Energy Data Analytics for Smart Meter Data

Energy Data Analytics for Smart Meter Data

Editors

Andreas Reinhardt
Lucas Pereira

MDPI • Basel • Beijing • Wuhan • Barcelona • Belgrade • Manchester • Tokyo • Cluj • Tianjin

Editors
Andreas Reinhardt　　　　　　　Lucas Pereira
Technische Universität Clausthal　Técnico Lisboa
Germany　　　　　　　　　　　Portugal

Editorial Office
MDPI
St. Alban-Anlage 66
4052 Basel, Switzerland

This is a reprint of articles from the Special Issue published online in the open access journal *Energies* (ISSN 1996-1073) (available at: https://www.mdpi.com/journal/energies/special_issues/Smart_Meter).

For citation purposes, cite each article independently as indicated on the article page online and as indicated below:

LastName, A.A.; LastName, B.B.; LastName, C.C. Article Title. *Journal Name* **Year**, *Volume Number*, Page Range.

ISBN 978-3-0365-2016-2 (Hbk)
ISBN 978-3-0365-2017-9 (PDF)

© 2021 by the authors. Articles in this book are Open Access and distributed under the Creative Commons Attribution (CC BY) license, which allows users to download, copy and build upon published articles, as long as the author and publisher are properly credited, which ensures maximum dissemination and a wider impact of our publications.

The book as a whole is distributed by MDPI under the terms and conditions of the Creative Commons license CC BY-NC-ND.

Contents

About the Editors vii

Preface to "Energy Data Analytics for Smart Meter Data" ix

Andreas Reinhardt and Lucas Pereira
Special Issue: "Energy Data Analytics for Smart Meter Data"
Reprinted from: *Energies* **2021**, *14*, 5376, doi:10.3390/en14175376 1

Patrick Huber, Alberto Calatroni, Andreas Rumsch and Andrew Paice
Review on Deep Neural Networks Applied to Low-Frequency NILM
Reprinted from: *Energies* **2021**, *14*, 2390, doi:10.3390/en14092390 5

**Christos Athanasiadis, Dimitrios Doukas, Theofilos Papadopoulos
and Antonios Chrysopoulos**
A Scalable Real-Time Non-Intrusive Load Monitoring System for the Estimation of Household Appliance Power Consumption
Reprinted from: *Energies* **2021**, *14*, 767, doi:10.3390/en14030767 39

Veronica Piccialli and Antonio M. Sudoso
Improving Non-Intrusive Load Disaggregation through an Attention-Based Deep Neural Network
Reprinted from: *Energies* **2021**, *14*, 847, doi:10.3390/en14040847 63

Anthony Faustine and Lucas Pereira
Multi-Label Learning for Appliance Recognition in NILM Using Fryze-Current Decomposition and Convolutional Neural Network
Reprinted from: *Energies* **2020**, *13*, 4154, doi:10.3390/en13164154 79

Zhengwei Qu, Hongwen Li, Yunjing Wang, Jiaxi Zhang, Ahmed Abu-Siada and Yunxiao Yao
Detection of Electricity Theft Behavior Based on Improved Synthetic Minority Oversampling Technique and Random Forest Classifier
Reprinted from: *Energies* **2020**, *13*, 2039, doi:10.3390/en13082039 97

Xiaofeng Feng, Hengyu Hui, Ziyang Liang, Wenchong Guo, Huakun Que, Haoyang Feng, Yu Yao, Chengjin Ye and Yi Ding
A Novel Electricity Theft Detection Scheme Based on Text Convolutional Neural Networks
Reprinted from: *Energies* **2020**, *13*, 5758, doi:10.3390/en13215758 117

**Augustyn Wójcik, Piotr Bilski, Robert Łukaszewski, Krzysztof Dowalla
and Ryszard Kowalik**
Identification of the State of Electrical Appliances with the Use of a Pulse Signal Generator
Reprinted from: *Energies* **2021**, *14*, 673, doi:10.3390/en14030673 135

Samira Ortiz, Mandoye Ndoye and Marcel Castro-Sitiriche
Satisfaction-Based Energy Allocation with Energy Constraint Applying Cooperative Game Theory
Reprinted from: *Energies* **2021**, *14*, 1485, doi:10.3390/en14051485 161

Jana Huchtkoetter, Marcel Alwin Tepe and Andreas Reinhardt
The Impact of Ambient Sensing on the Recognition of Electrical Appliances
Reprinted from: *Energies* **2021**, *14*, 188, doi:10.3390/en14010188 179

Benjamin Völker, Marc Pfeifer, Philipp M. Scholl and Bernd Becker
A Framework to Generate and Label Datasets for Non-Intrusive Load Monitoring
Reprinted from: *Energies* **2021**, *14*, 75, doi:10.3390/en14010075 . . . . **203**

Douglas Paulo Bertrand Renaux, Fabiana Pottker, Hellen Cristina Ancelmo, André Eugenio Lazzaretti, Carlos Raimundo Erig Lima, Robson Ribeiro Linhares, Elder Oroski, Lucas da Silva Nolasco, Lucas Tokarski Lima, Bruna Machado Mulinari, José Reinaldo Lopes da Silva, Júlio Shigeaki Omori, and Rodrigo Braun dos Santos
A Dataset for Non-Intrusive Load Monitoring: Design and Implementation
Reprinted from: *Energies* **2020**, *13*, 5371, doi:10.3390/en13205371 . . . . **229**

Manu Lahariya, Dries F. Benoit and Chris Develder
Synthetic Data Generator for Electric Vehicle Charging Sessions: Modeling and Evaluation Using Real-World Data
Reprinted from: *Energies* **2020**, *13*, 4211, doi:10.3390/en13164211 . . . . **265**

Xiao-Yu Zhang, Stefanie Kuenzel, José-Rodrigo Córdoba-Pachón and Chris Watkins
Privacy-Functionality Trade-Off: A Privacy-Preserving Multi-Channel Smart Metering System
Reprinted from: *Energies* **2020**, *13*, 3221, doi:10.3390/en13123221 . . . . **283**

Benjamin Völker, Andreas Reinhardt, Anthony Faustine and Lucas Pereira
Watt's up at Home? Smart Meter Data Analytics from a Consumer-Centric Perspective
Reprinted from: *Energies* **2021**, *14*, 719, doi:10.3390/en14030719 . . . . **313**

About the Editors

Andreas Reinhardt is head of the energy informatics lab at TU Clausthal, Germany. Prior to establishing his research group, he was a Vice-Chancellor's Postdoctoral Research Fellow at the University of New South Wales in Sydney, Australia. He completed his PhD and MSc degrees in Electrical Engineering and Information Technology at TU Darmstadt, Germany.

Lucas Pereira received his PhD in Computer Science from the University of Madeira, Portugal, in 2016. Since then, he is at ITI, LARSyS, where he leads the Further Energy and Environment research Laboratory (FEELab). Since 2019, he has been a research fellow at Técnico Lisboa, of the University of Lisbon, Portugal.

Preface to "Energy Data Analytics for Smart Meter Data"

The roll-out of smart electricity meters is a cornerstone for the realization of next-generation electrical power grids. In addition to measuring electrical consumption data at much greater temporal and amplitude resolutions than offered by traditional metering devices, smart meters can communicate collected data to external service providers and thus enable the creation of novel energy data-based services that extend far beyond their usage for billing purposes. These include, but are not limited to, the enablement of ambient-assisted living, generation, and demand forecasting, or the provision of recommendations on how to save energy.

A fundamental research challenge, still unresolved today, is how to fully explore and exploit the information content of smart meter data. This is a challenge pertaining not only to data processing but equally to their collection, transmission, and ensuring the appropriate security and privacy protection.

This book includes twelve research works and two review papers covering the entire lifecycle of smart meter data. The first set of contributions is centered around providing new solutions to the processing of smart meter data. These include innovative approaches for non-intrusive load monitoring (NILM) and the identification of electricity theft, exploiting novel techniques, and partially based on parameters beyond smart meter data alone. These contributions are followed by novel approaches to other data-driven services, including the privacy-preserving collection and synthetic generation of electricity consumption data. Two review papers cover the most up-to-date research efforts in deep learning for NILM, as well as a survey of viable options to streamline the provision of smart meter data applications from a consumer-centric perspective.

Andreas Reinhardt, Lucas Pereira
Editors

Editorial

Special Issue: "Energy Data Analytics for Smart Meter Data"

Andreas Reinhardt [1,*] and Lucas Pereira [2,*]

1. Department of Informatics, TU Clausthal, 38678 Clausthal-Zellerfeld, Germany
2. ITI, LARSyS, Técnico Lisboa, 1049-001 Lisboa, Portugal
* Correspondence: andreas.reinhardt@tu-clausthal.de (A.R.); lucas.pereira@tecnico.ulisboa.pt (L.P.)

Citation: Reinhardt, A.; Pereira, L. Special Issue: "Energy Data Analytics for Smart Meter Data". *Energies* **2021**, *14*, 5376. https://doi.org/10.3390/en14175376

Received: 25 August 2021
Accepted: 26 August 2021
Published: 30 August 2021

Publisher's Note: MDPI stays neutral with regard to jurisdictional claims in published maps and institutional affiliations.

Copyright: © 2021 by the authors. Licensee MDPI, Basel, Switzerland. This article is an open access article distributed under the terms and conditions of the Creative Commons Attribution (CC BY) license (https://creativecommons.org/licenses/by/4.0/).

Smart electricity meters are a cornerstone for the realization of next-generation electrical power grids. In addition to measuring electrical consumption data at much greater temporal and amplitude resolutions than offered by traditional metering devices, smart meters can communicate collected data to external service providers and, thus, enable the creation of novel energy data-based services. Such services include the enablement of ambient-assisted living, generation and demand forecasting, or the provision of recommendations on how to save energy. A fundamental research challenge, still unresolved as of today, is how to fully explore and exploit the information content of smart meter data—a challenge pertaining not only to data processing but equally to their collection, transmission, and security and privacy protection. This Special Issue includes twelve research works and two review papers covering the entire lifecycle of smart meter data.

The first paper in this Special Issue is a comprehensive survey of the state of the art in non-intrusive load monitoring (NILM) by Huber et al. [1]. Disaggregating the power consumption data captured by a metering device has seen a tremendous increase in research interest over the last years. As such, the comprehensive overview of existing technologies, including a comparison of their performance, is an excellent guide for researchers planning to join this research field. Nevertheless, even experienced NILM researchers are likely to discover new methods in this review paper, which may serve as baselines for comparative performance evaluations in their present and future work.

The review paper is followed by a technical NILM contribution by Athanasiadis et al. [2], which differs from existing work in that it is specifically designed to operate on real-time data streams. This not only renders large-scale buffering of data unnecessary but also shows remarkable results on the used dataset. Another interesting observation made during their analysis is the fact that the usage of data sampled at 100 Hz yields a remarkable performance improvement over methods using a baseline dataset, which was collected at 0.167 Hz.

Another technical NILM contribution, leveraging on the application of attention-based mechanisms in deep-neural networks (DNNs), is provided by Piccialli and Sudoso [3]. The proposed approach not only outperforms other DNN alternatives when applied to the two employed datasets, but it is also shown how attention-based mechanisms improve the underlying neural networks' ability to extract and exploit information that would otherwise be ignored (e.g., signal sections with high power consumption).

Although most of the proposed NILM algorithms attempt to disaggregate one load at a time, some authors are also exploring the possibility of disaggregating multiple appliances in parallel. This is the case of the work presented in Faustine and Pereira [4], where a convolutional neural network-based multi-label learning approach is proposed. Experimental results using one public dataset show that the proposed approach significantly outperforms state-of-the-art alternatives to multi-label learning in NILM.

A different application that relies on smart meter data is the detection of electricity theft. A corresponding contribution by Qu et al. [5] is presented next. Theft detection in electrical power grids is often challenging due to the strong imbalance in training data. Usually, only very few consumption traces that contain characteristic patterns of electricity theft are available. The authors have tackled the binary classification problem to discriminate between legitimate consumption patterns and electricity through a combination clustering and minority oversampling. The proposed method is shown to outperform two existing methods in terms of its theft detection accuracy.

Following up next, a second work on theft detection is provided by Feng et al. [ ] By applying text convolutional neural networks, anomalies in power consumption data can be located quickly and efficiently. By combining data augmentation techniques with realistic datasets, both intraday characteristics and diurnal periodicities are considered the training process, and remarkable results are reported.

The identification of devices within a building is another exciting challenge that relies on captured electrical voltage and current signals. Wójcik et al. [7] introduce a solution that applies a pulsed voltage signal to attached electrical appliances in order to record the transient responses. Through the creation of a dictionary of transients, their recognition in aggregate electrical load data is facilitated, and the potentials and limitations of the proposed method are discussed.

Another important application of smart meter data, particularly when disaggregated by individual load, is the definition of optimal schedules to optimize energy consumption. With this respect, in Ortiz et al. [8] the authors propose a satisfaction-based energy allocation algorithm that leverages cooperative game theory concepts to determine in which hour the energy should be allocated to maximize energy satisfaction while minimizing the power consumption. Experimental results using a publicly available dataset show that the proposed approach reduced energy consumption by 75%, while increasing the user satisfaction by 40%.

The use of sensor modalities beyond electrical data alone has been investigated by Huchtkoetter et al. [9]. A test bench, on which different electrical appliances-under-test can be operated, has been instrumented with sensors to capture eight ambient parameters in addition to measuring electrical voltage, current, and power. By analyzing the feature importance levels for different categories of appliance types, guidelines for selecting sensor types to increase appliance recognition accuracy were derived and confirmed to contribute to the unambiguous identification of appliances.

Most research works in the domain of smart meter data analytics rely on the use of real-world datasets to evaluate the performance of newly proposed methods and algorithms. limitation of most existing datasets, however, is the scarcity of annotations and metadata This limitation can sometimes be remedied through augmenting existing datasets with manual annotations; in other cases, the required aspects must already be considered during the data collection. Both aspects have been tackled by Völker et al. [10]: A new dataset with plentiful annotations, called FIRED, is presented in this work. Moreover, a semi-automated trace annotation tool is presented, which can be used to enrich the annotations of existing datasets. Based on these contributions, more sophisticated evaluations of smart meter data analytics methods become possible.

Another labeled electricity consumption dataset is presented in Renaux et al. [1] Three different subsets of data comprise the LIT dataset: (1) synthetic, where a controller is used to automatically switch appliances ON or OFF; (2) simulated, where appliance models are used to generate artificial appliance consumption profiles; and (3) natural where the appliance consumption is monitored and recorded in a real-world environment In addition to extensive details of the data collection procedures, the manuscript also thoroughly analyzes the consumption profiles in each of the LIT subsets.

Electric vehicle (EV) charging stations have become prominent in electricity grids the past few years. However, the availability of data regarding EV charging sessions currently very limited, which poses a significant hurdle to further research in the field. The challenge was investigated by the authors in Lahariya et al. [12], who propose a synthetic data generator to create samples of realistic EV session data, where each session is defined by an arrival time, a departure time, and the required energy. The proposed data generator was trained using real-world EV sessions, and the generated synthetic samples were shown to be statistically indistinguishable from real-world data.

Although electricity smart meters can promote the realization of several services, they can also pose a significant intrusion into a household's privacy. This issue has been investigated by Zhang et al. [13], who propose a privacy-preserving smart metering system that combines data aggregation and data down-sampling mechanisms. Of particular interest is the fact that the proposed system can be adapted to provide different levels of privacy, depending on the requirements and preferences.

The last paper in this Special Issue is a review by Völker et al. [14] that surveys a range of end-user services that can be enabled through smart meter data analytics. The reviewed services include, among others, end-user feedback, anomaly detection, and forecasting of load demand and renewable energy generation. The paper moreover provides a review of the technological foundations and open research challenges in this domain.

The contributed papers that are included in this Special Issue offer new and valuable insights that can stimulate ongoing research activities in the field. As guest editors, we would like to thank all authors who have submitted their contributions to this Special Issue.

Funding: This work was supported by the Portuguese Foundation for Science and Technology grants CEECIND/01179/2017 and UIDB/50009/2020.

Conflicts of Interest: The authors declare no conflict of interest.

References

Huber, P.; Calatroni, A.; Rumsch, A.; Paice, A. Review on Deep Neural Networks Applied to Low-Frequency NILM. *Energies* **2021**, *14*, 2390. [CrossRef]

Athanasiadis, C.; Doukas, D.; Papadopoulos, T.; Chrysopoulos, A. A Scalable Real-Time Non-Intrusive Load Monitoring System for the Estimation of Household Appliance Power Consumption. *Energies* **2021**, *14*, 767. [CrossRef]

Piccialli, V.; Sudoso, A.M. Improving Non-Intrusive Load Disaggregation through an Attention-Based Deep Neural Network. *Energies* **2021**, *14*, 847. [CrossRef]

Faustine, A.; Pereira, L. Multi-Label Learning for Appliance Recognition in NILM Using Fryze-Current Decomposition and Convolutional Neural Network. *Energies* **2020**, *13*, 4154. [CrossRef]

Qu, Z.; Li, H.; Wang, Y.; Zhang, J.; Abu-Siada, A.; Yao, Y. Detection of Electricity Theft Behavior Based on Improved Synthetic Minority Oversampling Technique and Random Forest Classifier. *Energies* **2020**, *13*, 2039. [CrossRef]

Feng, X.; Hui, H.; Liang, Z.; Guo, W.; Que, H.; Feng, H.; Yao, Y.; Ye, C.; Ding, Y. A Novel Electricity Theft Detection Scheme Based on Text Convolutional Neural Networks. *Energies* **2020**, *13*, 5758. [CrossRef]

Wójcik, A.; Bilski, P.; Łukaszewski, R.; Dowalla, K.; Kowalik, R. Identification of the State of Electrical Appliances with the Use of a Pulse Signal Generator. *Energies* **2021**, *14*, 673. [CrossRef]

Ortiz, S.; Ndoye, M.; Castro-Sitiriche, M. Satisfaction-Based Energy Allocation with Energy Constraint Applying Cooperative Game Theory. *Energies* **2021**, *14*. [CrossRef]

Huchtkoetter, J.; Tepe, M.A.; Reinhardt, A. The Impact of Ambient Sensing on the Recognition of Electrical Appliances. *Energies* **2021**, *14*, 188. [CrossRef]

Völker, B.; Pfeifer, M.; Scholl, P.M.; Becker, B. A Framework to Generate and Label Datasets for Non-Intrusive Load Monitoring. *Energies* **2021**, *14*, 75. [CrossRef]

Renaux, D.P.B.; Pottker, F.; Ancelmo, H.C.; Lazzaretti, A.E.; Lima, C.R.E.; Linhares, R.R.; Oroski, E.; Nolasco, L.d.S.; Lima, L.T.; Mulinari, B.M.; et al. A Dataset for Non-Intrusive Load Monitoring: Design and Implementation. *Energies* **2020**, *13*, 5371. [CrossRef]

Lahariya, M.; Benoit, D.F.; Develder, C. Synthetic Data Generator for Electric Vehicle Charging Sessions: Modeling and Evaluation Using Real-World Data. *Energies* **2020**, *13*, 4211. [CrossRef]

Zhang, X.Y.; Kuenzel, S.; Córdoba-Pachón, J.R.; Watkins, C. Privacy-Functionality Trade-Off: A Privacy-Preserving Multi-Channel Smart Metering System. *Energies* **2020**, *13*, 3221. [CrossRef]

Völker, B.; Reinhardt, A.; Faustine, A.; Pereira, L. Watt's up at Home? Smart Meter Data Analytics from a Consumer-Centric Perspective. *Energies* **2021**, *14*, 719. [CrossRef]

Review

Review on Deep Neural Networks Applied to Low-Frequency NILM

Patrick Huber *, Alberto Calatroni, Andreas Rumsch and Andrew Paice

iHomeLab, Engineering and Architecture, Lucerne University of Applied Sciences and Arts, 6048 Horw, Switzerland; alberto.calatroni@hslu.ch (A.C.); andreas.rumsch@hslu.ch (A.R.); andrew.paice@hslu.ch (A.P.)
* Correspondence: patrick.huber@hslu.ch

Abstract: This paper reviews non-intrusive load monitoring (NILM) approaches that employ deep neural networks to disaggregate appliances from low frequency data, i.e., data with sampling rates lower than the AC base frequency. The overall purpose of this review is, firstly, to gain an overview on the state of the research up to November 2020, and secondly, to identify worthwhile open research topics. Accordingly, we first review the many degrees of freedom of these approaches, what has already been done in the literature, and compile the main characteristics of the reviewed publications in an extensive overview table. The second part of the paper discusses selected aspects of the literature and corresponding research gaps. In particular, we do a performance comparison with respect to reported mean absolute error (MAE) and F_1-scores and observe different recurring elements in the best performing approaches, namely data sampling intervals below 10 s, a large field of view, the usage of generative adversarial network (GAN) losses, multi-task learning, and post-processing. Subsequently, multiple input features, multi-task learning, and related research gaps are discussed, the need for comparative studies is highlighted, and finally, missing elements for a successful deployment of NILM approaches based on deep neural networks are pointed out. We conclude the review with an outlook on possible future scenarios.

Keywords: non-intrusive load monitoring; load disaggregation; NILM; review; deep learning; deep neural networks; machine learning

Citation: Huber, P.; Calatroni, A.; Rumsch, A.; Paice, A. Review on Deep Neural Networks Applied to Low-Frequency NILM. *Energies* **2021**, *14*, 2390. https://doi.org/10.3390/en14092390

Received: 9 February 2021
Accepted: 16 April 2021
Published: 23 April 2021

Publisher's Note: MDPI stays neutral with regard to jurisdictional claims in published maps and institutional affiliations.

Copyright: © 2021 by the authors. Licensee MDPI, Basel, Switzerland. This article is an open access article distributed under the terms and conditions of the Creative Commons Attribution (CC BY) license (https://creativecommons.org/licenses/by/4.0/).

1. Introduction

Non-Intrusive Load Monitoring (NILM)—equally referred to as load disaggregation—aims to identify the individual power consumption or on/off state of electrical loads by relying exclusively on the measurement of the aggregated consumption of these loads. The term was coined by Hart in their seminal works [1,2], that initiated the NILM research field. As the term non-intrusive suggests, NILM is motivated by applications where metering of single appliances is necessary, but not feasible with conventional measurement devices, e.g., because of cost considerations or obtrusiveness. Potential NILM applications include, for example, user feedback for energy saving purposes, overseeing activities of daily living for the health assessment of elderly people, or demand management (see, e.g., [3–6]).

Before proceeding, it is essential to clarify some nomenclature typical of the NILM jargon and how this relates to the data used. Electric meters internally sample voltage and current signals at frequencies significantly greater than the base frequency of alternate current (AC). Meters can either output these raw data directly or averaged values such as, e.g., root mean square (RMS) voltage, current, power, or total harmonic distortion (THD) are calculated and output at lower frequencies. In the NILM literature, the following terms are used (We are aware that the given definition is slightly diverging from the conventionally used threshold of 1 Hz (e.g., [3]). However, we feel 1 Hz is somewhat arbitrary, whereas the provided definition seems to be a natural splitting point.):

- Low frequency approaches are those which use data (i.e., features) produced at rates lower than the AC current base frequency;
- High frequency approaches are those which use raw data, sampled at rates higher than the AC current base frequency.

The advantage of using high frequency data should be quite obvious, since they preserve the entire signals and therefore allow us to extract the maximum information content. This is confirmed by various works, e.g., ref. [7] showed that, using raw data sampled at 1 MHz, it is even possible to distinguish between two appliances of the same type. Nevertheless, the cost of gathering high frequency data constitutes a major obstacle. Dedicated infrastructure is needed, and this is expensive both in terms of the hardware itself and the extra installation effort. On the other hand, the loss in information intrinsic in low frequency features is offset by the tremendous ease with which those data can be collected. Indeed, around the year 2010, the European Union and the US started to mandate and actually roll-out smart meters [8]. The advanced versions of these meters can expose low frequency data to the outside (e.g., the meter E450 from Landis+Gyr has this capability). The pervasive roll-out of smart meters will therefore unlock all those applications which can benefit from NILM and which can be tackled using low frequency approaches.

Since its inception, the NILM research field has become quite diverse. The classic approach to low frequency NILM—as proposed in [1,2]—is event based. Simply stated, this means that the aggregate power series is first analyzed to find raises or drops that indicate device switching events, and these events are subsequently assigned to certain appliances. Later, one major avenue of NILM research used different variants of Hidden Markov Models, e.g., [9–12]. Over the years, the set of methods that have been applied to NILM has become extremely rich. A comprehensive overview is given in [13]. With the recent enormous success that deep learning has found in the vision and natural language processing domains, it was only a matter of time until deep neural networks (DNN) were also applied for the first time to NILM; this started in 2015 [14,15]. Since then, the number of DNN approaches to solve the NILM problem increased rapidly, as can be seen in Table 1.

Table 1. Number of DNN-NILM publications based on low frequency data per year. Numbers are compiled based on Table 2. That means the number for the year 2020 corresponds to the publications until end of November, see Section 1.3.

Year	2015	2016	2017	2018	2019	2020
Count	2	6	4	21	36	30

Over the years, different publications have surveyed the NILM literature from various angles: In [3], the authors focused on NILM feature types for low and high frequency data and touched algorithms and evaluation metrics. The authors of [16] proposed a taxonomy of appliance features and compared the reported performance of six classes of supervised and unsupervised NILM approaches. The authors of [17] surveyed unsupervised NILM algorithms and discussed the reported performance of eleven approaches. A very comprehensive review on inductive algorithms, employed feature sets, and the state of multi-label NILM classification approaches has been compiled by [13]. More recently, the authors of [18] discussed available public NILM datasets, and employed NILM performance metrics, tools and frameworks, and corresponding limitations and challenges. Ref. [19] touched on dataset complexity and compared the reported performance of eight NILM approaches under that viewpoint. Three of these approaches employed DNN approaches on low-frequency data. Finally, the authors of [20] summarized advancements on HMM and DNN approaches. With respect to DNN approaches, they summarized the work performed in [21,22]. While the previously mentioned survey papers compared performance as it was reported by the original authors, very recently, two works compared classical and DNN approaches under identical conditions [23,24]. The authors of [23] presented an extension to the NILM tool kit (NILMTK) library [25,26], with the exact purpose of simplifying direct comparisons, and [24] used this new functionality for an extensive comparison of eight available algorithms.

1.1. Contributions

Some of the mentioned review papers include DNN-based NILM approaches. However, what is missing so far is a comprehensive and structured overview on the ideas and findings in the field of DNNs applied to low frequency NILM and, based on that, a discussion on the usefulness of DNN architectural elements, input features or multi-task learning as well as research gaps particularly for applied deep learning-based NILM. In this paper, we intend to fill this gap. Therefore, the main contributions of this work are:

- A comprehensive review of NILM approaches based on low frequency data that employ DNNs, see Section 3. In particular, within Section 3, we discuss various options available for these approaches and provide a structured overview of the main characteristics of all reviewed approaches in Table 2.
- A discussion of selected aspects and corresponding research gaps in Section 4. In particular:
    - We compare the performance of approaches and extract common features of best performing approaches in Section 4.1;
    - We discuss the possible role of multiple input features and multi-task learning on NILM performance in Sections 4.2 and 4.3, respectively;
    - We illustrate the importance of parameter studies in Section 4.4, and
    - We outline major research gaps concerning the application of deep learning for NILM in Section 4.5.

Thereby, we hope that the interested reader will quickly identify the relevant literature for their own research and that our contributions will inspire new research activities, and thus ultimately advance the entire research field.

1.2. Scope

The scope of this review are NILM approaches based on DNNs using low frequency data. In the remainder of this text, we use the term DNN-NILM to designate the corresponding approaches. The choice to focus on low frequency data in our review is motivated by our strong belief that many applications could benefit from NILM, coupled with our observation that low frequency data will most likely be the only one available at scale in the near future. In our vision, all households equipped with smart meters will soon be able to become fully energy aware, informing their inhabitants of which appliances are being used, how they are being used, and even whether they are behaving abnormally or about to fail. This latter point is known as predictive maintenance and is currently applied in industrial settings, but being able to detect billions of appliances which consume an abnormal amount of power would have a beneficial impact of our society and its carbon footprint. With our review, we therefore try to make a contribution to push forward the development and understanding of low frequency NILM.

The focus on DNNs is motivated firstly by their proven success in other domains, and secondly by their good performance in the NILM domain: Recently, traditional and DNN-NILM approaches have been compared under identical conditions in two works [23,24]. The authors found that each of the compared DNN approaches—with few exceptions—performed better than each of the classical approaches. In particular for multi-state appliances, the performance gap was found to be "rather discernible" [24]. Publications that use shallow neural networks with only a single hidden layer such as, e.g., [27–29], are not included in our review. We restricted ourselves to approaches that train neural networks with back-propagation, excluding alternative approaches such as, e.g., [30,31]. Since the scope involves DNNs and NILM, we assume that the reader is familiar with the general concepts of the two fields, and we will merely introduce the basic NILM problem formulation in Section 2.1. With respect to DNNs and deep learning, we will refer the reader to comprehensive books on the topic in Section 2.2.

Table 2. Reviewed references. Publications are sorted by year. Except for the starred publications, the sorting within a year is arbitrary. The table is available in Excel format on our GitHub account, see 'Supplementary Materials' for the link. Explanations with respect to specific columns follow. **Best**: Best performing, according to Section 4.1. **Datasets**: See Table 3 for details. **Setting**: $R \rightarrow$ residential, $I \rightarrow$ industrial, $C \rightarrow$ commercial. The columns FGE to WDR indicate if the specific appliance has been disaggregated in the reference. **FGE**: fridge, **DWE**: dishwasher, **MWV**: microwave, **WME**: washing machine, **KET**: kettle, **SOC**: stove/oven/cooker, **TDR**: tumble dryer, **HPE**: heat pump, **WDR**: washer-dryer. **Further Appliances**: Additional appliances not listed in the previous columns. **E.Sce.**: Evaluation Scenarios; $sn \rightarrow$ only seen scenario evaluated, $usn \rightarrow$ additionally unseen scenario evaluated, $ctl \rightarrow$ also cross-domain transfer learning evaluated. **Aug.**: Data Augmentation; $da \rightarrow$ use synthetic training data, $yes \rightarrow$ data augmentation employed. **Input**; $P \rightarrow$ active power, $Q \rightarrow$ reactive power, $I \rightarrow$ current, $S \rightarrow$ apparent power, $P_{2D} \rightarrow$ active power window transformed into 2D representation, $P\text{-}S \rightarrow$ difference between active and apparent power, $\Delta P \rightarrow$ first-order difference of the active power signal, $PF \rightarrow$ power factor, $TofD \rightarrow$ time of day, $WE \rightarrow$ week or weekend day, $DofW \rightarrow$ day of week, $MofY \rightarrow$ month of year, $T_{out} \rightarrow$ outdoor temperature, $P_{var} \rightarrow$ variant power signature [32,33], $na \rightarrow$ see Section 3.3.1. **DNN Elements**: See Section 3.3.1. Comma separated descriptions refer to different trained models. Comma separated descriptions refer to different types. Similarly, the subscript m means that the DNN provides the identical output for *multiple* different appliances. Elements connected with an & indicate that a DNN has several outputs of a different type. Please refer to Section 3.5 for details concerning P_{app}, P_{total}, P_{rest}, *location*, and *stateChange*. **Code**: (electronic version) link to code repository as indicated in the reference or found through a very shallow google search.

Ref.	Best	Year	Dataset(s)	Setting	DWE	FGE	MWV	WME	KET	SOC	TDR	HVAC	WDR	HPE	Light	Further Appliances	E.Sce.	Aug.	Input	DNN Elements	Output	Code
[34]	*	2020	UK-DALE, REFIT	R	x	x	x	x	x								usn	no	P	CNN-dAE-GAN	P	https://github.com/DLZRMR/seq2subseq
[35]	*	2020	UK-DALE, REDD	R	x	x	x	x	x								usn	yes	P	CNN-att-GAN	on/off & P	-
[36]	*	2020	UK-DALE, ECO	R	x	x	x	x	x	x						TV	usn	no	P, Q, S, I, PF	CNN-biLSTM	P	-
[37]	*	2020	UK-DALE	R	x	x	x	x	x								usn	no	P	CNN-dAE	on/off$_m$	https://github.com/lmssdd/TPNILM
[38]		2020	UK-DALE, REFIT, SynD	R	x	x	x	x	x								sn	no	P_{2D}	CNN-dAE	P	https://github.com/BHafsa/image-nilm
[39]		2020	UK-DALE, REDD, REFIT	R	x		x	x									ctl	no	P	CNN-GAN, CNN-dAE-GAN	P	-
[40]		2020	REDD, Enertalk	R		x	x	x								TV, rice cooker	sn	no	P	CNN-biLSTM	P	-
[41]		2020	REDD	R		x	x	x									sn	no	P	CNN-dAE	P	-
[42]		2020	AMPds, REFIT	R	x			x		x	x						ctl	no	P	biLSTM	P	-
[43]		2020	AMPds, REFIT	R	x			x		x	x						sn	no	P	CNN-dAE-GAN	P	-
[44]		2020	proprietary	R												water heater	usn	yes	P	CNN-LSTM	P	-
[5]		2020	REFIT	R	x	x	x	x	x							toaster	sn	yes	P	CNN-biGRU	on/off$_m$ & P	-
[45]		2020	proprietary	I													sn	no	p	CNN-biGRU, LSTM	on/off	-
[46]		2020	UK-DALE	R	x	x	x	x	x								sn	no	P	CNN-dAE	on/off$_m$ & P_m	https://github.com/sambaiga/UNETNiLM

Table 2. *Cont.*

Ref.	Best	Year	Dataset(s)	Setting	DWE	FGE	MWA	MWE	KET	SOC	TDR	HVAC	WDR	HPE	Light	Further Appliances	E.Sce.	Aug.	Input	DNN Elements	Output	Code
[47]		2020	UK-DALE, REFIT, HES	R	x			x		x	x						usn	no	P	CNN-s2p	P	https://github.com/JackBarber98/pruned-nilm
[48]		2020	REFIT	R	x				x								usn	no	P	-	P	https://github.com/EdgeNILM/EdgeNILM
[49]		2020	UK-DALE, REDD	R	x	x		x								toaster	usn	no	P	CNN-s2p	P, P_m	-
[50]		2020	REFIT	R	x	x		x	x								usn	no	P	-	P	-
[51]		2020	UK-DALE, REDD, DRED	R	x	x	x	x	x	x			x		x	computer	sn	no	P	-	P	-
[52]		2020	DRED	R	x	x	x	x									sn	no	P	CNN-dAE	P	-
[53]		2020	REFIT	R		x		x									usn	no	P	LSTM	P	-
[54]		2020	REDD	R	x	x	x	x	x								usn	no	P		P	-
[55]		2020	UK-DALE, REDD	R	x	x	x	x	x				x		x		usn	no	P	att	P	https://github.com/Yueeeeeeee/BERT4NILM
[56]		2020	REDD	R											x	bathroom, heater, kitchen outlet	sn	no	P	CNN	on/off$_m$	-
[57]		2020	REDD, dataport	R	x	x	x	x				x	x		x		sn	no	P, ΔP	CNN-LSTM	P_m	-
[21,22]		2020	summary of [21,22]																			
[58]		2020	UK-DALE	R	x	x	x	x	x								usn	yes	P	CNN-s2p	P	-
[59]		2020	REFIT	R	x	x		x								TV	sn	no	P	CNN-GRU	P_{app} & P_{total} & P_{rest}	-
[60]		2020	AMPds	R	x	x	x		x						x		sn	dn	P, I	LSTM	on/off	-
[61]		2020	proprietary (dc)	R									x			dc appliances	sn	no	I	FE, LSTM	P_{class}	-
[62]	*	2019	REFIT	R	x		x	x	x								usn	no	P	CNN-wn	P, on/off	https://github.com/jiejiang-jojo/fast-seq2point
[63]	*	2019	UK-DALE, REDD, REFIT	R	x		x	x									ctl	no	P	biGRU, CNN-s2p	on/off & P	-
[64]	*	2019	REDD	R	x	x	x	x									usn	no	P	CNN, CNN-LSTM, LSTM	P	-
[65]		2019	UK-DALE	R	x	x	x	x									usn	no	S	CNN-s2sub	P	-
[66]		2019	proprietary	R	x	x										bottle warmer	sn	yes	P	AE/Kmeans-dAE	P, P_m	-
[67]		2019	REDD	R	x	x	x			x			x				sn	no	P	CNN-s2p	on/off	-
[68]		2019	AMPds	R	x	x		x		x	x			x			sn	no	P, MofY, DofW, TofD	LSTM-FF	P_{class}	-

Table 2. Cont.

Ref.	Best	Year	Dataset(s)	Setting	DWE	FGE	MWV	WME	KET	SOC	TDR	HVAC	WDR	HPE	Light	Further Appliances	E.Sce.	Aug.	Input	DNN Elements	Output	Code
[69]		2019	UK-DALE	R	x		x	x	x	x			x			toaster coffee machine, hair dryer, rice cooker, toaster, blender, iron, disposer	sn	yes	P	CNN-s2p	P	-
[70]		2019	proprietary	R													-	no	P	CNN-dAE	location	https://people.csail.mit.edu/cyhsu/sapple/
[71]		2019	UK-DALE, REDD	R		x			x	x	x						ctl	no	P_{2D}	CNN	on/off	https://github.com/LampriniKyrk/Imaging-NILM-time-series
[72]		2019	UK-DALE, AMPds	R	x	x	x	x	x	x	x			x			usn	yes	P	CNN-att-biLSTM	P, on/off	-
[73]		2019	proprietary	R			x		x			x			x	computer, fan, hair dryer, printer, TV, water dispenser	?	?	I	CNN	on/off$_m$	-
[74]		2019	ECO	R		x		x								computer, freezer	sn	no	P, P-S, ToD, WE	FF	P	-
[75]		2019	Enertalk	R			x	x	x	x	x					rice cooker, TV	usn	no	P, Q	CNN-s2p	P, on/off	https://github.com/ch-shin
[76]		2019	dataport	R	x	x	x	x	x	x	x	x					sn	no	P	CNN-GRU	P	-
[77]		2019	PLAID	R		x	x					x			x	computer, fan, hair dryer, heater, vacuum cleaner	usn	yes	na	CNN, FF	on/off$_m$	-
[78]		2019	REDD	R	x	x	x		x	x			x		x	bathroom, kitchen outlet	usn	no	P	RNN	P, on/off	https://github.com/nilmtk/nilmtk/
[23]		2019	dataport	R												air, furnace	usn	no	P	-	P	https://github.com/nilmtk/nilmtk-contrib
[79]		2019	AMPds	R	x	x	x	x	x	x	x						sn	no	P	LSTM-FF, GRU-FF	P	-
[80]		2019	AMPds	R		x	x			x	x			x		furnace, TV & entertainment	sn	no	na	CNN	P	-
[81]		2019	UK-DALE, REDD	R	x	x	x	x									sn	no	P	CNN, biLSTM	on/off$_m$	-

Table 2. Cont.

Ref.	Best	Year	Dataset(s)	Setting	DWE	FGE	MWV	WME	KET	SOC	TDR	HVAC	WDR	HPE	Light	Further Appliances	E.Sce.	Aug.	Input	DNN Elements	Output	Code
[82]		2019	proprietary	R		×				×						furnace	sn	no	na	FF	on/off$_m$	-
[83]		2019	ECO	R	×	×			×							freezer, home theater	sn	no	P	RNN	on/off$_m$	-
[84]		2019	UK-DALE, dataport	R	×	×	×	×	×			×					sn	no	P	CNN-s2s	P	-
[85]		2019	UK-DALE, dataport	R	×	×	×	×	×		×	×					sn	no	P	CNN	P, on/off	-
[86]		2019	UK-DALE	R	×	×	×	×	×								sn	no	P	CNN-LSTM	P$_m$	-
[87]		2019	proprietary	R												coffee filter, coffee machine, TV	sn	yes	P	RNN, CNN-LSTM, CNN-dAE	P	-
[88]		2019	AMPds	R	×	×		×			×			×			sn	dn	P, Q, S, I	LSTM-dAE	P, on/off	-
[89]		2019	AMPds	R	×					×	×			×			sn	no	P	CNN-s2s	P	-
[90]		2019	AMPds	R	×					×	×			×			sn	no	P	biLSTM	P	-
[91]		2019	UK-DALE, REDD, REFIT	R	×	×	×	×	×								ctl	no	P	CNN-s2p	P	-
[92]		2019	dataport	R	×	×	×	×	×		×	×					sn	no	P$_{2D}$	RNN, CNN-dAE	P	https://github.com/MingjunZhong/transferNILM
[93]		2019	UK-DALE, REDD	R	×	×	×	×	×				×				sn	no	P	-	P	https://github.com/villingia/TreeCNN-for-Energy-Breakdown
[94]		2019	AMPds2	R	×	×	×	×	×		×					kitchen outlet	sn	no	P, Q, S, I	CNN-wn	P$_m$	https://gitlab.com/a3LabShares/A3NeuralNILM
[95]		2019	REDD	R	×	×		×	×		×		×	×			sn	no	ΔP	FF, biGRU	stateChange	https://github.com/picagrad/WaveNILM
[96]		2019	REDD, dataport	R	×	×	×	×	×		×					air, furnace, kitchen outlet	usn	no	P	VRNN	P	https://bitbucket.org/gissemari/disaggregation-vrnn
[97]	*	2018	UK-DALE, REDD	R	×	×	×	×	×	×							usn	no	P	CNN-s2sub	on/off & P	https://github.com/ch-shin
[98]	*	2018	IDEAL	R	×	×	×	×	×	×						shower	usn	no	S	CNN-s2sub	P	-
[99]		2018	dataport	R								×					sn	no	P	LSTM	P, on/off	-
[100]		2018	UK-DALE	R		×											?	no	P	CNN-biLSTM	on/off	-
[101]		2018	proprietary	C	×	×	×	×	×	×			×		×	computer	sn	no	P	CNN-s2p	P	-
[102]		2018	UK-DALE	R	×	×	×	×	×	×							sn	no	P, P$_{nbr}$, IofD	CNN	on/off$_m$	-
[33]		2018	UK-DALE	R	×	×	×	×	×	×			×				usn	yes	P	CNN-s2p	P, P$_{class}$	-
[103]		2018	proprietary	R	×	×	×	×				×				bottle warmer, TV	?	no	P	dAE	P$_m$	-

11

Table 2. *Cont.*

Ref.	Best	Year	Dataset(s)	Setting	DWE	FGE	MWM	WME	KET	SOC	TDR	HVAC	WDR	HPE	Light	Further Appliances	E.Sce.	Aug.	Input	DNN Elements	Output	Code
[104]		2018	UK-DALE	R	x	x	x	x	x								usn	no	P	CNN-VAE	P	https://github.com/KaibinBao/ [1]
[105]		2018	UK-DALE	R		x	x	x									usn	yes	P	GAN	P	-
[106]		2018	AMPds2	R										x		home office	sn	no	P, T_{out}	LSTM	on/off & P	https://github.com/nlaptev
[107]		2018	dataport	R	x	x	x		x	x			x			~30 more	sn	no	P_m	dAE-LSTM	P, P_m	-
[108]		2018	UK-DALE	R	x	x	x		x								sn	no	P	biLSTM, biGRU	P	-
[109]		2018	IMD	I													sn	no	P	CNN-wn	P	[2]
[22]		2018	UK-DALE, AMPds2	R	x	x	x	x	x		x						usn	yes	P, Q	CNN-dAE	P	-
[10]		2018	REDD	R	x	x	x	x	x		x				x		usn	dn	P	CNN-s2sub	P	-
[111]		2018	UK-DALE	R	x	x	x	x	x	x							usn	no	P	biLSTM, biGRU, CNN-s2p	P	https://github.com/OdysseasKr/online-nilm
[112]		2018	UK-DALE	R	x	x	x	x	x			x				TV	usn	no	P	CNN-dAE	on/off	-
[113]		2018	proprietary, UK-DALE	C										x	x		usn	yes	P	CNN-dAE	P	-
[21]		2018	REDD, AMPds	R	x	x	x	x	x	x	x			x			usn	no	P	CNN-dAE	P	-
[114]		2018	REDD	R	x	x		x		x	x	x					sn	no	na	CNN	on/off_m	-
[115]		2017	proprietary	R			x	x				x				computer, heater	usn	yes	P, P_{rar}	dAE	P	-
[32]		2017	UK-DALE, REDD	R	x		x	x		x	x	x				dehumidifier, toaster, TV	sn	no	P, P_{rar}	LSTM	on/off_m	-
[16]		2017	REDD	R	x	x							x				sn	no	na	CNN, CNN-dAE	on/off_m	-
[17]		2017	proprietary	R		x	x						x				usn	no	P, Q, S	CNN, RCNN, biLSTM, biGRU	P & Q	-
[18]	*	2016	REDD	R	x		x	x	x								sn	yes	P	CNN-s2s, CNN-s2p	P_class	-
[19]		2016	UK-DALE, REDD	R	x	x	x	x	x								usn	no	P	RNN, GRU	P	https://github.com/MinguiZhong/NeuralNetNilm
[120]		2016	UK-DALE	R					x								sn	no	P	CNN-dAE	on/off	-
[121]		2016	UK-DALE	R	x	x	x	x									sn	no	P	CNN-LSTM	P	-
[122]		2016	REDD	R	x	x	x									kitchen outlet	sn	dn	P	HMM-DNN	P	-
[123]		2016	dataport	R	x	x	x					x					usn	no	P_{2D}	CNN	P	-
[5]		2015	REDD	R	x	x	x			x							usn	dn	P	biLSTM	P	-
[14]		2015	UK-DALE	R	x	x	x	x	x								usn	yes	P	CNN-dAE, CNN-biLSTM, CNN-FF	P	https://github.com/JackKelly/neuralnilm

[1] GitHub page of first author; [2] Experimental framework available upon request.

As the DNN-NILM literature reviewed contains only three publications using data from non-domestic settings (two commercial, one industrial), this distribution means that our review concentrates mainly on domestic NILM.

1.3. Methodology

Publications in the scope of this work have been collected in the following ways: Firstly, by systematically checking conference proceedings from the bi-annually 'International NILM Workshop' 2020 to 2016 (nilmworkshop.org, accessed on 11 January 2021) and from the 'ACM International Conference on Systems for Energy-Efficient Buildings, Cities, and Transportation (BuildSys)' 2020 to 2015 (buildsys.acm.org, accessed on 11 January 2021). The first conference is specifically dedicated to NILM, and the second in 2020 featured a dedicated NILM track. Secondly, by searching on Google Scholar and IEEE Xplore® for keyword combinations of 'DNN', 'deep learning', 'NILM', 'non-intrusive', 'load monitoring', and 'load disaggregation'. This search has been done on several occasions and by different persons. Thirdly, we checked very thoroughly all the references in identified papers for anything not yet on our list. While this approach might have missed a few recent publications, we are fairly sure that the survey is quite complete for the past years because of the systematic checking of references. The last iteration of our search process has been performed at the end of November 2020. We resulted with the DNN-NILM publications listed in Table 2, which reflects accordingly the body of work this survey paper is based on. The literature review, discussion, and all conclusions are deduced solely from these publications.

2. Fundamentals

As we assume that the reader is knowledgeable about both NILM and Deep Neural Networks, the following sections only skim the corresponding subjects and the reader is referred to relevant literature.

2.1. The Disaggregation Problem

The aggregate active power x_t^a of a set of appliances measured at time t can be formally defined as:

$$x_t^a = \sum_{m=1}^{M} y_t^m + \underbrace{\sum_{k=1}^{K} w_t^k + \epsilon_t}_{=e_t} \qquad (1)$$

where y_t^m are the contributions of individual appliances m that have been metered at the time of data acquisition, and M is their total number. The sum over k corresponds to the contribution of K further appliances w_t^k not sub-metered during the measurement campaign. ϵ_t is a noise term originating from the measurement equipment. In the literature, the NILM problem is typically stated such that the noise term e_t includes the sum over non measured equipment. We explicitly separate the two contributions, as their nature is quite different. We can assume that the measurement noise ϵ_t is well behaved, i.e., it follows approximately a standard distribution and is small compared to the actual signal. On the contrary, no such assumption can be made about the term $\sum w_t^k$. The contribution from non sub-metered appliances w_t^k typically amounts to a major part of x_t^a and the power distribution is non-Gaussian. From the point of view of disaggregation, the sum over m denotes the appliances that are disaggregated, and the sum over k consists of all the remaining appliances in the aggregate signal. If only a single appliance y_t^m is disaggregated, then $M = 1$.

One goal of energy disaggregation is to determine the individual y_t^m only based on the measurement of the aggregate signal. If machine learning or in particular deep learning is used to solve the problem, this leads to a so-called regression problem. While many authors work with the active power component x_t^a only, other information from the aggregate signal such as, e.g., apparent power, reactive power, or the current can also be used to solve the disaggregation challenge. In the particular case of countries where the residential power supply is fed on three phases, features from the aggregate power can even be available on all of these three phases.

A second, slightly less challenging goal of energy disaggregation is to find the on/off state s_t^m of appliance m at time t from the aggregate signal. If machine learning is used, this leads to a so-called binary classification problem. In this problem formulation, only the state of the machine will be output. After recognizing the on and off states, the run-time of an appliance can be calculated. By multiplying the run-time by the average energy consumption of a machine, one can still obtain an energy estimation. Such an estimate will be more in line with use cases that require only the average consumption over a certain time period.

2.2. Deep Neural Networks

Deep neural networks are a vast subject, and the focus of this review is merely their application to NILM. In this text, we therefore refrain from giving an introduction on DNNs and refer the reader to the following books:

- The book [124] (www.deeplearningbook.com accessed on 11 January 2021) is a very comprehensive resource on the topic, covering the basics up to research topics.
- The book [125] has been written by the initial author of Keras [126], a high level deep learning library. Accordingly, the book gives an applied introduction to deep learning. A lot of emphasis is put on the intuition behind concepts, and the text is interwoven with code examples based on the Keras library.

The references mentioned are the books we found useful in our work. The selection is of course a small subset of the many excellent resources available on the topic.

3. Literature Review

When applying DNNs to NILM, many options are available: For example, which data to use, what DNN architecture to employ, how to evaluate the results, and so on. An illustration of these 'degrees of freedom' is given in Figure 1. The subsections below roughly follow the grouping done there. The aim of this section is to provide the reader with an overview on what has already been done in the literature in the scope of this review.

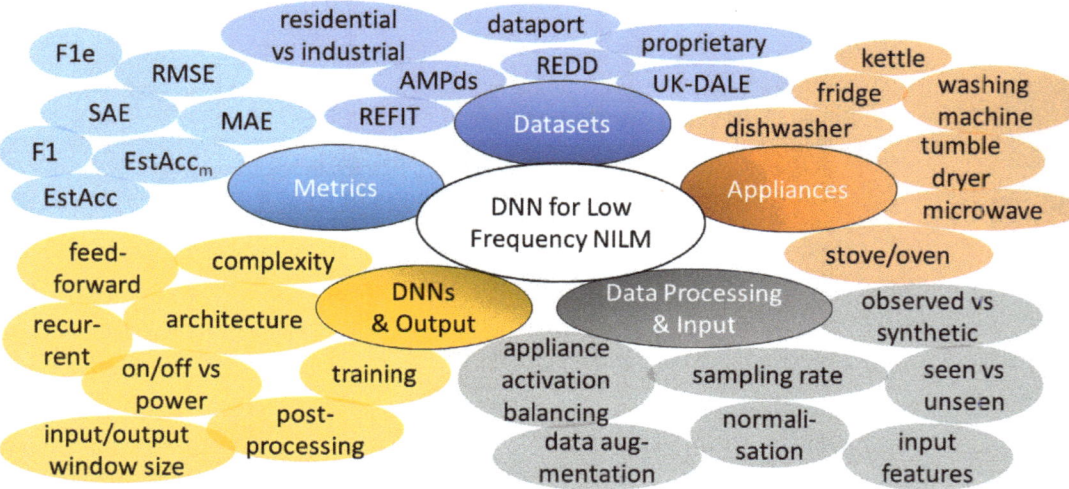

Figure 1. Illustration of the main degrees of freedom for DNN-NILM research. Colors indicate a loose grouping and should not be understood as a taxonomy.

3.1. Datasets and Appliances

DATASETS: The number of NILM datasets has been increasing over the last years, see [127,128] for recent overviews and [129–131] for the most recent published datasets we are aware of. In Table 3, we characterize only the publicly available datasets that have been used in the reviewed studies. The datasets at the beginning of the table are those more frequently used: Both UK-DALE and REDD were employed in approximately 40 and 30 studies, respectively, followed by AMPds, REFIT, and dataport each employed around 10 times. The ECO dataset is used three times, and other datasets are only used once or twice. The Industrial Machines Dataset (IMD) is to our knowledge the only available open industrial dataset. All remaining open datasets were measured in a residential setting. There are also a number of studies based on proprietary datasets measured in different settings: Nine residential, two commercial, one industrial, and one with dc-appliances. While not explicitly the scope of this review, the distribution of the employed datasets means that our review concerns mostly domestic NILM. Table 2 lists the datasets employed by each reviewed publication.

APPLIANCES: Appliances that have been disaggregated in the corresponding publications are listed in Table 2. The most investigated residential appliances in decreasing order are: dishwasher, fridge, microwave, washing machine, kettle, stove/oven/cooker, tumble dryer, HVAC, washer-dryer, heat pump, and light. Further electrical loads that appear fewer than ten times in the literature are given in the column 'Further Appliances' of Table 2. A few publications concentrated either on commercial or industrial applications using mostly proprietary datasets. These publications are marked in the column 'Setting' of Table 2.

Table 3. Main characteristics of the open datasets used in the reviewed DNN-NILM literature, see Table 2. Datasets closer to the top have been employed in more studies. *Type* indicates the type of the dataset: $R \to$ residential, $R_s \to$ synthetic residential, $I \to$ industrial. IMD is, to our knowledge, the only publicly available industrial dataset. '#H' and '#A' mean number of houses and appliances, respectively. '*Agg*' and '*Appl*' stand for 'aggregate' and 'appliance', respectively. For the IDEAL dataset, available information has been extracted from [98]. The authors plan to release the dataset.

Name	Country Code	Year	Type	#H	#A	Summed up Duration [d]	Agg Sampling	Appl Sampling
UK-DALE [132]	GBR	2017	R	5	109	2247	6 s, 16 kHz	6 s
REDD [133]	USA	2011	R	6	92	119	1 Hz, 15 kHz	$\frac{1}{3}$ Hz
AMPds(2) [134–137]	CAN	2016	R	1	20	730	1 min	1 min
REFIT [138]	GBR	2016	R	20	177	14,600	8 s	8 s
dataport [139]	USA	2015	R	1200+	8598	1,376,120	1 Hz, 1 min	1 Hz, 1 min
ECO [140]	CHE	2016	R	6	45	1227	1 Hz	1 Hz
DRED [141]	NLD	2014	R	1	12	183	1 Hz	1 Hz
Enertalk [127]	KOR	2019	R	22	75	1714	15 Hz	15 Hz
HES [142]	GBR	2010	R	251	5860	15,976	2–10 min	2–10 min
IDEAL	GBR	-	R	-	-	-	1 Hz	1 or 5 Hz
IMD [143]	BRA	2020	I	1	8	111	1 Hz	1 Hz
PLAID [144]	USA	2014	R	65	1876	1–20 s	-	30 kHz
SynD [130]	AUT	2020	R_s	1	21	180	5 z	5 Hz

3.2. Data Processing

Raw datasets can be employed differently for training and evaluating DNN-NILM approaches. Below, we review different aspects.

3.2.1. Training and Evaluation Scenarios

Training and evaluation of NILM algorithms can be done under different scenarios. Typical scenarios appearing in the literature are defined in the following.

OBSERVED VS. SYNTHETIC: In a *synthetic* scenario, the term $\sum w_t^k$ in Equation (1) is set to zero.

Corresponding data are typically created by summing up the power consumption from individual appliances. Only the measurement noise ϵ_t is therefore included in the noise term e_t. In an *observed* scenario, the noise term e_t also includes further appliances that have not been measured individually, i.e., $\sum w_t^k \neq 0$. The *synthetic* scenario can be considered a laboratory setting for a basic assessment of algorithms. Data in a real scenario will typically be *observed*.

We use here the terms *observed* and *synthetic* scenario equivalently to *noised* and *denoised* scenario, as these scenarios are commonly referred to in the literature. We introduce the new nomenclature because we believe that (i) the original terms are rather misleading for readers with less experience in the field, and (ii) the proposed terms express the essential difference between the two scenarios much more precisely.

SEEN VS. UNSEEN VS. CROSS-DOMAIN TRANSFER: The terms seen and unseen are used in the context of the evaluation of NILM algorithms. In the *seen* case, an algorithm is evaluated on new data from households that it has already been trained on. The resulting score gives, therefore, an indication on how well the trained algorithm can detect a particular appliance. *Unseen* means, that the algorithm is evaluated on data from a new household that was not available in the training data. This scenario tests the capability of algorithms to detect an appliance type [145]. Corresponding test results indicate the performance of a pre-trained model that is deployed on data from houses previously not seen during training. For the *cross-domain transfer learning* [91] scenario, the unseen house is taken from a different dataset. This scenario tests the transferability of the tested approach to an even more diverse setting as in the unseen case: Data could have been metered by different electrical meters or could originate from a different country. To our best knowledge, this scenario has only been investigated in [39,42,63,71,91]. The different scenarios are illustrated in Figure 2. The column 'Evaluation Scenario' in Table 2 lists the scenarios employed for the reviewed references.

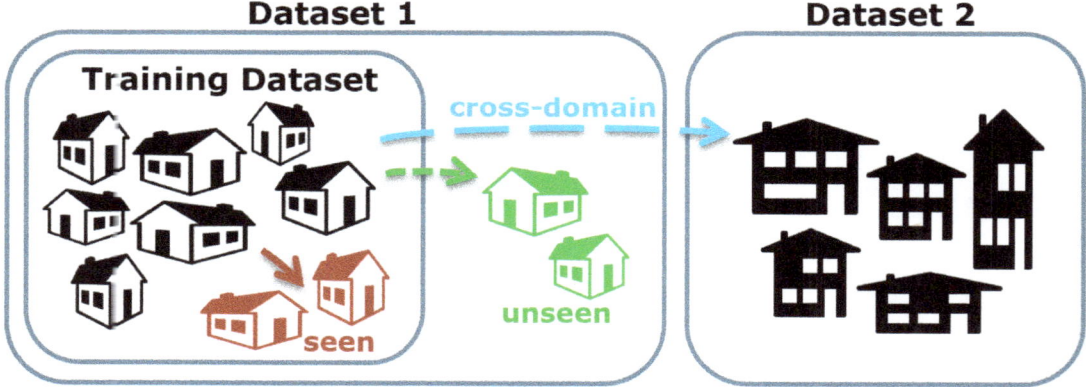

Figure 2. Different NILM evaluation scenarios: *seen*: the algorithm is evaluated on new data from a house that was already available during training; *unseen*: the algorithm is evaluated on data from a house not seen during training; *cross-domain transfer learning*: the algorithm is evaluated on data from a different dataset.

3.2.2. Preprocessing

Before data can be used by the DNNs, the raw data are transformed. Below we discuss typical data transformation steps employed in the literature.

RESAMPLING, FORWARD-FILLING, AND CLIPPING: The sampling frequencies of the published datasets are given in Table 3. As datasets exhibit missing values due to measurement or transmission equipment failures and a jitter in the timestamps, resampling is used to obtain evenly sampled data. While the range of sampling frequencies in the reviewed literature extends from $\frac{1}{3600}$ Hz [92,107] to 10 Hz [75], the large majority of the reviewed works employ either $\frac{1}{60}$ Hz or values between 1 and $\frac{1}{10}$ Hz. It is noteworthy that in two cases, data were upsampled to have a higher frequency than the original dataset [36,112]. Results on the influence of the sampling frequency on disaggregation results are presented in different studies [51,58,75,77]. Most of these studies find a marked dependence on the device [51,58,75]. This can be attributed to certain devices exhibiting more frequent fluctuations that get lost at lower resolution. Ref. [75] analyzes sampling rates from 10 Hz down to 0.03 Hz for on/off classification and energy estimation for TV, washing machine, rice cooker. They find that to prevent performance loss for the classification and regression tasks, the sampling rates should be at least 1 Hz and 3 Hz, respectively. Ref. [58] compares results obtained with 10 s and 1 min sampling intervals. The authors find "that the performance for dishwashers remains comparable while the performance for washing machine and washer dryer deteriorates dramatically". The publication [51] focuses exclusively on the influence of the sampling rate on the performance. The authors conclude that data sampled at 1/30 Hz might be sufficient to run NILM at high accuracy. It is important to note, that [51] did, contrary to [58,75], fix the number of inputs to the DNNs instead of the temporal window. Consequently, the temporal window seen by the network differs in this study depending on the sampling rate. Finally, [77] investigates the influence of the sampling rate in case of appliance on-event detection.

Short spans of missing data attributed to WiFi connectivity problems are forward-filled by many authors with the last available measurement. Typically, up to three minutes of missing data are filled in this manner [14]. In case of measurements exceeding the rating of the employed meter, values are clipped.

NORMALIZATION In the DNN-NILM literature, the input normalization for the DNNs comes in two main flavors:

$$x_{stdScaled} = \frac{x - \bar{x}}{\sigma(x)} \qquad (2)$$

$$x_{minmaxScaled} = \frac{x - x_{min}}{x_{max} - x_{min}} \qquad (3)$$

where x and x_{Scaled} are the input windows (see Section 3.3.1) before and after normalization. $\bar{x}$ corresponds to a mean value over the input. Different strategies have been employed: Most approaches calculate the mean over the complete training set so that the training data are centered. Other strategies center the data per house (see, e.g., [75]) or per input window (see, e.g., [14,107]). $\sigma(x)$ denotes the standard deviation, which is typically calculated on the complete training set. Alternatively, each input window was divided by the standard deviation from a random subset of the training data [14]. x_{max} and x_{min} correspond to maximal and minimal values. These values can be maximal or minimal values of the training dataset, parameters fixed by the authors [53], or quantile values [40]. In order to make the statistics of the data less sensitive to outliers, [44] transformed them with an *arcsinh* before normalizing. Some authors also normalized the target values for the training of the DNNs. While some publications mention that different normalization strategies were tried out, only two studies report on the influence of normalization strategies on training efficiency and testing performance: [34] finds that instance normalization [146] performs better than batch normalization [40,147] concludes that L_2-normalization works best.

3.2.3. Activation Balancing

In NILM literature, the time interval between an appliance being switched on and off is referred to as an *activation*. Domestic appliances exhibit typically one, up to several activations per day. Usually, the run-time of appliances is low compared to the time they are switched off. For the training of machine learning algorithms, one is consequently faced with a skewed dataset that contains only a few samples of the running appliance. To compensate, several authors balance samples with and without a (partial) activation during training [14,22,34,35,39,58,60,63,75,89,105,110,112,119,123]. The majority of works nevertheless train the models using the available data, without taking care of the class imbalance. In the scope of this review, we are only aware of [34], which investigates the effect of the ratio between samples with and without an activation on training results. They found that in case of batch normalization [147], the accuracy strongly decreased at a ratio of one to five, whereas for instance normalization [146], the performance increased slightly up to the largest tested ratio of one to seven. In general, it remains unclear how exact activation balancing influences the disaggregation quality and model convergence speed.

3.2.4. Data Augmentation

A common strategy in deep learning to overcome few labeled data samples or underrepresented classes employs data augmentation. It describes the process of transforming existing measured data or creating synthetic new data in order to achieve DNNs that generalize better. Recent overviews for data augmentation in the domains of computer vision and time series can be found in [148–150].

In the reviewed DNN-NILM literature, we see different data augmentation variants. First, some approaches train in a synthetic scenario, see Section 3.2.1. Only synthetic data consisting of the summed up loads from appliance sub-meters are used to train and test the algorithms. Such publications are denoted with a 'dn' in the column 'data augmentation' Table 2. A second group of publications train on measured aggregate data but add synthetic data—also created by summing up sub-metered load curves—to increase the size of the training set. Four authors added individual activations from appliances to a measured aggregate [6,58,66,115]. Finally, some authors employed specialized strategies: The authors of [35] found that by adding varying offsets specifically to the on state of the fridge, they were able to greatly enhance the corresponding disaggregation performance. So-called 'background filtering' has been proposed by [69] to remove all windows in the aggregate load curve that contain the target appliance. Activations from the target appliance are then added randomly to the filtered aggregate to create synthetic data for training. The authors of [44] use data obtained from SMACH [151], a tool that generates synthetic data based on time of use surveys and real appliances signatures. They compare scenarios with different amounts synthetic data and find good generalization performance for models trained only on synthetic data. We are not aware of any study that compares different data augmentation strategies.

3.3. Input

3.3.1. Shape

The vast majority of approaches take as a continuous regularly sampled window from the time series of the aggregate measurement data input for the DNNs. The range of employed window lengths extends from 90 s [75] to around 9 h [94,98] or even 24 h [66,107]. It is important to note that the number of input samples to the neural networks, i.e., the number of neurons in the first layer, depends on the sampling rate. Extreme values for the size of the input layer are 5 [100], 7 [83], and 10800 [112]. The influence of the window length on the disaggregation performance at a fixed sampling rate is investigated in [52,62,97]. While the investigations have been done on different datasets and sampling rates ([97] → UK-DALE, REDD at 3 and 6 Hz, [62] → REFIT at 10 Hz, and [52] → DRED 1 Hz), all authors find that the optimal window length depends on the appliance.

Few authors transform the time series data into a two-dimensional representation before feeding it into a DNN. Corresponding publications are marked with 'P_{2D}' in the column 'Input' of Table 2: [38,71,123] use the Grammian Angular Field (GAF) [152] to transform continuous part of the aggregate measurement into a two-dimensional representation which is then fed to a convolutional neural network (CNN). Ref. [38] additionally compares the performance of the GAF with the Markov Transition Field [152] and a Recurrence Plot [153] and finds that the GAF outperforms the other imaging techniques in the vast majority experiments. Ref. [92], on the other hand, arranged hourly consumption readings into two dimensions by setting the hour of the day as the x-coordinate and the day as the y-coordinate. The authors found that the optimal size of the first filter amounts to 7 × 7, allowing the filter to learn weekly correlations.

A minority of works use a DNN to classify events extracted by a previous detection stage [77,80,114,116] or to classify the on–off status directly from a single time step [82]. These publications are marked with 'na' in the column 'Input' of Table 2.

3.3.2. Features

The active power from the aggregate measurement is the only input for most of the reviewed works. There are, however, a number of papers that extended the input to further features: Reactive and apparent power, current, first-order difference of the active power signal, power factor, the variant power signature, and different time-based features have been used additionally. A noteworthy case is the input of the aggregate power from multiple neighboring buildings which, according to [107], lead to considerable performance improvements. Input features of the reviewed publications are marked in the column 'Input' of Table 2. See also Section 4.2 for a discussion of the benefits of multiple input features.

3.4. Deep Neural Networks

3.4.1. Architectures Elements

In Table 2, column 'DNN Elements', we summarize proposed DNN architectures based on a set of DNN building blocks or elements. Naturally, this attempt can only be a coarse approximation of the encountered diversity. It still provides a high-level view on what has been tried out. We mention only original architectures proposed by the authors. Models from earlier authors or baselines are not listed. As a consequence the column is empty in several cases, e.g., where the authors compare previous works. Looking at Table 2, we observe that starting with the year 2018, feedforward elements—in particular convolutional elements—gained in popularity. These elements are used roughly twice as much as recurrent elements. In the same time span, advanced DNN elements such as generative adversarial networks (GAN) and attention were also adapted to NILM.

Below, we give a description of the DNN elements used in Table 2 to describe the proposed models:

- FF → Feedforward network, see, e.g., [154]. We use the abbreviation in case a simple multilayer feedforward network is a major component of the network. It is not used for dAE, CNNs, or LSTM networks that contain, e.g., only a final feedforward layer for classification.
- dAE → Denoising autoencoder [155]. We use the abbreviation for architectures made up of encoder-decoder architectures. In the context of NILM, denoising AE are used to separate the appliance's signal from the rest of the aggregate signal.
- VAE → Variational autoencoder. Special variant of the AE that encodes the latent variables as distributions [156,157].
- CNN → Convolutional neural network, see, e.g., [154]. We use the abbreviation for networks that employ CNN layers. s2p, s2sub, s2s, and wn are more detailed subclasses.
- s2s → The abbreviation is used for networks that map from an input sequence to an output sequence with identical length. It is only used if the output of the network is active power or the on/off state of an appliance.
- s2sub → The abbreviation is used for networks that map from an input sequence to an output sub-sequence, i.e., the output length is smaller that the input length. It is only used if the output of the network is active power or the on/off state of an appliance.
- s2p → The abbreviation is used for networks that map from an input sequence to a single output value. It is only used if the output of the network is active power or the on/off state of an appliance.
- wn → Wavenet [158] inspired architectures. In particular, dilated convolutions and gating mechanisms are important elements of these architectures.
- att → This abbreviation subsumes different variants of attention mechanisms. Specifically, ref. [55] used the mechanism defined in [159], ref. [35] employed attention from [160], and [72] from [161].

- RNN → The abbreviation is used for networks that employ vanilla recurrent neural networks, see, e.g., [154].
- (bi)LSTM → The abbreviation is used for networks that employ (bidirectional) long short-term memory cells [162,163].
- (bi)GRU → (bidirectional) gated recurrent unit [164].
- GAN → The abbreviation is used for networks that employ elements from generative adversarial networks [165].
- RCNN → recurrent CNN [166].
- HMM-DNN → combination of Hidden Markov Model with a classic feed-forward network [122].
- VRNN → recurrent variational neural network [167].

In case different elements have been combined, they are joined with a hyphen '-'. That means, e.g., CNN-dAE corresponds to a dAE that includes convolutional layers.

3.4.2. Training and Loss Functions

DNN gradient descent has its own set of hyperparameters such as the type of optimizer, number of training epochs, or early stopping criterion. As these elements are not specific to the NILM problem, we just mention a few specific points in this section. There have been several authors that optimized (training) parameters of the proposed networks by automatic means. For example, grid search [72,92], Hill Climbing [41], and Bayesian optimization [42,75,90,102] have been employed. [118] investigated different variants of curriculum learning [168]. In this type of learning, samples are not randomly presented to the DNN, but organized in a meaningful order, the intuition being that humans learn by mastering concepts with increasing difficulty. Contrary to this intuition, ref. [118] finds that easy samples hinder training, and the author used synthetic training data composed from sets of more than 7 appliance sub-meters.

A key element of DNN optimization is the loss function that guides the optimization process. The vast majority of works employ either the mean absolute error (MAE) or the mean squared error (MSE) in case of power disaggregation and the cross entropy loss for on/off classification. Recent works also investigate alternative loss functions: Quantile regression [169] was employed by [46,59]. The authors of [59] found that their proposed loss increased the performance of two state-of-the-art models compared to the MSE loss. Some works [34,35,39,43] employ GAN loss functions—called 'adversarial loss' in [35]—that classify if the output of the regression DNN is a real or fake appliance load curve. This loss should make outputs more realistic and help especially in case of datasets of limited size. Finally, ref. [55] introduced a loss composed from four different terms: In addition to the MSE, a Kullback–Leibler divergence loss, a soft-margin loss, and the MAE are used. To our knowledge, a systematic comparison of loss functions for DNN-NILM approaches has not been published.

3.5. Output

With respect to their output, we observe four different dimensions along which DNN-NILM approaches can be distinguished: A first dimension is the number of time steps that are disaggregated by the DNN models in a single go, be it a sequence, subsequence, or a single value. This information is available through the abbreviations s2, s2sub, and s2p in the column 'DNN Elements' of Table 2, see also Section 3.4.1. The second dimension concerns the type of inferred output. With the exception of [70], where location data (*location*) are combined with the aggregated power consumption, and [95], where DNN infers state changes in the aggregate power (*stateChange*), the goal of DNN-NILM approaches is either to infer the on/off state or the energy usage of an appliance. We mark this information in column 'Output' of Table 2 with the abbreviations *on/off* and *E* respectively. Naturally, this dimension is mostly coupled with the third distinction, whether the learning problem is formulated as a classification or regression task. However, there are four works [33,61,68,118] where power values are clustered into groups, and the power regression problem is recast into a classification task. These references are marked with P_{class}. Lastly, we can distinguish between approaches that learn on a single task and those learning on multiple tasks simultaneously, i.e., perform multi-task learning. The majority of approaches train one model for each appliance to be disaggregated. A sizable number of approaches infer, however, the on/off state or power disaggregation of multiple appliances simultaneously. These cases are marked with a subscript m in the 'Output' column of Table 2. Where multi-task learning is done on different modes, the corresponding outputs are jointed with an '&' in Table 2: [35,63,97,106] trained networks on both on/off and active power data. Ref. [117] used both active and reactive power of an appliance as target, and [59] used three targets, i.e., the aggregate power (P_{total}), the appliance power (P_{app}), and the difference between the two (P_{rest}). Finally, ref. [46] took multi-task learning furthest by simultaneously learning on both on/off states and active power of multiple appliances.

3.5.1. Post-Processing

Different variants to process the output from the DNN-NILM approaches have been proposed in the literature. In cases where the DNN is of type *s2s* or *s2sub* and disaggregation is done by moving the input window a single time step at a time, the network will deliver n predictions for each time step, where n is the length of the output window. In order to obtain a single prediction, many authors used the mean, e.g., [14,22,35,89,119] or the median, e.g., [21,72]. In [21], the authors find that networks underestimate the power of appliances if activations are only partially in the input window. As a consequence, the mean also underestimates the ground truth and [21] proposes to use the median instead, which is less impacted by this problem. As the authors of [39] use a GAN, they only use the disaggregated signal for which the discriminator outputs the highest probability of being a true sample.

Some authors note that the models produce noisy output, e.g., in the form of sporadic activations either too short or too frequent for the target appliance. Ref. [37] filters out such events with the same approach as for activation detection in the ground truth data. Similarly, [36] removes all activations of an appliance that are shorter than those found in the ground truth data. Depending on the metric, the reported improvement ranges from 28% to 54%. Refs. [58,89] go one step further and train a second DNN to suppress spurious activations. According to [58], the additional DNN leads to "significant performance boosts".

3.6. Evaluation Metrics

The performance of NILM algorithm is assessed in various ways. The interested reader is referred to [18,145]: Ref. [18] provides a comprehensive review and discussion of employed metrics, and ref. [145] proposes a set of metrics to assess the generalization ability of NILM aglorithms. In the following, we only repeat the definition of the mean absolute error (MAE) and the F_1-score. In the reviewed literature, these were the most encountered metrics to assess the estimated energy consumption and on/off status of an appliance, and we use them for our comparison in Section 4.1.

$$\text{MAE} = \sum_t^T \frac{|y_t - \hat{y}_t|}{T} \tag{4}$$

where the sum goes over T time steps, and y_t, $\hat{y}_t$ correspond to the measured and estimated power consumption, respectively. In this publication, we use Watts as the unit for the MAE.

$$F_1 = 2 \cdot \frac{P \cdot R}{P + R} \tag{5}$$

where precision $P = TP/(TP + FP)$ and recall $R = TP/(TP + FN)$ with TP, FP, and FN denoting true positives, false positives, and false negatives, respectively [18].

4. Discussion and Current Research Gaps

The following sections discuss different aspects of the reviewed literature. Each section concludes with a paragraph on current research gaps we see concerning the discussed topic.

4.1. Performance Comparison

One of the basic questions that accompanied us throughout this literature review was "What is the most promising approach or classes of approaches?" As the last section hints, there is no straightforward answer to that. Too many degrees of freedom (see Figure) make the approaches differ in so many ways that a comparison based solely on the results presented in the publications can only give indications. For that purpose, the MAE and F_1 score were extracted from the reviewed publications wherever available. (The data available on our GitHub account. The link is provided in the 'Supplementary Materials'. These two metrics are the most applied performance measures in case of energy estimation and on/off state classification, respectively. Figures 3 and 4 each display the best reported results split up by dataset and appliance. Only results from the *observed, unseen* evaluation scenario are given. This scenario was selected as it is closest to an actual application DNN-NILM algorithms, see Section 3.2.1. The graphs only include results from approaches proposed in the corresponding publications: Results from baselines, or approaches from earlier work that were used for comparison, are not included. Appliances with a single result in the displayed range are excluded. We observe that the results for kettle and microwave are overall better and not as distributed as those from the other displayed appliances. We believe this is because of their simpler nature: Both kettle and microwave are appliances with only two states whereas dishwasher, washing machine, and fridge (to a certain degree) have a more diverse load signature.

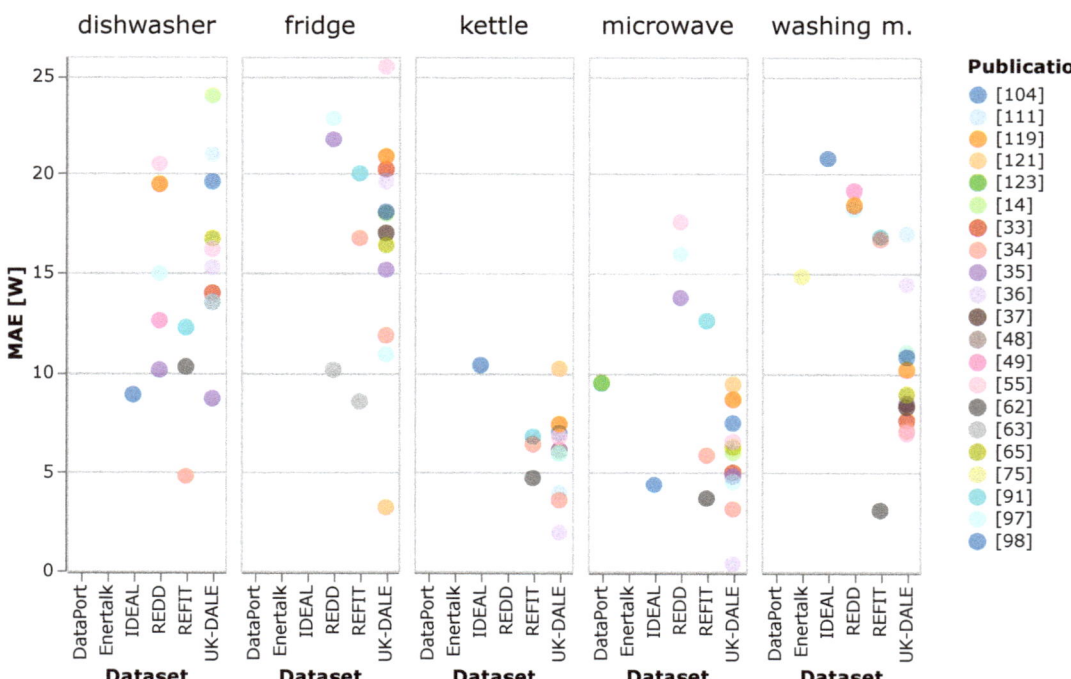

Figure 3. Minimal reported MAE for the corresponding dataset and appliance. Only results from the *observed, unseen* evaluation scenario have been included. Only approaches proposed by the authors in the corresponding publications are taken into consideration (i.e., no baselines or models from the state-of-the-art). Results have been split according to the appliance and employed dataset. Please note that appliances with a single result in the selected range are not shown.

We caution the reader *not to interpret the displayed values as the result of a direct comparison under identical conditions*. Results have been generated under broadly differing settings, see Table 2. One key difference is that evaluation data varied strongly between publications. While results in Figures 3 and 4 are not directly comparable, we try to identify common elements of successful approaches. For that purpose, we sorted the results for each appliance (irrespective of the dataset) and took the top quarter of the results. Depending on the appliance, a quarter consisted of four to six results in case of the MAE and two to four in case of the F_1-score. We then evaluated the number of times a publication appears in these results. Those with *more than one count* are [34,35] (five times), [62] (four times), and [36,63,97,98] (two times) in case of MAE, and [63] (six times), [118] (three times), and [36,37,64] (two times) for the F_1-score. These publications have been marked in column 'Best' of Table 2.

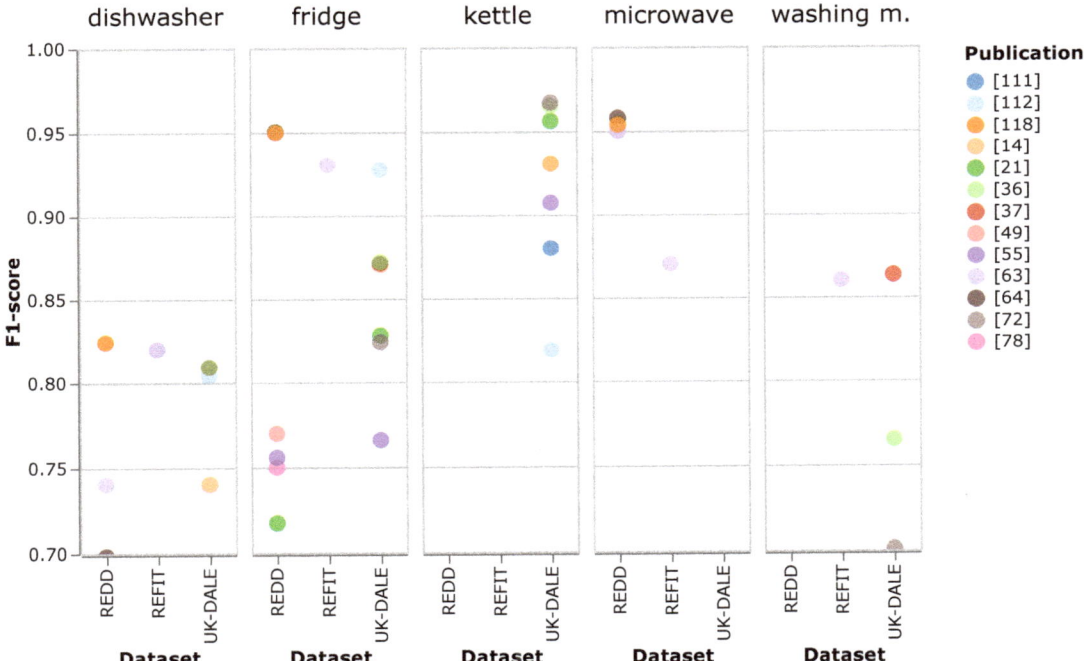

Figure 4. Maximal reported F_1-score for the corresponding dataset and appliance. Only results from the *observed, unseen* evaluation scenario have been included. Only approaches proposed by the authors in the corresponding publications are taken into consideration (i.e., no baselines or models from the state-of-the-art). Results have been split according to the appliance and employed dataset. Please note that appliances with a single result in the selected range are not shown.

Based on these publications, we make the following observations:

- With the exception of [37], who used 60 s intervals, the best results are based on data with sampling intervals up to 10 s.
- The architectures of [34,35,37,62,98] contain all elements that allow an output neuron to have a large field of view. Refs. [35,62,98] use dilated convolutions that allow for an exponential increase in the field of view of deeper network layers, see, e.g., WaveNet [158] that served as template in case of [62]. The models from [35,37,62] all concatenate and process at some stage in the network output from multiple layers that each have widely varying fields of view. This is called 'scale-awareness' by [35] and 'temporal pooling' by [37]. Ref. [34] adopts the U-net [170], an encoder–decoder architecture originally proposed for image segmentation.

- The approaches of [34,35] employ a GAN setup. This means that the network does not (only) learn based on a static loss function, but it rather exploits the 'feedback' of second network that classifies if the output of the first one is real or fake.
- Four approaches [35,37,63,97] use multi-task learning for network training: ref. [37] trains the model to classify the on/off state of multiple appliances at the same time whereas [35,63,97] simultaneously train their networks to estimate the energy usage and the on/off state of single appliances. In the latter three cases, this is done with networks that have one branch for each objective. The final result is then obtained by combining the branches.
- Refs. [36,37] use very similar post-processing schemes that remove sporadic activations, see Section 3.5.1. Ref. [36] reports improvements in the range of 28% to 54% compared to results without post-processing.

We see the following limitations in the previously performed comparison of the MAE and F_1-scores: As already pointed out, results have been obtained through wildly different procedures. It is also questionable if these metrics are the most relevant, they have simply been chosen as the ones mostly provided in the reviewed literature. That means also that publications reporting other metrics do not appear in the evaluation. With these words of caution said, we still believe the previous observations provide some value.

With respect to the initial question "How does the performance of approaches compare?" and the performed literature review, we identify several challenges worth addressing by the research community: We observe that the experiments performed in the reviewed literature are not always well specified with respect to the 'degrees of freedom' mentioned in Section 3. We hope future works profit from our listing of available options and specify their decisions clearly. Besides, we see several additional steps at different levels that could lead to a better comparability:

- Beside the metrics motivated by the intended use case, published approaches should report a set of standard metrics. Simply based on their availability in the reviewed literature, we propose to use the MAE and F_1-score in case of energy estimation and on/off state classification tasks, respectively.
- We see a great potential in defining a standard evaluation protocol that defines training and testing folds for cross-validation of models per dataset. Of course it should respect particularities of the NILM setting such as the evaluation scenarios, and it would ideally be in a machine readable form such as proposed for the ExperimentAPI of NILMTK [23].
- Authors would ideally publish the code for the employed models and experiments as already done by several authors, see Table 2. Based on that code, retraining for comparison with new approaches is greatly simplified.
- Beside the model specification as code, trained models could also be released as is done in the vision community, see, e.g., [171]. We are only aware of the trained models of [49] that have been publicly released.
- As has already been noted by other authors [19,140], a comparison of NILM approaches on the same terms is very beneficial for NILM practitioners. In case of DNN-NILM approaches, we feel that this work can serve as a good basis for such an undertaking providing a comprehensive overview on the relevant literature and published code. A Python framework for comparison has already been published in [23] and applied to the analysis of four DNN-NILM (beside classical) approaches in [24]. A possible route to stimulate such comparisons would also be to organize challenges like the one which took place in 2020 [172]. In the computer vision domain, for example, the ImageNet challenge was a great driver for innovation.

4.2. Multiple Input Features

While the reactive power has already been used in the founding works of NILM [1,2], most authors take the active power as the only input for disaggregation, see Section 3.3.

Therefore, we raise the question: "Can we find evidence that multiple input features benefit DNN-NILM performance?"

As can be seen in the overview Table 2, different authors employed alternative input features. The following authors report results from a comparison of input features [22,36,89,94,106,107,117]. Ref. [117] was the first DNN-NILM approach we are aware of that used multiple input features. Unfortunately, these results do not allow separation of the influence of input features from multi-task learning because the two are always used in conjunction. Ref. [22] explicitly exploits reactive power (Q). The authors evaluate the impact of Q on the F_1-score within the AMPds and UK-DALE datasets. Over the investigated appliances, they find an average improvement of around 12.5% in the seen and 8% in the unseen evaluation scenario. (Average F_1-scores for P in seen and unseen scenarios are 0.68 and 0.58, respectively.) Interestingly, the reported improvement is small or negative for purely resistive loads, such as kettle and electric oven. We hypothesize that in such cases, the reactive power provides no information, but purely noise. Refs. [89,94,106] all worked with the AMPds dataset. This data set contains measurements from a single house, thus all results stem from seen evaluation scenarios. Ref. [94] compares two feature sets with the help of the estimation accuracy: For the combination P and Q versus P alone, the authors find an improvement of 6%. For the combination P, Q, current (I), and apparent power (S), the improvement is slightly higher at 7%. (The estimation accuracy in case of P alone amounts to 0.83.) Ref. [89] investigates the same feature set P, Q, I, S versus P based on three different performance measures, namely the MAE, root mean square error (RMSE), and the normalized RMSE. In this work, the improvements with the additional features are much larger: For all measures, the average improvement is around 40% to 50%. (The MAE, RMSE, and normalized RMSE averaged over all investigated appliances and models for P alone amount to 36.7 W, 122.8 W, and 0.75, respectively.) The temperature as a supplementary feature has been used by [106]. The authors find that the disaggregation of 'heat pump' and 'home office' works 3% and 4% better based on the F_1-score and estimation accuracy, respectively. (The F_1-score and estimation accuracy averaged over the two appliances in case of P alone amount to 0.87 and 0.91, respectively.) The authors of [107] find that providing the aggregate electrical consumption from neighbors as additional features leads to performance improvements of 17% and 31% with and without multi-task learning, respectively. (The symmetric mean absolute percentage error (SMAPE) for P alone amounts to 23% and 38% in the respective cases.) Finally, ref. [36] calculates the mutual information between P, Q, S, I, voltage, the power factor of the aggregate measurement, and P of the appliance as a feature selection step. Voltage is the least informative feature, and is therefore dropped for the subsequent evaluations.

Previously mentioned improvements have been calculated from the original values reported by the authors according to the following formula

$$Improvement = \frac{Perf_{addFeatures} - Perf_{onlyP}}{Perf_{onlyP}}. \qquad (6)$$

where $Perf_{onlyP}$ and $Perf_{addFeatures}$ correspond to the performance (measured in any measure) of the approach based only on P and additional features, respectively. In cases where smaller values indicate better performance of a measure (e.g., MAE), we swapped $Perf_{addFeatures}$ and $Perf_{onlyP}$ to always result with a positive value for improved results.

Based on the results presented, we conclude that features beside P can improve disaggregation performance. No conclusions about the amount of improvement can, however, be made, as the spread of the results is quite broad. For the time being, we can only speculate about possible reasons. It might be a worthwhile investigation to examine what kind of factors, e.g., architectures, can make the most out of the information from features beside P. With the exception of those in [22], all results originate from seen evaluation scenarios. That effectively means that additional features help to estimate the power usage of a particular appliance. However, it is unclear how much they help to disaggregate an appliance *type*, see Section 3.2.1. Non DNN-NILM approaches already employed a very broad feature set [13]. Compared to this breadth, DNN-NILM approaches tested a very limited set of options. It would be interesting to see a systematic investigation on a broader feature set.

4.3. Multi-Task Learning

If a machine learning model trains on separate but related tasks, this process is referred to as multi-task learning. For a good introduction and overview on the topic with respect to deep learning, the reader is referred to [173]. The NILM problem is suitable to be framed as a multi-task learning problem: The column 'Output' in Table 2 lists the different variants that have been employed in the reviewed literature. We asked ourselves: "Can we find evidence that multi-task learning leads to superior performance compared to single task learning in the case of the DNN-NILM approaches?"

Based on the literature review, we found the following: Ref. [91] trains a CNN on the washing machine, freezes the parameters of the convolutional layers, and retrains afterwards only the final, fully connected layer for other appliances. The authors find that the results of this approach are comparable to standard training. This finding suggests that the learned features of different appliances are similar and can be shared between appliances. Simultaneous learning on different appliances could therefore make features more robust and lower the requirements on the amount of training data. A large improvement from joint learning on multiple appliances is also reported by [107] and, as was already mentioned in Section 4.1, four of the best approaches [35,37,63,97] use multi-task learning for network training. Only the authors of [49] report a general decrease in performance of multi-task learning models with respect to their single-task counterparts. They propose to employ a different architecture or share less layers between appliances as a remedy. Due to the presented observations and the general benefits of multi-task learning presented in [173], we conclude that multi-task learning is beneficial for DNN-NILM approaches. As has also been noted by [49], we see the additional benefit of multi-task learning in a reduced computational burden for edge devices because a major amount of computation for disaggregation can be shared between several applications.

4.4. Parameter Studies

As visualized in Figure 1, there are many degrees of freedom for DNN-NILM approaches. In Section 3, we listed the many options for the corresponding aspects that have already been tried out. Looking at the literature, however, we see a lack of understanding of the influence of the available options. Therefore, we want to stress the need and value of parameter studies for future research activities in the DNN-NILM field.

For example, in case of the data *sampling rate* and *window length*, several authors looked at the influence of these two parameters on the models performance, see Sections 3.2.2 and 3.3.1. There exists, however, no study that jointly investigates the two tightly connected parameters (maybe even on different datasets and based on different models). Similarly, we see potential in a systematic comparison between different normalization (Section 3.2.2), activation balancing (Section 3.2.3), data augmentation (Section 3.2.4), and post-processing (Section 3.5.1) strategies, as well as loss functions (Section 3.4.2).

4.5. Applied DNN-NILM

The best results of current DNN-NILM approaches are very promising, see Section 4. However, there are different aspects of relevance for an actual deployment of DNN-NILM approaches that have not yet been well investigated in the literature. In the following subsections, we motivate corresponding aspects and subsequently point out connected research gaps.

4.5.1. Data Scarcity

For a practical application of NILM, we see one of the main challenges in the scarcity of labeled data. While this challenge is not specific to DNN approaches, we think that recent developments in semi-supervised deep learning might be adaptable to NILM and could then be a great opportunity to tackle the problem of data scarcity for practical applications. In the following, we will first detail the challenge and subsequently formulate possible future research directions.

Net2Grid is a company providing NILM services to utilities. In a presentation, they stressed the point that "accurate NILM requires [...] a lot of high-quality data" [174]. Specifically, the company bases their NILM service on data from *hundreds* of houses. They also emphasized the point that machines with different programs or settings exhibit very variable load patterns and therefore need many observed cycles. The authors of [75] investigate how the disaggregation error of a DNN-NILM approach depends on the number of distinct houses used for its training. In agreement with Net2Grid, they find that for the washing machine, an appliance with a variety of programs, the disaggregation error steadily decreases with each house added to the training dataset without any sign of saturation until 40 houses, which was the maximum used for the investigation. Thus, both sources indicate that complex machines require a large variability in the training data to successfully generalize on unseen data. This observation is at the core of what we call 'data scarcity'.

These findings indicate that a company that wants to start a NILM service first has to obtain data from hundreds of houses. A possible source are public NILM datasets including aggregate and appliance consumption, see, e.g., [18,127] for an overview. While for some simple appliances these public datasets are certainly sufficient, the data are too restricted if we want to disaggregate appliances with variable load patterns [75,174]. Furthermore, appliances such as heat pumps or charging stations for electric vehicles are almost absent from public datasets. The only alternative is to engage in a measurement campaign involving *hundreds* of houses. This is an expensive and time consuming undertaking, as the metering, appliance-specific sub-metering, and the corresponding infrastructure has to be installed and maintained. It is also worth noting that, even if large datasets covering a large variety of appliances would be recorded, new devices come to the market continuously. This means that the effort for data collection is actually a recurrent one.

As DNNs are particularly data hungry, the problem of a shortage of labeled data has recently obtained a lot of attention by the research community in the computer vision and natural language processing (NLP) domain. A promising remedy to the situation is semi-supervised deep learning [175]. We see a lot of potential transferring these developments to the NILM domain as unlabeled data—the data obtained from the smart meter—are relatively easy to access compared to sub-metered ground truth data.

In the reviewed DNN-NILM literature, semi-supervised deep learning techniques have so far been employed by the following authors: Ref. [103] used the ladder network [176]. The presented results give no clear indication that unsupervised data actually improved the performance. These results might be caused by the relatively simple DNN. The authors of [66] trained an autoencoder on unlabeled data to subsequently use the learned embedding in a supervised training setting. This work was done on aggregate data with 15 min resolution that naturally led to large estimation errors. Ref. [81] derived their classification approach from the mean teacher approach [177], while ref. [67] adopted virtual adversarial training [178]. Both works present evidence that the disaggregation performance of the semi-supervised approaches improves compared to a strictly supervised settings. However, experiments were only conducted on data from houses already seen during training, and no conclusion can be drawn about improved generalization on previously unseen houses.

There is an increasing field of newer semi-supervised DNN approaches from the vision domain ready to be adapted to the NILM problem [175]. A particularly successful [179] semi-supervised strain of research is called *consistency learning* [175]. The method's main assumption is that a small perturbation or realistic transformation applied to a data point should not have an influence on the prediction. DNNs are then trained to provide a consistent output for an unlabeled data point and its perturbed version. Most recent publications demonstrate that for image classification it is feasible to get close to the performance of supervised approach with one order of magnitude fewer labeled samples [180,181]. While some of the consistency learning approaches seem to be well adaptable to NILM, there remain many open questions—to name a few: What type of consistency loss should be used in case of NILM? What types of data augmentation strategies should be employed? The last question is of particular interest because [180] demonstrated that the 'quality' of data transformation is the key for significant performance gains. While data augmentation was used in various DNN-NILM approaches, see Table 2, we are not aware of any work that did a detailed investigation of this aspect.

In summary, we see in the application of semi-supervised DNN approaches many worthwhile research questions and a great opportunity to tackle the problem of data scarcity for practical NILM applications.

4.5.2. NILM on Embedded Systems

If one imagines a NILM deployment at scale, the amount of data to transfer, store and process becomes an important factor. In the reviewed DNN-NILM literature, different aspects of such a deployment have been addressed. The work of [93] investigated data reduction policies: Different sampling strategies for data compression ($\frac{1}{4}$ to $\frac{1}{20}$ compression) in combination with DNN inference are tested. The authors found that the best sub-sampling policies outperform results with original sampling rates. Another option is to process the data directly in or close to the electric meter and only relay disaggregated high level information. That means that the DNN-NILM inference has to work on an embedded system, even though that can be quite challenging in terms of computational, storage, and energy resources. This direction has been investigated by [47–49,65,106]: Ref. [106] is to our knowledge the first to publish the implementation of DNN-NILM inference on an embedded device. Both [106] and later [65] used for that purpose a Raspberry Pi computer. Ref. [47] uses an efficient MobileNet [182,183] inspired DNN for disaggregation and compresses it by lowering the precision from 32 to bit floating point (used for training) to 8-bit integer representation by means of the TensorFlow Lite library. The resulting model was then evaluated with the Android SDK. The authors report that "disaggregation accuracy deviates up to ≈9.4% from original disaggregation model, but, on average, remains satisfactory". Both refs. [48] and [49] investigate different pruning methods based on the network from [119]. Pruning methods aim at reducing neurons in the network that contribute little to the final output. The final goal is to result with sparse networks that have lower storage and computational requirements but similar performance compared to the original networks. Both publications found that networks can be heavily pruned with only a slight decrease in performance: Ref. [48] reports a reduction of the number of network weights by 87% and [49] reports a 100-fold reduction in model size and a 25-fold reduction in inference times. Ref. [49] additionally investigates multi-task learning and vector decomposition as further paths towards efficient computation in embedded systems.

While the DNN-NILM community has taken first steps towards an implementation on embedded devices, the corresponding research field for DNNs in the vision and speech domain is vast, see, e.g., [184]. There remain therefore a multitude of research questions in this direction. From our perspective, an interesting question would be to see how best performing approaches (see Section 4.1) could be adapted to embedded devices because the architectures of these approaches are more elaborate than the ones used by [47–49,106].

4.5.3. 3-Phase Data

In some European countries, such as, e.g., Switzerland, residential power supply arrives in three phases at the master distribution board (breaker panel) and is then split into single phases. As a consequence, measurements from the electrical metering infrastructure are in principle also available on three phases. With respect to a practical NILM application, this additional information makes the problem at first glance easier to solve, as there are on average one third as many devices connected to each phase compared to households attached to a single phase. However, the challenge comes in the form of multi-phase appliances such as heat pumps, pool pumps, electrical heat storage radiators, or charging stations for electrical vehicles. These appliances require NILM algorithm to combine information from all three phases. When considering an approach that should perform on any households, the main challenge is that multi-phase devices can be connected in arbitrary permutations. Thus, the result of the DNN-NILM approach needs to be invariant to these permutations.

We are not aware of any DNN-NILM publication that works on 3-phase data and tackles the raised challenge (This might partially be because there are currently only few datasets with 3-phase information. We are aware of iAWE [185], ECO [140], and BLOND [186]). The desired permutation invariance is analog to the required rotational invariance in computer vision: An object needs to be recognized as such independently of its orientation in the image. This analogy points also to possible future research questions: Could permutation invariance be obtained by training a DNN with augmented data? Could the symmetry be directly anchored in the layer of the neural network via Group Equivariant Convolutions, see, e.g., [187,188]? These are convolutional layers specially designed to produce the same result for data subject to a group of symmetry operations. How do these two solution approaches compare to each other with respect to performance and complexity?

5. Outlook

Looking into the future, we can imagine different scenarios and directions for the (DNN-)NILM field. With the rapid development of the Internet of Things, we can well think of future appliances which are aware of their own current (and possibly future) energy consumption and feature a communication interface to relay this information to the outside world. In this scenario of energy-aware appliances, NILM would become obsolete. As this scenario would require a business case for appliance manufacturers and standards for interfaces and protocols, chances are good that this state will not be reached in the near future. We believe that the rapid increase of computing power in edge devices will have a much more immediate impact. Edge nodes will soon be able to perform DNN-NILM close to the meter without the need to transfer data to a cloud computing service. The culmination of this trend would be complex NILM algorithms that run directly on meter hardware, maybe even on the raw high frequency measurement data. Developing this scenario even further, one could imagine that NILM algorithms learn and improve on local data. For this to work, the learning problem will first have to be formulated in a way that the data available on the meter can be used for further improvements. A standard supervised training approach does not seem to be feasible. Furthermore, local improvements of the model will ideally also be made available to other smart meters. This concept of local learning with global exchange of improvements is a nascent research field called Federated Learning, see [189,190].

6. Conclusions

Summarizing, this publication presents a review on the DNN-NILM literature. The scope of this review comprises publications that employ deep neural networks to disaggregate appliances from low frequency data, i.e., data with sampling rates lower than the AC base frequency. Our motivation for the scope is our conviction that plenty of applications could benefit from NILM, coupled with the observation that low frequency data will most likely be available at scale in the near future and the enormous success of DNNs in other application domains. We systematically discuss the many degrees of freedom of these approaches and what has already been tested and tried out in the literature along these dimensions. One of the main contributions is Table 2, which gives a structured overview of the main characteristics of all reviewed DNN-NILM approaches. The review part is followed by a discussion of selected DNN-NILM aspects and corresponding research gaps.

We present a performance comparisons with respect to reported MAE and F_1-scores a[nd] observed different recurring elements in the best performing approaches, namely da[ta] sampling intervals below 10 s, a large field of view, the usage of GAN losses, multi-ta[sk] learning, and post-processing. Subsequently, the benefit of multiple input features a[nd] multi-task learning and related research gaps has been discussed, the need for comparati[ve] studies has been highlighted, and the missing elements for a successful deployment [of] DNN-NILM approaches have been pointed out. Finally, we also outline potential futu[re] scenarios for the NILM field. This contribution is currently missing in the literature, and ca[n] therefore be of value. We conclude that there remain many worthwhile research questio[ns] to be pursued.

Supplementary Materials: To facilitate future work based on the data collected for this publicati[on] we release Table 2 as a MS Excel file. We also provide data and code that was used to genera[te] Figures 3 and 4. All data and code is available at https://github.com/ihomelab/dnn4nilm_overvie[w] accessed on 11 January 2021.

Author Contributions: Conceptualization, P.H., A.R. and A.P.; data curation, P.H.; formal analys[is,] P.H.; funding acquisition, A.R. and A.P.;investigation, P.H.;project administration, A.R.;supervisi[on,] A.R. and A.P.;visualization, P.H.;writing, P.H.;writing—review and editing, P.H., A.C., A.R. and A[.P.] All authors have read and agreed to the published version of the manuscript

Funding: This research was funded by Innosuisse—Schweizerische Agentur für Innovatio[ns] förderung, grant number 36152.1 IP-EE and the Lucerne University of Applied Sciences and A[rts.] The APC was funded by the Lucerne University of Applied Sciences and Arts.

Institutional Review Board Statement: Not applicable.

Informed Consent Statement: Not applicable.

Data Availability Statement: The data presented in this study are available at https://github.co[m/] ihomelab/dnn4nilm_overview accessed on 11 January 2021.

Acknowledgments: We want to express our gratitude to Gianni Gugolz, who supported us comp[il]ing Table 3.

Conflicts of Interest: The authors declare no conflict of interest.

References

1. Hart, G.W. *Prototype Nonintrusive Appliance Load Monitor*; Technical Report 2; MIT Energy Laboratory and Electric Power Resear[ch] Institute: Concorde, MA, USA, 1985.
2. Hart, G.W. Nonintrusive Appliance Load Monitoring. *Proc. IEEE* **1992**, *80*, 1870–1891. [CrossRef]
3. Zeifman, M.; Roth, K. Nonintrusive Appliance Load Monitoring: Review and Outlook. *IEEE Trans. Consum. Electron.* **20**[11,] *57*, 76–84. [CrossRef]
4. Alcalá, J. Non-Intrusive Load Monitoring Techniques for Activity of Daily Living Recognition. Ph.D. Thesis, Universidad [de] Alcalá, Madrid, Spain, 2016.
5. Salani, M.; Derboni, M.; Rivola, D.; Medici, V.; Nespoli, L.; Rosato, F.; Rizzoli, A.E. Non Intrusive Load Monitoring for Deman[d] Side Management. *Energy Inform.* **2020**, *3*, 25. [CrossRef]
6. Çimen, H.; Çetinkaya, N.; Vasquez, J.C.; Guerrero, J.M. A Microgrid Energy Management System Based on Non-Intrusive Lo[ad] Monitoring via Multitask Learning. *IEEE Trans. Smart Grid* **2020**. [CrossRef]
7. Gupta, S.; Reynolds, M.S.; Patel, S.N. ElectriSense: Single-Point Sensing Using EMI for Electrical Event Detection and Classificati[on] in the Home. In Proceedings of the 12th ACM International Conference on Ubiquitous Computing, UbiComp '10, Copenhage[n,] Denmark, 26–29 September 2010; Association for Computing Machinery: New York, NY, USA, 2010; pp. 139–148. [CrossRef]
8. Uribe-Pérez, N.; Hernández, L.; de la Vega, D.; Angulo, I. State of the Art and Trends Review of Smart Metering in Electric[al] Grids. *Appl. Sci.* **2016**, *6*, 68. [CrossRef]
9. Kim, H.; Marwah, M.; Arlitt, M.; Lyon, G.; Han, J. Unsupervised Disaggregation of Low Frequency Power Measurements. *Proceedings of the 2011 SIAM International Conference on Data Mining*; Liu, B., Liu, H., Clifton, C., Washio, T., Kamath, C., Ed[s.;] Society for Industrial and Applied Mathematics: Philadelphia, PA, USA, 2011; pp. 747–758. [CrossRef]
10. Kolter, J.Z.; Jaakkola, T.S. Approximate Inference in Additive Factorial HMMs with Application to Energy Disaggregatio[n.] *AISTATS* **2012**, *22*, 1472–1482.

1. Parson, O.; Ghosh, S.; Weal, M.; Rogers, A. An Unsupervised Training Method for Non-Intrusive Appliance Load Monitoring. *Artif. Intell.* **2014**, *217*, 1–19. [CrossRef]
2. Makonin, S.; Popowich, F.; Bajić, I.V.; Gill, B.; Bartram, L. Exploiting HMM Sparsity to Perform Online Real-Time Nonintrusive Load Monitoring. *IEEE Trans. Smart Grid* **2016**, *7*, 2575–2585. [CrossRef]
3. Tabatabaei, S.M.; Dick, S.; Xu, W. Toward Non-Intrusive Load Monitoring via Multi-Label Classification. *IEEE Trans. Smart Grid* **2017**, *8*, 26–40. [CrossRef]
4. Kelly, J.; Knottenbelt, W. Neural NILM: Deep Neural Networks Applied to Energy Disaggregation. In Proceedings of the 2nd ACM International Conference on Embedded Systems for Energy-Efficient Built Environments, BuildSys '15, Seoul, Korea, 4–5 November 2015; Association for Computing Machinery: New York, NY, USA, 2015; pp. 55–64. [CrossRef]
5. Mauch, L.; Yang, B. A New Approach for Supervised Power Disaggregation by Using a Deep Recurrent LSTM Network. In Proceedings of the 2015 IEEE Global Conference on Signal and Information Processing (GlobalSIP), Orlando, FL, USA, 14–16 December 2015; IEEE: Orlando, FL, USA, 2015; pp. 63–67. [CrossRef]
6. Zoha, A.; Gluhak, A.; Imran, M.A.; Rajasegarar, S. Non-Intrusive Load Monitoring Approaches for Disaggregated Energy Sensing: A Survey. *Sensors* **2012**, *12*, 16838–16866. [CrossRef]
7. Bonfigli, R.; Squartini, S.; Fagiani, M.; Piazza, F. *Unsupervised Algorithms for Non-Intrusive Load Monitoring: An up-to-Date Overview*. In Proceedings of the 2015 IEEE 15th International Conference on Environment and Electrical Engineering (EEEIC), Rome, Italy, 10–13 June 2015; pp. 1175–1180. [CrossRef]
8. Pereira, L.; Nunes, N. Performance Evaluation in Non-Intrusive Load Monitoring: Datasets, Metrics, and Tools—A Review. *Wiley Interdiscip. Rev. Data Min. Knowl. Discov.* **2018**, *8*, e1265. [CrossRef]
9. Nalmpantis, C.; Vrakas, D. Machine Learning Approaches for Non-Intrusive Load Monitoring: From Qualitative to Quantitative Comparation. *Artif. Intell. Rev.* **2019**, *52*, 217–243. [CrossRef]
10. Bonfigli, R.; Squartini, S. *Machine Learning Approaches to Non-Intrusive Load Monitoring*; SpringerBriefs in Energy; Springer International Publishing: Cham, Switzerland, 2020; [CrossRef]
11. Bonfigli, R.; Felicetti, A.; Principi, E.; Fagiani, M.; Squartini, S.; Piazza, F. Denoising Autoencoders for Non-Intrusive Load Monitoring: Improvements and Comparative Evaluation. *Energy Build.* **2018**, *158*, 1461–1474. [CrossRef]
12. Valenti, M.; Bonfigli, R.; Principi, E.; Squartini, A.S. Exploiting the Reactive Power in Deep Neural Models for Non-Intrusive Load Monitoring. In Proceedings of the 2018 International Joint Conference on Neural Networks (IJCNN), Rio de Janeiro, Brazil, 8–13 July 2018; IEEE: Rio de Janeiro, Brazil, 2018; pp. 1–8. [CrossRef]
13. Batra, N.; Kukunuri, R.; Pandey, A.; Malakar, R.; Kumar, R.; Krystalakos, O.; Zhong, M.; Meira, P.; Parson, O. Towards Reproducible State-of-the-Art Energy Disaggregation. In Proceedings of the 6th ACM International Conference on Systems for Energy-Efficient Buildings, Cities, and Transportation, BuildSys '19, New York, NY, USA, 13–14 November 2019; Association for Computing Machinery: New York, NY, USA, 2019; pp. 193–202. [CrossRef]
14. Reinhardt, A.; Klemenjak, C. How Does Load Disaggregation Performance Depend on Data Characteristics? Insights from a Benchmarking Study. In Proceedings of the Eleventh ACM International Conference on Future Energy Systems, E-Energy '20, Virtual Conference, 22–26 June 2020; pp. 167–177. [CrossRef]
15. Batra, N.; Kelly, J.; Parson, O.; Dutta, H.; Knottenbelt, W.; Rogers, A.; Singh, A.; Srivastava, M. NILMTK: An Open Source Toolkit for Non-Intrusive Load Monitoring. In Proceedings of the 5th International Conference on Future Energy Systems, Cambridge, UK, 11–13 June 2014; ACM Press: New York, NY, USA, 2014; pp. 265–276. [CrossRef]
16. Kelly, J.; Batra, N.; Parson, O.; Dutta, H.; Knottenbelt, W.; Rogers, A.; Singh, A.; Srivastava, M. NILMTK v0.2: A Non-Intrusive Load Monitoring Toolkit for Large Scale Data Sets: Demo Abstract. In Proceedings of the 1st ACM Conference on Embedded Systems for Energy-Efficient Buildings, Memphis, TN, USA, 3–6 November 2014; ACM Press: New York, NY, USA, 2014; pp. 182–183. [CrossRef]
17. Roos, J.G.; Lane, I.E.; Botha, E.C.; Hancke, G.P. Using Neural Networks for Non-Intrusive Monitoring of Industrial Electrical Loads. In Proceedings of the 10th Anniversary, IMTC/94, Advanced Technologies in I M, 1994 IEEE Instrumentation and Measurement Technolgy Conference (Cat. No.94CH3424-9), Hamamatsu, Japan, 10–12 May 1994; Volume 3, pp. 1115–1118. [CrossRef]
18. Paradiso, F.; Paganelli, F.; Luchetta, A.; Giuli, D.; Castrogiovanni, P. ANN-Based Appliance Recognition from Low-Frequency Energy Monitoring Data. In Proceedings of the 2013 IEEE 14th International Symposium on "A World of Wireless, Mobile and Multimedia Networks" (WoWMoM), Madrid, Spain, 4–7 June 2013; pp. 1–6. [CrossRef]
19. Li, D.; Dick, S. Whole-House Non-Intrusive Appliance Load Monitoring via Multi-Label Classification. In Proceedings of the 2016 International Joint Conference on Neural Networks (IJCNN), Vancouver, BC, Canada, 24–29 July 2016; pp. 2749–2755. [CrossRef]
20. Salerno, V.M.; Rabbeni, G. An Extreme Learning Machine Approach to Effective Energy Disaggregation. *Electronics* **2018**, *7*, 235. [CrossRef]
21. Verma, S.; Singh, S.; Majumdar, A. Multi Label Restricted Boltzmann Machine for Non-Intrusive Load Monitoring. In Proceedings of the ICASSP 2019-2019 IEEE International Conference on Acoustics, Speech and Signal Processing (ICASSP), Brighton, UK, 12–17 May 2019; pp. 8345–8349.
22. Kim, J.; Le, T.T.H.; Kim, H. Nonintrusive Load Monitoring Based on Advanced Deep Learning and Novel Signature. *Comput. Intell. Neurosci.* **2017**, *2017*, 4216281. [CrossRef]

33. Al Zeidi, O. Deep Neural Networks for Non-Intrusive Load Monitoring. Master's Thesis, Monash University, Clayton, Australia, 2018.
34. Pan, Y.; Liu, K.; Shen, Z.; Cai, X.; Jia, Z. Sequence-To-Subsequence Learning With Conditional Gan For Power Disaggregation. In Proceedings of the ICASSP 2020—2020 IEEE International Conference on Acoustics, Speech and Signal Processing (ICASSP), Barcelona, Spain, 4–8 May 2020; pp. 3202–3206. [CrossRef]
35. Chen, K.; Zhang, Y.; Wang, Q.; Hu, J.; Fan, H.; He, J. Scale- and Context-Aware Convolutional Non-Intrusive Load Monitoring. *IEEE Trans. Power Syst.* **2020**, *35*, 2362–2373. [CrossRef]
36. Rafiq, H.; Shi, X.; Zhang, H.; Li, H.; Ochani, M.K. A Deep Recurrent Neural Network for Non-Intrusive Load Monitoring Based on Multi-Feature Input Space and Post-Processing. *Energies* **2020**, *13*, 2195. [CrossRef]
37. Massidda, L.; Marrocu, M.; Manca, S. Non-Intrusive Load Disaggregation by Convolutional Neural Network and Multilabel Classification. *Appl. Sci.* **2020**, *10*, 1454. [CrossRef]
38. Bousbiat, H.; Klemenjak, C.; Elmenreich, W. Exploring Time Series Imaging for Load Disaggregation. In Proceedings of the 7th ACM International Conference on Systems for Energy-Efficient Buildings, Cities, and Transportation, BuildSys '20, Virtual Event, 19–20 November 2020; Association for Computing Machinery: New York, NY, USA, 2020; pp. 254–257. [CrossRef]
39. Ahmed, A.M.A.; Zhang, Y.; Eliassen, F. Generative Adversarial Networks and Transfer Learning for Non-Intrusive Load Monitoring in Smart Grids. In Proceedings of the 2020 IEEE International Conference on Communications, Control, and Computing Technologies for Smart Grids (SmartGridComm), Tempe, AZ, USA, 11–13 November 2020; IEEE: Tempe, AZ, USA, 2020; pp. 1–7. [CrossRef]
40. Ayub, M.; El-Alfy, E.S.M. Impact of Normalization on BiLSTM Based Models for Energy Disaggregation. In Proceedings of the 2020 International Conference on Data Analytics for Business and Industry: Way Towards a Sustainable Economy (ICDABI), Sakheer, Bahrain, 26–27 October 2020; pp. 1–6. [CrossRef]
41. Chen, H.; Wang, Y.H.; Fan, C.H. A Convolutional Autoencoder-Based Approach with Batch Normalization for Energy Disaggregation. *J. Supercomput.* **2020**, *77*, 2961–2978. [CrossRef]
42. Kaselimi, M.; Doulamis, N.; Voulodimos, A.; Protopapadakis, E.; Doulamis, A. Context Aware Energy Disaggregation Using Adaptive Bidirectional LSTM Models. *IEEE Trans. Smart Grid* **2020**, *11*, 3054–3067. [CrossRef]
43. Kaselimi, M.; Voulodimos, A.; Protopapadakis, E.; Doulamis, N.; Doulamis, A. EnerGAN: A Generative Adversarial Network for Energy Disaggregation. In Proceedings of the ICASSP 2020—2020 IEEE International Conference on Acoustics, Speech and Signal Processing (ICASSP), Barcelona, Spain, 4–8 May 2020; pp. 1578–1582. [CrossRef]
44. Delfosse, A.; Hebrail, G.; Zerroug, A. Deep Learning Applied to NILM: Is Data Augmentation Worth for Energy Disaggregation? In Proceedings of the ECAI 2020, Santiago de Compostela, Spain, 29 August–8 September 2020; pp. 2972–2977. [CrossRef]
45. Yadav, A.; Sinha, A.; Saidi, A.; Trinkl, C.; Zörner, W. NILM Based Energy Disaggregation Algorithm for Dairy Farms. In Proceedings of the 5th International Workshop on Non-Intrusive Load Monitoring, NILM'20, Yokohama, Japan, 18 November 2020; Association for Computing Machinery: New York, NY, USA, 2020; pp. 16–19. [CrossRef]
46. Faustine, A.; Pereira, L.; Bousbiat, H.; Kulkarni, S. UNet-NILM: A Deep Neural Network for Multi-Tasks Appliances State Detection and Power Estimation in NILM. In Proceedings of the 5th International Workshop on Non-Intrusive Load Monitoring, NILM'20, Yokohama, Japan, 18 November 2020; Association for Computing Machinery: New York, NY, USA, 2020; pp. 84–88. [CrossRef]
47. Ahmed, S.; Bons, M. Edge Computed NILM: A Phone-Based Implementation Using MobileNet Compressed by Tensorflow Lite. In Proceedings of the 5th International Workshop on Non-Intrusive Load Monitoring, NILM'20, Yokohama, Japan, 18 November 2020; Association for Computing Machinery: New York, NY, USA, 2020; pp. 44–48. [CrossRef]
48. Barber, J.; Cuayáhuitl, H.; Zhong, M.; Luan, W. Lightweight Non-Intrusive Load Monitoring Employing Pruned Sequence-to-Point Learning. In Proceedings of the 5th International Workshop on Non-Intrusive Load Monitoring, NILM'20, Yokohama, Japan, 18 November 2020; Association for Computing Machinery: New York, NY, USA, 2020; pp. 11–15. [CrossRef]
49. Kukunuri, R.; Aglawe, A.; Chauhan, J.; Bhagtani, K.; Patil, R.; Walia, S.; Batra, N. EdgeNILM: Towards NILM on Edge Devices. In Proceedings of the 7th ACM International Conference on Systems for Energy-Efficient Buildings, Cities, and Transportation, BuildSys '20, Yokohama, Japan, 19–20 November 2020; Association for Computing Machinery: New York, NY, USA, 2020; pp. 90–99. [CrossRef]
50. Jones, R.; Klemenjak, C.; Makonin, S.; Bajić, I.V. Stop: Exploring Bayesian Surprise to Better Train NILM. In Proceedings of the 5th International Workshop on Non-Intrusive Load Monitoring, NILM'20, Yokohama, Japan, 18 November 2020; Association for Computing Machinery: New York, NY, USA, 2020; pp. 39–43. [CrossRef]
51. Huchtkoetter, J.; Reinhardt, A. On the Impact of Temporal Data Resolution on the Accuracy of Non-Intrusive Load Monitoring. In Proceedings of the 7th ACM International Conference on Systems for Energy-Efficient Buildings, Cities, and Transportation, BuildSys '20, Yokohama, Japan, 19–20 November 2020; Association for Computing Machinery: New York, NY, USA, 2020; pp. 270–273. [CrossRef]
52. Reinhardt, A.; Bouchur, M. On the Impact of the Sequence Length on Sequence-to-Sequence and Sequence-to-Point Learning for NILM. In Proceedings of the 5th International Workshop on Non-Intrusive Load Monitoring, NILM'20, Yokohama, Japan, 18 November 2020; Association for Computing Machinery: New York, NY, USA, 2020; pp. 75–78. [CrossRef]

Murray, D.; Stankovic, L.; Stankovic, V. Explainable NILM Networks. In Proceedings of the 5th International Workshop on Non-Intrusive Load Monitoring, NILM'20, Yokohama, Japan, 19–20 November 2020; Association for Computing Machinery: New York, NY, USA, 2020; pp. 64–69. [CrossRef]

Tongta, A.; Chooruang, K. Long Short-Term Memory (LSTM) Neural Networks Applied to Energy Disaggregation. In Proceedings of the 2020 8th International Electrical Engineering Congress (iEECON), Chiang Mai, Thailand, 4–6 March 2020; pp. 1–4. [CrossRef]

Yue, Z.; Witzig, C.R.; Jorde, D.; Jacobsen, H.A. BERT4NILM: A Bidirectional Transformer Model for Non-Intrusive Load Monitoring. In Proceedings of the 5th International Workshop on Non-Intrusive Load Monitoring, NILM'20, Yokohama, Japan, 18 December 2020; Association for Computing Machinery: New York, NY, USA, 2020; pp. 89–93. [CrossRef]

Zhang, Y.; Yin, B.; Cong, Y.; Du, Z. Multi-State Household Appliance Identification Based on Convolutional Neural Networks and Clustering. *Energies* **2020**, *13*, 792. [CrossRef]

Xia, M.; Liu, W.; Wang, K.; Song, W.; Chen, C.; Li, Y. Non-Intrusive Load Disaggregation Based on Composite Deep Long Short-Term Memory Network. *Expert Syst. Appl.* **2020**, *160*, 113669. [CrossRef]

Kong, W.; Dong, Z.Y.; Wang, B.; Zhao, J.; Huang, J. A Practical Solution for Non-Intrusive Type II Load Monitoring Based on Deep Learning and Post-Processing. *IEEE Trans. Smart Grid* **2020**, *11*, 148–160. [CrossRef]

Gomes, E.; Pereira, L. PB-NILM: Pinball Guided Deep Non-Intrusive Load Monitoring. *IEEE Access* **2020**, *8*, 48386–48398. [CrossRef]

Liu, H. *Non-Intrusive Load Monitoring: Theory, Technologies and Applications*; Springer: Singapore, 2020; [CrossRef]

Quek, Y.T.; Woo, W.L.; Logenthiran, T. Load Disaggregation Using One-Directional Convolutional Stacked Long Short-Term Memory Recurrent Neural Network. *IEEE Syst. J.* **2020**, *14*, 1395–1404. [CrossRef]

Jiang, J.; Kong, Q.; Plumbley, M.; Gilbert, N. Deep Learning Based Energy Disaggregation and On/Off Detection of Household Appliances. *arXiv* **2019**, arXiv:1908.00941.

Murray, D.; Stankovic, L.; Stankovic, V.; Lulic, S.; Sladojevic, S. Transferability of Neural Network Approaches for Low-Rate Energy Disaggregation. In Proceedings of the ICASSP 2019—2019 IEEE International Conference on Acoustics, Speech and Signal Processing (ICASSP), Brighton, UK, 12–17 May 2019; pp. 8330–8334. [CrossRef]

Çavdar, İ.; Faryad, V. New Design of a Supervised Energy Disaggregation Model Based on the Deep Neural Network for a Smart Grid. *Energies* **2019**, *12*, 1217. [CrossRef]

Santos, E.G.; Freitas, C.G.S.; Aquino, A.L.L. A Deep Learning Approach for Energy Disaggregation Considering Embedded Devices. In Proceedings of the 2019 IX Brazilian Symposium on Computing Systems Engineering (SBESC), Natal, Brazil, 19–22 November 2019; pp. 1–8. [CrossRef]

Chang, F.Y.; Ho, W.J. An Analysis of Semi-Supervised Learning Approaches in Low-Rate Energy Disaggregation. In Proceedings of the 2019 3rd International Conference on Smart Grid and Smart Cities (ICSGSC), Berkeley, CA, USA, 25–28 June 2019; pp. 145–150. [CrossRef]

Miao, N.; Zhao, S.; Shi, Q.; Zhang, R. Non-Intrusive Load Disaggregation Using Semi-Supervised Learning Method. In Proceedings of the 2019 International Conference on Security, Pattern Analysis, and Cybernetics (SPAC), Guangzhou, China, 20–23 December 2019; pp. 17–22. [CrossRef]

Wang, J.; Kababji, S.E.; Graham, C.; Srikantha, P. Ensemble-Based Deep Learning Model for Non-Intrusive Load Monitoring. In Proceedings of the 2019 IEEE Electrical Power and Energy Conference (EPEC), Montreal, QC, Canada, 16–18 October 2019; pp. 1–6. [CrossRef]

Cui, G.; Liu, B.; Luan, W.; Yu, Y. Estimation of Target Appliance Electricity Consumption Using Background Filtering. *IEEE Trans. Smart Grid* **2019**, *10*, 5920–5929. [CrossRef]

Hsu, C.Y.; Zeitoun, A.; Lee, G.H.; Katabi, D.; Jaakkola, T. Self-Supervised Learning of Appliance Usage. In Proceedings of the International Conference on Learning Representations—ICLR 2020, Virtual Conference, 26 April–1 May 2019.

Kyrkou, L.; Nalmpantis, C.; Vrakas, D. Imaging Time-Series for Nilm. In Proceedings of the 20th International Conference on Engineering Applications of Neural Networks (EANN 2019), Crete, Greece, 24–26 May 2019; Springer: Cham, Switzerland, 2019; pp. 188–196. [CrossRef]

Sudoso, A.M.; Piccialli, V. Non-Intrusive Load Monitoring with an Attention-Based Deep Neural Network. *arXiv* **2019**, arXiv:1912.00759.

Min, C.; Wen, G.; Yang, Z.; Li, X.; Li, B. Non-Intrusive Load Monitoring System Based on Convolution Neural Network and Adaptive Linear Programming Boosting. *Energies* **2019**, *12*, 2882. [CrossRef]

Jasiński, T. Modelling the Disaggregated Demand for Electricity at the Level of Residential Buildings with the Use of Artificial Neural Networks (Deep Learning Approach). *MATEC Web Conf.* **2019**, *282*, 02077. [CrossRef]

Shin, C.; Rho, S.; Lee, H.; Rhee, W. Data Requirements for Applying Machine Learning to Energy Disaggregation. *Energies* **2019**, *12*, 1696. [CrossRef]

Cao, H.; Weng, L.; Xia, M.; Zhang, D. Non-Intrusive Load Disaggregation Based on Residual Gated Network. *IOP Conf. Ser. Mater. Sci. Eng.* **2019**, *677*, 032092. [CrossRef]

Davies, P.; Dennis, J.; Hansom, J.; Martin, W.; Stankevicius, A.; Ward, L. Deep Neural Networks for Appliance Transient Classification. In Proceedings of the ICASSP 2019—2019 IEEE International Conference on Acoustics, Speech and Signal Processing (ICASSP), Brighton, UK, 12–17 May 2019; pp. 8320–8324. [CrossRef]

78. Linh, N.V.; Arboleya, P. Deep Learning Application to Non-Intrusive Load Monitoring. In Proceedings of the 2019 IEEE Milan PowerTech, Milan, Italy, 23–27 June 2019; IEEE: Milan, Italy, 2019; pp. 1–5. [CrossRef]
79. Gopu, R.; Gudimallam, A.; Thokala, N.; Chandra, M.G. On Electrical Load Disaggregation Using Recurrent Neural Networks. In Proceedings of the 6th ACM International Conference on Systems for Energy-Efficient Buildings, Cities, and Transportation, BuildSys '19, New York, NY, USA, 19–20 November 2019; Association for Computing Machinery: New York, NY, USA, 2019; pp. 364–365. [CrossRef]
80. Li, C.; Zheng, R.; Liu, M.; Zhang, S. A Fusion Framework Using Integrated Neural Network Model for Non-Intrusive Load Monitoring. In Proceedings of the 2019 Chinese Control Conference (CCC), Guangzhou, China, 27–30 July 2019; IEEE: Guangzhou, China, 2019; pp. 7385–7390. [CrossRef]
81. Yang, Y.; Zhong, J.; Li, W.; Gulliver, T.A.; Li, S. Semi-Supervised Multi-Label Deep Learning Based Non-Intrusive Load Monitoring in Smart Grids. *IEEE Trans. Ind. Inform.* **2019**. [CrossRef]
82. Buchhop, S.J.; Ranganathan, P. Residential Load Identification Based on Load Profile Using Artificial Neural Network (ANN). Proceedings of the 2019 North American Power Symposium (NAPS), Wichita, KS, USA, 13–15 October 2019; IEEE: Wichita, KS, USA, 2019; pp. 1–6. [CrossRef]
83. Hosseini, S.; Henao, N.; Kelouwani, S.; Agbossou, K.; Cardenas, A. A Study on Markovian and Deep Learning Based Architectures for Household Appliance-Level Load Modeling and Recognition. In Proceedings of the 2019 IEEE 28th International Symposium on Industrial Electronics (ISIE), Vancouver, BC, Canada, 12–14 June 2019; IEEE: Vancouver, BC, Canada, 2019; pp. 35–. [CrossRef]
84. Xia, M.; Liu, W.; Wang, K.; Zhang, X.; Xu, Y. Non-Intrusive Load Disaggregation Based on Deep Dilated Residual Network. *Electr. Power Syst. Res.* **2019**, *170*, 277–285. [CrossRef]
85. Xia, M.; Liu, W.; Xu, Y.; Wang, K.; Zhang, X. Dilated Residual Attention Network for Load Disaggregation. *Neural Comput. Appl.* **2019**. [CrossRef]
86. Yu, H.; Jiang, Z.; Li, Y.; Zhou, J.; Wang, K.; Cheng, Z.; Gu, Q. A Multi-Objective Non-Intrusive Load Monitoring Method Based on Deep Learning. *IOP Conf. Ser. Mater. Sci. Eng.* **2019**, *486*, 012110. [CrossRef]
87. Popa, D.; Pop, F.; Serbanescu, C.; Castiglione, A. Deep Learning Model for Home Automation and Energy Reduction in a Smart Home Environment Platform. *Neural Comput. Appl.* **2019**, *31*, 1317–1337. [CrossRef]
88. Wang, T.S.; Ji, T.Y.; Li, M.S. A New Approach for Supervised Power Disaggregation by Using a Denoising Autoencoder and Recurrent LSTM Network. In Proceedings of the 2019 IEEE 12th International Symposium on Diagnostics for Electrical Machines, Power Electronics and Drives (SDEMPED), Toulouse, France, 27–30 August 2019; IEEE: Toulouse, France, 2019; pp. 507–5. [CrossRef]
89. Kaselimi, M.; Protopapadakis, E.; Voulodimos, A.; Doulamis, N.; Doulamis, A. Multi-Channel Recurrent Convolutional Neural Networks for Energy Disaggregation. *IEEE Access* **2019**, *7*, 81047–81056. [CrossRef]
90. Kaselimi, M.; Doulamis, N.; Doulamis, A.; Voulodimos, A.; Protopapadakis, E. Bayesian-Optimized Bidirectional LSTM Regression Model for Non-Intrusive Load Monitoring. In Proceedings of the ICASSP 2019—2019 IEEE International Conference on Acoustics, Speech and Signal Processing (ICASSP), Brighton, UK, 12–17 May 2019; IEEE: Brighton, UK, 2019; pp. 2747–27. [CrossRef]
91. DIncecco, M.; Squartini, S.; Zhong, M. Transfer Learning for Non-Intrusive Load Monitoring. *arXiv* **2019**, arXiv:1902.08835.
92. Jia, Y.; Batra, N.; Wang, H.; Whitehouse, K. A Tree-Structured Neural Network Model for Household Energy Breakdown. Proceedings of the World Wide Web Conference, San Francisco, CA, USA, 13–17 May 2019; Association for Computing Machinery: New York, NY, USA, 2019; pp. 2872–2878. [CrossRef]
93. Fagiani, M.; Bonfigli, R.; Principi, E.; Squartini, S.; Mandolini, L. A Non-Intrusive Load Monitoring Algorithm Based on Non-Uniform Sampling of Power Data and Deep Neural Networks. *Energies* **2019**, *12*, 1371. [CrossRef]
94. Harell, A.; Makonin, S.; Bajić, I.V. Wavenilm: A Causal Neural Network for Power Disaggregation from the Complex Power Signal. *arXiv* **2019**, arXiv:1902.08736.
95. Xiao, P.; Cheng, S. Neural Network for NILM Based on Operational State Change Classification. *arXiv* **2019**, arXiv:1902.02675.
96. Bejarano, G.; DeFazio, D.; Ramesh, A. Deep Latent Generative Models for Energy Disaggregation. In Proceedings of the AAAI Conference on Artificial Intelligence, Honolulu, HI, USA, 27 January–1 February 2019; Volume 33. [CrossRef]
97. Shin, C.; Joo, S.; Yim, J.; Lee, H.; Moon, T.; Rhee, W. Subtask Gated Networks for Non-Intrusive Load Monitoring. *arXiv* **2018**, arXiv:1811.06692.
98. Brewitt, C.; Goddard, N. Non-Intrusive Load Monitoring with Fully Convolutional Networks. *arXiv* **2018**, arXiv:1812.03915.
99. Cho, J.; Hu, Z.; Sartipi, M. Non-Intrusive A/C Load Disaggregation Using Deep Learning. In Proceedings of the 2018 IEEE/PES Transmission and Distribution Conference and Exposition, Denver, CO, USA, 16–19 April 2018; pp. 1–5. [CrossRef]
100. Kundu, A.; Juvekar, G.P.; Davis, K. Deep Neural Network Based Non-Intrusive Load Status Recognition. In Proceedings of the 2018 Clemson University Power Systems Conference (PSC), Charleston, SC, USA, 4–7 September 2018; pp. 1–6. [CrossRef]
101. Van Zaen, J.; El Achkar, C.M.; Carrillo, R.E.; Hutter, A. Detection and Classification of Refrigeration Units in a Commercial Environment: Comparing Neural Networks to Unsupervised Clustering. In Proceedings of the 4th International Workshop on Non-Intrusive Load Monitoring, Austin, TX, USA, 7–8 March 2018; Nilmworkshop Org: Austin, TX, USA, 2018.

2. Dash, P.; Naik, K. A Very Deep One Dimensional Convolutional Neural Network (VDOCNN) for Appliance Power Signature Classification. In Proceedings of the 2018 IEEE Electrical Power and Energy Conference (EPEC), Toronto, ON, Canada, 10–11 October 2018; pp. 1–6. [CrossRef]
3. Chang, F.Y.; Chen, C.; Lin, S.D. An Empirical Study of Ladder Network and Multitask Learning on Energy Disaggregation in Taiwan. In Proceedings of the 2018 Conference on Technologies and Applications of Artificial Intelligence (TAAI), Taichung, Taiwan, 30 November–2 December 2018; IEEE: Taichung, Taiwan, 2018; pp. 86–89. [CrossRef]
4. Sirojan, T.; Phung, B.T.; Ambikairajah, E. Deep Neural Network Based Energy Disaggregation. In Proceedings of the 2018 IEEE International Conference on Smart Energy Grid Engineering (SEGE), Oshawa, ON, Canada, 12–15 August 2018; pp. 73–77. [CrossRef]
5. Bao, K.; Ibrahimov, K.; Wagner, M.; Schmeck, H. Enhancing Neural Non-Intrusive Load Monitoring with Generative Adversarial Networks. *Energy Inform.* **2018**, *1*, 18. [CrossRef]
6. Harell, A.; Makonin, S.; Bajic, I.V. A Recurrent Neural Network for Multisensory Non-Intrusive Load Monitoring on a Raspberry Pi. In Proceedings of the IEEE 20th International Workshop on Multimedia Signal Processing (MMSP 2018), Vancouver, BC, Canada, 29–31 August 2018.
7. Laptev, N.; Ji, Y.; Rajagopal, R. Using the Wisdom of Neighbors for Energy Disaggregation from Smart Meters. In Proceedings of the 4th International Workshop on Non-Intrusive Load Monitoring, Austin, TX, USA, 7–8 March 2018; pp. 1–5.
8. Rafiq, H.; Zhang, H.; Li, H.; Ochani, M.K. Regularized LSTM Based Deep Learning Model: First Step towards Real-Time Non-Intrusive Load Monitoring. In Proceedings of the 2018 IEEE International Conference on Smart Energy Grid Engineering (SEGE), Oshawa, ON, Canada, 12–15 August 2018; pp. 234–239. [CrossRef]
9. Martins, P.B.M.; Gomes, J.G.R.C.; Nascimento, V.B.; de Freitas, A.R. Application of a Deep Learning Generative Model to Load Disaggregation for Industrial Machinery Power Consumption Monitoring. In Proceedings of the 2018 IEEE International Conference on Communications, Control, and Computing Technologies for Smart Grids (SmartGridComm), Aalborg, Denmark, 29–31 October 2018; IEEE: Aalborg, Denmark, 2018; pp. 1–6. [CrossRef]
10. Chen, K.; Wang, Q.; He, Z.; Chen, K.; Hu, J.; He, J. Convolutional Sequence to Sequence Non-Intrusive Load Monitoring (arXiv Version). *arXiv* **2018**, arXiv:1806.02078.
11. Krystalakos, O.; Nalmpantis, C.; Vrakas, D. Sliding Window Approach for Online Energy Disaggregation Using Artificial Neural Networks. In Proceedings of the 10th Hellenic Conference on Artificial Intelligence, SETN '18, Patras, Greece, 9–12 July 2018; Association for Computing Machinery: New York, NY, USA, 2018; pp. 7:1–7:6. [CrossRef]
12. Barsim, K.S.; Yang, B. On the Feasibility of Generic Deep Disaggregation for Single-Load Extraction. *arXiv* **2018**, arXiv:1802.02139.
13. Tsai, C.W.; Yang, C.W.; Ho, W.J.; Yin, Z.X.; Chiang, K.C. Using Autoencoder Network to Implement Non-Intrusive Load Monitoring of Small and Medium Business Customer. In Proceedings of the 2018 International Conference on Applied System Invention (ICASI), Chiba, Japan, 13–17 April 2018; pp. 433–436. [CrossRef]
14. de Paiva Penha, D.; Castro, A.R.G. Home Appliance Identification for NILM Systems Based on Deep Neural Networks. *Int. J. Artif. Intell. Appl. (IJAIA)* **2018**, *9*, 69–80. [CrossRef]
15. Garcia, F.C.C.; Creayla, C.M.C.; Macabebe, E.Q.B. Development of an Intelligent System for Smart Home Energy Disaggregation Using Stacked Denoising Autoencoders. *Procedia Comput. Sci.* **2017**, *105*, 248–255. [CrossRef]
16. de Paiva Penha, D.; Castro, A.R.G. Convolutional Neural Network Applied to the Identification of Residential Equipment in Non-Intrusive Load Monitoring Systems. *Comput. Sci. Inf. Technol.* **2017**, *7*, 11–21. [CrossRef]
17. Morgan, E.S. Applications of Deep Learning for Load Disaggregation in Residential Environments. Bachelor's Thesis, Universidade Federal do Rio de Janeiro, Rio de Janeiro, Brazil, 2017.
18. Nascimento, P.P.M. Applications of Deep Learning Techniques on Nilm. Master's Thesis, Universidade Federal do Rio de Janeiro, Rio de Janeiro, Brazil, 2016.
19. Zhang, C.; Zhong, M.; Wang, Z.; Goddard, N.; Sutton, C. Sequence-to-Point Learning with Neural Networks for Nonintrusive Load Monitoring. *arXiv* **2016**, arXiv:1612.09106.
20. Le, T.T.H.; Kim, J.; Kim, H. Classification Performance Using Gated Recurrent Unit Recurrent Neural Network on Energy Disaggregation. In Proceedings of the 2016 International Conference on Machine Learning and Cybernetics (ICMLC), Jeju, Korea, 10–13 July 2016; Volume 1, pp. 105–110. [CrossRef]
21. He, W.; Chai, Y. An Empirical Study on Energy Disaggregation via Deep Learning. In Proceedings of the 2016 2nd International Conference on Artificial Intelligence and Industrial Engineering (AIIE 2016), Beijing, China, 20–21 November 2016; Volume 133, pp. 338–342. [CrossRef]
22. Mauch, L.; Yang, B. A Novel DNN-HMM-Based Approach for Extracting Single Loads from Aggregate Power Signals. In Proceedings of the 2016 IEEE International Conference on Acoustics, Speech and Signal Processing (ICASSP), Shanghai, China, 20–25 March 2016; pp. 2384–2388. [CrossRef]
23. Mottahedi, M.; Asadi, S. Non-Intrusive Load Monitoring Using Imaging Time Series and Convolutional Neural Networks. In Proceedings of the ICCCBE, Osaka, Japan, 6–8 July 2016.
24. Goodfellow, I.; Bengio, Y.; Courville, A. *Deep Learning*; MIT Press: Cambridge, MA, USA, 2016.
25. Chollet, F. *Deep Learning with Python*; Manning Publications Co.: Shelter Island, NY, USA, 2018.
26. Chollet, F. Keras: The Python Deep Learning API. 2015. Available online: https://keras.io (accessed on 11 January 2021).

127. Shin, C.; Lee, E.; Han, J.; Yim, J.; Rhee, W.; Lee, H. The ENERTALK Dataset, 15 Hz Electricity Consumption Data from 22 Houses in Korea. *Sci. Data* **2019**, *6*, 193. [CrossRef] [PubMed]
128. Himeur, Y.; Alsalemi, A.; Bensaali, F.; Amira, A. Building Power Consumption Datasets: Survey, Taxonomy and Future Directions. *Energy Build.* **2020**, *227*, 110404. [CrossRef]
129. Huber, P.; Ott, M.; Friedli, M.; Rumsch, A.; Paice, A. Residential Power Traces for Five Houses: The iHomeLab RAPT Dataset. *Data* **2020**, *5*, 17. [CrossRef]
130. Klemenjak, C.; Kovatsch, C.; Herold, M.; Elmenreich, W. A Synthetic Energy Dataset for Non-Intrusive Load Monitoring in Households. *Sci. Data* **2020**, *7*, 108. [CrossRef]
131. Völker, B.; Pfeifer, M.; Scholl, P.M.; Becker, B. FIRED: A Fully-Labeled hIgh-fRequency Electricity Disaggregation Dataset. In Proceedings of the 7th ACM International Conference on Systems for Energy-Efficient Buildings, Cities, and Transportation, Yokohama, Japan, 19–20 November 2020; Association for Computing Machinery: New York, NY, USA, 2020; pp. 294–297. [CrossRef]
132. Kelly, J.; Knottenbelt, W. The UK-DALE Dataset, Domestic Appliance-Level Electricity Demand and Whole-House Demand from Five UK Homes. *Sci. Data* **2015**, *2*, 150007. [CrossRef] [PubMed]
133. Kolter, J.Z.; Johnson, M.J. REDD: A Public Data Set for Energy Disaggregation Research. In *Workshop on Data Mining Applications in Sustainability (SIGKDD)*; Citeseer: San Diego, CA, USA, 2011; Volume 25, pp. 59–62.
134. Makonin, S.; Popowich, F.; Bartram, L.; Gill, B.; Bajic, I. AMPds: A Public Dataset for Load Disaggregation and Eco-Feedback Research. In Proceedings of the 2013 IEEE Electrical Power Energy Conference (EPEC), Halifax, NS, Canada, 21–23 August 2013; pp. 1–6. [CrossRef]
135. Makonin, S. *AMPds: Almanac of Minutely Power Dataset (R2013)*; Harvard Dataverse: Cambridge, MA, USA, 2015. [CrossRef]
136. Makonin, S.; Ellert, B.; Bajić, I.V.; Popowich, F. Electricity, Water, and Natural Gas Consumption of a Residential House in Canada from 2012 to 2014. *Sci. Data* **2016**, *3*, 160037. [CrossRef]
137. Makonin, S. *AMPds2: The Almanac of Minutely Power Dataset (Version 2)*; Harvard Dataverse: Cambridge, MA, USA, 2016; [CrossRef]
138. Murray, D.; Stankovic, L.; Stankovic, V. An Electrical Load Measurements Dataset of United Kingdom Households from Two-Year Longitudinal Study. *Sci. Data* **2017**, *4*, 160122. [CrossRef]
139. Parson, O.; Fisher, G.; Hersey, A.; Batra, N.; Kelly, J.; Singh, A.; Knottenbelt, W.; Rogers, A. Dataport and NILMTK: A Building Data Set Designed for Non-Intrusive Load Monitoring. In Proceedings of the 2015 IEEE Global Conference on Signal and Information Processing (GlobalSIP), Orlando, FL, USA, 14–16 December 2015; pp. 210–214. [CrossRef]
140. Beckel, C.; Kleiminger, W.; Cicchetti, R.; Staake, T.; Santini, S. The ECO Data Set and the Performance of Non-Intrusive Load Monitoring Algorithms. In Proceedings of the 1st ACM Conference on Embedded Systems for Energy-Efficient Buildings, BuildSys '14, Memphis, TN, USA, 5–6 November 2014; Association for Computing Machinery: New York, NY, USA, 2014; pp. 80–89. [CrossRef]
141. Uttama Nambi, A.S.; Reyes Lua, A.; Prasad, V.R. LocED: Location-Aware Energy Disaggregation Framework. In Proceedings of the 2nd ACM International Conference on Embedded Systems for Energy-Efficient Built Environments, BuildSys '15, Seoul, Korea, 4–5 November 2015; Association for Computing Machinery: New York, NY, USA, 2015; pp. 45–54. [CrossRef]
142. Zimmermann, J.P.; Evans, M.; Griggs, J.; King, N.; Harding, L.; Roberts, P.; Evans, C. *Household Electricity Survey: A Study of Domestic Electrical Product Usage*; Intertek Report R66141; Intertek Testing & Certification Ltd.: Milton Keynes, UK, 2012; p. 600.
143. Martins, P.B.M.; Nascimento, V.B.; de Freitas, A.R.; Bittencourt e Silva, P.; Guimarães Duarte Pinto, R. *Industrial Machines Dataset for Electrical Load Disaggregation*; IEEE Dataport: New York, NY, USA, 2018. [CrossRef]
144. Gao, J.; Giri, S.; Kara, E.C.; Bergés, M. PLAID: A Public Dataset of High-Resoultion Electrical Appliance Measurements for Load Identification Research: Demo Abstract. In Proceedings of the 1st ACM Conference on Embedded Systems for Energy-Efficient Buildings, BuildSys '14, Memphis, TN, USA, 5–6 November 2014; Association for Computing Machinery: New York, NY, USA, 2014; pp. 198–199. [CrossRef]
145. Klemenjak, C.; Faustine, A.; Makonin, S.; Elmenreich, W. On Metrics to Assess the Transferability of Machine Learning Models in Non-Intrusive Load Monitoring. *arXiv* **2019**, arXiv:1912.06200.
146. Ulyanov, D.; Vedaldi, A.; Lempitsky, V. Instance Normalization: The Missing Ingredient for Fast Stylization. *arXiv* **2016**, arXiv:1607.08022.
147. Ioffe, S.; Szegedy, C. Batch Normalization: Accelerating Deep Network Training by Reducing Internal Covariate Shift. *arXiv* **2015**, arXiv:1502.03167.
148. Shorten, C.; Khoshgoftaar, T.M. A Survey on Image Data Augmentation for Deep Learning. *J. Big Data* **2019**, *6*, 60. [CrossRef]
149. Iwana, B.K.; Uchida, S. An Empirical Survey of Data Augmentation for Time Series Classification with Neural Networks. *arXiv* **2020**, arXiv:2007.15951.
150. Wen, Q.; Sun, L.; Song, X.; Gao, J.; Wang, X.; Xu, H. Time Series Data Augmentation for Deep Learning: A Survey. *arXiv* **2020**, arXiv:2002.12478.
151. Reynaud, Q.; Haradji, Y.; Sempé, F.; Sabouret, N. Using Time Use Surveys in Multi Agent Based Simulations of Human Activity. In Proceedings of the 9th International Conference on Agents and Artificial Intelligence, Porto, Portugal, 24–26 February 2017; SCITEPRESS—Science and Technology Publications: Porto, Portugal, 2017; Volume 1, pp. 67–77. [CrossRef]

2. Wang, Z.; Oates, T. Encoding Time Series as Images for Visual Inspection and Classification Using Tiled Convolutional Neural Networks. In Proceedings of the Workshops at the Twenty-Ninth AAAI Conference on Artificial Intelligence, Austin, TX, USA, 25–30 January 2015; AAAI Publications: Austin, TX, USA, 2015; Volume 1.
3. Eckmann, J.P.; Kamphorst, S.O.; Ruelle, D. Recurrence Plots of Dynamical Systems. *Europhys. Lett. (EPL)* **1987**, *4*, 973–977. [CrossRef]
4. LeCun, Y.; Bengio, Y.; Hinton, G. Deep Learning. *Nature* **2015**, *521*, 436–444. [CrossRef]
5. Vincent, P.; Larochelle, H.; Bengio, Y.; Manzagol, P.A. Extracting and Composing Robust Features with Denoising Autoencoders. In Proceedings of the 25th International Conference on Machine Learning, ICML '08, Helsinki, Finland, 5–9 July 2008; Association for Computing Machinery: New York, NY, USA, 2008; pp. 1096–1103. [CrossRef]
6. Kingma, D.P.; Welling, M. Auto-Encoding Variational Bayes. *arXiv* **2014**, arXiv:1312.6114.
7. Rezende, D.J.; Mohamed, S.; Wierstra, D. Stochastic Backpropagation and Approximate Inference in Deep Generative Models. *arXiv* **2014**, arXiv:1401.4082.
8. van den Oord, A.; Dieleman, S.; Zen, H.; Simonyan, K.; Vinyals, O.; Graves, A.; Kalchbrenner, N.; Senior, A.; Kavukcuoglu, K. WaveNet: A Generative Model for Raw Audio. *arXiv* **2016**, arXiv:1609.03499.
9. Devlin, J.; Chang, M.W.; Lee, K.; Toutanova, K. Bert: Pre-Training of Deep Bidirectional Transformers for Language Understanding. *arXiv* **2018**, arXiv:1810.04805.
10. Zhang, H.; Goodfellow, I.; Metaxas, D.; Odena, A. Self-Attention Generative Adversarial Networks. In Proceedings of the 36th International Conference on Machine Learning, Long Beach, CA, USA, 9–15 June 2019; PMLR: Long Beach, CA, USA, 2019; Volume 97, pp. 7354–7363.
11. Raffel, C.; Ellis, D.P.W. Feed-Forward Networks with Attention Can Solve Some Long-Term Memory Problems. *arXiv* **2016**, arXiv:1512.08756.
12. Hochreiter, S.; Schmidhuber, J. Long Short-Term Memory. *Neural Comput.* **1997**, *9*, 1735–1780. [CrossRef]
13. Graves, A.; Schmidhuber, J. Framewise Phoneme Classification with Bidirectional LSTM and Other Neural Network Architectures. *Neural Netw.* **2005**, *18*, 602–610. [CrossRef] [PubMed]
14. Cho, K.; van Merrienboer, B.; Gulcehre, C.; Bahdanau, D.; Bougares, F.; Schwenk, H.; Bengio, Y. Learning Phrase Representations Using RNN Encoder-Decoder for Statistical Machine Translation. *arXiv* **2014**, arXiv:1406.1078.
15. Goodfellow, I.; Pouget-Abadie, J.; Mirza, M.; Xu, B.; Warde-Farley, D.; Ozair, S.; Courville, A.; Bengio, Y. Generative Adversarial Nets. In *Advances in Neural Information Processing Systems*; Ghahramani, Z., Welling, M., Cortes, C., Lawrence, N., Weinberger, K.Q., Eds.; Curran Associates, Inc.: Red Hook, NY, USA, 2014; Volume 27, pp. 2672–2680.
16. Liang, M.; Hu, X. Recurrent Convolutional Neural Network for Object Recognition. In Proceedings of the IEEE Conference on Computer Vision and Pattern Recognition, Boston, MA, USA, 7–12 June 2015; pp. 3367–3375.
17. Chung, J.; Kastner, K.; Dinh, L.; Goel, K.; Courville, A.C.; Bengio, Y. A Recurrent Latent Variable Model for Sequential Data. In *Advances in Neural Information Processing Systems*; Curran Associates, Inc.: Red Hook, NY, USA, 2015; Volume 28.
18. Bengio, Y.; Louradour, J.; Collobert, R.; Weston, J. Curriculum Learning. In Proceedings of the 26th Annual International Conference on Machine Learning, ICML'09, Montreal, QC, Canada, 14–18 June 2009; Association for Computing Machinery: New York, NY, USA, 2009; pp. 41–48. [CrossRef]
19. Koenker, R.; Bassett, G. Regression Quantiles. *Econometrica* **1978**, *46*, 33–50. [CrossRef]
20. Ronneberger, O.; Fischer, P.; Brox, T. U-Net: Convolutional Networks for Biomedical Image Segmentation. In *Medical Image Computing and Computer-Assisted Intervention–MICCAI 2015*; Navab, N., Hornegger, J., Wells, W.M., Frangi, A.F., Eds.; Springer International Publishing: Cham, Switzerland, 2015; Volume 9351, pp. 234–241. [CrossRef]
21. TensorFlow Hub—A Repository of Trained Machine Learning Models. Available online: Www.Tensorflow.Org/Hub (accessed on 11 January 2021).
22. 14th ACM International Conference on Distributed and Event-Based Systems DEBS 2020—Challenge (2020.Debs.Org/Call-for-Grand-Challenge-Solutions/). 2020. Available online: https://2020.debs.org/call-for-grand-challenge-solutions/ (accessed on 11 January 2021).
23. Ruder, S. An Overview of Multi-Task Learning in Deep Neural Networks. *arXiv* **2017**, arXiv:1706.05098.
24. Doukas, D. EU NILM Workshop 2019-NILM Datasets, Benchmarking and Evaluation. 2019. Available online: www.youtube.com/watch?v=v5XoLtQH9Uw&list=PLJrF-gxa0ImryGeNtil-s9zPJOaV4w-Vy&index=22&t=0s (accessed on 10 October 2020).
25. Ouali, Y.; Hudelot, C.; Tami, M. An Overview of Deep Semi-Supervised Learning. *arXiv* **2020**, arXiv:2006.05278.
26. Rasmus, A.; Valpola, H.; Honkala, M.; Berglund, M.; Raiko, T. Semi-Supervised Learning with Ladder Networks. *arXiv* **2015**, arXiv:1507.02672.
27. Tarvainen, A.; Valpola, H. Mean Teachers Are Better Role Models: Weight-Averaged Consistency Targets Improve Semi-Supervised Deep Learning Results. In Proceedings of the 31st International Conference on Neural Information Processing Systems, NIPS'17, Long Beach, CA, USA, 4–9 December 2017; Curran Associates Inc.: Red Hook, NY, USA, 2017; pp. 1195–1204.
28. Miyato, T.; Maeda, S.I.; Koyama, M.; Ishii, S. Virtual Adversarial Training: A Regularization Method for Supervised and Semi-Supervised Learning. *IEEE Trans. Pattern Anal. Mach. Intell.* **2018**, *41*, 1979–1993. [CrossRef]
29. Kuo, C.W.; Ma, C.Y.; Huang, J.B.; Kira, Z. FeatMatch: Feature-Based Augmentation for Semi-Supervised Learning. *arXiv* **2020**, arXiv:2007.08505.

180. Xie, Q.; Dai, Z.; Hovy, E.; Luong, M.T.; Le, Q.V. Unsupervised Data Augmentation for Consistency Training. *arXiv* 201 arXiv:1904.12848.
181. French, G.; Oliver, A.; Salimans, T. Milking CowMask for Semi-Supervised Image Classification. *arXiv* **2020**, arXiv:2003.1202:
182. Howard, A.G.; Zhu, M.; Chen, B.; Kalenichenko, D.; Wang, W.; Weyand, T.; Andreetto, M.; Adam, H. MobileNets: Efficie Convolutional Neural Networks for Mobile Vision Applications. *arXiv* **2017**, arXiv:1704.04861.
183. Howard, A.; Zhmoginov, A.; Chen, L.C.; Sandler, M.; Zhu, M. Inverted Residuals and Linear Bottlenecks: Mobile Networks f Classification, Detection and Segmentation. In Proceedings of the CVPR 2018, Salt Lake City, UT, USA, 18–22 June 2018.
184. Sze, V. Efficient Processing of Deep Neural Networks: From Algorithms to Hardware Architectures. In Proceedings of t NeurIPS 2020, Virtual Conference, 6–12 December 2019.
185. Batra, N.; Gulati, M.; Singh, A.; Srivastava, M.B. It's Different: Insights into Home Energy Consumption in India. In Proceedin of the 5th ACM Workshop on Embedded Systems For Energy-Efficient Buildings, BuildSys'13, Roma, Italy, 13–14 Novemb 2013; Association for Computing Machinery: New York, NY, USA, 2013; pp. 1–8. [CrossRef]
186. Kriechbaumer, T.; Jacobsen, H.A. BLOND, a Building-Level Office Environment Dataset of Typical Electrical Appliances. S *Data* **2018**, *5*, 1–14. [CrossRef]
187. Cohen, T.; Welling, M. Group Equivariant Convolutional Networks. In Proceedings of the 33rd International Conference o Machine Learning, New York, NY, USA, 19–24 June 2016; PMLR: New York, NY, USA, 2016; Volume 48, pp. 2990–2999.
188. Zaheer, M.; Kottur, S.; Ravanbakhsh, S.; Poczos, B.; Salakhutdinov, R.R.; Smola, A.J. Deep Sets. In *Advances in Neural Informati Processing Systems*; Curran Associates, Inc.: Long Beach, CA, USA, 2017; Volume 30, pp. 3391–3401.
189. McMahan, B.; Moore, E.; Ramage, D.; Hampson, S.; y Arcas, B.A. Communication-Efficient Learning of Deep Networks fro Decentralized Data. In Proceedings of the 20th International Conference on Artificial Intelligence and Statistics, Lauderdale, F USA, 20–22 April 2017; PMLR: Fort Lauderdale, FL, USA, 2017; Volume 54, pp. 1273–1282.
190. Kairouz, P.; McMahan, H.B.; Avent, B.; Bellet, A.; Bennis, M.; Bhagoji, A.N.; Bonawitz, K.; Charles, Z.; Cormode, G.; Cu mings, R.; et al. Advances and Open Problems in Federated Learning. *arXiv* **2019**, arXiv:1912.04977.

Article

A Scalable Real-Time Non-Intrusive Load Monitoring System for the Estimation of Household Appliance Power Consumption

Christos Athanasiadis [1,2], Dimitrios Doukas [2,*], Theofilos Papadopoulos [1] and Antonios Chrysopoulos [2,3]

1. Department of Electrical and Computer Engineering, Democritus University of Thrace, 67100 Xanthi, Greece; cathanas@ee.duth.gr (C.A.); thpapad@ee.duth.gr (T.P.)
2. NET2GRID BV, Krystalli 4, 54630 Thessaloniki, Greece; antonios@net2grid.com
3. School of Electrical and Computer Engineering, Aristotle University of Thessaloniki, 54124 Thessaloniki, Greece
* Correspondence: dimitrios@net2grid.com

Abstract: Smart-meter technology advancements have resulted in the generation of massive volumes of information introducing new opportunities for energy services and data-driven business models. One such service is non-intrusive load monitoring (NILM). NILM is a process to break down the electricity consumption on an appliance level by analyzing the total aggregated data measurements monitored from a single point. Most prominent existing solutions use deep learning techniques resulting in models with millions of parameters and a high computational burden. Some of these solutions use the turn-on transient response of the target appliance to calculate its energy consumption, while others require the total operation cycle. In the latter case, disaggregation is performed either with delay (in the order of minutes) or only for past events. In this paper, a real-time NILM system is proposed. The scope of the proposed NILM algorithm is to detect the turning-on of a target appliance by processing the measured active power transient response and estimate its consumption in real-time. The proposed system consists of three main blocks, i.e., an event detection algorithm, a convolutional neural network classifier and a power estimation algorithm. Experimental results reveal that the proposed system can achieve promising results in real-time, presenting high computational and memory efficiency.

Keywords: convolutional neural network; energy consumption; energy data analytics; energy disaggregation; machine learning; non-intrusive load monitoring; real-time; smart meter data; smart meters; transient load signature

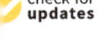

1. Introduction

Nowadays, the amount of data that is generated almost continuously is enormous. Once analyzed, they can reveal useful information in many different disciplines; economy, healthcare, and e-commerce, to name a few. In this context, the energy sector could not have been an exception. Traditionally, energy data was acquired at a few critical points of the power grid, usually at the transmission level, but the landscape has changed due to the advance in smart-metering technologies. Thousands of internet-of-things (IoT) endpoints are placed within the smart grid, providing energy utilities access to valuable data; thus, new opportunities have been created for energy services and data-driven business models [1–5]. Energy disaggregation is an example of such a service.

Energy disaggregation is the process of consumption breakdown at appliance or activity level for residential or commercial-industrial (C&I) users; in other words, it estimates the individual power consumption for all appliances contributing to the total mains power. This process can help energy utilities reveal useful information to support load forecasting and demand-side management programs. Regarding residential consumers, it can be used

to provide accurate billing and meaningful feedback regarding their energy consumption as well as to improve the appliance efficiency (e.g., by detecting old devices and replacing them with more efficient ones) [6].

There are two main possible energy disaggregation solutions: (a) Intrusive Load Monitoring (ILM) and (b) Non-Intrusive Load Monitoring (NILM). In ILM, i.e., a hardware based approach, power meters are attached behind each target appliance. The large number of hardware devices required for ILM makes the installation process difficult and cost inefficient but results in very accurate power estimates. On the other hand, NILM is a software-based approach. It requires a single meter for the total aggregated power, thus the installation process is simplified and the corresponding cost is reduced. However, since there is no information about the aggregated power appliances, appropriate algorithms should be created to perform energy decomposition.

The utilities should perform a large-scale deployment to support thousands of consumers to benefit as much as possible from energy disaggregation services; only then it is possible to extract useful information for business models. This large-scale deployment makes NILM far more favorable than ILM due to the low cost, installation simplicity and minimum hardware requirements. However, in many cases, NILM algorithms present high computational complexity and significant memory requirements. In this sense, utilities should either use high-end smart meters—or extra hardware attached to them—with powerful central processing units (CPUs) and sufficient memory. Alternatively, energy disaggregation must be performed in cloud services. In the latter case, the cost of cloud services increases with the number of consumers. To this end, utilities must adopt scalable solutions. Scalability can be more critical even than disaggregation accuracy. As it is realized, low computational and memory requirements are necessary to run the service on the edge with conventional microprocessors or minimize the cost of needed cloud services. Furthermore, to improve user experience, minimum feedback must be required; thus the necessity for pre-trained generic appliance models is of utmost importance.

Several approaches have been proposed to cope with the NILM problem [7,8]. It was first introduced by Hart [9]. Hart's approach was based on monitoring power changes (corresponding to the appliance turning-on/off events) of both active and reactive power signals. These power changes are grouped into clusters, with each cluster representing a state change of a target appliance. Since then, several works have investigated the NILM problem utilizing different sampling rates and techniques. Earlier approaches employed sampling rates lower than 1 Hz, where event detection (appliances turning on/off) is impractical and probabilistic models, such as variants of hidden Markov models (HMM) were examined [10–17]. HMMs yield promising results but present disadvantages, e.g., high computational complexity increases when the number of appliances increases and difficulty in classifying appliances that present similar power consumption [18]. Due to these disadvantages, researchers have turned to alternative methods, including machine learning and deep learning techniques [18–38]. NILM approaches can be generally categorized as event-based and state-based.

Event-based solutions [33–39] leverage the information-rich transient response of an appliance turning-on. Specifically, they consist of two modules: (a) an event detection algorithm for discovering power changes corresponding to an appliance turning-on and (b) a classifier for identifying the appliance that caused the power change. This approach is based on the fact that turn-on transient responses contain more information regarding the operating device than steady states. However, in order to obtain this transient state information, high-resolution data is vital [33–38]. One widespread event-based method is the V-I trajectory, utilizing high-resolution voltage and current measurements. In [37,38] useful features are extracted from the V-I trajectories and neural networks are trained for classification. Other researchers depict the trajectories as binary images [34–36]. This visual representation solves the appliance recognition problem by exploiting computer vision techniques. Furthermore, transfer learning techniques have been investigated, as in [36] where an image classifier has been implemented based on AlexNet [40]. An important advantage of event-based approaches is the low complexity; only a few time instants corresponding to on/off events are processed. Furthermore, such approaches can detect an appliance turn-on event in real-time since information only from the transient state is required. However, high-resolution data of several kHz is essential, applying mainly to detect appliance turn-on/off events, without calculating power consumption.

On the other hand, state-based approaches [19–27] mainly require lower frequency data. These approaches do not detect state transitions. On the contrary, they parse all available data of a time-series, even if no events occur. In such approaches, the appliance must operate for at least some minutes to determine if it is on [20,22]. There are even cases where the appliance end-use has to be fully completed to estimate the power consumption [19,24]. In [19], three different neural network architectures were presented, i.e., (a) long short-term memory (LSTM) networks, (b) stacked denoising autoencoders, and (c) a regression algorithm to forecast the start time, stop time and average power demand of devices. In [21], a bidirectional LSTM cell was used; in [26] a deep convolutional neural network (CNN) that uses as input a time window of active power consumption and predicts the active power in the center of the window. In [23], the authors feed their network with active, reactive, and apparent power and current data. Furthermore, they use mainly CNN blocks in order to create a recurrent property similar to LSTMs. Finally, in [24], an attention-based deep neural network is introduced, inspired by deep learning techniques used in Natural Language Processing (NLP). State-based approaches use low-resolution data and predict the power consumption per appliance. However, they present higher computational complexity since all available data are used, thus cannot detect in real-time an appliance being turned-on/off.

The scope of this paper is to present a real-time event-based NILM methodology to detect an appliance turn-on event and calculate its power consumption in real-time. The proposed NILM design is built on top of three main blocks, i.e., an event detector, a CNN classifier and a power estimation algorithm. The main strengths of the proposed NILM system rely on the following:

- The proposed system can identify when an appliance is turned-on in real-time, based on its active power transient response sampled at 100 Hz; processing data of the total appliance operational duration is not required, as in [19,24].
- The proposed system is delay-free; once the appliance has been turned-on, the system can calculate its power in real-time.
- The combination of a machine learning model to detect appliance turning-on and a heuristic algorithm to estimate the power in real-time constitute a system lightweight, presenting less memory and CPU requirements than end-to-end deep learning models [19,23,24].
- The proposed NILM algorithm is automatic, thus, no feedback is required by the user.
- Data sampling rate of 100 Hz for active power measurements is used, contrary to several kHz in relevant works [33–35,41,42].

Generally, as it can be suggested from the above analysis, the proposed system constitutes a real-time scalable solution presenting minimum hardware requirements; thus, it can be integrated into low-cost chip-sets and, consequently, run on the edge.

The paper is structured as follows: In Section 2, the proposed methodology is presented. In Section 3, the dataset and the metrics used for evaluation are described. In Section 4, experimental validation results from real-life installations are analyzed and the performance of the system is compared to other state-of-the-art approaches. In Section 5, an industrial perspective regarding scalable real-time NILM services is discussed. Finally, Section 6 concludes the paper.

2. Proposed System

The proposed methodology comprises of three main parts: (a) an event-detection system to find active power changes corresponding to turn-on events, (b) a CNN binary classifier to determine if the turn-on event was caused by a specific target appliance or

not, and (c) a power estimation algorithm to calculate in real-time the appliance power p
second and consequently the energy consumption. An overview of the system in flowcha
form is illustrated in Figure 1.

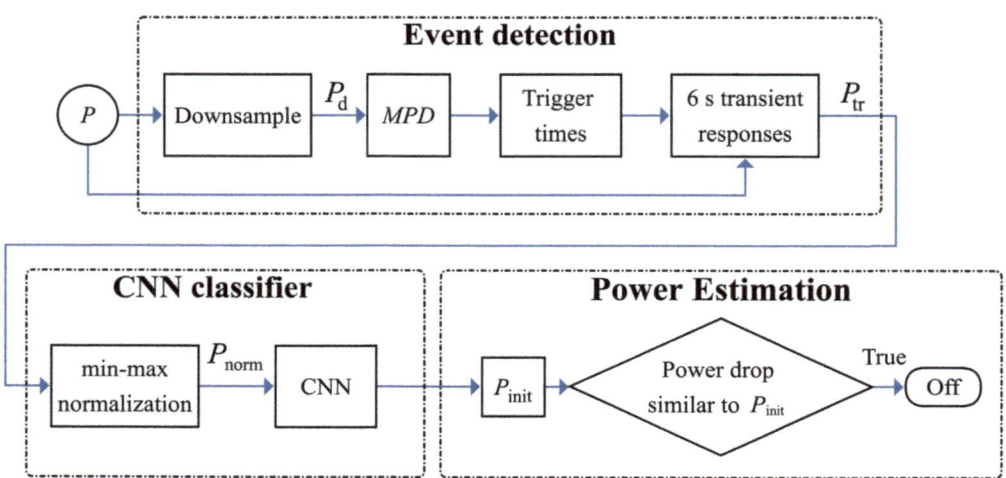

Figure 1. Proposed system flowchart.

2.1. Event Detection

The event detection algorithm is used to identify the time instant (trigger time) when sudden increase of active power occurs, indicating a possible turn-on event. The advantag of the proposed event detection algorithm are its simplicity and the fact that no pre-trainir is required.

Let us assume that the aggregated active power time-series at 100 Hz is P. The origin signal P is down-sampled at 1 Hz by means of averaging, resulting into signal P_d. Dow sampling is applied for two main reasons: (a) the event detection algorithm becom simpler, presenting less computational burden and (b) most of power changes are st easily identifiable assuming an 1 Hz sampling frequency. However, if two or more even occur almost simultaneously, e.g., in a period of less than a second, the algorithm detec these events as a single one. Considering that the probability of this scenario is very lo the frequency of 1 Hz has been selected. Next, the maximum power difference (MPD) fc each second n is calculated as:

$$MPD(n) = \max(P_d(n+1), P_d(n+2), P_d(n+3)) - \min(P_d(n-3), P_d(n-2), P_d(n-1)).$$

MPD shows the maximum difference in active power in a region around n, i.e., tl maximum power during the first three seconds after n, minus the minimum power of tl three first seconds before n. In this sense, the transient onset can be accurately determine since the real power increase may not appear immediately, but some seconds after To determine the trigger time candidates, MPD is compared with a threshold, P_{th}, which determined in terms of the appliance rating power. This means that, at time instant n a event occurs if

$$MPD(n) > P_{th}.$$

At this point, it should be mentioned that trigger time candidates close in time a merged. For each trigger time, a 6 s window of the captured transient response, P_{tr}, generated from P (100 × 6 = 600 samples). The pseudo-code for the process described is presented in Algorithm 1.

Algorithm 1: Event detection.

Input: P, P_{th}
Output: list of captured transient responses
$P_d = P$ down-sampled at 1 Hz;
Initialize an empty list L;
Initialize an empty list transients;
for each second n **do**
    max_after = max($P_d(n + 1)$, $P_d(n + 2)$, $P_d(n + 3)$);
    min_before = min($P_d(n-3)$, $P_d(n-2)$, $P_d(n-1)$);
    MPD = max_after − min_before;
    if $MPD \geq P_{th}$ **then**
        | Append n to L;
    end
end
Merge consecutive seconds in L;
for t in L **do**
    $P_{tr} = P(t - 300:t + 299)$;
    Append P_{tr} to transients;
end
return transients;

2.2. CNN Classifier

In the proposed methodology, the transient response generated by an appliance's turning-on is used as the load signature for appliance classification [7]. Whenever a target appliance is turned on, a transient response can be detected in the aggregated active power waveform. Besides appliance classification, this load signature presents two additional advantages. Firstly, for a given appliance, the turn-on transient response pattern is unique and relates only to the operational characteristics of the appliance [43]. Consequently, the identification algorithm's performance is independent of the simultaneous operation of other types of appliances, even when a large number of devices is considered [7,44]. Secondly, the proposed algorithm can successfully treat various types of appliances, even though presenting similar consumption levels at steady-state, since classification is performed based on the unique appliance transient characteristics instead of calculating steady-state features.

The same principle can detect specific operational states by identifying transient responses caused by a state transition regarding multi-state appliances. For example, for a washing machine or a dishwasher, the water heating process's transient response can be used to identify this specific state, being of primary interest as the most energy-intensive process during an operation cycle.

In order to associate a given transient response, P_{tr}, with a specific target appliance behavior, a CNN classifier is utilized. In this sense, for each target appliance, a dedicated CNN classifier is used, identifying P_{tr} as positive when related to the target appliance or negative otherwise.

Different types of appliances generate transient responses with distinct characteristics, primarily when a high sampling frequency, e.g., at 100 Hz, is used. Suppose a user was initially given an example of such a response corresponding to a specific appliance. In that case, he/she could later recognize a new response of the same appliance by simple visual inspection. However, the implementation of such a recognition algorithm is not an easy task.

Inspired from the area of computer vision, where CNN models are used for image recognition, and classification [45], a similar approach has been adopted in this paper.

Convolutional layers can automatically extract useful features from the input data without user supervision [45]. Thus, there is no need to implement specific algorithms; instead, by training a CNN model, the classification problem can be successfully solved. A block diagram of the proposed CNN architecture is depicted in Figure 2.

Initially, min-max normalization is applied to P_{tr} by means of (3); the resulting normalized vector, P_{norm}, is forwarded as input to the CNN model.

$$P_{norm} = \frac{P_{tr} - \min(P_{tr})}{\max(P_{tr}) - \min(P_{tr})}$$

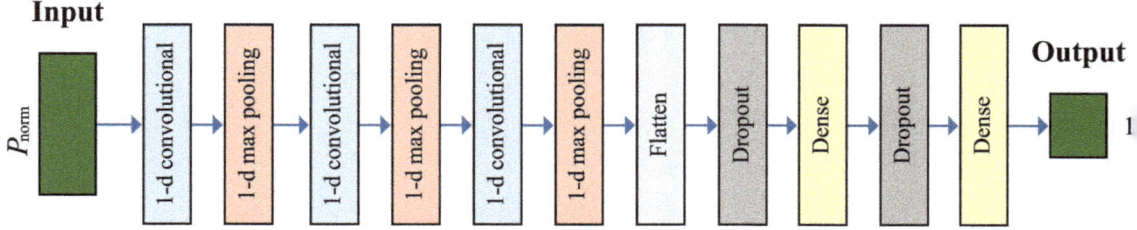

Figure 2. Convolutional neural network (CNN) block diagram.

Since the CNN input model is an one-dimensional signal, one-dimensional convolutional layers are used to extract the useful features from P_{norm}. In particular, three consecutive 1-d convolutional layers are used in combination with an 1-d max-pooling layer. All convolutional layer parameters have been set to 32 filters, kernel size equal to 3, strides equal to 1, 'same' padding, and rectified linear unit (ReLU) activation function. The ReLU function is defined as

$$\text{ReLU}(x) = \max(x, 0)$$

for $x \in \mathbb{R}$. For max-pooling layers, the pool size was set to 2. Generally, at each 1-d convolutional layer, a number of filters is applied to the corresponding input, $\mathbf{x}_{conv}$. Assuming that the size of $\mathbf{x}_{conv}$ is $M_{conv} \times N_{conv}$ and a single filter, $\mathbf{f}$, is of $3 \times N_{conv}$, the output of the convolution between $\mathbf{x}_{conv}$ and $\mathbf{f}$ will be a $M_{conv} \times 1$ matrix. The resulting $\mathbf{y}_{conv}$ is calculated as

$$\mathbf{y}_{conv}(m) = \max \left(\sum_{t=1}^{3} \sum_{n=1}^{N_{conv}} \mathbf{x}_{conv}(m+t-2, n) \, \mathbf{f}(t,n), 0 \right)$$

for $m \in [1, ..., M_{conv}]$, where $\mathbf{x}_{conv}(0, n)$ and $\mathbf{x}_{conv}(M_{conv}+1, n)$ are considered zero for any $n \in [1, ..., N_{conv}]$ as a result of zero-padding. In our case, where 32 filters are used in convolutional layer results $\mathbf{y}_{conv}$, are stacked as columns, forming a $M_{conv} \times 32$ matrix.

Each layer is followed by a max-pooling layer to down-sample the extracted features of the input signal. In this sense, a summarized version of the extracted features (half the size) is created, maintaining the most important features and is further used as input to the next layer. Assuming $\mathbf{x}_{pool}$, with size $M_{pool} \times N_{pool}$ is the max-pooling layer input matrix, the output, $\mathbf{y}_{pool}$, has a size of $(M_{pool}/2) \times N_{pool}$ and is calculated as

$$\mathbf{y}_{pool}(m,n) = \max \left(\mathbf{x}_{pool}(2m-1, n), \mathbf{x}_{pool}(2m, n) \right)$$

for $m \in [1, ..., M_{pool}/2]$ and $n \in [1, ..., N_{pool}]$.

Following the three convolutional/pooling pairs, a flattening layer is applied, transforming its input to a single vector by column-wise stacking. Finally, two dense layers are used of 20 and 1 output nodes, respectively. For the first dense layer, the ReLU activation function is applied; for the last layer, the sigmoid activation function defined in (7) for

$x \in \mathbb{R}$ is used to compute the probability of the transient response to correspond to the positive class.

$$S(x) = \frac{1}{1+e^{-x}} \qquad (7)$$

Generally speaking, a dense layer with M_{dense} input nodes and K output nodes includes two trainable parameters, i.e., a weight matrix, $\mathbf{w}$, with size $M_{\text{dense}} \times K$ and a bias vector, $\mathbf{b}$, with size K. Given an input vector, $\mathbf{x}_{\text{dense}}$, with M_{dense} elements, the output $\mathbf{y}_{\text{dense}}$ of size K is calculated as

$$\mathbf{y}_{\text{dense}}(k) = F\left(\sum_{m=1}^{M_{\text{dense}}} \mathbf{x}_{\text{dense}}(m)\, \mathbf{w}(m,k) + \mathbf{b}(k) \right) \qquad (8)$$

for $k \in [1, ..., K]$, where F is the corresponding activation function. Before each dense layer, a dropout layer [46] is used. Its value is set to 0.2 to prevent model over-fitting.

A standard backpropagation algorithm is used during training to optimize the binary cross-entropy loss between the predicted probabilities and the actual labels. Assuming that the predicted probabilities are $p_1, p_2, ..., p_B$ for B samples and the actual labels are $q_1, q_2, ..., q_B$, the binary cross-entropy loss is

$$L = -\frac{1}{B} \sum_{b=1}^{B} [q_b \log_2 p_b + (1-q_b) \log_2 (1-p_b)]. \qquad (9)$$

The CNN classifier is trained for a maximum of 50 iterations. The Adamax optimizer [47] was selected assuming an initial learning rate of 0.01 and batch size 32. In order to avoid over-fitting, early stopping with patience is used. The training process stops once the validation accuracy does not improve after five consecutive iterations.

2.3. Consumed Energy Estimation Algorithm

The last module is related to the real-power estimation of the target appliance. The implemented algorithm considers the appliance end-uses as pulses of constant power; this approximation is well-suited for single-state appliances such as microwave oven, kettle or toaster. In the case of appliances with operating cycles comprising of multiple pulses, the algorithm considers each pulse as a new appliance end-use and not as a single end-use event of several pulses. An example is the oven turning-on and off controlled by a thermostat and the dishwasher, where several water heating pulses may occur depending on the selected program. In this sense, the proposed algorithm performance may degrade for multi-state appliances. They are characterized by varying power consumption and cannot be approximated with a constant power pulse. However, such appliances present a predominant energy-intensive process during a full operating cycle while the rest operating states are less critical regarding the total energy consumption. For example, washing machine or dishwasher cycles include energy-intensive water heating processes and low energy-consuming processes, e.g., water pumping. Therefore, regarding multi-state appliances, the proposed power estimation algorithm focuses on the estimation of the energy-intensive processes neglecting the effect of the minor consuming ones.

When the CNN classifies a transient response as positive, it is implied that the appliance has been turned-on. The calculated power increase, P_{init} is considered equal to the appliance power consumption and assumed constant during the total time of operation of the appliance. When a power decrease between two consecutive seconds in P_d inside the interval [0.8 P_{init}, 1.2 P_{init}] is detected, the appliance is considered to be turned-off. The pseudo-code of the energy consumption estimation algorithm is shown in Algorithm 2, having as input the time (in seconds), t, when the target appliance is turned-on and P_d.

Algorithm 2: Energy consumption estimation.

Input: t, P_d
Output: turn-off time, P_{init}
$P_{init} = P_d(t+2) - P_d(t-2)$;
for each second $n \geq t$ **do**
　if $0.8\, P_{init} \leq (P_d(n-1) - P_d(n)) \leq 1.2\, P_{init}$ **then**
　　Assume target appliance is turned-off;
　　break;
　else
　　Assume power consumption equal to P_{init};
　end
end
return n, P_{init};

3. Evaluation Methodology

3.1. Dataset

The proposed NILM system is based on the fact that each household appliance presents a transient response pattern with distinct characteristics, becoming more noticeable as the sampling frequency increases. In this paper, the selected sampling frequency is 100 Hz; at this frequency the transient characteristics are captured in contrast to low sampling rates where such information may be lost. In Figure 3, turn-on transient responses at 100 Hz and 1 Hz for five appliances are depicted. It is evident that the frequency of 100 Hz reveals unique details that are lost when sampling at 1 Hz. More specifically, Figure 3a presents the turn-on response of a high-power consumption (~1.2 kW) fridge compressor with a duration of fewer than two seconds. Figure 3b visualizes the water heating process of a washing machine, which corresponds to a steep power step-up. Next, Figure 3c illustrates the transient response of a microwave oven as a high-power spike followed by a smooth power increase. In Figure 3d, a stove turn-on presenting a smooth and convex power increase is shown, and finally, in Figure 3e visualizes the transient response from a heat pump dryer appliance, including a high-power spike at motor starting time.

An extensive set of transient responses for each target appliance is required to train the CNN classifier. For this purpose, a private dataset that includes transient responses of different household appliances sampled at 100 Hz from different installations is used. The type of appliance and the number of samples for each case are summarized in Table. Note that, the duration of the transient responses contained in the dataset ranges from 12 to 1 min.

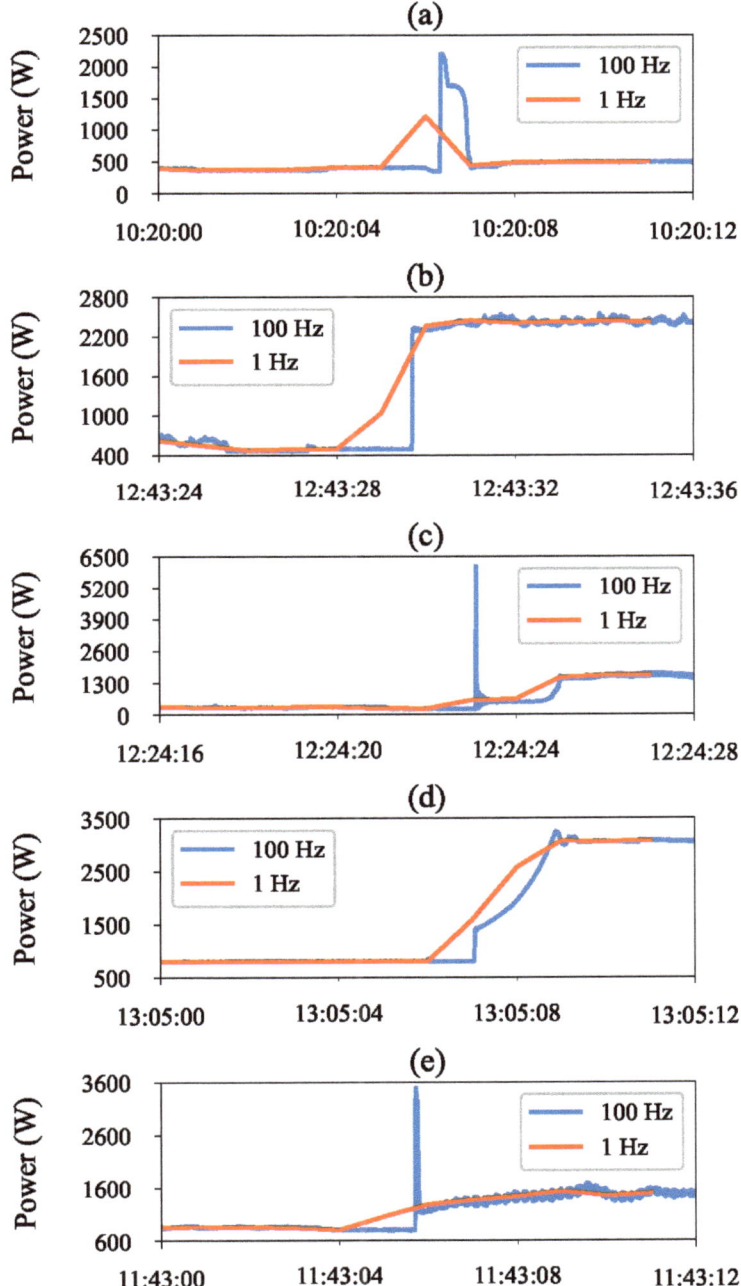

Figure 3. Comparison of the turn-on transient response with sampling frequency at 100 Hz and 1 Hz for (**a**) fridge, (**b**) washing machine, (**c**) microwave oven, (**d**) stove and (**e**) heat pump dryer.

Table 1. Private dataset: Number of transient responses for the appliances of interest.

Appliance	Number of Transient Responses
Fridge	132
Dishwasher	171
Heat pump	202
Washing machine	135
Oven	82
Stove	148
Heat pump dryer (drum spinning)	54
Heat pump dryer (heating)	42
Microwave	290

In this study, three appliances are selected to test the proposed methodology's performance, i.e., fridge, washing machine, and microwave oven. Pulses can approximate the end-use of these appliances without significant error in power estimation. Furthermore, such appliances are considered typical for most households, corresponding to substantial total energy consumption. The selected appliances represent a larger group of appliances since both single-state and multi-state appliances are considered. Additionally, detailed results regarding the analysis of such appliances can be found in several relevant works [19,20,22,25–27]; thus, a comprehensive comparative analysis can be performed. Finally, low energy-consuming appliances such as game consoles and phone chargers have not been investigated, being of trivial importance and hard to be identified in terms of NILM algorithm application [19].

For each target appliance, a binary classifier is implemented and trained. During training, the transient responses of the appliance under consideration are labeled positive; the responses corresponding to a different appliance are labeled negative. Balancing of the positive and negative classes is performed in order to prevent bias towards the class with the most samples; the number of negative responses is the same as the number of positive ones. A training/validation/testing split is used assuming a ratio of 60%/20%/20% to avoid over-fitting for each class separately.

However, because the number of samples per appliance is small, augmentation techniques are used. These techniques aim to increase the number as well as the diversity of the training samples by artificially introducing variations in existing transient responses. Specifically, for each transient response, 15 samples with the required length of 6 s are created. Assuming that the time-series that contains a response is z, each one of the 15 samples is generated by means of the following steps:

1. Considering that the transient response starts at index s of z, a random number u in the interval [s − 500, s − 100] is selected, following uniform distribution. The selected sample is equal to z from index u to index u + 599.
2. White Gaussian noise with mean value (μ) equal to 0 and standard deviation (σ) equal to 1 is added to the sample; 10 W maximum power is considered.

The number of samples for training, validation and testing the sets per appliance is shown in Table 2.

Table 2. Number of positive samples per set.

Appliance	Training	Validation	Testing
Fridge	2400	780	780
Washing machine	2430	810	810
Microwave	5220	1740	1740

3.2. Performance Metrics

The proposed methodology is evaluated in terms of the event detection algorithm, the CNN classifiers as well as the overall system performance. For each case, different metrics are used.

3.2.1. Metrics for Event Detection Evaluation

For the event detection algorithm the true positive rate ($TPR = TP/(TP + FN)$), the false positive rate ($FPR = FP/(FP + TN)$) and false negative rate ($FNR = FN/(TP + FN)$) are calculated; TP, FN, FP and TN are the number of true positives, false negatives, false positives and true negatives, respectively. Here, a sample (a time instant) is positive if it is an actual event (there is an appliance turning-on or off) and negative if not.

3.2.2. Metrics for Classifier Evaluation

To evaluate the classifier, the most common metrics used in classification and NILM problems are adopted [18,29,32,34,35]. Specifically, the accuracy, precision, recall and F_1-score, defined in (10)–(13), respectively, are calculated

$$accuracy = \frac{TP + TN}{TP + TN + FP + FN} \qquad (10)$$

$$precision = \frac{TP}{TP + FP} \qquad (11)$$

$$recall = \frac{TP}{TP + FN} \qquad (12)$$

$$F_1 = 2 \cdot \frac{precision \cdot recall}{precision + recall}. \qquad (13)$$

In this context, for a transient response classifier, a sample (i.e., transient response of 6 s) is positive if the transient response corresponds to the target appliance. Otherwise, it is assumed negative.

3.2.3. Metrics for Overall NILM System Evaluation

The overall proposed NILM system is tested by using the same metrics as previously, i.e., accuracy, precision, recall, and F_1-score to evaluate the predicted status of the appliance (ON or OFF). Thus, a sample (i.e., a time instant) is considered positive if the appliance is ON and negative if not. It should be mentioned that an appliance is considered turned-on if the measured active power is higher than 5 W. Additionally, for energy estimation, the mean absolute error (MAE) and the root mean square error ($RMSE$) in (14) and (15), respectively, are computed

$$MAE = \frac{1}{N} \sum_{n=1}^{N} |y[n] - \hat{y}[n]| \qquad (14)$$

$$RMSE = \sqrt{\frac{1}{N} \sum_{n=1}^{N} (y[n] - \hat{y}[n])^2} \qquad (15)$$

where $y[n]$ and $\hat{y}[n]$ is the original and the estimated power response with N samples. Moreover, the relative error in total energy (RE), defined in (16), is calculated

$$RE = \frac{|E - \hat{E}|}{max(E, \hat{E})}$$

where E and $\hat{E}$ is the original and the estimated total energy consumption of the appliance.

4. Results

In this section, experimental validation results are analyzed considering data from real world installations. The Building-Level fully-labeled dataset for Electricity Disaggregation (BLUED) [48] is used to test the applicability of the proposed event-detection algorithm. Energy consumption data from three household installations are also used to evaluate the performance of the proposed methodology; common metrics are employed and results are compared with those obtained from other state-of-the-art methods proposed in the literature. Finally, the computational and memory efficiency of the proposed system is discussed.

4.1. Event Detection Evaluation

The BLUED dataset contains aggregate voltage/current and active power data, sampled at 12 kHz and 60 Hz, respectively, from a 2-phase household in Pittsburgh, USA. The recording duration is eight days. The time instants when a turn-on or turn-off event occurred are also reported in the dataset. In particular, for testing the proposed event detection algorithm, the active power measurements of phase A, at 1 Hz, from 11:58:3 20 October 2011, to 09:29:55 21 October 2011, are used. In fact, during this period, 125 events have occurred, including six pairs of simultaneous events. The proposed algorithm detects the simultaneous events as well as two near-simultaneous turn-off events as single events, respectively. Finally, one false event is detected; an appliance power drop, was incorrectly identified as an appliance turning-off, while the appliance being still in operation. In summary, 118 out of the 125 events have been correctly detected by the proposed event-detection algorithm. In Figure 4, the active power and the detected events for the period from 18:30:00 to 20:30:00 are shown.

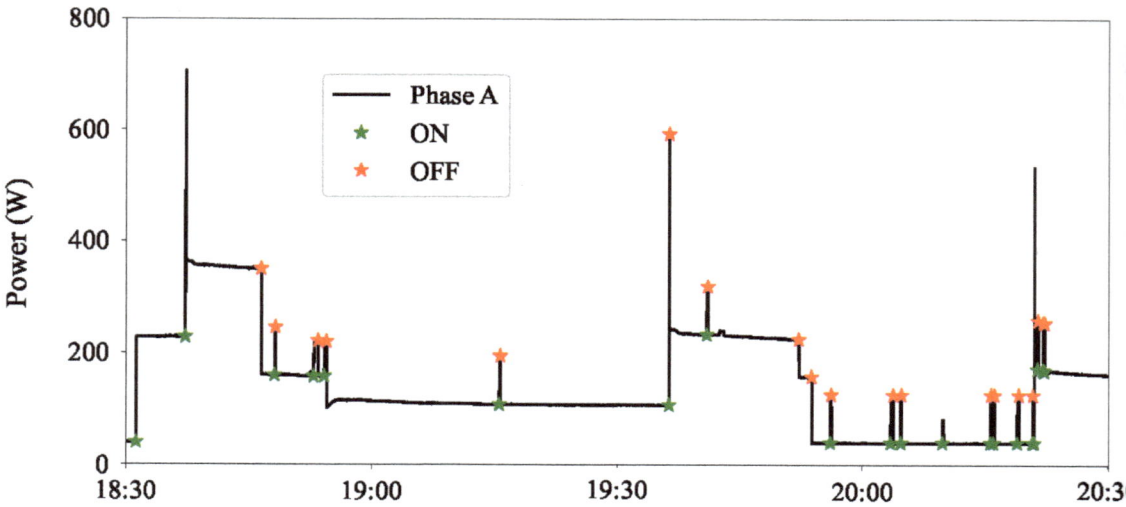

Figure 4. Event detection in a real household on a given day.

The TPR, FPR and FNR metrics are calculated and compared to other more complex solutions [32,49,50] in Table 3. It can be seen that the proposed algorithm can achieve good results while being simple and computationally efficient.

Table 3. Event detection evaluation.

Reference	TPR	FPR	FNR
Proposed	94.400%	0.003%	5.600%
[32]	94.000%	0.088%	6.000%
[49]	96.700%	0.810%	3.300%
[50]	94.130%	0.260%	5.870%

4.2. Classification Evaluation

To evaluate the classifiers performance regarding the three target appliances, the private testing sets mentioned in Section 3.1 are used. The calculated accuracy, precision, recall and F_1-score results are summarized in Table 4.

Table 4. Classification results.

Appliance	Accuracy	Precision	Recall	F_1-Score
Fridge	0.978	0.984	0.972	0.978
Washing machine	0.872	0.875	0.867	0.871
Microwave	0.992	0.986	0.999	0.992

It is evident that the proposed classification algorithm presents high performance regarding the microwave and the fridge. These appliances are related to transient response patterns presenting specific characteristics, thus can be identified with high confidence. However, this is not the case for the washing machine, since the turn-on transient response is a simple steep step-up waveform. Similar patterns are also related to the heating processes of most of the household appliances, e.g., dishwasher, oven and generally appliances that use resistive elements for heating as shown in Figure 5. This illustrates the relatively lower scores obtained for the washing machine metrics compared to the other appliances.

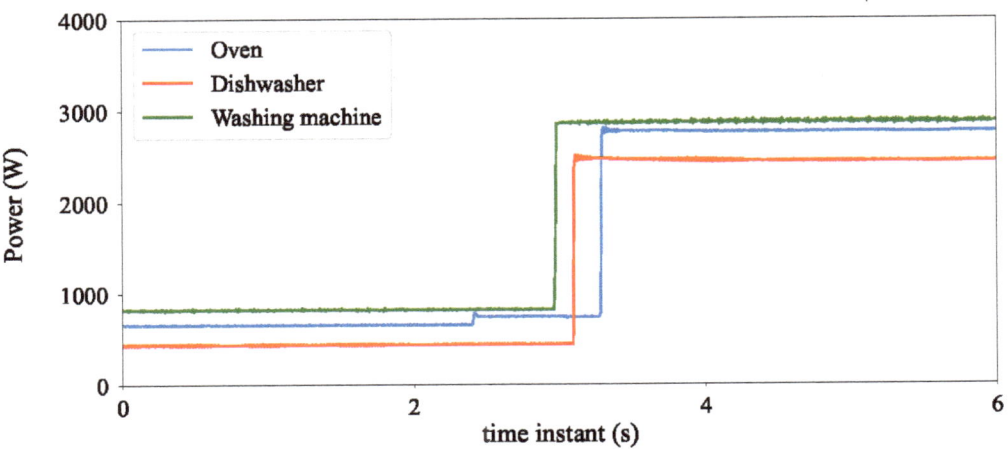

Figure 5. Turn-on transient responses from different household appliances.

4.3. Application on Residential Households

The overall performance of the proposed methodology is tested on a private dataset. This dataset includes three 3-phase power supply households located in the Netherlands. For each household, aggregated active power per phase was measured at 100 Hz along with power consumption of selected appliances for 15 days. For evaluation purposes, the proposed NILM system is applied only when the target appliance is connected.

Figure 6 presents the results for each target appliance, assuming an operational duration of four hours. Specifically, the aggregated power is colored in blue. The actual target appliance power measured with plugwise meters is colored in red. The target appliance power, as estimated by the proposed methodology, is colored in green.

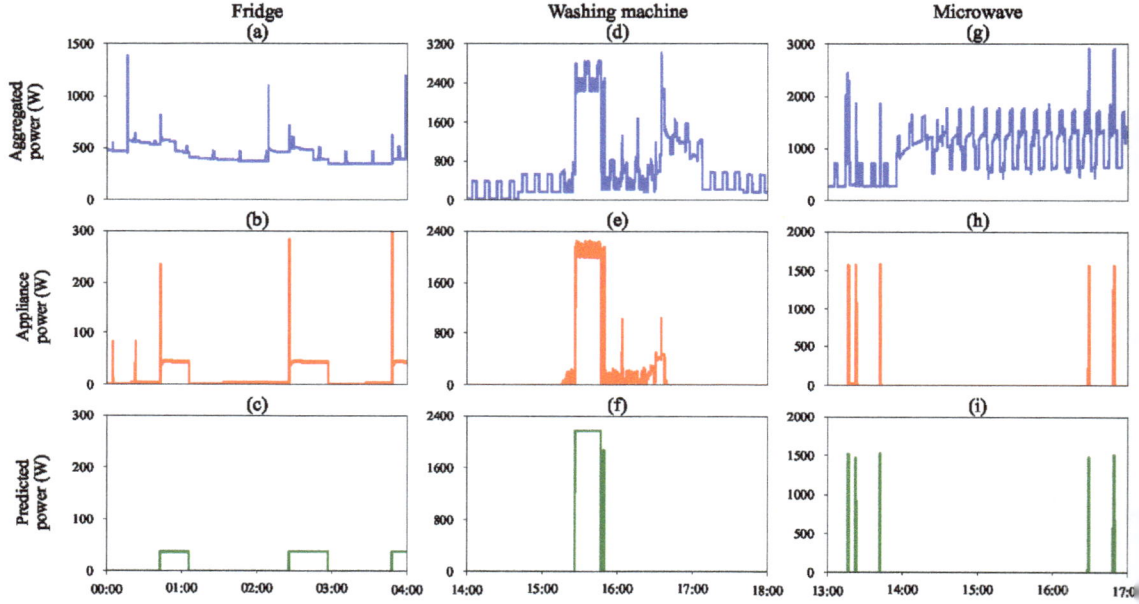

Figure 6. Power estimation for the selected appliances in real households. Time-series of (**a**) aggregated power, (**b**) actual target appliance power, (**c**) estimated power for fridge; (**d**) aggregated power, (**e**) actual target appliance power, (**f**) estimated power for washing machine; (**g**) aggregated power, (**h**) actual target appliance power, (**i**) estimated power for microwave.

The accuracy, precision, recall, F_1-score, MAE, RMSE and RE are calculated as well as their average considering the three households for 15 days. Results for the fridge, washing machine and microwave oven are shown in Tables 5–7, respectively. It can be generally observed that the proposed algorithm presents high accuracy regarding the power and energy estimates of the fridge and the microwave. On the contrary, the microwave oven recall metric is low. This can be attributed to the fact that the proposed methodology considers this appliance standby mode of operation as OFF. In fact, the power consumption during this period is low, thus, of trivial importance regarding energy consumption calculation. Regarding the washing machine results, the NILM system is designed to detect only the most energy-intensive process during the washing machine operation cycle, i.e., water heating mode of operation. For the rest of the operational cycles (non-detected), i.e., water pumping, drum spinning, rinsing, the appliance status is assumed OFF. The partial detection of the washing machine appliance is evident in Figure 6, resulting into low recall score. Moreover, in the third household, the calculated low precision is due to the operation of appliances presenting similar transient response patterns, being misclassified as washing machine end-uses.

Table 5. Results for fridge.

House	Accuracy	Precision	Recall	F$_1$-Score	MAE (W)	RMSE (W)	E (kWh)	Ê (kWh)	RE
1	0.86	0.94	0.77	0.85	12.97	38.67	14.26	11.74	0.18
2	0.89	0.96	0.74	0.84	7.05	17.14	6.19	4.24	0.32
3	0.98	1.00	0.90	0.95	3.68	12.89	4.74	5.05	0.06
Average	0.91	0.97	0.80	0.88	7.90	22.90	-	-	0.19

Table 6. Results for washing machine.

House	Accuracy	Precision	Recall	F$_1$-Score	MAE (W)	RMSE (W)	E (kWh)	Ê (kWh)	RE
1	0.94	0.99	0.24	0.39	20.86	168.96	19.85	12.83	0.35
2	0.97	0.94	0.24	0.38	10.02	116.83	10.44	8.44	0.19
3	0.93	0.42	0.13	0.20	24.27	183.42	7.27	12.61	0.42
Average	0.95	0.78	0.20	0.32	18.38	156.40	-	-	0.32

Table 7. Results for microwave.

House	Accuracy	Precision	Recall	F$_1$-Score	MAE (W)	RMSE (W)	E (kWh)	Ê (kWh)	RE
1	1.00	0.94	0.52	0.67	1.28	36.55	2.16	2.13	0.01
2	1.00	0.99	0.46	0.63	1.08	37.63	2.10	1.90	0.10
3	0.99	0.82	0.47	0.60	2.61	57.27	2.19	2.57	0.15
Average	1.00	0.92	0.48	0.63	1.66	43.82	-	-	0.09

4.4. Comparison with Other Methods

The performance of the proposed methodology is compared to other NILM-based energy consumption estimation systems. The average MAE, RE, precision, recall, F$_1$-score and accuracy calculations obtained by the proposed method are summarized in Tables 8–10 regarding the fridge, washing machine and microwave, respectively. The corresponding results (where available) reported in the relevant literature are also presented as well as the associated NILM technique, sampling frequency, and testing dataset. Note that, most of the literature state-of-the-art methods have been tested by using the well-known UK Domestic Appliance-Level Electricity (UK-DALE) [51] dataset. This dataset includes aggregated active power and appliance measurements of 0.167 Hz for several months, recorded for a small number of household installations. Moreover, the Reference Energy Disaggregation Data Set (REDD) [52] has been used in [21] to evaluate the LSTM algorithm performance; the sampling frequency is 1 Hz for mains and 0.333 Hz for the appliances. The proposed NILM system is tested by using an 100-Hz private dataset, since high-frequency sampling data are not provided in the above mentioned public datasets. It is important to stress out that in order to conduct a fair comparison between the different approaches, all metrics should be taken into consideration. However, this is not possible, since results for all metrics calculations are not always provided in the corresponding literature. Therefore, a direct comparison should be carried out with caution.

Table 8. Comparison results among existing non-intrusive load monitoring (NILM) solutions for fridge identification and energy consumption estimation.

Reference	Method	Sampling Frequency	Dataset	MAE	RE	Precision	Recall	F_1-Score	Accuracy
Proposed		100 Hz	private	7.90	0.19	0.97	0.80	0.88	0.91
[19]	Autoencoder	0.167 Hz	UK-DALE	26.00	0.38	0.85	0.88	0.87	0.90
[19]	CNN	0.167 Hz	UK-DALE	18.00	0.13	0.79	0.86	0.82	0.87
[19]	LSTM	0.167 Hz	UK-DALE	36.00	0.25	0.72	0.77	0.74	0.81
[20]	LSTM	0.167 Hz	UK-DALE	51.00	0.21	0.45	0.51	0.47	0.60
[20]	GRU	0.167 Hz	UK-DALE	51.00	0.26	0.46	0.75	0.57	0.60
[20]	seq2point	0.167 Hz	UK-DALE	51.00	0.29	0.42	0.74	0.53	0.54
[21]	LSTM	0.333 Hz	REDD	-	-	0.91	0.96	0.93	-
[22]	WGRU	0.167 Hz	UK-DALE	28.46	0.13	-	-	0.82	-
[22]	SAEDdot	0.167 Hz	UK-DALE	35.25	0.60	-	-	0.62	-
[22]	SAEDadd	0.167 Hz	UK-DALE	32.31	0.65	-	-	0.66	-
[25]	PCNN AE	0.167 Hz	UK-DALE	3.46	-	-	-	-	-
[25]	PCNN LSTM	0.167 Hz	UK-DALE	3.22	-	-	-	-	-
[26]	seq2seq	0.167 Hz	UK-DALE	24.49	-	-	-	-	-
[26]	seq2point	0.167 Hz	UK-DALE	20.89	-	-	-	-	-
[27]	UNet	0.167 Hz	UK-DALE	15.12	-	-	-	-	-

Note: GRU stands for gated recurrent units, seq2point/seq2seq for sequence-to-point/sequence-to-sequence, WGRU for window GRU, SAEDdot/SAEDadd for self-attentive energy disaggregation with 'additive'/'dot' attention mechanism, PCNN AE for parallel CNN autoencoder.

Table 9. Comparison results among existing NILM solutions for washing machine identification and energy consumption estimation.

Reference	Method	Sampling Frequency	Dataset	MAE	RE	Precision	Recall	F_1-score	Accuracy
Proposed		100 Hz	private	18.38	0.32	0.78	0.20	0.32	0.95
[19]	Autoencoder	0.167 Hz	UK-DALE	24.00	0.48	0.07	1.00	0.13	0.82
[19]	CNN	0.167 Hz	UK-DALE	11.00	0.74	0.29	0.24	0.27	0.98
[19]	LSTM	0.167 Hz	UK-DALE	109.00	0.91	0.01	0.73	0.03	0.23
[20]	LSTM	0.167 Hz	UK-DALE	25.00	0.35	0.16	0.56	0.24	0.95
[20]	GRU	0.167 Hz	UK-DALE	30.00	0.58	0.22	0.54	0.31	0.96
[20]	seq2point	0.167 Hz	UK-DALE	17.00	0.28	0.26	0.55	0.35	0.97
[22]	WGRU	0.167 Hz	UK-DALE	10.45	0.43	-	-	0.34	-
[22]	SAEDdot	0.167 Hz	UK-DALE	13.10	0.34	-	-	0.30	-
[22]	SAEDadd	0.167 Hz	UK-DALE	22.01	0.53	-	-	0.30	-
[25]	PCNN AE	0.167 Hz	UK-DALE	83.40	-	-	-	-	-
[25]	PCNN LSTM	0.167 Hz	UK-DALE	73.16	-	-	-	-	-
[26]	seq2seq	0.167 Hz	UK-DALE	10.15	-	-	-	-	-
[26]	seq2point	0.167 Hz	UK-DALE	12.66	-	-	-	-	-
[27]	UNet	0.167 Hz	UK-DALE	11.51	-	-	-	-	-

Table 10. Comparison results among existing NILM solutions for microwave identification and energy consumption estimation.

Reference	Method	Sampling Frequency	Dataset	MAE	RE	Precision	Recall	F_1-Score	Accuracy
Proposed		100 Hz	private	1.66	0.09	0.92	0.48	0.63	1.00
[19]	Autoencoder	0.167 Hz	UK-DALE	9.00	0.73	0.15	0.94	0.26	0.99
[19]	CNN	0.167 Hz	UK-DALE	6.00	0.50	0.14	0.40	0.21	0.99
[19]	LSTM	0.167 Hz	UK-DALE	20.00	0.88	0.07	0.99	0.13	0.98
[20]	LSTM	0.167 Hz	UK-DALE	86.00	0.10	0.01	0.45	0.02	0.93
[20]	GRU	0.167 Hz	UK-DALE	97.00	0.07	0.02	0.75	0.04	0.93
[20]	seq2point	0.167 Hz	UK-DALE	103.00	0.16	0.01	0.79	0.03	0.91
[21]	LSTM	0.333 Hz	REDD	-	-	0.50	0.05	0.09	-
[22]	WGRU	0.167 Hz	UK-DALE	4.36	0.25	-	-	0.44	-
[22]	SAEDdot	0.167 Hz	UK-DALE	5.97	0.19	-	-	0.25	-
[22]	SAEDadd	0.167 Hz	UK-DALE	5.98	0.17	-	-	0.26	-
[25]	PCNN AE	0.167 Hz	UK-DALE	27.50	-	-	-	-	-
[25]	PCNN LSTM	0.167 Hz	UK-DALE	9.42	-	-	-	-	-
[26]	seq2seq	0.167 Hz	UK-DALE	13.62	-	-	-	-	-
[26]	seq2point	0.167 Hz	UK-DALE	8.67	-	-	-	-	-
[27]	UNet	0.167 Hz	UK-DALE	6.48	-	-	-	-	-

From the results of Table 8 it can be seen that the proposed algorithm presents a high performance on most metrics. In particular, the method presents the third-best MAE, being inferior only to PCNN AE and PCNN LSTM. Regarding energy estimation, the RE metric is low (equal to 0.19), thus the proposed method is outperformed only by the CNN [19] and the WGRU [22] algorithms. Finally, the proposed solution presents the highest precision in terms of status estimation. In particular, the fridge status has been falsely identified as ON (real status was OFF) for the minimum of cases from all examined NILM solutions. On the other hand, the proposed method presents moderate performance in terms of recall (0.80), since the Autoencoder, CNN [19] and LSTM [21] algorithms achieve better results. This is mainly attributed to the proposed power estimation algorithm design. The fridge status may be falsely considered OFF prior to an actual turning-off, due to similar power step-down recordings, caused by appliances different from the target one. A possible solution is to determine the fridge duration pulse. However, this is practically infeasible since the fridge duration pulse varies significantly due to temperature difference inside and outside the appliance. Finally, by ranking all methods in terms of the F_1-score and accuracy, it can be realized that the proposed method is the second-best and first, respectively, among all examined solutions (where the corresponding metrics were available).

By analysing the washing machine results in Table 9, it can be observed that the proposed method presents relatively high MAE; seven out of the fourteen examined methods perform better. Regarding energy estimation the proposed method can be considered as the second-best in terms of RE, following the seq2point implementation [20]. Moreover, the proposed method presents the highest precision and the lowest recall among the examined solutions. This is due to the fact that the proposed system is specifically designed to detect the most energy-intensive and lower-duration process of the appliance, i.e., heating. The rest of the washing machine operation cycles, e.g., drum-spinning and rinsing are not taken into account as low energy-consumption longer-duration processes; thus, being of less importance. This implies that the proposed NILM system can accurately estimate the washing-machine energy consumption (low RE value) but predicts the appliance idle status (no water heating process) as OFF, resulting into low recall and high MAE. Some of the current state-of-the-art NILM systems can indeed detect these low energy-intensive processes. However, this results into an increased number of FP and consequently a low precision. Note that, the low precision (although the highest among the examined solutions) is attributed to the fact that the transient response of the heating process is similar to that of other household appliances; thus, may lead to an increased number of FP predictions. Finally, the F_1-score and accuracy metrics set the proposed method as the third- and fourth-best, respectively, among the examined solutions (where metrics were available).

Finally, regarding the microwave oven (Table 10), the proposed method outperforms the examined NILM methods presenting the lowest MAE and RE as well as the highest precision, F_1-score and accuracy. Better results by other methods are observed only in terms of recall. This is due to the fact that the proposed system can not detect the microwave oven standby mode of operation. However, the power consumption during this period can be considered negligible. It is also important to note, that in NILM and from a user-experience point of view, precision is considered more important than recall; missing an appliance event is preferable than detecting an appliance event that has not actually occurred. In this sense, missing standby modes is more favored than predicting false microwave end-uses. The superiority of the proposed method for the analysis of the microwave oven is based on the following: (a) the microwave transient response pattern is unique, thus, it can be easily identified, and (b) the microwave oven end-use duration is short, varying from few seconds to minutes; thus, the number of the possible turning-off events caused from other appliances that may degrade the power estimation algorithm performance is very limited.

4.5. Computational and Memory Efficiency

The proposed methodology is designed to be memory and computationally efficient. The first part, i.e., the event detector, calculates the power difference over time. The second part, i.e., the classifier, is triggered only when a significant power step-up is detected. If the classifier detects a target appliance, the power estimation algorithm is enabled. This event based approach can be considered computationally efficient compared to other solutions operating continuously, i.e., even no turning-on event occurs. Furthermore, the transient response classifier consists of 54,377 parameters. This is a small number compared to other end-to-end deep learning models requiring a number of parameters in the order of millions, e.g., the model parameters proposed in [19] range from 1 million to more than 150 million parameters. Therefore, the proposed NILM system can be considered as memory efficient.

The only drawback is the use of 100 Hz active power data to recognize appliance turning-on when a transient occurs. However, this feature is important to enable the real-time application of the proposed NILM system, contrary to other approaches requiring power data of more extended periods (minutes to hours) in order to identify which appliance is operating. Moreover, it must be noted that the 100 Hz time-series is used only when an event is detected, and only a 6 s window is extracted. Based on the above, it is evident that the proposed system can operate on the edge without the need of high-end microprocessors.

5. Discussion—Towards Scalable Real-Time NILM Services

As already reported, the proposed methodology is implemented as a real-time scalable solution with minimum hardware requirements, thus allowing utilities to perform a large-scale deployment. However, some criteria need to be met from an industry perspective before massively adopting such a service. Coming up with the correct blend of characteristics is not a trivial issue. So, it is no surprise that no real-time NILM solution based on sub-second energy data resolution has been rolled out in scale (>50 K end-users) globally yet. In this section, four necessary criteria are investigated and we examine if the proposed methodology meets them or not.

1. First of all, as expected, comes the accuracy metric. Accuracy usually refers to weights-based combination of (i) correctly detected events, (ii) precise energy con

sumption estimation for the detected appliance events and (iii) minimized FP. Energy companies and electricity consumers usually trust a NILM service when its accuracy exceeds 90% and when they are not receiving reports for appliances/activities never actually occurred.
2. Second comes the data resolution and as a result the data volumes required for an accurate NILM output. As mentioned above in Section 3, for real-time appliance identification sub-second data granularity is needed. Note that, most of the solutions presented in literature deal with kHz or even MHz of data. Considering as a rule of thumb that 1-s resolution data from separate phases in a 3-phase installation result in almost 1 GB of data being produced per year, we realize that moving into the kHz resolution areas makes data parsing, storing and analysing a rather complicated, costly and therefore non-scalable option.
3. Next in the list comes the computational/RAM efficiency of such a service. Although the recent trend was to move everything to the cloud, now NILM vendors and energy companies realise that such a decision is not always the most cost-effective; the opposite actually. Running for example the whole service for ~100 k end users on the cloud can increase cloud operation costs that much, that there is no business case that can be built on top of a NILM layer, no matter how accurate that is. So, the key to unlock scalability opportunities here is to built a system that is so efficient that can run on the edge instead of the cloud.
4. Strictly connected to the hardware constraints of the previous point comes the hardware cost. Traditionally sub-second data can be acquired only via a din meter hardware installed in the metering cabinet (it's only recently that a few smart-meter manufacturers make >1 Hz resolutions available through their S1 port [53]). On the other hand, utilities and energy retail companies see NILM as a great customer engagement tool on top of which they can build value-added services and they usually tend to offer that as a freemium service. As a result, hardware cost has to be as low as possible and ideally within the companies customer retention and acquisition budgets.

In Figure 7, three of the criteria mentioned above are analyzed for the proposed system, i.e., accuracy, sampling frequency, and computational burden. As we can see, scalability-related criteria #1 and #2 are met; the proposed system presents accuracy higher than 90% in all examined cases by utilizing the sampling frequency of 100 Hz (see results in Section 4). Although this frequency is high, it is still considerably lower than a resolution of several kHz used in most state-of-the-art real-time implementations [33–35,39]. To that end, it is an excellent "do a lot with a little" decision to take. Regarding criterion #3, i.e., computational and memory efficiency, as demonstrated in Section 4.5, optimized design can efficiently run on the edge and even on low-cost chip-sets. Specifically, in Figure 7, it is assumed that the "High" value refers to expensive algorithms, incorporating several parameters that cannot be easily integrated into a low-cost microprocessor. On the other hand, the "Very Low" value refers to low computational complexity algorithms that can be integrated and run in a low-cost microprocessor. The proposed system is between the "Low" and "Very Low" area. Criterion #4 is expected to be met as a consequence of #3. However, such an investigation falls out of the scope of this paper.

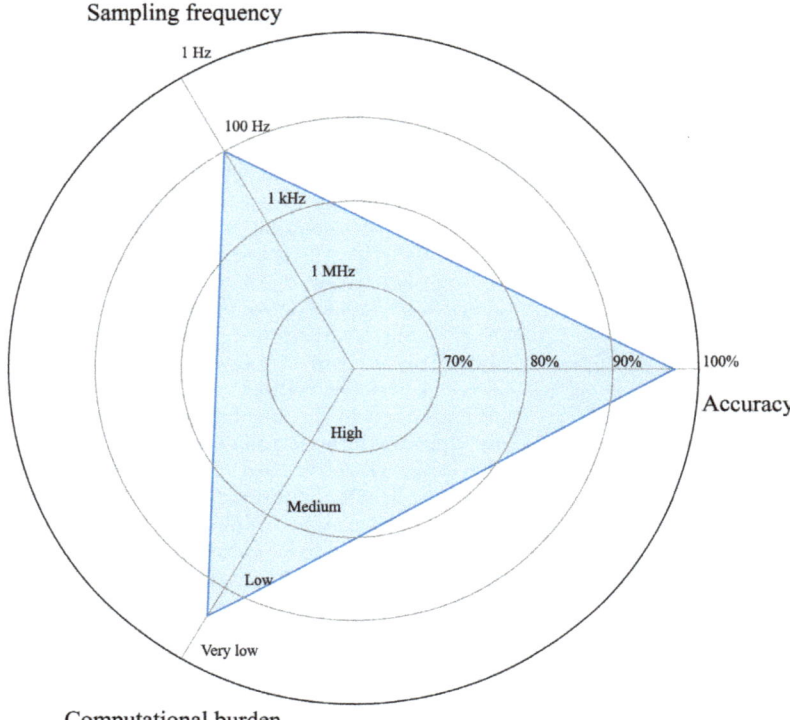

Figure 7. Scalability evaluation of the proposed implementation.

6. Conclusions

In this paper, a novel real-time event-based energy disaggregation methodology introduced. Initially, a simple event-detection algorithm is proposed to find time instan when an appliance is turned-on and extract transient responses at 100 Hz. Next, a conv lutional neural network classifier identifies if a transient response was caused by a targ appliance. Finally, a power estimation algorithm is implemented considering applian end-uses as pulses of constant power. Experimental results show a promising performan for specific appliances.

Unlike most relevant papers in the literature, the proposed non-intrusive load mo itoring system can identify in real-time when an appliance is turned-on based on i information-rich transient response sampled at 100 Hz. Furthermore, it is delay-free sin once a target appliance has been turned-on, the active power can be directly calculate Moreover, the system is computational and memory-efficient and can be integrated in smart meters.

The proposed approach can be used for a significant number of appliances wi negligible error. However, energy consumption of specific appliances, e.g., heat pum tumble dryer, including many states of operation, cannot be calculated by the propose methodology. For such cases, dedicated algorithms should be implemented. As futu steps, a more robust power estimation algorithm will be examined for multi-state applian uses. Additionally, the proposed methodology will be tested on more types of applianc

Author Contributions: Conceptualization, methodology, software, visualization, formal analys writing—original draft preparation, C.A.; validation, investigation, writing—review and editir C.A., D.D., T.P. and A.C.; resources, C.A., D.D. and T.P.; data curation, C.A. and A.C.; supervisic

D.D. and T.P.; project administration, funding acquisition, D.D. All authors have read and agreed to the published version of the manuscript.

Funding: This research has been co–financed by the European Regional Development Fund of the European Union and Greek national funds through the Operational Program Competitiveness, Entrepreneurship and Innovation, under the call RESEARCH–CREATE–INNOVATE (project: T2EDK-03898).

Data Availability Statement: The public available BLUED [48] dataset as well as a private dataset obtained from NET2GRID BV that is not public available have been used in this study.

Conflicts of Interest: The authors declare no conflict of interest.

Abbreviations

The following abbreviations are used in this manuscript:

AE	autoencoder
BLUED	Building-Level fUlly-labeled dataset for Electricity Disaggregation
C&I	commercial-industrial
CNN	convolutional neural network
CPU	central processing unit
FN	false negative
FNR	false negative rate
FP	false positive
FPR	false positive rate
GRU	gated recurrent unit
HMM	hidden Markov model
ILM	intrusive load monitoring
IoT	internet-of-things
LSTM	long short-term memory
MAE	mean absolute error
MPD	maximum power difference
NILM	non-intrusive load monitoring
NLP	natural language processing
PCNN	parallel CNN
RE	relative error in total energy
REDD	Reference Energy Disaggregation Data Set
ReLU	rectified linear unit
RMSE	root mean square error
SAEDadd	self-attentive energy disaggregation with 'dot' attention mechanism
SAEDdot	self-attentive energy disaggregation with 'additive' attention mechanism
seq2point	sequence-to-point
seq2seq	sequence-to-sequence
TN	true negative
TP	true positive
TPR	true positive rate
UK-DALE	UK Domestic Appliance-Level Electricity
WGRU	window GRU

References

Wang, Y.; Chen, Q.; Hong, T.; Kang, C. Review of Smart Meter Data Analytics: Applications, Methodologies, and Challenges. *IEEE Trans. Smart Grid* **2019**, *10*, 3125–3148. [CrossRef]

Hayes, B.P.; Prodanovic, M. State Forecasting and Operational Planning for Distribution Network Energy Management Systems. *IEEE Trans. Smart Grid* **2016**, *7*, 1002–1011. [CrossRef]

Huang, Y.; Wang, L.; Guo, W.; Kang, Q.; Wu, Q. Chance Constrained Optimization in a Home Energy Management System. *IEEE Trans. Smart Grid* **2018**, *9*, 252–260. [CrossRef]

Buzau, M.M.; Tejedor-Aguilera, J.; Cruz-Romero, P.; Gómez-Expósito, A. Detection of Non-Technical Losses Using Smart Meter Data and Supervised Learning. *IEEE Trans. Smart Grid* **2019**, *10*, 2661–2670. [CrossRef]

Chakraborty, S.; Das, S. Application of Smart Meters in High Impedance Fault Detection on Distribution Systems. *IEEE Trans. Smart Grid* **2019**, *10*, 3465–3473. [CrossRef]

6. Chatzigeorgiou, I.; Andreou, G. A systematic review on feedback research for residential energy behavior change through mobile and web interfaces. *Renew. Sust. Energ. Rev.* **2021**, *135*, 110187. [CrossRef]
7. Nalmpantis, C.; Vrakas, D. Machine learning approaches for non-intrusive load monitoring: From qualitative to quantitative comparation. *Artif. Intel. Rev.* **2018**, *52*. [CrossRef]
8. Ruano, A.; Hernandez, A.; Ureña, J.; Ruano, M.; Garcia, J. NILM Techniques for Intelligent Home Energy Management and Ambient Assisted Living: A Review. *Energies* **2019**, *12*, 2203. [CrossRef]
9. Hart, G.W. Non-intrusive appliance load monitoring. *Proc. IEEE* **1992**, *80*, 1870–1891. [CrossRef]
10. Kolter, J.Z.; Jaakkola, T. Approximate Inference in Additive Factorial HMMs with Application to Energy Disaggregation. In Proceedings of the Fifteenth International Conference on Artificial Intelligence and Statistics, La Palma, Canary Islands, 21–2 April 2012; Volume 22, pp. 1472–1482. Available online: http://proceedings.mlr.press/v22/zico12.html (accessed on 15 January 2021).
11. Zoha, A.; Gluhak, A.; Nati, M.; Imran, M.A. Low-power appliance monitoring using Factorial Hidden Markov Models. In Proceedings of the 2013 IEEE Eighth International Conference on Intelligent Sensors, Sensor Networks and Information Processing, Melbourne, VIC, Australia, 2–5 April 2013; pp. 527–532. [CrossRef]
12. Bonfigli, R.; Principi, E.; Fagiani, M.; Severini, M.; Squartini, S.; Piazza, F. Non-intrusive load monitoring by using active and reactive power in additive Factorial Hidden Markov Models. *Appl. Energy* **2017**, *208*, 1590–1607. [CrossRef]
13. Lu, T.; Xu, Z.; Huang, B. An Event-Based Nonintrusive Load Monitoring Approach: Using the Simplified Viterbi Algorithm. *IEEE Pervasive Comput.* **2017**, *16*, 54–61. [CrossRef]
14. Ji, T.Y.; Liu, L.; Wang, T.S.; Lin, W.B.; Li, M.S.; Wu, Q.H. Non-Intrusive Load Monitoring Using Additive Factorial Approximate Maximum a Posteriori Based on Iterative Fuzzy c-Means. *IEEE Trans. Smart Grid* **2019**, *10*, 6667–6677. [CrossRef]
15. Makonin, S.; Popowich, F.; Bajic, I.V.; Gill, B.; Bartram, L. Exploiting HMM Sparsity to Perform Online Real-Time Nonintrusive Load Monitoring. *IEEE Trans. Smart Grid* **2016**, *7*, 2575–2585. [CrossRef]
16. Kong, W.; Dong, Z.Y.; Hill, D.J.; Ma, J.; Zhao, J.H.; Luo, F.J. A Hierarchical Hidden Markov Model Framework for Home Appliance Modeling. *IEEE Trans. Smart Grid* **2018**, *9*, 3079–3090. [CrossRef]
17. Mueller, J.A.; Kimball, J.W. Accurate Energy Use Estimation for Nonintrusive Load Monitoring in Systems of Known Devices. *IEEE Trans. Smart Grid* **2018**, *9*, 2797–2808. [CrossRef]
18. Kim, J.; Le, T.T.H.; Kim, H. Non-intrusive Load Monitoring Based on Advanced Deep Learning and Novel Signature. *Comput. Intel. Neurosc.* **2017**, *2017*, 4216281. [CrossRef]
19. Kelly, J.; Knottenbelt, W. Neural NILM. In Proceedings of the 2nd ACM International Conference on Embedded Systems for Energy-Efficient Built Environments—BuildSys'15, Seoul, Korea, 4–5 November 2015. [CrossRef]
20. Krystalakos, O.; Nalmpantis, C.; Vrakas, D. Sliding Window Approach for Online Energy Disaggregation Using Artificial Neural Networks. In Proceedings of the 10th Hellenic Conference on Artificial Intelligence, SETN '18, Patra, Greece, 9–15 July 2018; Association for Computing Machinery: New York, NY, USA, 2018; doi:10.1145/3200947.3201011. [CrossRef]
21. Mauch, L.; Yang, B. A new approach for supervised power disaggregation by using a deep recurrent LSTM network. In Proceedings of the 2015 IEEE Global Conference on Signal and Information Processing (GlobalSIP), Orlando, FL, USA, 14–16 December 2015; pp. 63–67. [CrossRef]
22. Virtsionis Gkalinikis, N.; Nalmpantis, C.; Vrakas, D. Attention in Recurrent Neural Networks for Energy Disaggregation. In *Discovery Science*; Appice, A., Tsoumakas, G., Manolopoulos, Y., Matwin, S., Eds.; Springer International Publishing: Cham, Switzerland, 2020; pp. 551–565._36. [CrossRef]
23. Kaselimi, M.; Protopapadakis, E.; Voulodimos, A.; Doulamis, N.; Doulamis, A. Multi-Channel Recurrent Convolutional Neural Networks for Energy Disaggregation. *IEEE Access* **2019**, *7*, 81047–81056. [CrossRef]
24. Sudoso, A.M.; Piccialli, V. Non-Intrusive Load Monitoring with an Attention-based Deep Neural Network. *arXiv* 2019, arXiv:1912.00759.
25. He, W.; Chai, Y. An Empirical Study on Energy Disaggregation via Deep Learning. In Proceedings of the 2016 2nd International Conference on Artificial Intelligence and Industrial Engineering (AIIE 2016), Beijing, China, 20–21 November 2016; Atlantis Press: Amsterdam, The Netherlands, 2016; pp. 338–342. [CrossRef]
26. Zhang, C.; Zhong, M.; Wang, Z.; Goddard, N.; Sutton, C. Sequence-to-point learning with neural networks for non-intrusive load monitoring. *arXiv* **2016**, arXiv:1612.09106.
27. Faustine, A.; Pereira, L.; Bousbiat, H.; Kulkarni, S. UNet-NILM: A Deep Neural Network for Multi-Tasks Appliances State Detection and Power Estimation in NILM. In *Proceedings of the 5th International Workshop on Non-Intrusive Load Monitoring*; NILM'20; Association for Computing Machinery: New York, NY, USA, 2020; pp. 84–88. [CrossRef]
28. Devlin, M.A.; Hayes, B.P. Non-Intrusive Load Monitoring and Classification of Activities of Daily Living Using Residential Smart Meter Data. *IEEE Trans. Consum. Electron.* **2019**, *65*, 339–348. [CrossRef]
29. Penha, D.; Castro, A. Home Appliance Identification for NILM Systems Based on Deep Neural Networks. *Int. J. Artif. Intell. Appl.* **2018**, *9*, 69–80. [CrossRef]
30. Kong, W.; Dong, Z.Y.; Wang, B.; Zhao, J.; Huang, J. A Practical Solution for Non-Intrusive Type II Load Monitoring Based on Deep Learning and Post-Processing. *IEEE Trans. Smart Grid* **2020**, *11*, 148–160. [CrossRef]
31. Cui, G.; Liu, B.; Luan, W.; Yu, Y. Estimation of Target Appliance Electricity Consumption Using Background Filtering. *IEEE Trans. Smart Grid* **2019**, *10*, 5920–5929. [CrossRef]

Alcalá, J.; Ureña, J.; Hernández, Á.; Gualda, D. Event-Based Energy Disaggregation Algorithm for Activity Monitoring From a Single-Point Sensor. *IEEE Trans. Instrum. Meas.* **2017**, *66*, 2615–2626. [CrossRef]

Faustine, A.; Pereira, L. Improved Appliance Classification in Non-Intrusive Load Monitoring Using Weighted Recurrence Graph and Convolutional Neural Networks. *Energies* **2020**, *13*, 3374. [CrossRef]

Yang, D.; Gao, X.; Kong, L.; Pang, Y.; Zhou, B. An Event-Driven Convolutional Neural Architecture for Non-Intrusive Load Monitoring of Residential Appliance. *IEEE Trans. Consum. Electron.* **2020**, *66*, 173–182. [CrossRef]

De Baets, L.; Ruyssinck, J.; Develder, C.; Dhaene, T.; Deschrijver, D. Appliance classification using VI trajectories and convolutional neural networks. *Energ. Build.* **2018**, *158*, 32–36. [CrossRef]

Liu, Y.; Wang, X.; You, W. Non-Intrusive Load Monitoring by Voltage—Current Trajectory Enabled Transfer Learning. *IEEE Trans. Smart Grid* **2019**, *10*, 5609–5619. [CrossRef]

Hassan, T.; Javed, F.; Arshad, N. An Empirical Investigation of V-I Trajectory Based Load Signatures for Non-Intrusive Load Monitoring. *IEEE Trans. Smart Grid* **2014**, *5*, 870–878. [CrossRef]

Du, L.; He, D.; Harley, R.G.; Habetler, T.G. Electric Load Classification by Binary Voltage—Current Trajectory Mapping. *IEEE Trans. Smart Grid* **2016**, *7*, 358–365. [CrossRef]

Faustine, A.; Pereira, L. Multi-Label Learning for Appliance Recognition in NILM Using Fryze-Current Decomposition and Convolutional Neural Network. *Energies* **2020**, *13*, 4154. [CrossRef]

Krizhevsky, A.; Sutskever, I.; Hinton, G. ImageNet Classification with Deep Convolutional Neural Networks. *Neural Inf. Process Syst.* **2012**, *25*. [CrossRef]

Chang, H.H.; Lin, L.S.; Chen, N.; Lee, W.J. Particle Swarm Optimization based non-intrusive demand monitoring and load identification in smart meters. In Proceedings of the 2012 IEEE Industry Applications Society Annual Meeting, Las Vegas, NV, USA, 7–11 October 2012; pp. 1–8. [CrossRef]

Chang, H.H.; Chen, K.L.; Tsai, Y.P.; Lee, W.J. A New Measurement Method for Power Signatures of Nonintrusive Demand Monitoring and Load Identification. *IEEE Trans. Ind. Appl.* **2012**, *48*, 764–771. [CrossRef]

Zoha, A.; Gluhak, A.; Imran, M.; Rajasegarar, S. Non-Intrusive Load Monitoring Approaches for Disaggregated Energy Sensing: A Survey. *Sensors* **2012**, *12*, 16838–16866. [CrossRef]

Revuelta Herrero, J.; Lozano Murciego, A.; Barriuso, A.; Hernández de la Iglesia, D.; Villarrubia, G.; Corchado Rodríguez, J.; Carreira, R. Non Intrusive Load Monitoring (NILM): A State of the Art. In *Trends in Cyber-Physical Multi-Agent Systems*; Springer International Publishing: Cham, Switzerland, 2018; pp. 125–138._12. [CrossRef]

LeCun, Y.; Bengio, Y.; Hinton, G. Deep Learning. *Nature* **2015**, *521*, 436–44. [CrossRef]

Srivastava, N.; Hinton, G.; Krizhevsky, A.; Sutskever, I.; Salakhutdinov, R. Dropout: A Simple Way to Prevent Neural Networks from Overfitting. *J. Mach. Learn. Res.* **2014**, *15*, 1929–1958. Available online: http://jmlr.org/papers/v15/srivastava14a.html (accessed on 15 January 2021).

Kingma, D.P.; Ba, J. Adam: A Method for Stochastic Optimization. *arXiv* **2017**, arXiv:1412.6980.

Anderson, K.; Ocneanu, A.; Benitez, D.; Carlson, D.; Rowe, A.; Bergés, M. BLUED: A Fully Labeled Public Dataset for Event-Based Non-Intrusive Load Monitoring Research. In Proceedings of the 2nd KDD Workshop on Data Mining Applications in Sustainability (SustKDD), Beijing, China, 12 August 2012.

Lu, M.; Li, Z. A Hybrid Event Detection Approach for Non-Intrusive Load Monitoring. *IEEE Trans. Smart Grid* **2020**, *11*, 528–540. [CrossRef]

Bergés, M.; Goldman, E.; Matthews, H.; Soibelman, L.; Anderson, K. User-Centered Nonintrusive Electricity Load Monitoring for Residential Buildings. *J. Comput. Civ. Eng.* **2011**, *25*, 471–480. [CrossRef]

Kelly, J.; Knottenbelt, W. The UK-DALE dataset, domestic appliance-level electricity demand and whole-house demand from five UK homes. *Sci. Data* **2015**, *2*. [CrossRef]

Kolter, J.; Johnson, M. REDD: A Public Data Set for Energy Disaggregation Research. In Proceedings of the SustKDD Workshop on Data Mining Applications in Sustainability, San Diego, CA, USA, 21 August 2011.

Fluvius. *User Ports of Smart Meters Explanatory Note on the Possibilities for Product Developers*; Technical Report; Melle, Belgium, 2019. Available online: https://www.fluvius.be/sites/fluvius/files/2019-07/1901-fluvius-technical-specification-user-ports-digital-meter.pdf (accessed on 15 January 2021).

Article

Improving Non-Intrusive Load Disaggregation through an Attention-Based Deep Neural Network

Veronica Piccialli *,† and Antonio M. Sudoso †

Department of Civil and Computer Engineering, University of Rome Tor Vergata, 00133 Rome, Italy; antonio.maria.sudoso@uniroma2.it
* Correspondence: veronica.piccialli@uniroma2.it
† Both authors contributed equally to this work.

Abstract: Energy disaggregation, known in the literature as Non-Intrusive Load Monitoring (NILM), is the task of inferring the power demand of the individual appliances given the aggregate power demand recorded by a single smart meter which monitors multiple appliances. In this paper, we propose a deep neural network that combines a regression subnetwork with a classification subnetwork for solving the NILM problem. Specifically, we improve the generalization capability of the overall architecture by including an encoder–decoder with a tailored attention mechanism in the regression subnetwork. The attention mechanism is inspired by the temporal attention that has been successfully applied in neural machine translation, text summarization, and speech recognition. The experiments conducted on two publicly available datasets—REDD and UK-DALE—show that our proposed deep neural network outperforms the state-of-the-art in all the considered experimental conditions. We also show that modeling attention translates into the network's ability to correctly detect the turning on or off an appliance and to locate signal sections with high power consumption, which are of extreme interest in the field of energy disaggregation.

Keywords: attention mechanism; deep neural network; energy disaggregation; non-intrusive load monitoring

1. Introduction

Non-Intrusive Load Monitoring (NILM) is the task of estimating the power demand of each appliance given aggregate power demand signal recorded by a single electric meter monitoring multiple appliances [1]. In the last years, machine learning and optimization played a significant role in the research on NILM [2,3]. In the literature, solutions based on k-Nearest Neighbor(k-NN), Support Vector Machine (SVM), Matrix Factorization have been proposed [4,5]. A practical approach to NILM has to handle real power measurements sampled at intervals of seconds or minutes. In this setting, one of the most popular approaches is based on the Hidden Markov Model (HMM) [6], because of its ability to model transitions in consumption levels of real energy consumption for target appliances. Some successive papers focused on enhancing the expressive power of this class of methods [7,8]. Recently, the energy disaggregation problem has been reformulated as a multi-label classification problem [9]. In order to detect the active appliances at each time step, the idea is to associate each value of the main power to a vector of labels of length equal to the number of appliances, that are set to 1 if the appliance is active and 0 otherwise. The reformulated problem has been solved with different approaches [10–12]. However, there is no direct way to derive the power consumption for each appliance at that time step. During the last years, approaches based on deep learning have received particular attention as they exhibited noteworthy disaggregation performance. Deep Neural Networks (DNNs) have been successfully applied for the first time to NILM by Kelly and Knottenbelt in [13], who coined the term "Neural NILM". Neural NILM is a nonlinear regression problem that consists of training a neural network for each appliance in order to predict a time window of the appliance load given the corresponding time window of aggregated data. Kelly and Knottenbelt proposed three different neural networks to perform NILM with high-frequency time series data: a recurrent neural network (RNN) using Long Short-Term Memory units (LSTM); a Denoising Autoencoder (DAE); and a regression model that predicts the start time, end time, and average power demand of each appliance. The capability of LSTMs to successfully learn long-range temporal dependencies of time series data makes it a suitable candidate for NILM. Their first approach is based on stacked layers of LSTM units combined with a Convolutional Neural Network (CNN) at the beginning of the network to automatically extract features from the raw data. In the same paper, NILM is treated as a noise reduction problem, in which the disaggregated load represents the clean signal, and the aggregated signal is considered corrupted by the presence of the remaining appliances and by the measurement noise. For this purpose, noise reduction is performed by means of a DAE composed of convolutional layers and fully connected layers. In the experiments conducted by the authors, the DAE network outperforms the LSTM-based architecture and the other approaches frequently employed for this problem, such as HMMs and Combinatorial Optimization. In [14], an empirical investigation of deep learning methods is conducted by using two types of neural network architectures for NILM. The first neural network solves a regression problem which estimates the transient power demand of a single appliance given the whole series of the aggregate power. The second type of network is a multi-layer RNN using LSTM units, which is similar to the structure used in [13]. Zhang et al. [15] proposed instead a sequence-to-point learning for energy disaggregation where the midpoint of an appliance window is treated as classification output of a neural network with the aggregate window being the input. Bonfigli et al. [16] proposed different algorithmic and architecture improvements to the DAE for NILM, showing that the Neural NILM approach improves on the best known NILM approaches not based on DNNs like Additive Factorial Approximate Maximum A Posteriori estimation (AFAMAP) by Kolter and Jaakkola [6]. Compared to the work in [13], their DAE approach is improved by introducing pooling and upsampling layers in the architecture and a median filter in the disaggregation phase to reconstruct the output signal from the overlapped portions of the disaggregated signal. Shin et al. [17] proposed a subtask gated network (SGN) that combines two CNNs, namely, a regression subnetwork and a classification subnetwork. The building block of the two subnetworks is the sequence-to-sequence CNN proposed in [15]. In their work, the regression subnetwork is used to infer the initial power consumption, whereas the classification subnetwork focuses on the binary classification of the appliance state (on/off). The final estimate of the power consumption is obtained by multiplying the regression output with the probability classification output. In the experiments conducted by the authors, the SGN outperforms HHMs and state-of-the-art CNN architectures that have been proposed recently [13,15]. Chen et al. [18] adopted the structure of the SGN proposed in [17] and added to the SGN backbone a Generative Adversarial Network (GAN). In their model, the disaggregator for a given appliance is followed by a generator that produces the load pattern for the appliance. They show that adding the adversarial loss can help the model to produce more accurate result with respect to the basic SGN architecture. None of these state-of-the art deep learning models use RNNs. In fact, in the NILM literature, CNNs have always shown better performance than RNNs, even though RNNs are still widely employed for sequence modeling tasks. In [19], a CNN-based DNN has been combined with data augmentation and an effective postprocessing phase, improving its ability to correctly detect the activation of each appliance with a small amount of data available. The attention mechanism applied to NILM is a relatively new idea [20]. The DNN in [20] remarkably improves over Kelly's DAE when trained and tested on the same house. On the other hand, the generalization capability on houses not seen during the training is modest. Moreover, training and testing for the NILM task are time-consuming as they used the same architecture proposed in [21] for machine translation which consists of RNN layers in both the encoder and the decoder.

In this paper, we propose a RNN-based encoder–decoder model to extract appliance specific power usage from the aggregated signal and we enhance it with a scalable and lightweight attention mechanism designed for the energy disaggregation task. More in detail, we substantially improve the generalization capability of the SGN by Shin et al. by encapsulating our model in the regression subnetwork and by combining it with the classification subnetwork. The implemented attention mechanism has the function to strengthen the representational power of the neural network to locate the positions of the input sequence where the relevant information is present. The intuition is that our attention-based model could help the energy disaggregation task by assigning importance to each position of the aggregated signal which corresponds to the position of a state change of the target appliance. This feature is crucial for developing appliance models that generalize well on buildings not seen during the training.

The proposed DNN is tested on two publicly available datasets—REDD and UK-DALE—and the performance is evaluated using different metrics. The obtained results show that our algorithm outperforms state-of-the-art DNNs in all the addressed experimental conditions. The paper is organized as follows. Section 2 describes the NILM problem. Section 3 presents our DNN architecture. Section 4 describes the experimental procedure and the obtained results. Finally, Section 5 concludes the paper.

2. NILM Problem

Given the aggregate power consumption $(x_1, \ldots, x_T)$ from N active appliances at the entry point of the meter, the task of the NILM algorithm is to deduce the contribution $(y_1^i, \ldots, y_T^i)$ of appliance $i = 1, \ldots, N$, such that at time $t = 1, \ldots, T$, the aggregate power consumption is given by the sum of the power consumption of all the known appliances plus a noise term. The energy disaggregation problem can be stated as

$$x_t = \sum_{i=1}^{N} y_t^i + \epsilon_t, \qquad (1)$$

where x_t is the aggregated active power measured at time t, y_t^i is the individual contribution of appliance i, N is the number of appliances, and ϵ_t is a noise term. In a denoised scenario, there is no noise term, whereas in a noised scenario, ϵ_t is given by the total contribution from appliances not included and the measurement noise. Similarly to the work in [13], we refer to the power over a complete cycle of an appliance as an appliance activation.

3. Encoder–Decoder with Attention Mechanism

In this section, we describe the adopted attention mechanism and DNN architecture for solving the NILM problem.

3.1. Attention Mechanism

In the classical setting, a sequence-to-sequence network is a model consisting of two components called the encoder and decoder [22]. The encoder is an RNN that takes an input sequence of vectors $(x_1, \ldots, x_T)$, where T is the length of input sequence, and encodes the information into fixed length vectors $(h_1, \ldots, h_T)$. This representation is expected to be a good summary of the entire input sequence. The decoder is also an RNN which is initialized with a single context vector $c = h_T$ as its inputs and generates an output sequence $(y_1, \ldots, y_N)$ vector by vector, where N is the length of output sequence. At each time step, h_t and σ_t denote the hidden states of the encoder and decoder, respectively. There are two well-known challenges with this traditional encoder–decoder framework. First, a critical disadvantage of single context vector design is the incapability of the system to remember long sequences: all the intermediate states of the encoder are eliminated and only the final hidden state vector is used to initialize the decoder. This technique works only for small sequences, however, as the length of the sequence increases, the vector becomes a bottleneck and may lead to loss of information [23]. Second, it is unable to capture the need of alignment between input and output sequences, which is an essenti aspect of structured output tasks such as machine translation or text summarization [2 The attention mechanism, first introduced for machine translation by Bahdanau et al. [2 was born to address these problems. The novelty in their approach is the introductio of an alignment function that finds for each output word significant input words. In th way, the neural network learns to align and translate at the same time. The central ide behind the attention is not to discard the intermediate encoder states but to combine ar utilize all the states in order to construct the context vectors required by the decoder generate the output sequence. This mechanism induces attention weights over the inp sequence to prioritize the set of positions where relevant information is present. Followir the definition from Bahdanau et al., attention-based models compute a context vector c_t f each time step as the weighted sum of all hidden states of the encoder. Their correspondir attention weights are calculated as

$$e_{tj} = f(\sigma_{t-1}, h_j), \quad \alpha_{tj} = \frac{exp(e_{tj})}{\sum_{k=1}^{T} exp(e_{tk})}, \quad c_t = \sum_{j=1}^{T} \alpha_{tj} h_j,$$

where f is a learned function that computes a scalar importance value for h_j given the valu of h_j and the previous state σ_{t-1} and each attention weight α_{tj} determines the normalize importance score for h_j. As shown in Figure 1, the context vectors c_t are then used compute the decoder hidden state sequence, where σ_t depends on σ_{t-1}, c_t, and y_{t-} The attention weights can be learned by incorporating an additional feed-forward neur network that is jointly trained with encoder–decoder components of the architecture.

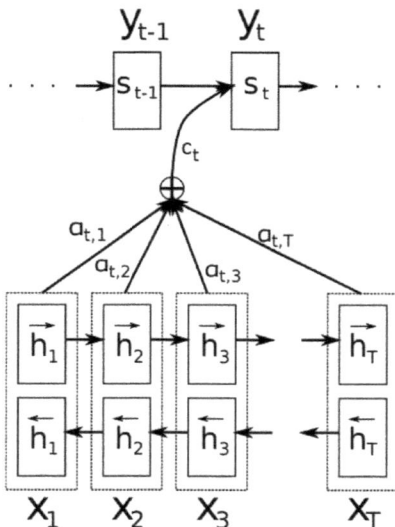

Figure 1. Original graphical representation of the attention model by Bahdanau et al. in [21].

The intuition is that an attention-based model could help in the energy disaggregatic task by assigning importance to each position of the aggregated signal which correspond to the position of an activation, or more generally, to a state change of the target applianc This allows the neural network to focus its representational power on selected time step of the target appliance in the aggregated signal, rather than on the activations of no target appliances, hopefully yielding more accurate predictions. In fact, some events (e.g turning on or off an appliance) or signal sections (e.g., high power consumption) are mo important than other parts within the input signal. For this reason, being able to correctly detect the corresponding time steps can play a key role in the disaggregation task. In neural machine translation, languages are typically not aligned because of the word ordering between the source and the target language. For the NILM problem, the aggregated power consumption is perfectly aligned with the load of the corresponding appliance and the alignment is known ahead of time. For this reason, to amplify the contribution of an appliance activation in the aggregated signal, we use a simplified attention model inspired by Raffel and Ellis [25], that combines all the hidden states of the encoder using their relative importance. The attention mechanism can be formulated as

$$e_t = a(h_t), \quad \alpha_t = \frac{exp(e_t)}{\sum_{j=1}^{T} exp(e_j)}, \quad c = \sum_{t=1}^{T} \alpha_t h_t, \qquad (3)$$

where a is a learnable function the depends only on the hidden state vector of the encoder h_t. The function a can be implemented with a feed-forward network that learns a particular attention weight α_t that determines the normalized importance score for h_j. This allows the network to recognize the time steps that are more important to the desired output as the ones having higher attention value.

3.2. Model Design

From a practical point of view, DNNs use partial sequences obtained with a sliding window technique. The duration of an appliance activation is used to determine the size of the window that selects the input and output sequences for the NILM modeling. To be precise, let $x_{t,L} = (x_t, \ldots, x_{t+L-1})$ and $y_{t,L}^i = (y_t^i, \ldots y_{t+L-1}^i)$ be, respectively, the partial aggregate and appliance sequences of length L starting at time t. In addition, we build the auxiliary state sequence $(s_1^i, \ldots, s_T^i)$, where $s_t^i \in \{0, 1\}$ represent the on/off state of the appliance i at time t. The state of an appliance is considered "on" when the consumption is greater than some threshold and "off" when the consumption is less or equal the same threshold. We use the notation $s_{t,L}^i = (s_t, \ldots, s_{t+L-1})$ for the partial state sequences of length L starting at time t. Our idea is to exploit the structure of the SGN architecture proposed in [17] as the building block of the model. This general framework uses an auxiliary sequence-to-sequence classification subnetwork that is jointly trained with a standard sequence-to-sequence regression subnetwork. The difference here is that we generate a more accurate estimate of the power consumption by performing the regression subtask with a scalable RNN-based encoder–decoder with attention mechanism. The intuition behind the proposed model is that the tailored attention mechanism allows the regression subnetwork to implicitly detect and assign more importance to some events (e.g., turning on or off of the appliance) and to specific signal sections (e.g., high power consumption), whereas the classification subnetwork helps the disaggregation process by enforcing explicitly the on/off states.

Differently from the DNN in [20], the scalability of the overall architecture is ensured by the regression subnetwork where no RNN is needed in the decoder. In fact, the adopted attention mechanism allows one to decouple the input representation from the output and the structure of the encoder from the structure of the decoder. We exploit these benefits and we design a hybrid encoder–decoder which is based on a combination of convolutional layers and recurrent layers for the encoder and fully connected layers for the decoder.

3.3. Network Topology

Let $f_{reg}^i : \mathbb{R}_+^L \to \mathbb{R}_+^L$ be the appliance power estimation model, then the regression subnetwork learns the mapping $\hat{p}_{t,L}^i = f_{reg}^i(x_{t,L})$. The topology of the regression subnetwork is as follows.

Encoder: The encoder network is composed by a CNN with 4 one-dimensional convolutional layers (Conv1D) with ReLU activation function that processes the input aggregated signal and extracts the appliance-specific signature as a set of feature maps. Finally, a RNN takes as input the set of feature maps and produces the sequence of the hidden states summarizing all the information of the aggregated signal. We use Bidirectional LSTM (BiLSTM) in order to get the hidden states h_t that summarize the information from both directions. A bidirectional LSTM is made up of a forward LSTM $\overrightarrow{g}$ that reads the sequence from left to right and a backward LSTM $\overleftarrow{g}$ that reads it from right to left. The final sequence of the hidden states of the encoder is obtained by concatenating the hidden state vectors from both directions, i.e., $h_t = [\overrightarrow{h_t}; \overleftarrow{h_t}]^T$.

Attention: The attention unit between the encoder and the decoder consists of a single layer feed-forward neural network that computes the attention weights and returns the context vector as a weighted average of the output of the encoder over time. Not all the feature maps produced by the CNN have equal contribution in the identification of the activation of the target appliance. Thus, the attention mechanism captures salient activations of the appliance, extracting more valuable feature maps than others for the disaggregation. The implemented attention unit is shown in Figure 2, and it is mathematically defined as

$$e_t = V_a^T tanh(W_a h_t + b_a),$$

$$\alpha_t = softmax(e_t),$$

$$c = \sum_{t=1}^{T} \alpha_t h_t,$$

where V_a, W_a, and b_a are the attentions parameters jointly learned with the other components of the architecture. The output of the attention unit is the context vector c that is used as the input vector for the following decoder.

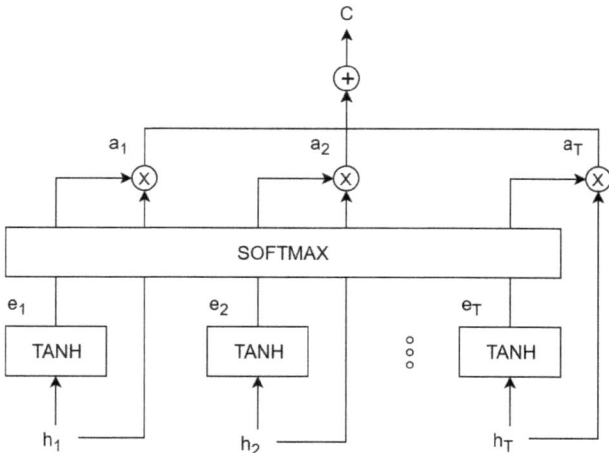

Figure 2. Graphical illustration of the implemented attention unit.

Decoder: The decoder network is composed by 2 fully connected layers (Dense). The second layer has the same number of units of the sequence length L.

The exact configuration of regression subnetwork is as follows:

1. Input (sequence length L determined by the appliance duration)
2. Conv1D (convolutional layer with F filters, kernel size K, stride 1, and ReLU activation function)
3. Conv1D (convolutional layer with F filters, kernel size K, stride 1, and ReLU activation function)
4. Conv1D (convolutional layer with F filters, kernel size K, stride 1, and ReLU activation function)
5. Conv1D (convolutional layer with F filters, kernel size K, stride 1, and ReLU activation function)
6. BiLSTM (bidirectional LSTM with H units, and tangent hyperbolic activation function)
7. Attention (single layer feed-forward neural network with H units, and tangent hyperbolic activation function)
8. Dense (fully connected layer with H units, and ReLU activation function)
9. Dense (fully connected layer with L units, and linear activation function)
10. Output (sequence length L)

Let $f_{reg}^i : \mathbb{R}_+^L \to [0,1]^L$ be the appliance state estimation model, then the classification subnetwork learns the mapping $\hat{s}_{t,L}^i = f_{cls}^i(x_{t,L})$. We use the sequence-to-sequence CNN proposed in [15] consisting of 6 convolutional layers followed by 2 fully connected layers. The exact configuration of the classification subnetwork is the following:

1. Input (sequence length L determined by the appliance duration)
2. Conv1D (convolutional layer with 30 filters, kernel size 10, stride 1, and ReLU activation function)
3. Conv1D (convolutional layer with 30 filters, kernel size 8, stride 1, and ReLU activation function)
4. Conv1D (convolutional layer with 40 filters, kernel size 6, stride 1, and ReLU activation function)
5. Conv1D (convolutional layer with 50 filters, kernel size 5, stride 1, and ReLU activation function)
6. Conv1D (convolutional layer with 50 filters, kernel size 5, stride 1, and ReLU activation function)
7. Conv1D (convolutional layer with 50 filters, kernel size 5, stride 1, and ReLU activation function)
8. Dense (fully connected layer with 1024 units, and ReLU activation function)
9. Dense (fully connected layer with L units, and sigmoid activation function)
10. Output (sequence length L)

The final estimate of the power consumption is obtained by multiplying the regression output with the probability classification output:

$$\hat{y}_{t,L}^i = f_{out}^i(x_{t,L}) = \hat{p}_{t,L} \odot \hat{s}_{t,L}, \tag{7}$$

where $\odot$ is the component-wise multiplication. The overall architecture is shown in Figure 3, and we call it LDwA, that is, Load Disaggregation with Attention.

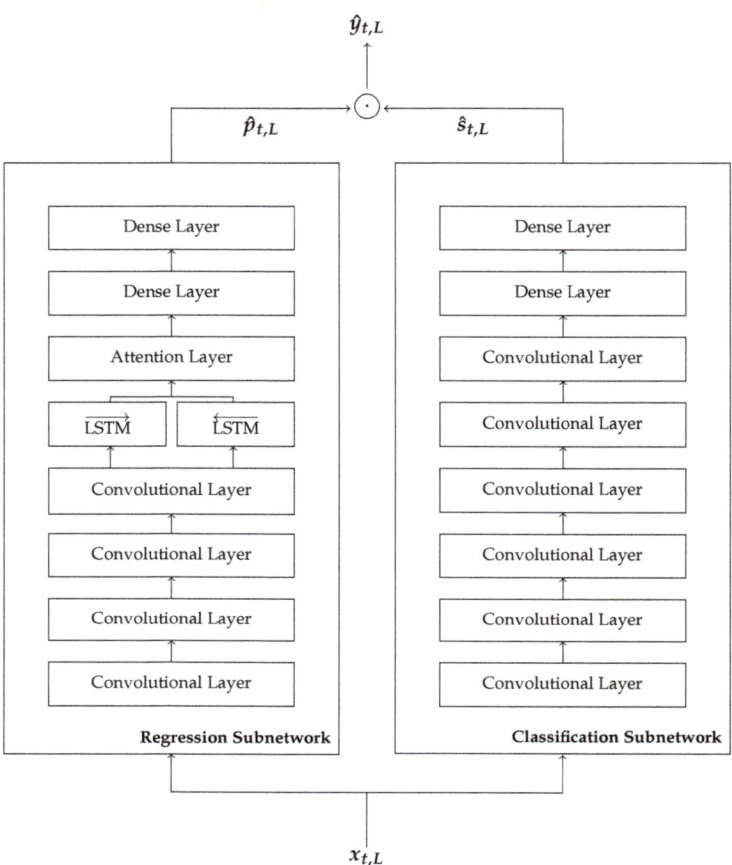

Figure 3. Proposed Load Disaggregation with Attention (LDwA) architecture used in our experiments.

4. Experiments

In this section, we show the experiments performed to evaluate our LDwA approach. First, we describe the datasets, the performance metrics, and the experimental procedure adopted. Then, we present and discuss the obtained results.

4.1. Datasets

In order to evaluate our algorithm and perform a fair comparison with state-of-the art methods, we choose two publicly available real-world datasets and adopt the same partition into training and test sets of the previous studies [15,17,18]. The Reference Energy Disaggregation Data Set (REDD) [26] contains data for six houses in the USA at 1 second sampling period for the aggregate power consumption, and at 3 s for the appliance power consumption. Following the previous studies, we consider the 3 top-consuming appliances: dishwasher (DW), fridge (FR), and microwave (MW). We use the data of houses 2–6 to build the training set, and house 1 as the test set. The preprocessed REDD dataset is provided by the authors of [17]. The second dataset, the Domestic Appliance-Level Electricity dataset UK-DALE [27], contains over two years of consumption profiles of five houses in UK, at 6 s sampling period. Here, the experiments are conducted using the 5 top-consuming appliances: dishwasher (DW), fridge (FR), kettle (KE), microwave (MW), and washing machine (WM). For evaluation, we use houses 1, 3, 4, and 5 for training and house 2 for testing as in the previous works [15,17,18]. The UK-DALE dataset has been preprocessed by the authors of [13]. We stress that for both datasets we consider the *unseen* setting in which we train and test on different households. In fact, the best way to test the generalization capability of a model is to use the model on a building not seen during the training. This is a particularly desirable property for a NILM algorithm since the unseen scenario is more likely in the real world application of the NILM service.

4.2. Metrics

In order to evaluate our NILM approach, we recall specific metrics that allow to capture significant performance of the algorithm. Following the previous studies in [15,17,18], we use the Mean Absolute Error (MAE) and the Signal Aggregate Error (SAE). Let $y_i(t)$ and $\hat{y}_i(t)$ be the true power and the estimated power at time t for the appliance i, respectively. The MAE for the appliance i is defined as

$$\text{MAE}_i = \frac{1}{T} \sum_{t=1}^{T} |y_i(t) - \hat{y}_i(t)|. \tag{8}$$

Give a predicted output sequence of length T, the SAE for the appliance i is defined as

$$\text{SAE}_i = \frac{1}{N} \sum_{\tau=1}^{N} \frac{1}{K} |r_i(\tau) - \hat{r}_i(\tau)|, \tag{9}$$

where N is the number of disjoint time periods of length K, $T = K \cdot N$, and $r_i(\tau)$ and $\hat{r}_i(\tau)$ represent the sum of the true power and the sum of the predicted power in the τth time period, respectively. In our experiments, we set $N = 1200$ which corresponds to a time period of approximately one hour for the REDD dataset and two hours for the UK-DALE dataset. For both metrics, lower values indicate better disaggregation performance.

In order to measure how accurately each appliance is running in on/off states, we use classification metrics such as the F1-score, that is, the harmonic mean of precision (P) and recall (R):

$$F_1 = \frac{2P \cdot R}{P + R}, \quad P = \frac{TP}{TP + FP}, \quad R = \frac{TP}{TP + FN}, \tag{10}$$

where TP, FP, and FP stand for true positive, false positive, and false negative, respectively. An appliance is considered "on" when the active power is greater than some threshold and "off" when it is less or equal the same threshold. The threshold is assumed to be the same value used for extracting the activations [17,18]. In our experiments, we use a threshold of 15 Watt for labeling the disaggregated loads. Precision, recall, and F1-score return a value between 0 and 1 where a higher number corresponds to better classification performance.

4.3. Network Setup

According to the Neural NILM approach, we train a network for each target appliance. A mini-batch of 32 examples is fed to each neural network, and mean and variance standardization is performed on the input sequences. For the target data, min-max normalization is used where minimum and maximum power consumption values of the related appliance are computed in the training set. The training phase is performed with a sliding window technique over the aggregated signal, using overlapped windows of length L with hop size equal to 1 sample. As stated in [13], the window size for the input and output pairs has to be large enough to capture an entire appliance activation, but not too large to include contributions of other appliances. In Table 1, we report the adopted window length L for each appliance that is related to the dataset sampling rate. The state classification label is generated by using a power consumption of 15 Watt as threshold. Each network is trained with the Stochastic Gradient Descent (SGD) algorithm with Nesterov momentum [28] set to 0.9. The loss function used for the joint optimization of the two subnetworks is given by $\mathcal{L} = \mathcal{L}_{out} + \mathcal{L}_{cls}$, where $\mathcal{L}_{out}$ is the Mean Squared Error (MSE) between the overall output of the network and the ground truth of a single appliance, and $\mathcal{L}_{cls}$ is the binary cross-entropy (BCE) that measures the classification error of the on/off state for the classification subnetwork. The maximum number of epochs is set to 100, the initial learning rate is set to 0.0 and it is reduced by a decay factor equal to 10^{-6} as the training progresses. Early stopping is employed as a form of regularization to avoid overfitting since it stops the training when the error on the validation set starts to grow [29]. For the classification subnetwork we adopt the hyperparameters from in [17] as our focus is only the effectiveness of the proposed components. The hyperparameter optimization of the regression subnetwork regards the number of filters (F), the size of each kernel (K), and the number of neurons in the recurrent layer (H). Grid search is used to perform the hyperparameter optimization, which is simply an exhaustive search through a manually specified subset of points in the hyperparameter space of the neural network where $F = \{16, 32, 64\}$, $K = \{4, 8, 16\}$ and $H = \{256, 512, 1024\}$. We evaluate the configuration of the hyperparameters on a held-out validation set and we choose the architecture achieving the highest performance on The disaggregation phase, also carried out with a sliding window over the aggregate signal with hop size equal to 1 sample, generates overlapped windows of the disaggregated signal. Differently from what proposed in [13], where the authors reconstruct the overlapped windows by aggregating their mean value, we adopt the strategy proposed in [16] in which the disaggregated signal is reconstructed by means of a median filter of the overlapped portions. The neural networks are implemented in Python with PyTorch, an open source machine learning framework [30] and the experiments are conducted on a cluster of NVIDIA Tesla K80 GPUs. The training time requires several hours for each architecture depending on the network dimension and on the granularity of the dataset.

Table 1. Sequence length (L) for the LDwA architecture.

Dataset	DW	FR	KE	MW	WM
REDD	2304	496	-	128	-
UK-DALE	1536	512	128	288	1024

4.4. Results

We compare our approach with the HMM implemented in [31] and the DNNs recently proposed: DAE, Seq2Point, S2SwA, SGN, and SCANet. We report the MAE, SAE, and F score for the REDD and the UK-DALE datasets in Tables 2 and 3, respectively. The results show that our approach turns out to be by far the best for both datasets. Apart from us, the two most competitive methods are SGN and SCANet, which share the same backbone we drew inspiration from. Results show that our network is better than both SGN and SCANet, implying that the differences introduced in our approach are significantly beneficial. In particular, our network outperforms SGN, showing that including our regression network significantly improves both the estimate of the power consumption and the load classification, and thus the overall disaggregation performance. More in detail, for the dataset REDD the improvement in terms of MAE (SAE) with respect to SGN ranges from a minimum of 24.13% (23.44%) on the fridge to a maximum of 45.15% (54.4%) on the dishwasher, with an average improvement of 32.64% (39.33%). As for the F1-score, the classification performance increase from a minimum of 6.67% on the fridge to a maximum of 24.03% on the microwave, with an average increase of 15.45%. For the UK-DALE dataset instead, the improvement in terms of MAE (SAE) with respect to SGN ranges from a minimum of 18.62% (8.93%) on the fridge to a maximum of 39.78% (50.25%) on the dishwasher, with an average improvement of 27.84% (30.65%). The F1-score increases from a minimum of 2.49% on the kettle to a maximum of 10.82% on the washing machine, with an average increment of the accuracy of 6.79%. Moreover, our method outperforms the more recent SCANet getting better disaggregation performance on all the appliances for both the datasets and both the metric. Looking at the tables, we see that for the REDD dataset the improvement in terms of MAE (SAE) with respect to SCANet ranges from a minimum of 9% (5.84%) on the fridge to a maximum of 18.76% (28.59%) on the microwave, with an average improvement of 13.21% (15.03%). The improvement of the F1-score ranges from a minimum of 3.64% on the fridge to maximum of 11.58% on the microwave, with an average increment of 6.81% of the accuracy. Finally, on the UK-DALE dataset, the improvement in terms of MAE (SAE) with respect to SCANet ranges from a minimum of 7.33% (7.2%) on the kettle to a maximum of 24.57% (19.55%) on the dishwasher, with an average improvement of 15.69% (14%). The F1-score increases from a minimum of 0.92% on the kettle to a maximum of 8.85% on the washing machine, with an overall improvement of 4.41%.

In order to evaluate the computational burden of the proposed LDwA, we report in Tables 4 and 5 the training time with respect to the most accurate DNNs. Clearly, LDwA is less efficient than SGN as LSTM layers have larger number of trainable parameters than the convolutional ones. However, the efficiency of our architecture with respect to the attention-based S2SwA is remarkable. This is explained by the presence of the tailored attention mechanism that does not require additional recurrent layers in the decoder. There is also a huge improvement in the training time with respect to the SCANet. We achieve better performance without the need to train a Generative Adversarial Network, that requires a significant amount of computational resources and has notorious convergence issues.

The profiles related to the dishwasher, microwave, fridge, and kettle are shown in Figures 4–7, respectively, where each appliance activation is successfully detected by the LDwA in the disaggregated trace.

Table 2. Disaggregation performance for the REDD dataset. We report in boldface the best approach.

Model	Metric	DW	FR	MW	Overall
FHMM [31]	MAE	101.30	98.67	87.00	95.66
	SAE	93.64	46.73	65.03	68.47
	F1 (%)	12.93	35.12	11.97	20.01
DAE [16]	MAE	26.18	29.11	23.26	26.18
	SAE	21.46	20.97	19.14	20.52
	F1 (%)	48.81	74.76	18.54	47.37
Seq2Point [15]	MAE	24.44	26.01	27.13	25.86
	SAE	22.87	16.24	18.89	19.33
	F1 (%)	47.66	75.12	17.43	46.74
S2SwA [20]	MAE	23.48	25.98	24.27	24.57
	SAE	22.64	17.26	16.19	18.69
	F1 (%)	49.32	76.98	19.31	48.57
SGN [17]	MAE	15.77	26.11	16.95	19.61
	SAE	15.22	17.28	12.49	15.00
	F1 (%)	58.78	80.09	44.98	61.28
SCANet [18]	MAE	10.14	21.77	13.75	15.22
	SAE	8.12	14.05	9.97	10.71
	F1 (%)	69.21	83.12	57.43	69.92
Proposed LDwA	MAE	**8.65**	**19.81**	**11.17**	**13.21**
	SAE	**6.94**	**13.23**	**7.12**	**9.10**
	F1 (%)	**74.41**	**86.76**	**69.01**	**76.73**

Table 3. Disaggregation performance for the UK-DALE dataset. We report in boldface the best approach.

Model	Metric	DW	FR	KE	MW	WM	Overall
FHMM [31]	MAE	48.25	60.93	38.02	43.63	67.91	51.75
	SAE	46.04	51.90	35.41	41.52	64.15	47.80
	F1 (%)	11.79	33.52	9.35	3.44	4.10	12.44
DAE [16]	MAE	22.18	17.72	10.87	12.87	13.64	15.46
	SAE	18.24	8.74	7.95	9.99	10.67	11.12
	F1 (%)	54.88	75.98	93.43	31.32	24.54	56.03
Seq2Point [15]	MAE	15.96	17.48	10.81	12.47	10.87	13.52
	SAE	10.65	8.01	5.30	10.33	8.69	8.60
	F1 (%)	50.92	80.32	94.88	45.41	49.11	64.13
S2SwA [20]	MAE	14.96	16.47	12.02	10.37	9.87	12.74
	SAE	10.68	7.81	5.78	8.33	8.09	8.14
	F1 (%)	53.67	79.04	94.62	47.99	45.79	64.22
SGN [17]	MAE	10.91	16.27	8.09	5.62	9.74	10.13
	SAE	7.86	6.61	5.03	4.32	7.14	6.20
	F1 (%)	60.02	84.43	96.32	58.55	61.12	72.09
SCANet [18]	MAE	8.71	15.16	6.14	4.82	8.48	8.67
	SAE	4.86	6.54	4.03	3.81	5.77	5.00
	F1 (%)	63.30	85.77	98.89	62.22	63.09	74.65
Proposed LDwA	MAE	**6.57**	**13.24**	**5.69**	**3.79**	**7.26**	**7.31**
	SAE	**3.91**	**6.02**	**3.74**	**2.98**	**4.87**	**4.30**
	F1 (%)	**68.99**	**87.01**	**99.81**	**67.55**	**71.94**	**79.06**

Table 4. Training time in hours for the REDD dataset.

Model	DW	FR	MW
S2SwA [20]	8.91	5.75	3.93
SGN [17]	3.31	2.43	0.57
SCANet [18]	6.44	3.24	2.68
Proposed LDwA	5.37	3.07	1.88

Table 5. Training time in hours for the UK-DALE dataset.

Model	DW	FR	KE	MW	WM
S2SwA [20]	14.53	9.42	5.02	6.21	6.12
SGN [17]	5.43	3.11	2.43	2.76	3.78
SCANet [18]	8.01	7.98	6.41	5.77	7.11
Proposed LDwA	6.21	4.93	3.65	3.37	5.87

The tailored attention mechanism inserted into the regression branch of the network allows us to correctly identify the relevant time steps in the signal and generalize well in unseen houses. Furthermore, modeling attention is particularly interesting from the perspective of the interpretability of deep learning models because it allows one to direct inspect the internal working of the architecture. The hypothesis is that the magnitude of the attention weights correlates with how relevant the specific region of the input sequence is, for the prediction of the output sequence. As shown in Figures 4–7, our network effective at predicting the activation of an appliance and the attention weights present peak in correspondence of the state change of that appliance.

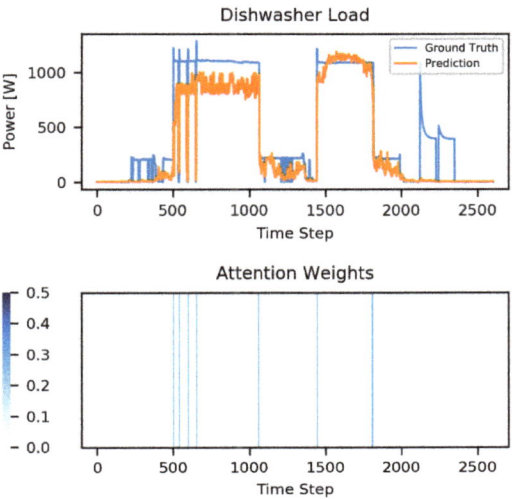

Figure 4. REDD dishwasher load and the heatmap of the attention weights at 3 s resolution.

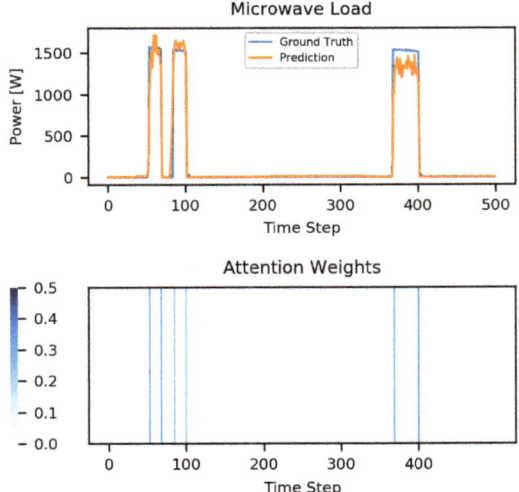

Figure 5. REDD microwave load and the heatmap of the attention weights at 3 s resolution.

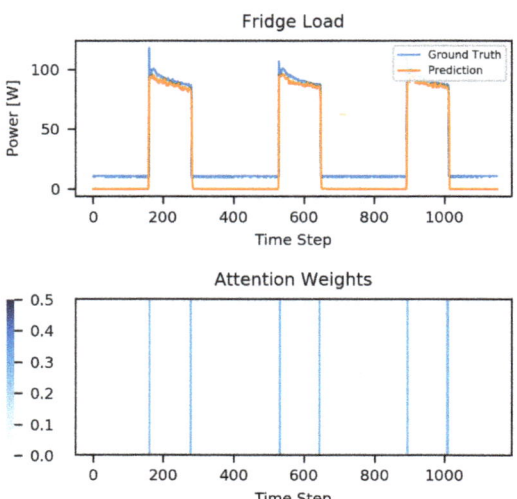

Figure 6. UK-DALE fridge load and the heatmap of the attention weights at 6 s resolution.

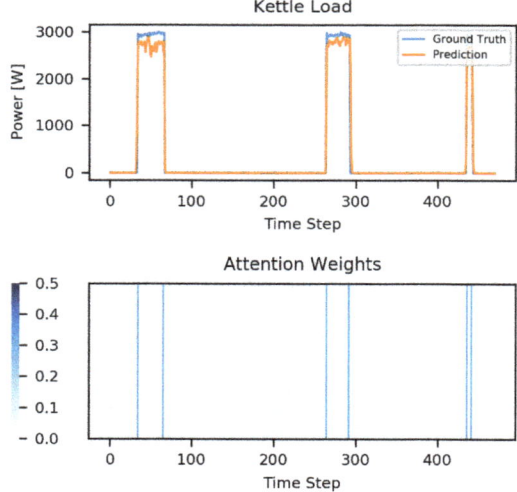

Figure 7. UK-DALE kettle load and the heatmap of the attention weights at 6 s resolution.

In conclusion, our approach does not only predict the correct disaggregation in term of scale, but is also successful at deciding if the target appliance is active in the aggregate load at a given time step.

5. Conclusions

This paper proposes LDwA, a new deep neural network architecture for the NILM problem that features a tailored attention mechanism with the encoder–decoder framework to extract appliance specific power usage from the aggregated signal. The integration of convolutional layers and recurrent layers in the regression subnetwork facilitates feature extraction and allows to build better appliance models where the locations of relevant features are successfully identified by the attention mechanism. The use of the proposed model for the regression subtask increases the network's ability to extract and exploit information dramatically. The proposed system is tested on two real-world datasets with different granularity, REDD and UK-DALE. The experimental results demonstrate that the proposed model significantly improves accuracy and generalization capability for load recognition of all the appliances for both datasets compared to the deep learning state-of-the-art.

Author Contributions: Conceptualization, A.M.S.; Funding acquisition, V.P.; Methodology, A.M.S.; Software, A.M.S.; Supervision, V.P.; Validation, V.P.; Writing—original draft, V.P. and A.M.S.; Writing—review and editing, V.P. All authors have read and agreed to the published version of the manuscript.

Funding: This research was funded by the grant provided to Veronica Piccialli by ENEA in the collaboration agreement "Clustering di tipologie di abitazione per scegliere modelli di disaggregazione di consumi elettrici realizzati tramite reti deep" within the PAR2018 project. The APC was funded by the Department of Civil and Computer Engineering of the University of Rome Tor Vergata.

Institutional Review Board Statement: Not applicable.

Informed Consent Statement: Not applicable.

Data Availability Statement: The pre-processed data used in this study are available on request from the corresponding author.

Conflicts of Interest: The authors declare no conflict of interest.

References

Hart, G.W. Nonintrusive appliance load monitoring. *Proc. IEEE* **1992**, *80*, 1870–1891. [CrossRef]
Ruano, A.; Hernandez, A.; Ureña, J.; Ruano, M.; Garcia, J. NILM techniques for intelligent home energy management and ambient assisted living: A review. *Energies* **2019**, *12*, 2203. [CrossRef]
de Souza, W.A.; Garcia, F.D.; Marafão, F.P.; Da Silva, L.C.P.; Simões, M.G. Load disaggregation using microscopic power features and pattern recognition. *Energies* **2019**, *12*, 2641. [CrossRef]
Faustine, A.; Mvungi, N.H.; Kaijage, S.; Michael, K. A survey on non-intrusive load monitoring methodies and techniques for energy disaggregation problem. *arXiv* **2017**, arXiv:1703.00785.
Rahimpour, A.; Qi, H.; Fugate, D.; Kuruganti, T. Non-Intrusive Energy Disaggregation Using Non-Negative Matrix Factorization With Sum-to-k Constraint. *IEEE Trans. Power Syst.* **2017**, *32*, 4430–4441. [CrossRef]
Kolter, J.Z.; Jaakkola, T. Approximate inference in additive factorial hmms with application to energy disaggregation. In Proceedings of the Fifteenth International Conference on Artificial Intelligence and Statistics, La Palma, Spain, 21–23 April 2012; pp. 1472–1482.
Makonin, S.; Popowich, F.; Bajić, I.V.; Gill, B.; Bartram, L. Exploiting HMM sparsity to perform online real-time nonintrusive load monitoring. *IEEE Trans. Smart Grid* **2015**, *7*, 2575–2585. [CrossRef]
Nashrullah, E.; Halim, A. Performance Evaluation of Superstate HMM with Median Filter For Appliance Energy Disaggregation. In Proceedings of the 2019 6th International Conference on Electrical Engineering, Computer Science and Informatics (EECSI), Bandung, Indonesia, 18–20 September 2019; pp. 374–379.
Basu, K.; Debusschere, V.; Bacha, S.; Maulik, U.; Bondyopadhyay, S. Nonintrusive load monitoring: A temporal multilabel classification approach. *IEEE Trans. Ind. Inform.* **2014**, *11*, 262–270. [CrossRef]
Singhal, V.; Maggu, J.; Majumdar, A. Simultaneous detection of multiple appliances from smart-meter measurements via multi-label consistent deep dictionary learning and deep transform learning. *IEEE Trans. Smart Grid* **2018**, *10*, 2969–2978. [CrossRef]
Singh, S.; Majumdar, A. Non-Intrusive Load Monitoring via Multi-Label Sparse Representation-Based Classification. *IEEE Trans. Smart Grid* **2019**, *11*, 1799–1801. [CrossRef]
Faustine, A.; Pereira, L. Multi-Label Learning for Appliance Recognition in NILM Using Fryze-Current Decomposition and Convolutional Neural Network. *Energies* **2020**, *13*, 4154. [CrossRef]
Kelly, J.; Knottenbelt, W. Neural nilm: Deep neural networks applied to energy disaggregation. In Proceedings of the 2nd ACM International Conference on Embedded Systems for Energy-Efficient Built Environments, Seoul, Korea, 4–5 November 2015; pp. 55–64.
He, W.; Chai, Y. An Empirical Study on Energy Disaggregation via Deep Learning. In Proceedings of the 2016 2nd International Conference on Artificial Intelligence and Industrial Engineering (AIIE 2016), Beijing, China, 20–21 November 2016.
Zhang, C.; Zhong, M.; Wang, Z.; Goddard, N.; Sutton, C. Sequence-to-point learning with neural networks for non-intrusive load monitoring. In Proceedings of the Thirty-Second AAAI Conference on Artificial Intelligence, New Orleans, LA, USA, 2–7 February 2018.

16. Bonfigli, R.; Felicetti, A.; Principi, E.; Fagiani, M.; Squartini, S.; Piazza, F. Denoising autoencoders for non-intrusive load monitoring: improvements and comparative evaluation. *Energy Build.* **2018**, *158*, 1461–1474. [CrossRef]
17. Shin, C.; Joo, S.; Yim, J.; Lee, H.; Moon, T.; Rhee, W. Subtask gated networks for non-intrusive load monitoring. In Proceedings the AAAI Conference on Artificial Intelligence, Honolulu, HI, USA, 27 January–1 February 2019; Volume 33, pp. 1150–1157.
18. Chen, K.; Zhang, Y.; Wang, Q.; Hu, J.; Fan, H.; He, J. Scale-and Context-Aware Convolutional Non-intrusive Load Monitoring. *IEEE Trans. Power Syst.* **2019**, *35*, 2362–2373. [CrossRef]
19. Kong, W.; Dong, Z.Y.; Wang, B.; Zhao, J.; Huang, J. A Practical Solution for Non-Intrusive Type II Load Monitoring Based on Deep Learning and Post-Processing. *IEEE Trans. Smart Grid* **2020**, *11*, 148–160. [CrossRef]
20. Wang, K.; Zhong, H.; Yu, N.; Xia, Q. Nonintrusive Load Monitoring based on Sequence-to-sequence Model With Attention Mechanism. *Zhongguo Dianji Gongcheng Xuebao/Proc. Chin. Soc. Electr. Eng.* **2019**, *39*, 75–83. [CrossRef]
21. Bahdanau, D.; Cho, K.; Bengio, Y. Neural machine translation by jointly learning to align and translate. *arXiv* **2014**, arXiv:1409.04.
22. Sutskever, I.; Vinyals, O.; Le, Q.V. Sequence to sequence learning with neural networks. In *Advances in Neural Information Processing Systems 27, Proceedings of the 27th International Conference on Neural Information Processing Systems, December 2014, Pages 3104–3112*; Montreal, Canada 2014; Ghahramani, Z., Welling, M., Cortes, C., Lawrence, N., Weinberger, K.Q., Eds.; MIT Press: Cambridge, MA, USA, 2014; pp. 3104–3112.
23. Cho, K.; Van Merriënboer, B.; Bahdanau, D.; Bengio, Y. On the properties of neural machine translation: Encoder-decoder approaches. *arXiv* **2014**, arXiv:1409.1259.
24. Young, T.; Hazarika, D.; Poria, S.; Cambria, E. Recent trends in deep learning based natural language processing. *IEEE Comp. Intell. Mag.* **2018**, *13*, 55–75. [CrossRef]
25. Raffel, C.; Ellis, D.P. Feed-forward networks with attention can solve some long-term memory problems. *arXiv* **2015**, arXiv:1512.08756.
26. Kolter, J.Z.; Johnson, M.J. REDD: A public data set for energy disaggregation research. In Proceedings of the Workshop on Data Mining Applications in Sustainability (SIGKDD), San Diego, CA, USA, 21–24 August 2011; Volume 25, pp. 59–62.
27. Kelly, J.; Knottenbelt, W. The UK-DALE dataset, domestic appliance-level electricity demand and whole-house demand from five UK homes. *Sci. Data* **2015**, *2*, 150007. [CrossRef] [PubMed]
28. Sutskever, I.; Martens, J.; Dahl, G.; Hinton, G. On the importance of initialization and momentum in deep learning. In Proceedings of the International Conference on Machine Learning, Atlanta, GA, USA, 16–21 June 2013; pp. 1139–1147.
29. Caruana, R.; Lawrence, S.; Giles, L. Overfitting in Neural Nets: Backpropagation, Conjugate Gradient, and Early Stopping. In *Proceedings of the 13th International Conference on Neural Information Processing Systems*; MIT Press: Cambridge, MA, USA, 2000; pp. 381–387.
30. Paszke, A.; Gross, S.; Chintala, S.; Chanan, G.; Yang, E.; DeVito, Z.; Lin, Z.; Desmaison, A.; Antiga, L.; Lerer, A. Automatic Differentiation in PyTorch. In Proceedings of the NIPS Autodiff Workshop, Long Beach, CA, USA, 9 December 2017.
31. Batra, N.; Kelly, J.; Parson, O.; Dutta, H.; Knottenbelt, W.; Rogers, A.; Singh, A.; Srivastava, M. NILMTK: An open source toolkit for non-intrusive load monitoring. In Proceedings of the 5th International Conference on Future Energy Systems, Cambridge, UK, 11–13 June 2014; pp. 265–276.

Article

Multi-Label Learning for Appliance Recognition in NILM Using Fryze-Current Decomposition and Convolutional Neural Network

Anthony Faustine [1],* and Lucas Pereira [2]

[1] Ireland's National Centre for Applied Data Analytics (CeADER) University College Dublin; Belfield Office Park, Unit 9, Clonskeagh, 4 Dublin, Ireland
[2] ITI, LARSyS, Té cnico Lisboa; Av. Rovisco Pais, 1000 268 Lisboa, Portugal; lucas.pereira@tecnico.ulisboa.pt
* Correspondence: sambaiga@gmail.com or anthony.faustine@ucd.ie; Tel: +32493972982

Received: 18 May 2020; Accepted: 6 July 2020; Published: 11 August 2020

Abstract: The advance in energy-sensing and smart-meter technologies have motivated the use of a Non-Intrusive Load Monitoring (NILM), a data-driven technique that recognizes active end-use appliances by analyzing the data streams coming from these devices. NILM offers an electricity consumption pattern of individual loads at consumer premises, which is crucial in the design of energy efficiency and energy demand management strategies in buildings. Appliance classification, also known as load identification is an essential sub-task for identifying the type and status of an unknown load from appliance features extracted from the aggregate power signal. Most of the existing work for appliance recognition in NILM uses a single-label learning strategy which, assumes only one appliance is active at a time. This assumption ignores the fact that multiple devices can be active simultaneously and requires a perfect event detector to recognize the appliance. In this paper proposes the Convolutional Neural Network (CNN)-based multi-label learning approach, which links multiple loads to an observed aggregate current signal. Our approach applies the Fryze power theory to decompose the current features into active and non-active components and use the Euclidean distance similarity function to transform the decomposed current into an image-like representation which, is used as input to the CNN. Experimental results suggest that the proposed approach is sufficient for recognizing multiple appliances from aggregated measurements.

Keywords: multi-label learning; Non-intrusive Load Monitoring; appliance recognition; fryze power theory; V-I trajectory; Convolutional Neural Network; distance similarity matrix; activation current

1. Introduction

Recently, most of the world has witnessed a rapid increase in energy use in buildings (residential and commercial). Residential and commercial buildings consume approximately 60% of the world's electricity (The United Nation's Environment Programme's Sustainable Building and Climate Initiative (UNEP-SBCI)). Energy efficiency and conservation in buildings can be generally achieved through replacing devices with more efficient ones, improving the efficiency of the building (for example, using better insulation), or optimizing energy usage through behavior changes and application of cost-effective technologies [1]. Unlike other strategies for building energy saving, optimizing energy use through behavior changes is very fast and highly profitable. With the application of cost-effective technologies, this strategy can also provide end-use appliances consumption to households that give insight into what appliances are used, when they are used, how much power they consume, and why such consumption [2]. The end-use appliances consumption is also useful for estimating the amount of energy demand at consumer premises [3]. It further increases the awareness about the energy consumption behavior of consumers.

The recent advance in energy-sensing and smart-meter technologies has led to the rise of Non-Intrusive Load Monitoring (NILM) [4,5]. NILM is a computational technique that uses aggregate power data monitored from a single point source such as a smart meter or current or voltage sensor-plug to infer the end-appliances running in the building and estimate their respective power consumption [6]. It relies on signal processing and machine learning techniques that analyze appliance patterns from aggregate power measurements. NILM provides households with cost-effective monitoring of appliance-specific energy consumption, and it can be easily integrated into existing buildings without causing any inconvenience to inhabitants. Several machine learning techniques have been proposed to address the energy-disaggregation [7–12].

Recognizing appliances from aggregated power measurements is one of the vital sub-tasks of NILM [11,13]. It uses machine learning techniques to analyze the pattern of the electrical features vector extracted from aggregated measurements and classifies them into the respective appliance category. Feature vectors are obtained after the state-transitions of appliances have been detected. These features are extracted at different sampling rates (high-frequency or low-frequency) depending on the measurements and electrical characteristics needed by the NILM algorithm [14]. The high sampling frequency offers the possibility to consider fine-grained features such as voltage-current (V-I) trajectory, harmonics, wavelet coefficients from steady-state, and transient behavior. As a result, several techniques for appliance recognition applying high-frequency features such as V-I trajectory have been proposed [15,16].

It has been demonstrated that transforming the V-I trajectory into image representation and feeding it as the input to machine learning classifiers improves classification performance [11,14,17–21]. However, the presented works use single-label learning, thus assuming that only one appliance is active at a time. This strategy ignores the fact that multiple devices can be active simultaneously as well as dependencies between appliance usage. It further requires a perfect event detector to extract the appliance features just after an event has been detected, particularly in aggregated measurements [19]. In contrast to single-label learning, multi-label learning links multiple appliances to an observed aggregate power signal [22,23].

Several studies have demonstrated that multi-label learning represents a viable alternative to conventional NILM approaches [23–27]. For example, the work by [25], investigated the possibility of applying a temporal multi-label classification approach in non-event based NILM where a novel set of meta-features was proposed. In [26] an extensive survey for the multi-label classification and the multi-label meta-classification framework for low-sampling power measurements is presented. Recent studies have explored deep neural networks for multi-label appliance recognition in NILM [23,24], yet these approaches also rely on low-frequency data.

Instead, this paper presents a CNN-based multi-label appliance classification approach. The proposed method uses the current waveform generated from the aggregated measurements, taken in brief windows of time containing one or more than one event. The underlying assumption of this method is that the extracted aggregated current will be the summation of active appliances. Therefore, by training a classification model to learn the patterns of different combinations, it is possible to successfully identify such appliances when these appear in a future time window.

To improve the discriminating power of our method, we apply the *Fryze power theory*, which enables the decomposition of the current waveform into active and non-active components in time-domain [20,28]. Our research hypothesis is that the sum of the active and non-active components will exhibit unique and consistent characteristics based on the appliances that are running simultaneously, hence providing a distinctive feature for multi-label classification.

The decomposed current is then transformed into an image representation using the Euclidean-distance-similarity matrix [29] and fed into the CNN for multi-label classification. The proposed approach is evaluated against the PLAID dataset [30], which consists of aggregated voltage and current measurements at a 30 kHz sampling rate. The source code used in our experiments is publicly available on a GitHub repository (https://github.com/sambaiga/MLCFCD).

The main contribution of this paper is a multi-label learning strategy for appliance recognition in NILM. The proposed approach associates multiple appliances to an observed aggregated current signal. Overall, this contribution folds into four sub-contributions

1. We first demonstrate that for aggregated measurements, the use of activation current as an input feature offers improved performance compared to the regularly used V-I binary image feature.
2. Second, we apply the Fryze power theory and Euclidean distance matrix as pre-processing steps for the multi-label classifier. This pre-processing step improves the appliance feature's uniqueness and enhances the performance of the multi-label classifier.
3. Third, we propose a CNN multi-label classifier that uses *softmax* activation to capture the relations between multiple appliances implicitly.
4. Fourth, we conduct an experimental evaluation of the proposed approach on an aggregated public dataset and compare the general and per-appliance performances. We also provide an in-depth error analysis and identified three types of errors for multi-label appliance recognition in NILM. Finally, a complexity analysis of the proposed approach method is also presented.

The remainder of this paper is organized as follows: Section 2 summarizes related works while Section 3 introduces the methods utilized in this work. Section 4 describes the experimental design. Section 5 presents the results and discussion of the performed evaluations. Finally, Section 6 summarises the contributions of this paper and suggests future research direction.

2. Related Works

The concept of multi-label classification for NILM has gained momentum recently, as the systematic review finds in [26]. Besides an extensive survey of the topic, the authors present the multi-label meta-classification framework (RAkEL) and the bespoke multi-label classification algorithm (MLkNN), where both employ time-domain and wavelet-domain feature sets. Other approaches to multi-label NILM comprise restricted Boltzmann machines [31], and multi-target classification [32].

In [33], the authors present an algorithm that uses Sparse Representation based classification for multi-label NILM. Furthermore, the authors compare their algorithm to other cutting edge multi-label NILM approaches such as classification based on extreme learning machines (ELM) [34], graph-based semi-supervised learning [35], and an approach based on deep dictionary learning and deep transform learning [36]. Nalmpantis and Vrakas [37] present a multi-label NILM based on the Signal2Vec algorithm that maps any time series into a vector space. A deep neural network (DNN) based multi-label NILM applying active power features at low-sampling frequency is proposed in [23,24]. In [23], the authors propose an approach that builds on Temporal Convolutional Networks (TCNN). At the same time, Massidda et al. [24] applied Fully Convolutional Networks (FCNN) for multi-label-learning in NILM, adopting some methods used in semantic segmentation.

Even though multi-label learning was found to be competitive with state-of-the-arts NILM algorithms, none of the previous works have considered the V-I trajectory-based features for multi-label-classification. Existing NILM methods that use V-I based features for appliance classification uses single-label learning [11,14–21,38]. The use of V-I based features for appliance classification was first introduced in [15], where shape-based features extracted from V-I (e.g., number of self-interceptions) were used as input to a machine learning classifier. A review and performance evaluation of the seven load wave-shape is presented in [39]. The shape-based feature was found to have a direct correspondence to operating characteristics of appliances as contained in the current wave-shape. Several other features such as asymmetry, mean line and self-intersection assessment extracted from V-I waveforms were used to classify appliances in [16]. However, this approach compresses the information in the V-I-trajectory into a limited amount of features extracted solely based on deep engineering knowledge.

Against this background, other researchers demonstrated that transforming the V-I trajectory into a binary image representation improves classification performance [17,18] by leveraging on state-of-the-art deep-learning algorithms for image recognition. For example, De Baets et al. [14,19] transforms the V-I trajectory into weighted pixelated V-I images and uses a CNN classifier. In another work, a hardware implementation of the appliance recognition system based on V-I curves and a CNN classifier is also proposed [21]. The work by [20,28] demonstrated that applying the Fryze power theory to decompose the current into active and non-active components could enhance the uniqueness of the V-I binary image and consequently improve classification performance. The work by Teshome et al. [28] applied the non-active current and non-active voltage (V-I_f) for appliance recognition. Liu et al. [20] further demonstrated that the visual representation of (V-I_f) is robust enough to be used in Transfer learning. Recently it has been shown that transforming the V-I into compressed distance similarity matrix consistently improves the appliance classification performance compared to the commonly used V–I image representation [11,13].

Motivated by these two works, we apply decomposed currents as input features for recognizing multiple running appliances. Still, unlike Liu et al. [20], and Teshome et al. [28], we transform the decomposed current into a 2D Euclidean-distance similarity matrix, which is later used as the input to the CNN model.

3. Proposed Methods

The goal of appliance recognition in NILM is to identify appliance states s_t^m from the aggregate measurements x_t composed of individual appliance measurements y_t^m such that $m = 1, 2, \ldots M$, where M indicates the number of appliances such that

$$x_t = \sum_{m=1}^{M} y_t^m \cdot s_t^m + \epsilon_t \tag{1}$$

where ϵ_t represents both any contribution from appliances not accounted for and measurement noise [40]. We refer to this as a multi-label NILM problem, where given observed aggregate measurements, the unobserved appliance states s_t of electrical appliances are estimated.

Specifically, the problem is formulated as follows: Let $\mathbf{X} \in \mathbb{R}^{T \times d} = \{x_1 \ldots x_T\}$ denote a set of input features derived from the aggregate measurements of M appliances and $\mathbf{Y} \in \mathbb{R}^{T \times M}$ indicates the associated appliance measurements, where each appliance has k states denoted as $\mathbf{s}^m = \{s_1^m, \ldots s_k^m\}$ such that $s_k^m \in \{0, 1\}$. The matrix $\mathbf{S} \in \mathbb{R}^{T \times M}$ indicates the associated multi-label states. Thus, given $D = \{\mathbf{x}_t, \mathbf{s}_t | t = 1, \ldots, T\}$ datasets, the goal is to learn a multi-label classifier that predicts the state vector $\mathbf{s_t} = \{s_t^m, \ldots s_t^M\}$ from the input aggregate power feature vector $\mathbf{x}_t$.

The proposed approach is summarized in Figure 1.

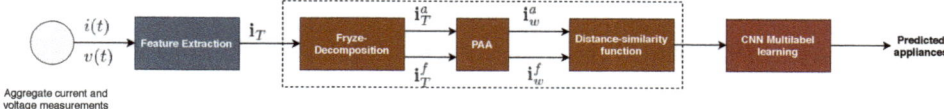

Figure 1. The block diagram of the proposed method. The dotted block is the pre-processing block where PAA stand for Piecewise Aggregate Approximation, a dimension reduction method for high-dimensional time series signal.

3.1. Feature Extraction from Aggregate Measurements

In this work, we consider the appliance feature extracted from high-frequency aggregate voltage and current measurements in brief windows of time. This feature contains one or more than one event and allows us to distinguish multiple appliances running simultaneously, as illustrated in Figure 2a. We define an activation current $\mathbf{i}$ and voltage $\mathbf{v}$ to be a one-cycle steady-state signal extracted from the aggregate current and voltage waveform.

To obtain the activation waveforms from aggregate measurements we measure $N_c = 20$ cycles of current and voltage $\{\mathbf{i}^{(a)}, \mathbf{v}^{(a)}\}$ after an event. As depicted in Figure 2b, the N_c circles correspond to steady-state behavior and is equivalent to $T_s \times N_s$ samples where $T_s = \frac{f_s}{f}$, f_s is sampling frequency and f is mains frequency. These cycles are aligned at the zero-crossing of the voltage and there-after one-cycle activation current **i** and voltage **v** with size T_s is extracted, as illustrated in Figure 2.

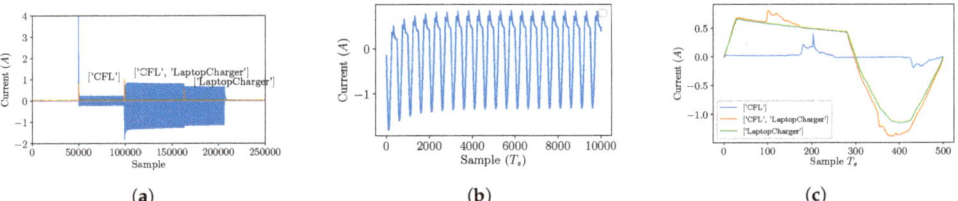

Figure 2. (**a**) Aggregated current signal for different events. (**b**) Twenty cycles of current after an event. (**c**) Extracted activation currents for different events. The activation current of the event at time **i** is the summation of the activation current for the all running appliances.

3.2. Feature Pre-Processing

As discussed in the related work section, the V-I binary trajectory mapping has been one of the favored features for appliance classification in *single-label learning*. However, in this work, we consider the features derived from source current $i(t)$ in recognizing multiple running appliances from aggregate measurements. Through experimentation, it was found that the aggregated activation voltage $v(t)$ has an almost identical pattern for most of the events, as illustrated in Figure 3. This suggests that the activation current i reflects the electrical properties of an appliance.

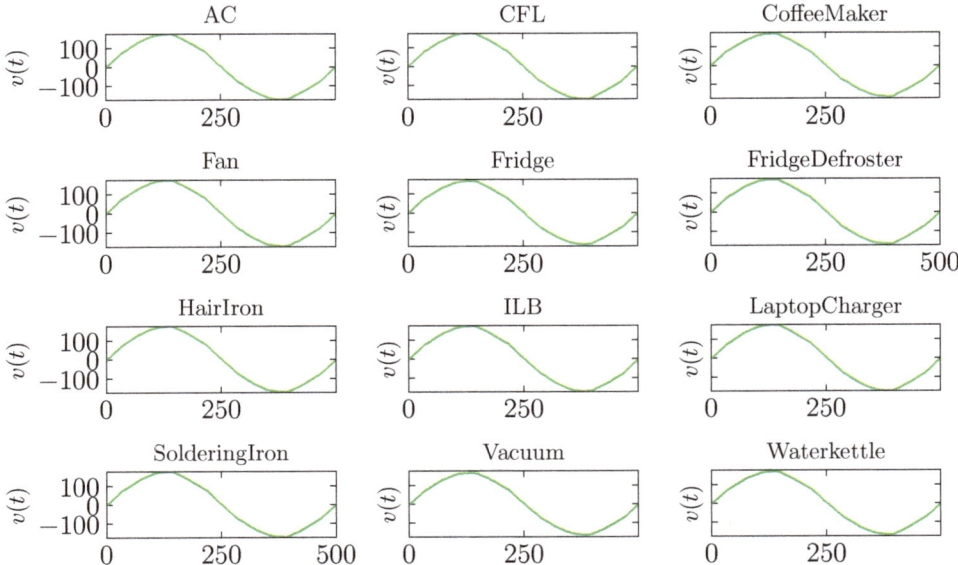

Figure 3. Activation voltage $v(t)$ for different appliances in the PLAID dataset. The voltage has an almost identical pattern for all the appliances.

Therefore, we propose the decomposed current features obtained by applying the Fryze power theory [41].

The Fryze power theory decomposes activation current into orthogonal components related to electrical energy in time-domain [41]. According to this theory, it is possible to decompose the activation current **i** into active $i(t)_a$ and non-active components $i(t)_f$, such that:

$$i(t) = i(t)_a + i(t)_f \tag{2}$$

The active current $i(t)_a$ is the current of the resistive load, having the same active power at the same activation voltage. In Fryze's theory, the active power is calculated as the average value of $i(t) \cdot v(t)$ over one fundamental cycle T_s defined as follows:

$$p_a = \frac{1}{T_s} \sum_{t=1}^{T_s} i(t) v(t) \tag{3}$$

The active current is therefore defined as

$$i_a(t) = \frac{p_a}{v_{rms}^2} v(t) \tag{4}$$

where v_{rms} is the rms voltage, expressed as follows:

$$v_{rms} = \sqrt{\frac{1}{T_s} \sum_{t=1}^{T_s} v(t)^2} \tag{5}$$

The current $i_a(t)$ represents the resistance information and is purely sinsoidal. The non-active component is then equal to

$$i(t)_f = i(t) - i(t)_a \tag{6}$$

Figure 4 presents the source currents and the corresponding active and non-active components for the twelve appliances in the PLAID dataset. It can be observed from Figure 4 that the active component approaches a pure sine wave even for non-periodic load currents like a Compact Fluorescent Lamp (CFL) and Laptop.

Once the activation-current has been decomposed, the Piece-wise Aggregate Approximation (PAA) is used to reduce the dimensional of the decomposed signal i_a and i_f from T_s to a predefined size w. PAA is a dimension reduction method for high-dimensional time series signal [42]. This is a crucial pre-processing step as it reduces the high-dimensionality of the extracted activation current feature with minimal information loss.

To further enhance the uniqueness of the decomposed-current feature, a Euclidean distance function $d_{u,v} = ||i(t)_u - i(t)_v||_2$ that measures how similar or related two data points are is applied on the active and non-active current. The distance similarity function is widely used as a pre-processing step for many of the machine learning approaches such as K-means clustering and K-nearest neighbor algorithms [29,43]. The distance similarity matrix $D_{w,w}$ for points $i(t)_1, i(t)_2, \ldots i(t)_w$ is the a matrix of squared euclidean distances representing the spacing of a set of w points in euclidean space [29] such that

$$D_{w,w} = \begin{bmatrix} 0 & d_{1,2} & \cdots & d_{1,w} \\ d_{2,1} & 0 & \cdots & d_{2,w} \\ \vdots & \vdots & \ddots & \vdots \\ d_{w,1} & d_{w,2} & \cdots & 0 \end{bmatrix} \tag{7}$$

Figure 5 depicts the activation current, its components and their corresponding distance similarity matrix when a CFL and a laptop charger are active.

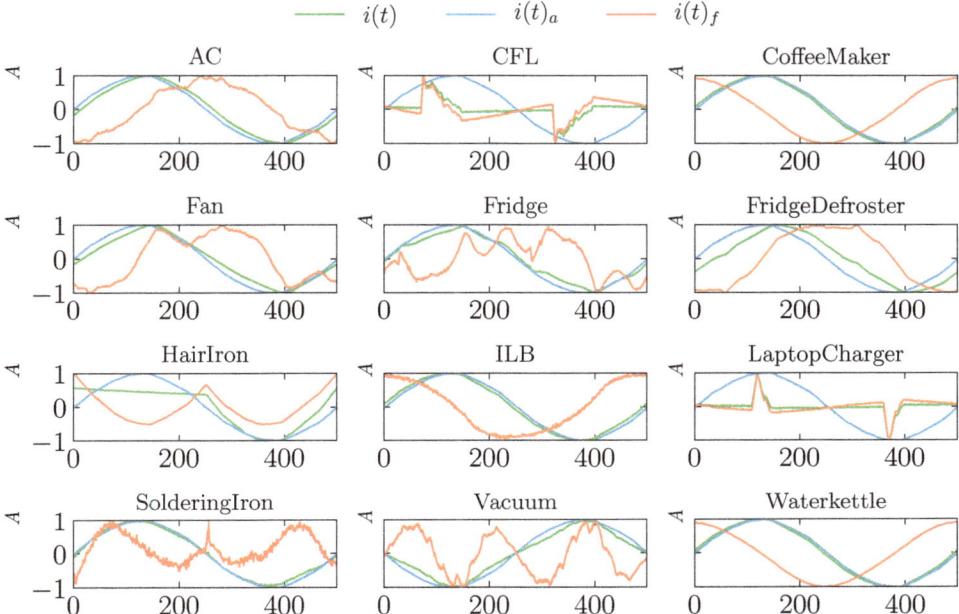

Figure 4. Normalized source current $i(t)$ and their respective active $i(t)_a$ and reactive components $i(t)_f$ after applying Fryze power theory. The current is normalized for visualization purposes.

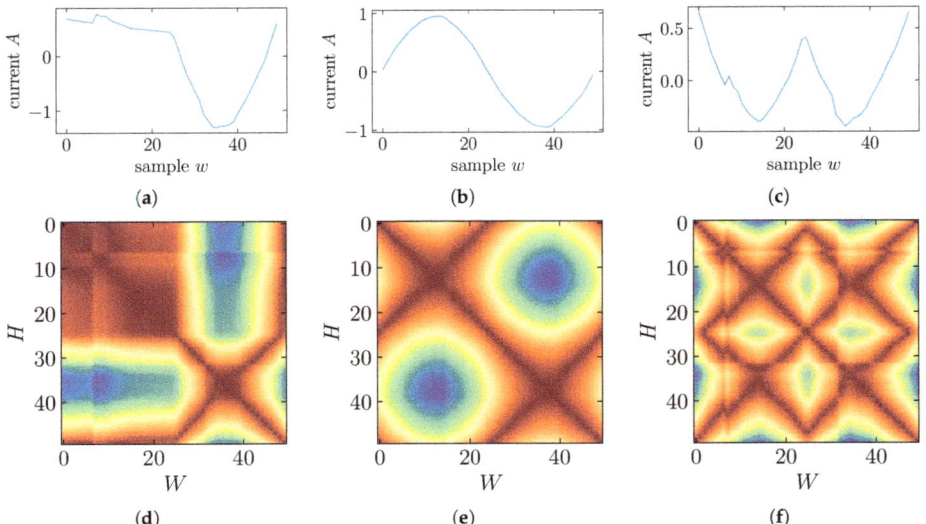

Figure 5. Currents and distance matrix when Compact Fluorescent Lamp (CFL) and laptop charger are active. (**a**) Source current $i(t)$. (**b**) Active current $i_a(t)$. (**c**) Non-active current $i_a(t)$. (**d**) Distance matrix for source current. (**e**) Distance matrix for active current (**f**) Distance matrix for non-active current.

3.3. Multi-Label Modeling

A common approach that extends neural networks to multi-label classification is to use one neural network to learn the joint probability of multiple labels conditioned to the input features representation. The final predicted multi-label is obtained by applying a *sigmoid* activation function [23]. This process requires an additional thresholding mechanism to transform the sigmoid probabilities to multi-label outputs. However, building such a threshold function is very challenging. Therefore a default threshold of 0.5 is often employed [44].

To address this challenge, we propose a CNN multi-label classifier that uses softmax to implicitly capture the relations between multiple labels. As shown in Figure 6, the proposed CNN multi-label classifier consists of a four-stage CNN layer each with 16, 32, 64, and 128 feature maps, 2 × 2 strides.

The first two CNN layers use a 5 × 5 filter size, while the last two layers use a 3 × 3 filter size. The four CNN layers are followed by a batch normalization layer and the ReLU activation function. The last CNN layer is followed by an adaptive average pooling layer with an output size of 1 × 1. The CNN layer takes current-based features as inputs and produces a latent-feature vector z_i.

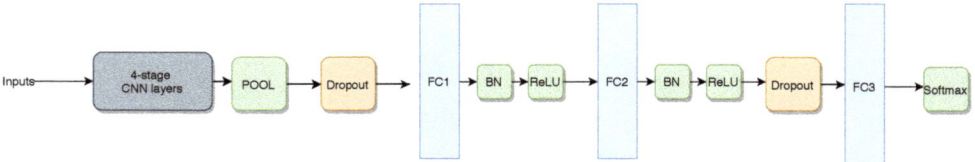

Figure 6. Block diagram of the Convolutional Neural Network (CNN) multi-label classifier. It consists of a CNN encoder to learn feature representation from the input feature, and the output layer to produce the predicted labels.

The output layer consists of three FC layers with a hidden size of 502, 1024, and 2M, respectively. The last layer is followed by an adaptive average pooling layer and three linear layers with a hidden size of 5012, 1024, and M, respectively. M is the maximum number of appliances available. This layer receives the output of the CNN layer to produce an output $\mathbf{O}_s$ of size $(2 \times M)$. The final predicted multi-label states, $\hat{\mathbf{s}}_t$, is obtained by applying the softmax activation function $\hat{\mathbf{s}}_t = \text{softmax}(\mathbf{O}_s)$. Thus, the proposed multi-label classifier learns the joint representation of multiple appliances states conditioned on activation-based input features.

To learn the model parameters, a standard back propagation is used to optimize the cross-entropy between the predicted softmax distribution and the multi-label target of each input feature.

$$\mathcal{L}(\hat{\mathbf{s}}, \mathbf{s}) = -\frac{1}{N} \sum_{t=1}^{N} \sum_{i=1}^{M} s_{ti} \cdot \log \frac{\exp(\hat{s}_{ti})}{\sum_j^2 \exp(\hat{s}_{tj})} \quad (8)$$

The joint cross-entropy loss implicitly captures the relations between labels.

The CNN multi-label classifier is trained for 500 iterations using the Adam optimizer with an initial learning rate of 0.001, betas of (0.9, 0.98), and a batch size of 16. A factor of 0.1 reduces the learning rate once the learning stagnates for 20 consecutive iterations. To avoid over-fitting, early stopping with patience is used where the training model terminates once the validation performance does not change after 50 iterations. The dropout is set to 0.25.

4. Evaluation Methodology

4.1. Dataset

The proposed method is evaluated on the PLAID dataset [30] that contains aggregate voltage and current measurements. The PLAID aggregated measurement data include measurements of more than one concurrently running appliances sampled at 30 kHz. It includes 1478 aggregated activations and deactivations for 12 different appliances measured at one location. Since we are interested in recognizing multiple active appliances, we only select activations and deactivations with at least one running appliance in the background resulting in 1154 samples. The distribution of the number of active appliances and appliances on the extracted 1154 activations is depicted in Figure 7.

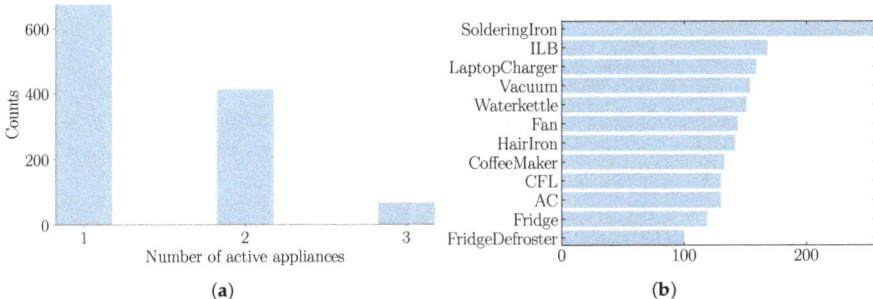

Figure 7. (**a**) Active appliances distributions. (**b**) Appliances distribution on the extracted 1154 activations. The soldering iron has large number of activations because it has two start-up events.

4.2. Performance Metrics

We quantitatively evaluate the classification performance with label-based and instance-based metrics. Label-based metrics work by evaluating each label separately and returning the average (micro or macro) value across all appliances. In contrast, the instance-based metrics evaluate bi-partition over all instances. To this end, two metrics, namely example-based F_1 (F_1-eb) and macro-averaged F_1 (F_1-macro) measures are used. Example-based F_1 (F_1-eb) is an instance-based metric that measures the ratio of correctly predicted labels to the sum of the total true and predicted labels such that:

$$F_1\text{-eb} = \frac{\sum_{i=1}^{M} 2 \cdot t_p}{\sum_{i=1}^{M} y_i + \sum_{i=1}^{M} \hat{y}_i} \qquad (9)$$

The F_1-macro is derived from F_1 score and measures the label-based F_1 score averaged over all labels and is defined as:

$$F_1\text{-macro} = \frac{1}{M} \sum_{i=1}^{M} \frac{2 \cdot t_{pi}}{2 \cdot t_{pi} + f_{pi} + f_{pi}} \qquad (10)$$

where t_p is true positive, f_p is false positive and f_n is false negative. High F_1-ma usually indicate high performance on less frequent labels [45].

4.3. Experiment Description

To benchmark our approach, we adopt multi-label stratified 10-fold cross-validation with random shuffle [46]. This evaluation approach provides stratified randomized folds for multi-label while preserving the label's percentage in each fold. We compare the performance of the proposed CNN model against the commonly used multi-label k-nearest-neighbor (MLkNN) [47] and Binary relevance k-nearest-neighbor (BRkNN) [48] model.

To evaluate the proposed activation current feature, we first establish a baseline in which the V-I binary image is used as an appliance feature. The VI binary image of size $w \times w$, is obtained by meshing the $V-I$ trajectory and assigning a binary value that denotes whether the trajectory traverses it as described in [14]. This experiment setup helps us to answer an essential question on whether the proposed approach is sufficient for recognizing multiple appliances from aggregated measurements. We analyze this by altering the type input features and compare the obtained performance. To gain more insight into the proposed approach, we further examine the individual appliance performance and misclassification errors.

To analyze the computational complexity of the proposed approach, we also assess the training and inference times as a function of the number of data samples. This was achieved by training the MLkNN baseline and CNN-based multi-label classifier while varying the training and testing size.

In each run, the model is trained on p samples of data for 100 iterations and tested on $(1-p)$ samples data where $p \in [0.1, 0.9]$.

Finally, we compare the appliance classification results with related state-of-art methods. However, we should emphasize that due to the difficulty in producing fair comparisons as a result of different experimental settings (e.g., sampling frequency, measurements, learning strategy, dataset, and the metrics) these comparisons are merely illustrative of the potential of the proposed method.

5. Results and Discussion

5.1. Comparison with Baseline

The results of the comparisons between the baselines and the proposed CNN multi-label learning for the V-I binary image and activation current feature are depicted in Figure 8a.

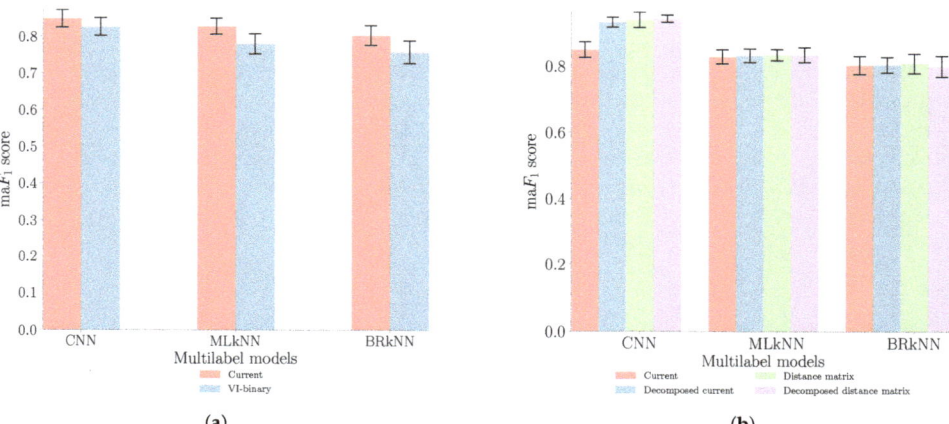

Figure 8. maF_1 score performance comparison between the proposed CNN model and the two baselines for different inputs features: (**a**) Comparison between voltage-current (V-I) binary image and current activation features; (**b**) Comparison between the activation current based features.

From Figure 8a, we see that the proposed CNN multi-label learning performs better than the baselines in both feature types. We also observe that compared to the current activation feature, the V-I binary feature representation yields low maF_1 scores in both the proposed CNN model and the two baseline algorithms. We see a slight increase in maF_1 score (from 0.826 ± 0.024 to 0.849 ± 0.024 for the CNN model and from 0.779 ± 0.028 to 0.827 ± 0.021 for the MLkNN model) when activation current is used as input features. This result suggests that features derived from activation current could be useful in recognizing appliances from total measurements.

We, therefore, analyzed three additional features derived from the activation current, namely decomposed current, current distance similarity matrix, and decomposed distance similarities. The results are presented in Figure 8b.

As it can be observed, the three current-based features significantly improve the classification performance in the CNN model, while achieving nearly the same performance on the two baselines. For the CNN model, the decomposed current feature attains an average 9.4% maF_1 score (from 0.849 ± 0.024 to 0.931 ± 0.015) increase over the activation current feature. This result is in line with the one obtained in [20], which suggested that decomposing the activation current into its active components enhances the uniqueness of the V-I trajectory. We also see about 10 percentage points increase in maF_1 (from 0.849 ± 0.024 to 0.94 ± 0.015) for the decomposed distance similarities. The decomposed current and current distance matrix features achieve comparable performance.

This result indicates that the decomposed current features could help increase the performance of appliance recognition in NILM.

Figure 9 presents the predicted multiple appliances from the CNN based classifier with different feature representation. We see that compared to the activation current in Figure 9a and the V-I image Figure 9c, the proposed Fryze's current-decomposition in Figure 9b,d is capable of detecting all multiple running appliances. This shows that the Fryze current decomposition-based feature alone is sufficient for the identification of multiple running appliances.

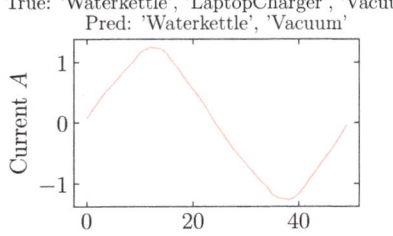
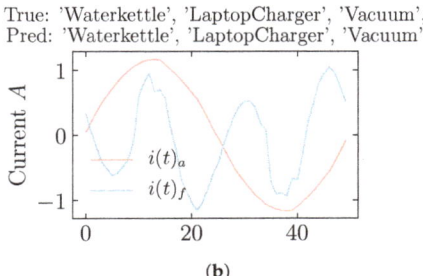

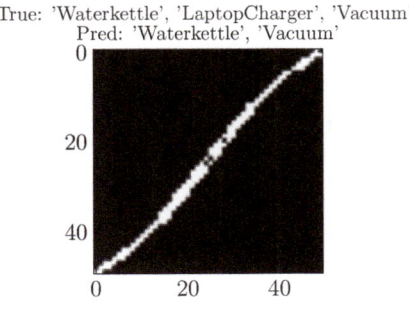
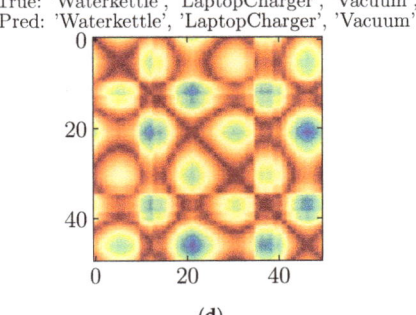

Figure 9. Prediction comparison for different feature representations with the proposed CNN multilabel classifier. (**a**) Action current. (**b**) Decomposed current. (**c**) V-I image. (**d**) Distance matrix.

To gain insights on the performance of individual appliances, we further analyze the per-appliance ebF_1 score for the MLkNN and the proposed CNN multi-label classifier, as depicted in Figure 10. The CNN model with decomposed current distance matrix feature obtains over 90% ebF_1 score for each appliance except for AC, ILB, and LaptopCharger. We also see that the MLkNN baseline with the same decomposed current distance matrix feature obtains over 90% ebF_1 score for only four appliances, namely FridgeDefroster, CoffeeMaker, Vacuum, and CFL. In both cases, we observe low scores for V-I binary features except for FridgeDefroster, CoffeeMaker, and Vacuum, which score above 90% ebF_1.

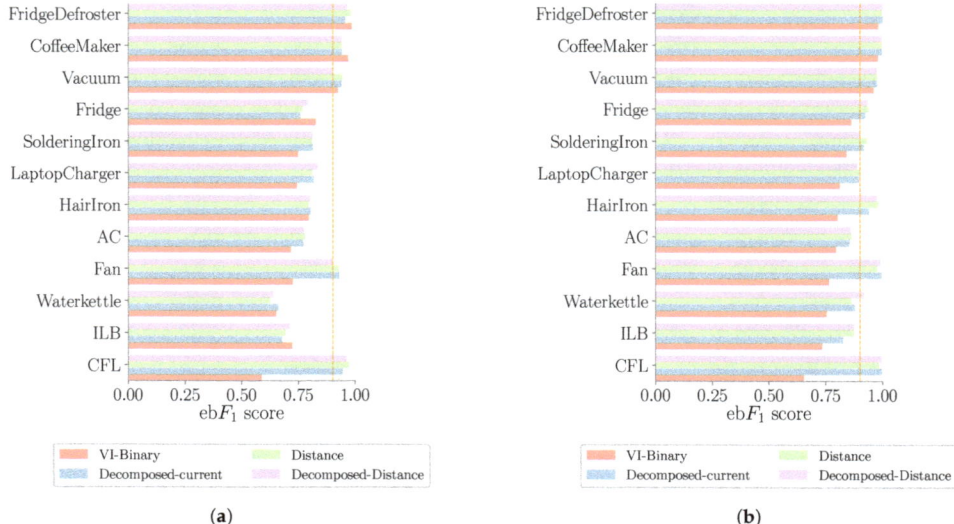

Figure 10. Per-appliance ebF_1 score on PLAID dataset. AC = air conditioning, CFL = compact fluorescent lamp, ILB = incandescent light bulb. (**a**) Multi-label k-nearest-neighbor (MLkNN) (**b**) CNN.

5.2. Error Analysis

We also analyze the miss-classification errors for the proposed CNN model. To this end, we identified three types of errors, namely *zero-error*, *one-to-one*, and *many-to-many errors*.

The *zero-type* mistake happens when a model predicts no appliance is running while there is at least one active appliance. It can be observed from Figure 11 that the number of zero-type mistakes is very low for the three feature types with decomposed-current distance making no such type of error.

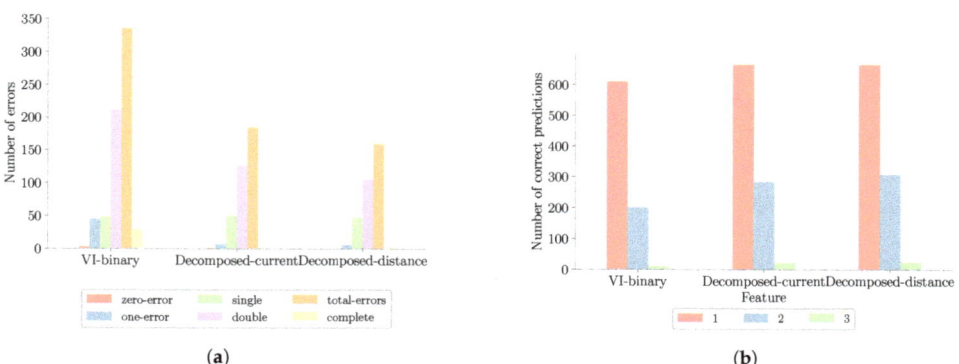

Figure 11. (**a**) Distributions of type errors the model makes. (**b**) Number of correct predictions for single, double and triple activations.

On the other hand, the *one-to-one* is the type of error that the model makes when there is only one active appliance running. We see from Figure 11 that the V-I binary image makes 45 one-type errors while the current based features reduce this to seven for the decomposed current, and six for the decomposed current distance feature. The low error rate when one appliance is running can be attributed to the high number of single activations, over 50%, as presented in Figure 7a. It further shows the effectiveness of the proposed CNN multi-label learning in recognizing individually operating appliances, with over 98% accuracy, as shown in Figure 11b.

The *many-to-many errors* are confusions that a model makes when several appliances are active. Since the PLAID dataset used in our experiment consists of up to three simultaneous active appliances, we further categorized *many-to-many errors* into *single*, *double*, or *complete-error*. A *single* error occurs when a model confuses only one appliance when two or three appliances are active, whereas in *double* fault, the model confuses two appliances when three appliances are active. The *complete-error* is the case when the model produces incorrect predictions for all the active appliances. It can be inferred from Figure 11 that the proposed CNN multi-label model makes a higher number of *double* errors for the three input feature types used. This is likely to be caused by the fewer numbers of samples with more than two appliances running simultaneously at about 5.8%, as depicted in Figure 7a.

5.3. Complexity Analysis

The results for the complexity analysis between the baseline and the proposed CNN multi-label learning are presented in Figure 12. As expected, since the proposed method is an eager learner (i.e., a model is created in the training phase), it takes significantly longer to train than the MLKNN baseline (Figure 12a). In contrast, the proposed method has a much shorter inference time since the model was already created in the training phase. Furthermore, from Figure 12b it can be observed that the proposed method achieves better performance even with less training data, which is positive if one considers that labeled data is scarce and often hard to acquire.

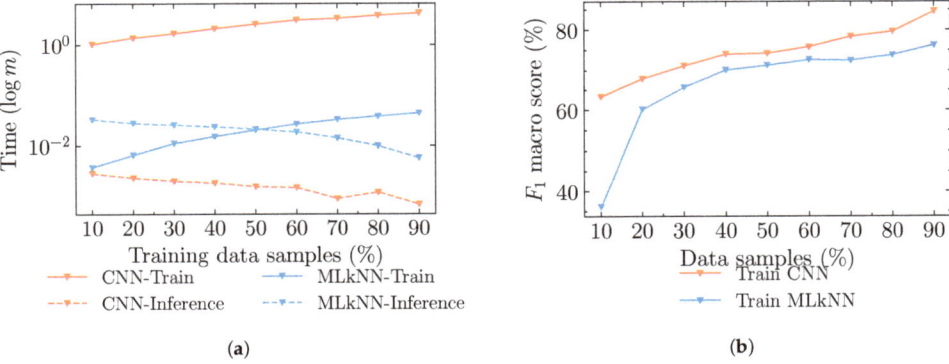

Figure 12. (**a**) Distributions of type errors the model makes. (**b**) Number of correct predictions for single, double and triple activations.

5.4. Comparison with State-of-the-Art Methods

Table 1 provides an overview of the results obtained in other related works. As it can be observed, there are many differences that make a fair and objective comparison impossible to achieve. For instance, while our approach uses current waveforms extracted from high-frequency power measurements, the results presented in [26] were obtained on low-frequency data, and on a different dataset. Yet, they also used the MLkNN multi-label classifier, achieving considerably lower results. Moreover, our results cannot be directly compared with the ones presented in [49], as these were obtained from a private dataset, besides the very different experimental settings including a different performance metric. In [23,37], the F_1 macro score for TCNN and FCNN DNN based multi-label classifiers are given; however, they use UK-DALE dataset making the comparison irrelevant.

An almost direct comparison is only possible between our method and the results from [19] who have used the same dataset and performance metric. Still, it should be stressed that the performance evaluation method was different since their work targets single-label classification. Yet, the results obtained with our approach are superior by six percentage points.

In short, for a fair comparison, we would have to re-implement all these approaches, which unfortunately is not always possible. Nevertheless, to make this task easier for other authors, we open-sourced the code necessary to replicate our experiments.

Table 1. Results comparison.

Approach	Learning Strategy	Model	Dataset	Sampling Frequency	Results (Metric)
De Baets et al. [19]	single	CNN	PLAID [30]	High	88.0% (F_1 macro)
Faustine et al. [13]	single	CNN	PLAID [30]	High	97.77% (F_1 macro)
Tabatabaei et al. [26]	multi	MLkNN	REDD-House1 [50]	Low	61.90% (F_1 macro)
Lai et al. [49]	multi	SVM/GMM	Private	-	90.72% (Accuracy)
Yang et al. [23]	multi	FCNN	UK-DALE-house 1 [51]	Low	93.8% (F_1 score)
Nalmpantis and Vrakas [37]	multi	TCNN	UK-DALE-house 1 [51]	Low	92.5% (F_1 score)
Proposed approach	**multi**	**CNN**	**PLAID [30]**	**High**	**94.0% (F_1 score)**

6. Conclusions and Future Work Directions

In this work, we have approached appliance recognition in NILM as a multi-label learning problem which links multiple appliances to an observed aggregate current signal. We first show that features derived from activation current alone could be useful in recognizing devices from total measurements. We later apply Fryze's power theory, which decomposes the current waveform into active and non-active components. The decomposed current signal was then transformed into an image-like representation using the Euclidean-distance-similarity function and fed into the CNN multi-label classifier. Experimental evaluation on the PLAID aggregated dataset shows that the proposed approach is very successful at recognizing multiple appliances from aggregated measurements with an overall 0.94 F-score.

We further show the effectiveness of the proposed CNN multi-label learning in recognizing a single running appliance with over 98% accuracy. We will investigate the use of Fryze's current decomposition and distance similarity matrix for single-label appliance recognition in future iterations of this work. Finally, we presented a detailed error analysis and identified three types of errors: *zero-error*, *one-to-one*, and *many-to-many* errors.

At this point, we acknowledge that the performance of the proposed approach is not yet satisfactory in detecting triple running appliances. A possible explanation for this issue is the small number of training samples with more than two running appliances. In the future, we would like to test our approaches against datasets with more training data. However, this may imply the development of such a dataset since the currently available ones are still scarce concerning high-frequency measurements [52,53].

Finally, it should be mentioned that the proposed method assumes that the appliance state transition (power event) is known in advance. However, in practice, this information has to be provided by an event detection algorithm (e.g., [54–56]). Therefore, future work should investigate how to integrate the proposed approach in the event-based NILM pipeline. Specifically, we plan to explore the use of the proposed Fryze current decomposition for event detection in multi-label appliance recognition.

Author Contributions: Conceptualization, A.F; data curation, A.F.; formal analysis, A.F. and L.P.; methodology, A.F. and L.P.; resources, A.F.; software, A.F.; supervision, L.P.; validation, L.P.; writing—original draft, A.F.; writing—review and editing, A.F. and L.P. All authors have read and agreed to the published version of the manuscript.

Funding: Lucas Pereira has received funding from the Portuguese Foundation for Science and Technology (FCT) under grants CEECIND/01179/2017 and UIDB/50009/2020.

Acknowledgments: The authors thank Christoph Klemenjak and Shridhar Kulkarni for providing insightful comments and advice towards the completion of this work.

Conflicts of Interest: The authors declare no conflict of interest.

References

1. Monacchi, A.; Versolatto, F.; Herold, M.; Egarter, D.; Tonello, A.M.; Elmenreich, W. An Open Solution to Provide Personalized Feedback for Building Energy Management. *CoRR* **2015**, *abs/1505.0*, 1–28. [CrossRef]
2. Batra, N.; Singh, A.; Whitehouse, K. If You Measure It, Can You Improve It? Exploring The Value of Energy Disaggregation. In Proceedings of the 2nd ACM International Conference on Embedded Systems for Energy-Efficient Built Environments—BuildSys '15, Seoul, Korea, 4–5 November 2015; pp. 191–200. [CrossRef]
3. Froehlich, J.; Larson, E.; Gupta, S.; Cohn, G.; Reynolds, M.; Patel, S. Disaggregated end-use energy sensing for the smart grid. *IEEE Pervasive Comput.* **2011**, *10*, 28–39. [CrossRef]
4. Reyes Lua, A. Location-aware Energy Disaggregation in Smart Homes. Master's Thesis, Delft University of Technology, Delft, Netherlands, 2015.
5. Klemenjak, C.; Jost, S.; Elmenreich, W. Yomopie: A user-oriented energy monitor to enhance energy efficiency in households. In Proceedings of the 2018 IEEE Conference on Technologies for Sustainability (SusTech), Long Beach, CA, USA, 11–13 November 2018; IEEE: Piscataway, NJ, USA, 2018; pp. 1–7.
6. Hart, G. Nonintrusive appliance load monitoring. *Proc. IEEE* **1992**, *80*, 1870–1891. [CrossRef]
7. Zeifman, M.; Roth, K. Nonintrusive appliance load monitoring: Review and outlook. *IEEE Trans. Consum. Electron.* **2011**, *57*, 76–84. [CrossRef]
8. Zoha, A.; Gluhak, A.; Imran, M.A.; Rajasegarar, S. Non-intrusive Load Monitoring approaches for disaggregated energy sensing: A survey. *Sensors* **2012**, *12*, 16838–16866. [CrossRef]
9. Klemenjak, C.; Elmenreich, W. On the applicability of correlation filters for appliance detection in smart meter readings. In Proceedings of the 2017 IEEE International Conference on Smart Grid Communications (SmartGridComm), Dresden, Germany, 23–27 October 2017; IEEE: Piscataway, NJ, USA 2017; pp. 171–176.
10. Nalmpantis, C.; Vrakas, D. Machine learning approaches for non-intrusive load monitoring: From qualitative to quantitative comparison. *Artif. Intell. Rev.* **2018**, *52*, 217–243. [CrossRef]
11. Faustine, A.; Pereira, L. Improved Appliance Classification in Non-Intrusive Load Monitoring Using Weighted Recurrence Graph and Convolutional Neural Networks 2019. *Energies* **2020**, *13*, 3374. [CrossRef]
12. Gomes, E.; Pereira, L. PB-NILM: Pinball Guided Deep Non-Intrusive Load Monitoring. *IEEE Access* **2020**, *8*, 48386–48398. [CrossRef]
13. Faustine, A, Pereira, L.; Klemenjak, C. Adaptive Weighted Recurrence Graphs for Appliance Recognition in Non-Intrusive Load Monitoring. *IEEE Trans. Smart Grid* **2020**, 1. [CrossRef]
14. De Baets, L.; Ruyssinck, J.; Develder, C.; Dhaene, T.; Deschrijver, D. Appliance classification using VI trajectories and convolutional neural networks. *Energy Build.* **2018**, *158*, 32–36. [CrossRef]
15. Lam, H.Y.; Fung, G.S.K.; Lee, W.K. A Novel Method to Construct Taxonomy Electrical Appliances Based on Load Signaturesof. *IEEE Trans. Consum. Electron.* **2007**, *53*, 653–660. [CrossRef]
16. Wang, A.L.; Chen, B.X.; Wang, C.G.; Hua, D. Non-intrusive load monitoring algorithm based on features of V–I trajectory. *Electr. Power Syst. Res.* **2018**, *157*, 134–144. [CrossRef]
17. Du, L.; He, D.; Harley, R.G.; Habetler, T.G. Electric Load Classification by Binary Voltage–Current Trajectory Mapping. *IEEE Trans. Smart Grid* **2016**, *7*, 358–365. [CrossRef]
18. Gao, J.; Kara, E.C.; Giri, S.; Bergés, M. A feasibility study of automated plug-load identification from high-frequency measurements. In Proceedings of the 2015 IEEE Global Conference on Signal and Information Processing (GlobalSIP), Orlando, FL, USA, 14–16 December 2015; pp. 220–224. [CrossRef]
19. De Baets, L.; Dhaene, T.; Deschrijver, D.; Develder, C.; Berges, M. VI-Based Appliance Classification Using Aggregated Power Consumption Data. In Proceedings of the 2018 IEEE International Conference on Smart Computing (SMARTCOMP), Sicily, Italy, 18–20 June 2018; pp. 179–186. [CrossRef]
20. Liu, Y.; Wang, X.; You, W. Non-Intrusive Load Monitoring by Voltage–Current Trajectory Enabled Transfer Learning. *IEEE Trans. Smart Grid* **2019**, *10*, 5609–5619. [CrossRef]

21. Baptista, D.; Mostafa, S.; Pereira, L.; Sousa, L.; Morgado, D.F. Implementation Strategy of Convolution Neural Networks on Field Programmable Gate Arrays for Appliance Classification Using the Voltage and Current (V-I) Trajectory. *Energies* **2018**, *11*, 2460. [CrossRef]
22. Yeh, C.K.; Wu, W.C.; Ko, W.J.; Wang, Y.C.F. Learning Deep Latent Spaces for Multi-Label Classification. In Proceedings of the Thirty-First AAAI Conference on Artificial Intelligence (AAAI-17), San Francisco, CA, USA, 4–9 February 2017; pp. 2838–2844.
23. Yang, Y.; Zhong, J.; Li, W.; Gulliver, T.A.; Li, S. Semi-Supervised Multi-Label Deep Learning based Non-intrusive Load Monitoring in Smart Grids. *IEEE Trans. Ind. Inform.* **2019**, *10*, 1. [CrossRef]
24. Massidda, L.; Marrocu, M.; Manca, S. Non-Intrusive Load Disaggregation by Convolutional Neural Network and Multilabel Classification. *Appl. Sci.* **2020**, *10*, 1454. [CrossRef]
25. Basu, K.; Debusschere, V.; Bacha, S.; Maulik, U.; Bondyopadhyay, S. Nonintrusive Load Monitoring: A Temporal Multilabel Classification Approach. *IEEE Trans. Ind. Inform.* **2015**, *11*, 262–270. [CrossRef]
26. Tabatabaei, S.M.; Dick, S.; Xu, W. Toward Non-Intrusive Load Monitoring via Multi-Label Classification. *IEEE Trans. Smart Grid* **2016**, *8*, 26–40. [CrossRef]
27. Buddhahai, B.; Wongseree, W.; Rakkwamsuk, P. A non-intrusive load monitoring system using multi-label classification approach. *Sustain. Cities Soc.* **2018**, *39*, 621–630. [CrossRef]
28. Teshome, D.F.; Huang, T.D.; Lian, K. Distinctive Load Feature Extraction Based on Fryze's Time-Domain Power Theory. *IEEE Power Energy Technol. Syst. J.* **2016**, *3*, 60–70. [CrossRef]
29. Dokmanic, I.; Parhizkar, R.; Ranieri, J.; Vetterli, M. Euclidean Distance Matrices: Essential theory, algorithms, and applications. *IEEE Signal Process. Mag.* **2015**, *32*, 12–30. [CrossRef]
30. Medico, R.; De Baets, L.; Gao, J.; Giri, S.; Kara, E.; Dhaene, T.; Develder, C.; Bergés, M.; Deschrijver, D. A voltage and current measurement dataset for plug load appliance identification in households. *Sci. Data* **2020**, *7*, 49. [CrossRef] [PubMed]
31. Verma, S.; Singh, S.; Majumdar, A. Multi Label Restricted Boltzmann Machine for Non-intrusive Load Monitoring. In Proceedings of the ICASSP 2019—2019 IEEE International Conference on Acoustics, Speech and Signal Processing (ICASSP), Brighton, UK, 12–17 May 2019; IEEE: Piscataway, NJ, USA, 2019; pp. 8345–8349.
32. Buddhahai, B.; Wongseree, W.; Rakkwamsuk, P. An Energy Prediction Approach for a Nonintrusive Load Monitoring in Home Appliances. *IEEE Trans. Consumer Electron.* **2019**, *66*, 96–105. [CrossRef]
33. Singh, S.; Majumdar, A. Non-intrusive Load Monitoring via Multi-label Sparse Representation based Classification. *IEEE Trans. Smart Grid* **2019**, *11*, 1799–1801. [CrossRef]
34. Kongsorot, Y.; Horata, P. Multi-label classification with extreme learning machine. In Proceedings of the 2014 6th International Conference on Knowledge and Smart Technology (KST), Chonburi, Thailand, 30–31 January 2014; IEEE: Piscataway, NJ, USA, 2014; pp. 81–86.
35. Li, D.; Dick, S. Residential household non-intrusive load monitoring via graph-based multi-label semi-supervised learning. *IEEE Trans. Smart Grid* **2018**, *10*, 4615–4627. [CrossRef]
36. Singhal, V.; Maggu, J.; Majumdar, A. Simultaneous detection of multiple appliances from smart-meter measurements via multi-label consistent deep dictionary learning and deep transform learning. *IEEE Trans. Smart Grid* **2018**, *10*, 2269–2987. [CrossRef]
37. Nalmpantis, C.; Vrakas, D. On time series representations for multi-label NILM. *Neural Comput. Appl.* **2020**. [CrossRef]
38. Li, L.; Zhao, Y.; Jiang, D.; Zhang, Y.; Wang, F.; Gonzalez, I.; Valentin, E.; Sahli, H. Hybrid Deep Neural Network–Hidden Markov Model (DNN-HMM) Based Speech Emotion Recognition. In Proceedings of the 2013 Humaine Association Conference on Affective Computing and Intelligent Interaction, Geneva, Switzerland, 2–5 September 2013; pp. 312–317. [CrossRef]
39. Hassan, T.; Javed, F.; Arshad, N. An Empirical Investigation of V-I Trajectory Based Load Signatures for Non-Intrusive Load Monitoring. *IEEE Trans. Smart Grid* **2014**, *5*, 870–878. [CrossRef]
40. Klemenjak, C.; Makonin, S.; Elmenreich, W. Towards Comparability in Non-Intrusive Load Monitoring: On Data and Performance Evaluation. In Proceedings of the 2020 IEEE Power & Energy Society Innovative Smart Grid Technologies Conference (ISGT), The Hague, The Netherlands, 25–28 October 2020; IEEE: Piscataway, NJ, USA, 2020.

41. Staudt, V. Fryze-Buchholz-Depenbrock: A time-domain power theory. In Proceedings of the 2008 International School on Nonsinusoidal Currents and Compensation, Lagow, Poland, 10–13 June 2008. [CrossRef]
42. Keogh, E.J.; Pazzani, M.J. Scaling Up Dynamic Time Warping for Datamining Applications. In *Proceedings of the 6th ACM SIGKDD International Conference on Knowledge Discovery and Data Mining*; ACM: New York, NY, USA, 2000; KDD '00, pp. 285–289. [CrossRef]
43. Ontañón, S. An overview of distance and similarity functions for structured data. *Artif. Intell. Rev.* **2020**. [CrossRef]
44. Mahajan, D.; Girshick, R.; Ramanathan, V.; Paluri, M.; Li, Y.; Bharambe, A.; Maaten, L.v.d. Exploring the Limits of Weakly Supervised Pretraining. In *Computer Vision—ECCV 2018*; Springer: Berlin/Heidelberg, Germany, 2018. [CrossRef]
45. Lanchantin, J.; Sekhon, A.; Qi, Y. Neural Message Passing for Multi-Label Classification. In *Machine Learning and Knowledge Discovery in Databases*; Brefeld, U., Fromont, E., Hotho, A., Knobbe, A., Maathuis, M., Robardet, C., Eds.; ECML PKDD 2019, Lecture Notes in Computer Science; Springer: Cham, Switzerland; Volume 11907. [CrossRef]
46. Sechidis, K.; Tsoumakas, G.; Vlahavas, I. On the Stratification of Multi-label Data. In *Machine Learning and Knowledge Discovery in Databases*; Gunopulos, D., Hofmann, T., Malerba, D., Vazirgiannis, M., Eds.; Springer: Berlin/Heidelberg, Germany, 2011; pp. 145–158.
47. Zhang, M.L.; Zhou, Z.H. ML-KNN: A lazy learning approach to multi-label learning. *Pattern Recognit.* **2007**, *40*, 2038–2048. [CrossRef]
48. Spyromitros, E.; Tsoumakas, G.; Vlahavas, I. An Empirical Study of Lazy Multilabel Classification Algorithms. In *Artificial Intelligence: Theories, Models and Applications*; Springer: Berlin/Heidelberg, Germany; pp. 401–406. [CrossRef]
49. Lai, Y.X.; Lai, C.F.; Huang, Y.M.; Chao, H.C. Multi-appliance recognition system with hybrid SVM/GMM classifier in ubiquitous smart home. *Inf. Sci.* **2013**, *230*, 39–55. [CrossRef]
50. Kolter, J.Z.; Johnson, M.J. REDD : A Public Data Set for Energy Disaggregation Research. In Proceedings of the 1st KDD Workshop on Data Mining Applications in Sustainability (SustKDD'11), San Diego, CA, USA, 21 August 2011; pp. 1–6.
51. Kelly, J.; Knottenbelt, W. The UK-DALE dataset, domestic appliance-level electricity demand and whole-house demand from five UK homes. *Sci. Data* **2015**, *2*, 150007. [CrossRef]
52. Pereira, L.; Nunes, N. Performance evaluation in non-intrusive load monitoring: Datasets, metrics, and tools—A review. *Wiley Interdiscip. Rev. Data Min. Knowl. Discov.* **2018**, *8*, e1265. [CrossRef]
53. Klemenjak, C.; Reinhardt, A.; Pereira, L.; Makonin, S.; Bergés, M.; Elmenreich, W. Electricity Consumption Data Sets: Pitfalls and Opportunities. In Proceedings of the 6th ACM International Conference on Systems for Energy-Efficient Buildings, Cities, and Transportation,BuildSys '19, New York, NY, USA, 13–14 November 2019; ACM: New York, NY, USA, 2019. [CrossRef]
54. Pereira, L. Developing and evaluating a probabilistic event detector for non-intrusive load monitoring. In *2017 Sustainable Internet and ICT for Sustainability (SustainIT)*; IEEE: Funchal, Portugal, 2017; pp. 1–10. [CrossRef]
55. De Baets, L.; Ruyssinck, J.; Develder, C.; Dhaene, T.; Deschrijver, D. On the Bayesian optimization and robustness of event detection methods in NILM. *Energy Build.* **2017**, *145*, 57–66. [CrossRef]
56. Houidi, S.; Auger, F.; Sethom, H.B.A.; Fourer, D.; Miègeville, L. Multivariate event detection methods for non-intrusive load monitoring in smart homes and residential buildings. *Energy Build.* **2020**, *208*, 109624. [CrossRef]

© 2020 by the authors. Licensee MDPI, Basel, Switzerland. This article is an open access article distributed under the terms and conditions of the Creative Commons Attribution (CC BY) license (http://creativecommons.org/licenses/by/4.0/).

Article

Detection of Electricity Theft Behavior Based on Improved Synthetic Minority Oversampling Technique and Random Forest Classifier

Zhengwei Qu [1,*], Hongwen Li [1], Yunjing Wang [1], Jiaxi Zhang [1], Ahmed Abu-Siada [2] and Yunxiao Yao [3]

1. Key Laboratory of Power Electronics for Energy Conservation and Drive Control, Yanshan University, Qinhuangdao 066004, China; lhw@stumail.ysu.edu.cn (H.L.); wyj@ysu.edu.cn (Y.W.); zjx@stumail.ysu.edu.cn (J.Z.)
2. School of Electrical Engineering Computing and Mathematical Sciences, Curtin University, Perth WA 6102, Australia; A.AbuSiada@curtin.edu.au
3. State Grid Hubei DC Operation and Maintenance Company, Yichang 443008, China; yunx_yao@163.com
* Correspondence: zhengwei.qu@ysu.edu.cn; Tel.: +86-0335-807-4883

Received: 13 March 2020; Accepted: 15 April 2020; Published: 19 April 2020

Abstract: Effective detection of electricity theft is essential to maintain power system reliability. With the development of smart grids, traditional electricity theft detection technologies have become ineffective to deal with the increasingly complex data on the users' side. To improve the auditing efficiency of grid enterprises, a new electricity theft detection method based on improved synthetic minority oversampling technique (SMOTE) and improve random forest (RF) method is proposed in this paper. The data of normal and electricity theft users were classified as positive data (PD) and negative data (ND), respectively. In practice, the number of ND was far less than PD, which made the dataset composed of these two types of data become unbalanced. An improved SOMTE based on K-means clustering algorithm (K-SMOTE) was firstly presented to balance the dataset. The cluster center of ND was determined by K-means method. Then, the ND were interpolated by SMOTE on the basis of the cluster center to balance the entire data. Finally, the RF classifier was trained with the balanced dataset, and the optimal number of decision trees in RF was decided according to the convergence of out-of-bag data error (OOB error). Electricity theft behaviors on the user side were detected by the trained RF classifier.

Keywords: smart grid; nontechnical losses; electricity theft detection; synthetic minority oversampling technique; K-means cluster; random forest

1. Introduction

Power losses are usually divided into technical losses (TLs) and nontechnical losses (NTLs) [1]. NTLs refer to the power loss during the transformation, transmission, and distribution process and are mainly caused by electricity theft on the user side [2]. In most countries, electricity theft losses (ETLs) account for the predominant part of the overall electricity losses [3], and are mainly taking place in the medium- and low-voltage power grids. ETLs can cause serious problems, such as loss of revenue of power suppliers, reducing the stability, security, and reliability of power grids and increasing unnecessary resources consumption. In India, ETLs were valued at about US$4.5 billion [4], which is still rising year by year. ETLs are reported to reach up to 40% of the total electricity losses in countries such as Brazil, Malaysia, and Lebanon [5]. ETLs of some provinces in China reached about 200 million kWh, with an overall cost of 100 million yuan. As reported in [6], the losses due to electricity theft reached about 100 million Canadian dollars every year with a power loss that could have supplied 77,000 families for one year. The annual income loss caused by electricity theft in the United States accounted for 0.5% to 3.5% of the total income [7,8]. Therefore, the research on advancing electricity theft detection techniques has become essential due to its significance for energy saving and consumption reductions [9].

In the past an electric meter packaging, professional electric meters, and bidirectional metering conventional methods were adopted to deal with electricity theft [10,11]. Today, electricity theft detection methods rely on classifying the data collected by smart meter measurement system. Classification of the electricity theft and normal behaviors is conducted through data analysis [12].

The modern methods for electricity theft detection mainly include state-based analysis, game theory, and classification [13].

State-based detection schemes employ specific devices to provide high detection accuracy. A novel hybrid intrusion detection system framework that incorporates power information and sensor placement has been developed in [14] to detect malicious activities such as consumer attacks. In [15], an integrated intrusion detection solution (AMIDS), was presented to identify malicious energy theft attempts in advanced metering infrastructures. AMIDS makes use of different information sources to gather a sufficient amount of evidence about an ongoing attack before marking an activity as a malicious energy theft. In [16], state estimation was used to determine electricity theft users. When there was a difference between the voltage of the state estimation and the voltage of the measured node, the breadth-first search was conducted from the root node of the distribution network, and the magnitude of the difference at the same depth was compared to locate electricity theft users. In [17], in order to detect and localize the occurrence of theft in grid-tied microgrids, A Stochastic Petri Net (SPN) with a low sampling rate was used to first detect the random occurrence of theft and then localize it. The detection was based on determining the accurate line losses through (Singular Value Decomposition) SVD, which led to the recognition of theft in grid-tied MGs. State-based detection schemes will bring additional investment required for monitoring systems, including equipment costs, system implementation costs, software costs, and operation/training costs. In [18], it investigated energy theft detection in microgrids, considering a realistic model for the microgrid's power system and the protection of users' privacy. It proposed two energy theft detection algorithms capable of successfully identifying energy thieves. One algorithm, called centralized state estimation algorithm based on the Kalman filter (SEK), employed a centralized Kalman filter. However, it could not protect users' privacy and did not have very good numerical stability in large systems with high measurement errors. The other one, called privacy-preserving bias estimation algorithm (PPBE), was based on two loosely coupled filters, and could preserve users' privacy by hiding their energy measurements from the system operator, other users, and eavesdroppers. However, state-based detection schemes employ specific devices to provide high detection accuracy, which, however, come with the price of extra investment required for the monitoring system including device cost, system implementation cost, software cost, and operating/training cost.

Another approach for theft detection is based on game theory. Reference [19] formulated the problem of theft detection as a game between an illegitimate user and a distributor. The distributor wants to maximize the probability of theft detection while illegitimate users or thieves want to minimize the likelihood of being caught by changing their Probability Density Functions (PDFs) of electricity usage.

Classification-based methods include expert systems and machine learning. Expert systems are based on computer models trained by human experts to deal with complex problems and draw the same conclusions as experts [20]. The expert system of electricity theft detection based on specific decision rules was initially used. With the rapid development of artificial intelligence technology, machine learning enables computers to learn decision rules from training. Therefore, in recent years, machine learning has become the main research direction of electricity theft detection [21]. In [22], it explored the possibilities that exist in the implementation of Machine-Learning techniques for the detection of nontechnical losses in customers. The analysis was based on the work done in collaboration with an international energy distribution company. It reported on how the success in detecting nontechnical losses can help the company to better control the energy provided to their customers, avoiding a misuse, and, hence, improving the sustainability of the service that the company provides. Reference [23] provides a novel knowledge-embedded sample model and deep semi-supervised learning algorithm to detect NTL by using the data in smart meter. It first analyzed the characteristic of realistic NTL, and designed a knowledge-embedded sample model referring to the principle of electricity measurement. Next, it proposed an autoencoder based semi-supervised learning model.

In [24], fuzzy logic and expert system were combined to integrate human expert knowledge into the decision-making process to identify electricity theft behavior. A grid-based local outliers algorithm was proposed in [25] to achieve unsupervised learning of abnormal behavior of power users. This method mapped variables features into two-dimensional plane by factor analysis (FA) and principal component analysis (PCA). The dimensionality of data and the operation cost of outlier factor algorithm were reduced by grid technology. In [26], electricity theft detection method based on probabilistic neural network was employed to detect two types of illegal consumption.

In [27], clustering analysis was carried out firstly to reduce the number of data to be analyzed, then the suspected users were found through neural network. In [28], the extreme learning machine (ELM) was used to identify the weight between the hidden and output layer, and electricity theft was detected through the measured data of the meter. In [29], a five-joint neural network was trained with power data comprising 20,000 customers and achieved considerable accuracy. SVM-FIS method was proposed in [30], which could reduce the calculation complexity and improve the detection accuracy by combining the fuzzy inference system with the SVM. In [31], a data-based method was proposed to detect sources of electricity theft and other commercial losses. Prototypes of typical consumption behavior were extracted through clustering the data collected from smart meters.

For an unbalanced dataset, intelligent algorithms tend to favor positive data (PD) in the training process and ignore the important information contained in a few negative data (ND), which may reduce the detection accuracy [32]. Therefore, optimizing the unbalance of the dataset plays an important role for improving the efficiency and accuracy of the algorithm. Data-oriented methods mostly rely on existing and validated cases of fraud either for training or validation. However, since frauds are scarce, it is difficult to obtain these samples, unless another Fraud detection methods such as unsupervised detection, or a manual inspection campaign are used [33].

The theory of unbalanced data processing has been widely used in the fields of network fraud identification, network intrusion detection, medical diagnosis, and text classification. However, it is still rarely used in electricity theft detection. Reference [34] introduced consumption pattern-based energy theft detector (CPBETD), a new algorithm for detecting energy theft in advanced metering infrastructure (AMI). CPBETD relies on the predictability of customers' normal and malicious usage patterns, and it addresses the problem of imbalanced data and zero-day attacks by generating a synthetic attack dataset, benefiting from the fact that theft patterns are predictable. In [35], a methodology was proposed to improve the performance and evaluation of supervised classification algorithms in the context of NTL detection with imbalanced data. The main contributions of our work lie in two aspects: (1) The strategies considered to counteract the effects of imbalanced classes, and (2) an extensive list of performance metrics detailed and tested in the experiments.

A comprehensive detection method for NTLs of unbalanced power data was proposed in [36], which contained three detection models (Boolean rule, fuzzy logic, and support vector machine). Reference [37] proposed two undersampling methods for the classification of unbalanced data, easy ensemble (EE) algorithm and balance cascade (BC) algorithm. The above two methods exhibited high computation and implementation complexity. In [38], a one-sided selection (OSS) method was proposed for dealing with unbalanced data. In [39], a KNN-near miss method based on the K-nearest neighbor (KNN) undersampling method was proposed. In [40], an oversampling method, called synthetic minority oversampling technique (SMOTE), was adopted, which achieved excellent results in the processing of unbalanced data and effectively solved the problem of excessive random sampling.

However, the algorithm had certain blindness in the selection of neighbors, did not consider the distribution of data when generating new data, and had strong marginality.

Reference [41] reported that, compared with single strong decision tree, weak decision tree had high computational efficiency. In addition, considering the weight sparsity of weak classifier, the recognition rate of the cluster could be further improved [42]. In [43], decision trees were used for NTL detecting and the algorithms were tested with real a database of Endesa Company. In addition, random forest classifier (RF) can save resources and computational time because the multiple decision trees run in parallel. Moreover, each decision tree can achieve random selection of data and attributes without overfitting [44].

In summary, considering the shortcomings of existing electricity theft detection methods and the unbalance of user data, a method for electricity theft behaviors detection was proposed based on improved SMOTE and random forest classifier in this paper. The main contributions of this paper can be listed as follows.

(1) Considering the high unbalance of power user-side dataset and the shortcomings of existing methods, a new K-SMOTE method was proposed to deal with the unbalanced initial datasets. The proposed method can reduce the impact of detection accuracy caused by unbalanced data.
(2) Combined with the unbalanced data, considering the limitation of setting decision tree in RF algorithm, the improved random forest classifier was applied to detect electricity theft behaviors. The efficiency of power theft detection could be greatly improved because multiple decision trees were running in parallel. Then, the improved RF algorithm and K-SMOTE oversampling algorithm were combined to establish the electricity theft detection system, which considered the unbalance of the users' electricity dataset.
(3) The detection method of this paper had higher detection accuracy and better stability compared with the existing methods.

This paper is organized as follows: Section 2 proposes the K-SMOTE. Section 3 describes the proposed detection method for electricity theft behaviors. In Section 4, simulation results are presented to verify the feasibility and superiority of the proposed method. Section 5 summarizes the main conclusion of this study. In addition, the nomenclature table is shown in Appendix A.

2. Proposed Algorithm

The detection of electricity theft behaviors is a binary classification problem which calls for distinguishing of normal and electricity theft users. If the electricity data of the user side are directly used by a classifier, unbalanced data may make the classifier more prone to PD and ignore the important information contained in ND, which may degrade the performance of the classifier substantially.

As shown in Figure 1, the triangle and circle represent two kinds of datasets. Respectively, the solid box represents the actual decision boundary of the two kinds of datasets, while the dotted box represents the possible learning decision boundary of the classification algorithm. The number of triangle data in Figure 1a is less than the circular data, so they represent an unbalanced dataset. From Figure 1a,b that shows the normal dataset, it can be seen that the decision boundary of the classification algorithm may be quite different from the real decision boundary if the dataset is unbalanced.

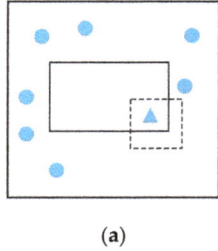 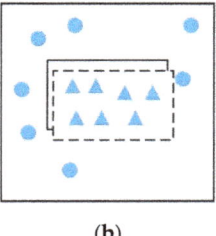

Figure 1. The schematic diagram of the impact of unbalanced data on the classification algorithm. (**a**) Unbalanced data, (**b**) normal data.

In the actual power consumption environment, the number of users stealing electricity is far less than normal users, so the users' electricity dataset is an unbalanced dataset. Unbalanced user data will make the classification algorithm more prone to normal user samples, thereby ignoring the important information contained in a small number of electricity theft user samples, making the decision boundary of the classifier and the actual decision boundary inconsistent, resulting in serious performance degradation of the classifier. Therefore, it was necessary to use an appropriate method to balance the dataset. The traditional SMOTE method was easy to cause data marginalization problems. If there are more PD between some ND, the artificial data generated around these ND will cause the problem of blurred boundaries of PD and ND.

In the field of the detection of electricity theft, the problem about the low detection accuracy due to the unbalance of the power consumption dataset on the user side needs to be solved. Based on a kind of unbalanced data processing method based on K-means clustering and SMOTE, named K-SMOTE, the problem of low electricity theft detection accuracy caused by unbalance electricity data is solved in this paper.

2.1. SMOTE

SMOTE is a classic oversampling algorithm normally used for solving data unbalance problems [45]. Compared to the random oversampling approach, SMOTE performance is better in preventing overfitting [40] by adding ND to achieve balancing distribution with PD. The basic idea is to perform linear interpolation between the existing ND and their neighbors. Specific steps of SMOTE are as follows:

(1) For a sample x_i in minority class samples set X, calculate the Euclidean distance from this sample to all other samples in the set, and get its k nearest neighbor, denoted as y_j ($j = 1, 2, \ldots, k$).
(2) Sampling rate is set according to the data unbalance ratio to determine the sampling magnification. For data x_i, n numbers are randomly selected from their K-nearest neighbors, and new data can be constructed as follows:

$$x_{new} = x_i + rand(0,1) \times (y_j - x_i) \tag{1}$$

where $x_j = 1, 2, \ldots, n$, $rand(0,1)$ represents a random number between 0 and 1.

New data synthesized by SMOTE is shown in Figure 2.

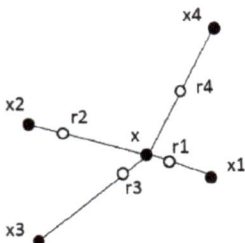

Figure 2. New data synthesized by SMOTE.

In Figure 2 x is the core data currently used to construct the new data: x_1, x_2, x_3, x_4 are the four nearest neighbor data of x; r_1, and r_2, r_3, r_4 are synthetic new data.

2.2. K-Means Clustering Algorithm

K-means clustering is a widely used algorithm that takes the distance between data points and cluster center as the optimization objective [46]. The algorithm would maximize the similarity of elements in clusters while minimizing the similarity between clusters. The K-means selects the desired cluster center, K, minimizes the variance of the whole cluster through continuous iteration and recalculation of the cluster center, and takes the relatively compact and mutually independent clusters as the ultimate goal.

The basic idea of the K-means is to determine the number of initial clusters centers, K, and randomly select K data as the center of the initial cluster in the given dataset D. Then, for each remaining data in D, calculate the Euclidean distance to each cluster center, divide it into the cluster class belonging to the nearest cluster center, and repeat the calculation to generate new cluster centers. The clustering process converges when cluster centers encountered no longer change or the number of iterations reaches the preset threshold limit.

Specific steps are as follows:

(1) Dataset D, denoted as $\{x_1, x_2, x_3, \ldots, x_n\}$, randomly select k initial cluster center as $\mu_1, \mu_2, \ldots, \mu_k \in D^n$.

(2) Calculate the Euclidean distance using Equation (2), that is, calculate the distance $d(x_i, \mu_j)$ between x_i to each cluster center, find the minimum d, and divide x_i into the same cluster as μ_i:

$$d(i, j) = \sqrt{(x_{i1} - x_{j1})^2 + (x_{i2} - x_{j2})^2 + \cdots + (x_{in} - x_{jn})^2} \qquad (2)$$

where $i = \{x_{i1}, x_{i2}, x_{i3}, \ldots, x_{in}\}$ and $j = \{x_{j1}, x_{j2}, x_{j3}, \ldots, x_{jn}\}$ are n-dimensional dataset.

(3) After all data have been calculated, the new clustering center of each class can be recalculated by Equation (3):

$$\mu_{j+1} = \frac{1}{N_j} \sum_{x_i \in S_j} x_i \qquad (3)$$

where N_j represents the number of data in class j.

(4) If the cluster to which each data belongs does not change with the increase of iteration process, go to Step 2. Otherwise, go to Step 5.

(5) Output clustering results.

2.3. K-SMOTE

This paper combined the K-means algorithm and SMOTE to balance the electricity data on the user side. The specific steps are as follows:

(1) Let M, O, P, and N represent the unbalanced electricity dataset on the user side, the PD, and the ND. T is the training set in P, S is the majority training set, and T and S constitute the total training set O. K is the number of the initial clusters, u_i is the cluster center, and X_{new} is the corresponding new interpolated data point set.
(2) Determine the number of initial clusters K.
(3) For T, K-means algorithm was used to perform clustering and record cluster centers. T was divided into K clusters, and the cluster center were $\{\mu_1, \mu_2, \mu_3, \ldots, \mu_k\}$.
(4) SMOTE was used in T to achieve data interpolation based on cluster centers $\{\mu_1, \mu_2, \mu_3, \ldots, \mu_k\}$, then the interpolated dataset X_{new} was obtained
(5) T, S, and X_{new} were combined to form new training set O'.

3. Random Forest Classification Based on K-SMOTE

RF, a statistical learning algorithm proposed by Breiman in 2001 [47], is essentially a combinatorial classifier containing multiple decision trees [48]. It mainly uses bagging method to generate bootstrap training datasets and classification and regression tree (CART) to generate pruning-free decision trees. As a new machine learning classification and prediction algorithm, random forest features the following advantages.

(1) Compared with existing classification algorithms, its average accuracy is at a high level [49].
(2) It can process input data with high dimensional characteristics without dimensionality reduction [50].
(3) An unbiased estimate of the internal generation error can be obtained during the generation process.
(4) It is robust to default value problems.
(5) Each decision tree in the random forest operates independently, realizing parallel operation of multiple decision trees and saving resources and computational time.
(6) Randomness is reflected in the random selection of data and attributes, even if each tree is not pruned, there will be no overfitting.

The electricity data of the grid user side includes various types, such as voltage, current, power consumption, user classification, etc. So the electricity theft users need to be detected quickly and accurately, so as to promptly notify the power department or relevant stakeholders to take timely and proper action.

On the other hand, the RF classifier has poor processing ability for unbalanced datasets, so in this paper it was combined with the K-SMOTE to detect electric power theft.

3.1. Decision Tree

Random forest is a single classifier composed of several decision trees. Decision trees can be regarded as a tree model including three kinds of nodes: Root, intermediate, and leaf nodes. Each node represents the attribute of the object, the bifurcation path from each node represents a possible attribute value, and each leaf node corresponds to the value of the object represented by the path from the root to the leaf node. The path, which starts from the root to the leaf node, represents a rule, and the whole tree represents a set of rules determined by the training dataset. The decision tree has only a single output, which starts from the root node, and only the unique leaf nodes can be reached. In other words, the rule is unique essentially. The classification idea of decision tree is a data mining process which is achieved by analyzing data with a series of generated rules.

Concept learning system (CLS), iterative dichotomiser 3 (ID3), classification 4.5 (C4.5), CART and other node-splitting algorithms can be used to generate the decision trees [51]. This paper selected the CART node-splitting algorithm because it can handle both continuous variables and discrete variables.

The principle of CART node-splitting algorithm is as follows.

Information entropy (IE) is the most commonly used indicator to measure the purity of a sample set. Assume that the proportion of the k-th samples in the set D is p_k ($k = 1, 2, \ldots, r$), then the information entropy of D ($Ent(D)$) is defined as:

$$Ent(D) = -\sum_{k=1}^{r} p_k \log_2 p_k. \quad (4)$$

The smaller the value of $Ent(D)$, the higher is the purity of D.

CART decision tree uses the Gini-index to select the partitioning attributes. Using the same sign as in Equation (9), the purity of the dataset D can be measured using the $Gini$ value, calculated as below:

$$Gini(D) = \sum_{k=1}^{r}\sum_{k' \neq k} p_k = 1 - \sum_{k=1}^{r} p_k^2. \quad (5)$$

Intuitively, $Gini(D)$ reflects the probability that two samples are randomly selected from the dataset D, and their class labels are inconsistent. Therefore, the smaller the $Gini(D)$, the higher is the purity of the dataset D.

Assume that the discrete attribute a has V possible values $\{a^1, a^1, \ldots, a^V\}$. If property a is used to partition the dataset D, there will be V branch nodes, in which the v node contains all the data with a^v value on property a in D, and is denoted as D^v. We can calculate the IE of D^v according to Equation (4). Considering that the number of samples contained in different branch nodes is different, then give each branch node a weight $|D^v|/|D|$, that is, the more samples of branch nodes, the greater the influence of branch nodes. Then, the $Gini$-index of the attribute a is defined as:

$$Gini(D,a) = \sum_{v=1}^{V} \frac{|D^v|}{|D|} Gini(D^v) \quad (6)$$

In the candidate attribute set A, select the attribute that minimizes the $Gini$-index after division as the optimal division attribute, and define the optimization attribute as a^*; then:

$$a_* = \min_{a \in A} Gini\_index(D,a) \quad (7)$$

3.2. Discretization of Continuous Variable

The continuous attribute in the decision tree needs to be discretized. The dichotomy method is used for node splitting of decision trees. The main idea of the method is to find the maximum and minimum values of a continuous variable, and set multiple equal breakpoints between them. These equal breakpoints divide the dataset into two small sets and calculate the information gain rate generated by each breakpoint. In CART decision tree, the steps of discretization of continuous variables are as follows.

(1) Sort the values of continuous variables to find the maximum (MAX) and minimum (MIN).
(2) If there are N values for continuous variables and each value is a breakpoint, the interval (MIN, MAX) is divided into N-1 intervals.
(3) For each breakpoint A_i ($i = 1, 2, \ldots, N$), $Gini$-index is calculated with A and B as intervals.
(4) Select a breakpoint A_i with the largest $Gini$-index coefficient as the best split point of the continuous attribute.

3.3. Random Forest

3.3.1. Bootstrap Random Sampling

Bootstrap random sampling algorithm is used to obtain different training datasets for training base classifiers.

The mathematical model of bootstrap is as follows: Assuming that there are n different data $\{x_1, x_2, x_3, \ldots, x_n\}$ in the dataset D, if any data is extracted from D and put back for n times to form a new set $D*$, then the probability that $D*$ does not contain the x_i ($i = 1, 2, \ldots, n$) is $(1-1/n)^n$. When $n \to \infty$, it can be launched:

$$\lim_{n \to \infty} \left(1 - \frac{1}{n}\right)^n = e^{-1} \approx 0.368. \tag{8}$$

Equation (8) indicates that approximately 36.8% of the original data are not extracted in each sampling. This part of the data are called out-of-bag (OOB) data.

3.3.2. OOB Error Estimate

OOB data are not fitted to the training set. However, OOB data can be used to test the generalization capabilities of the model. It has been proven that the error calculated by OOB, called OBB error, is an unbiased estimate of the true error of the random forest [52]. Therefore, OOB error can be used to evaluate the accuracy of the random forest algorithm.

The performance of the generated random forest can be tested with OOB data. The principle of OOB is shown in the Table 1, in the first column, where x_i represents the input sample and y_i represents the classification label corresponding to x_i. In the first row, T_i represents the decision tree constructed by RF. Yes "Y" indicates that the current sample participates in the classification of the corresponding decision tree, and No "N" indicates that the current sample does not participate in the classification of the corresponding decision tree. Therefore, it can be seen from Table 1 that (x_1, y_1) was not used for the construction of T_1, T_2, and T_3, so (x_1, y_1) was the OOB data of the decision trees T_1, T_2, and T_3. After RF model is trained, its performance can be tested by OOB dataset, and the test result is the OOB error. In addition, there is also a relationship between the number of decision trees and the OOB error; therefore, for a certain dataset, this relationship can be used to solve the optimal number of decision trees in RF.

Table 1. The schematic of OOB.

Data \ Decision Tree	T_1	T_2	T_3	...	T_n
(X_1, Y_1)	N	N	N	...	Y
(X_2, Y_2)	N	N	Y	...	Y
(X_3, Y_3)	N	N	Y	...	N
...	...	...	...	...	...
(X_4, Y_4)	Y	Y	N	...	N

Suppose the random forest consists of k decision trees. The OOB dataset is O and the OOB data of each decision tree are O_i ($i = 1, 2, \ldots, k$), bringing the OOB data into the corresponding decision tree for classification. The numbers of misclassifications of each decision tree are set to X_i ($i = 1, 2, \ldots, k$), and the error size of the OOB is calculated from:

$$\text{OOB Error} = \frac{1}{k} \sum_{i=1}^{k} \frac{X_i}{O_i}. \tag{9}$$

3.3.3. Random Forest

Random forest is a set of tree classifiers $\{h(x, \theta_k), k = 1, 2, \ldots, n)\}$ and $h(x, \theta_k)$ is the meta-classifier, which is a classification regression tree composed of CART algorithm. As an independent random vector, $h(x, \theta_k)$ determines the growth of each decision tree and x is the input vector of the classifier.

A schematic diagram of the random forest algorithm is shown in Figure 3.

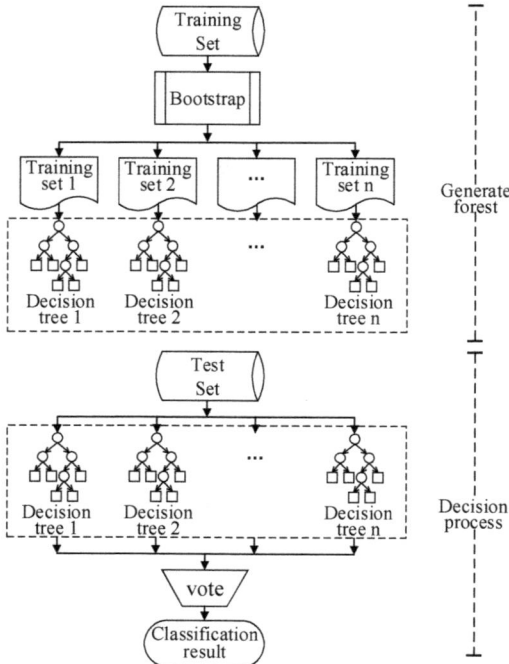

Figure 3. Schematic diagram of random forest algorithm.

Combined with the proposed oversampling method, the specific electricity theft detection steps are as follows:

(1) The unbalanced user-side dataset M is oversampled by K-SMOTE to obtain dataset M'.
(2) Divide the training set Tr and test set Te of random forest.
(3) Set the number of initial decision tree $nTree$.
(4) Use the bootstrap method to select training data for each decision tree. The total features in M are K. Selecting n features randomly, n is calculated using Equation (15). Then, use the CART algorithm to generate the unpruned decision trees.

$$n = \sqrt{K} \qquad (10)$$

(5) Input test set Te into each trained decision trees, and the classification result is determined according to the voting result of each decision tree. The voting classification formula is as follows:

$$f(Te_i) = MV\{h_t(Te_i)\}_{t=1}^{nTree} \qquad (11)$$

where Te_i ($i = 1, 2, \ldots, k$) represents each element in the test set, MV represents the majority vote, and $h_t(Te_i)$ represents the classification result of element Te_i in decision tree T.

(6) The current OOB error is calculated according to Equation (9). If the OOB error converges, then go to Step 7. If the error does not converge, update the decision tree number *nTree* according to Equation (11) and return to Step 4.
(7) Output classification result.

Based on the above theory and steps, the proposed electricity theft detection process was as shown in Figure 4.

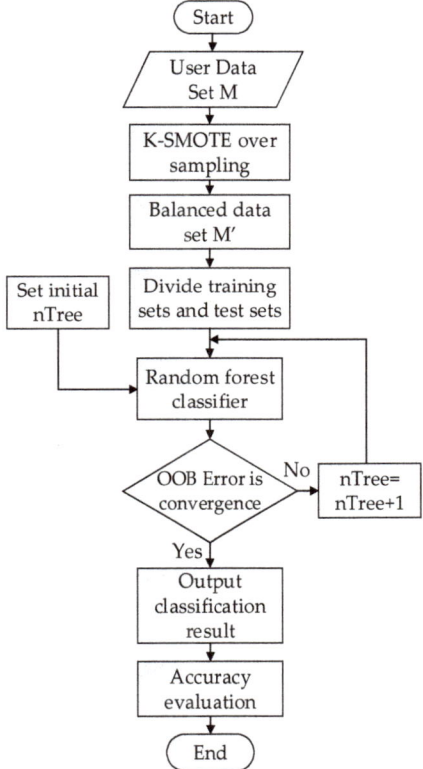

Figure 4. The proposed process of electricity theft detection.

4. Simulation Results

In order to verify the accuracy and effectiveness of the proposed detection method, three models, back-propagation neural network (BPN), support vector machine (SVM), and RF were established. The parameters of models are as follows.

(1) The neurons' number of input layer, hidden layer, and output layer in BPN were 20, 40 and 1, respectively. Learning rate was 0.3, momentum term was 0.3, batch volume was 100, and the maximum number of iterations was 50 [53].
(2) The kernel function of SVM qA radial basis function (RBF), the parameter coefficient of kernel function penalty g WAS 0.07, and the penalty factor coefficient c of PD and ND were 1 and 0.01, respectively [30].

This paper selected the short-term load data of 50 urban electricity users from 15 March 2018 to 16 May 2018 in Hebei province, China. The dataset included six data types (peak, flat, and valley active power; power factor; voltage; and current) and three user types (industrial, commercial, and residential). Data were sampled at intervals of 30 min through smart meters. Among them, the unbalance ratio was 16.47%.

4.1. Evaluation Indexes

After the classification of unbalanced data, all test sets were divided into four cases: TN (true negative), TP (true positive), FP (false positive), and FN (false negative). These indicators constituted a confusion matrix, as shown in Table 2. Confusion matrix is a way to evaluate the model performance, where the row corresponds to the category to which the object actually belongs and the column represents the category predicted by the model.

Table 2. Two-class confusion matrix.

Predicted / Actual	Predicted Positive	Predicted Negative
True Positive	TP	FN
True Negative	FP	TN

FP is the first type of error and FN is the second type of error. Through confusion matrix, multiple evaluation indexes can be extended.

(1) Accuracy (ACC): ACC is the ratio of the number of correct classifications to the total number of samples. The higher the value of ACC, the better is the performance of the detection algorithm. Mathematically ACC is defined as:

$$ACC = \frac{TP + TN}{TP + FP + TN + FN}. \quad (12)$$

(2) True Positive Rate (TPR): TPR describes the sensitivity of the detection model to PD. The higher the value of TPR, the better is the performance of the detection algorithm. TPR is defined as:

$$TPR = \frac{TP}{TP + FN}. \quad (13)$$

(3) False Positive Rate (FPR): FPR refers to the proportion of data in ND, which actually belongs to ND, and is wrongly judged as PD by the detection algorithm. FPR is defined as:

$$FPR = \frac{FP}{FP + TN}. \quad (14)$$

(4) True Negative Rate (TNR): TNR describes the sensitivity of the detection model to ND, which is defined as:

$$TNR = \frac{TN}{TN + FP}. \quad (15)$$

(5) G-mean index: G-mean index is used for the evaluation of classifier performance [54]. Large G-mean index reveals better classification performance. The value of G-mean depends on the square root of the product of the accuracy of PD and ND. G-mean can reasonably evaluate the overall classification performance of unbalanced dataset, and it can be expressed as:

$$G-mean = \sqrt{TPR * TNR}. \quad (16)$$

(6) Receiver operating characteristic (ROC) and area under the ROC curve (AUC): Receiver operator characteristic chive (ROC) was originally created to test the performance of a radar [55]. ROC curve

describes the relationship between the relative growth of FPR and TPR in the confusion matrix. For values output by the binary classification model, the closer the ROC curve is to the point (0, 1), the better the classification performance. Area under the ROC curve (AUC), is an index to evaluate the performance of the detection algorithm in the ROC curve. The AUC value of 1 corresponds to an ideal detection algorithm.

4.2. Unbalanced Processing of User-Side Data

Random oversampling, SMOTE, and K-SMOTE were used to oversample the datasets, and the results are shown in Figure 5, of which the black circle represents the normal users, the red asterisk represents the electricity theft users, and the blue box represents the data generated after oversampling.

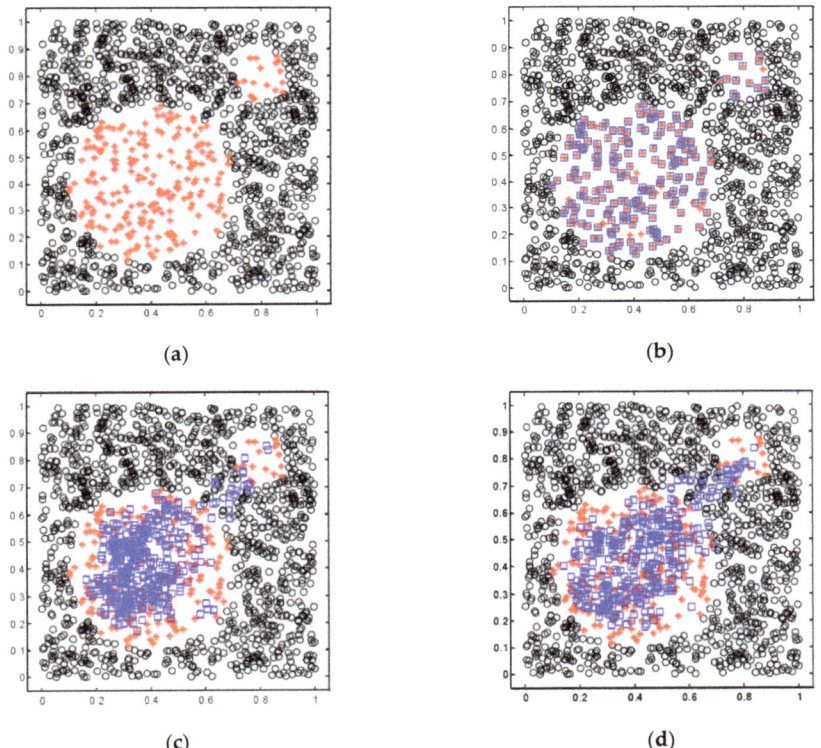

Figure 5. Schematic diagram of samples generated by different oversampling methods. (**a**) Raw data, (**b**) random oversampling, (**c**) synthetic minority oversampling technique, (**d**) improved SOMTE based on K-means. red: electricity theft users; Blue: data generated after oversampling; GREY: normal users.

In addition, Table 3 shows the repetition rate of artificial data and original data generated by several oversampling algorithms.

Table 3. The schematic of OOB.

Algorithm	Random Oversampling	SMOTE	K-SMOTE
Data Repetition Rate/%	95.02	30.50	15.84

It can be observed from Figure 5 that a large amount of duplicated data were included in the result of random oversampling algorithm, and some data were never selected. From Table 3, the data repetition rate of random oversampling was 95.02%, which indicates that the oversampling effect was not ideal. The data repetition rate of SMOTE was 30.5%, as can be seen from Figure 5. Data generated by SMOTE were scattered with other data and introduced noise points. The problem of data overlap still existed and could not be ignored. K-SMOTE can generate data near the center, and use representative points to limit the boundaries of the generated data to avoid introducing noise. Data generated by K-SMOTE generally follows the original distribution. Further, as shown in Table 3, the data repetition rate was only 15.84%.

4.3. Electricity Theft Detection Based on Improved RF

4.3.1. Determination of the Number of Decision Trees

The number of decision trees is relevant to the accuracy of the algorithm.

In this paper, 80% of the user data were set to form a training set and 20% to form a test set. The optimal number of decision trees can be determined by minimizing the OOB error. The relationship between the OOB error and the number of decision trees, *nTree*, is shown in Figure 6.

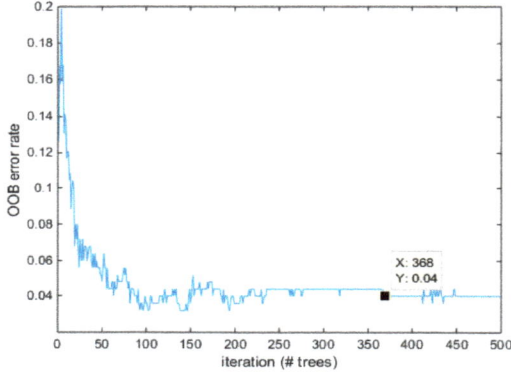

Figure 6. Value of OOB error varies with the number of decision trees.

It can be observed that when the decision tree number was larger than 368, the OBB error almost converged to a minimum level. If the number of decision trees was too small, the accuracy was low. On the other hand, too many decision trees did not improve the accuracy further and the algorithm burden was increasing. Therefore, the decision tree number was set to 368.

4.3.2. Detection Results of RF

The above-mentioned electricity users' dataset processed by K-SMOTE and not processed by K-SMOTE oversampling were, respectively, detected by RF. In order to make the simulation results more convincing and avoid randomness, three independent tests were carried out for each detection. ACC values of test data are listed in Table 4, and ROC curves are shown in Figures 7 and 8. In Figures 7 and 8, the three differently colored curves of red, green and blue represent the ROC curve of three independent tests. According to these results, when K-SMOTE was not used for unbalanced data processing, the mean value of ACC in RF was 85.53%, while the average value of ACC in RF after K-SMOTE was 94.53%. In addition, it can be concluded that ROC curve detected by RF with K-SMOTE was obviously closer to (0.1) than ROC curve detected RF algorithm without K-SMOTE. That is, the area under the former ROC curve was larger.

Moreover, the AUC index of the former was obviously better than that of the latter, which shows that it is necessary to use K-SMOTE to deal with unbalanced data before the detection of electricity theft behaviors. In addition, the detection performance of RF was also ideal.

Table 4. Accuracy value of random forest.

Simulation Times	K-SMOTE	ACC/%
1	with	96.02
	without	83.24
2	with	96.21
	without	82.01
3	with	81.32
	without	91.35
Mean	with	94.53
	without	85.53

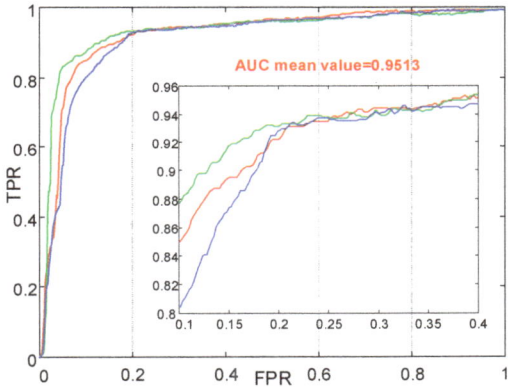

Figure 7. Receiver operating characteristic curve detected by RF with K-SMOTE.

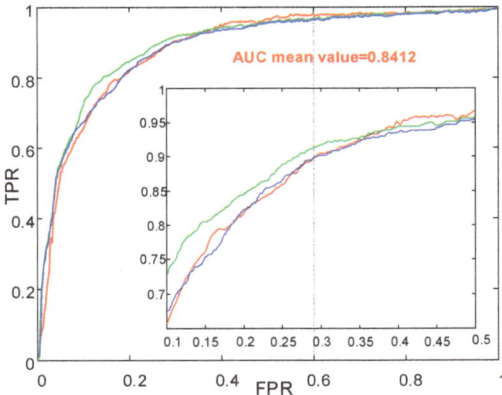

Figure 8. ROC curve detected RF algorithm without K-SMOTE.

4.3.3. Comparison of Detection Performance of Different Algorithms

The electricity users' data processed by K-SMOTE were tested by BPN and SVM. Again, in order to make the simulation results more convincing, the same dataset was used and three independent tests were performed. The testing results are shown in Table 5 and Figures 9 and 10. In Figures 9 and 10, the three differently colored curves of red, green and blue represent the ROC curve of three independent tests.

Table 5. Accuracy value of three algorithms.

Simulation Times	Algorithm	ACC/%
1	SVM	70.21
	BPN	85.24
2	SVM	72.54
	BPN	83.01
3	SVM	71.02
	BPN	86.35
Mean	SVM	71.26
	BPN	84.87

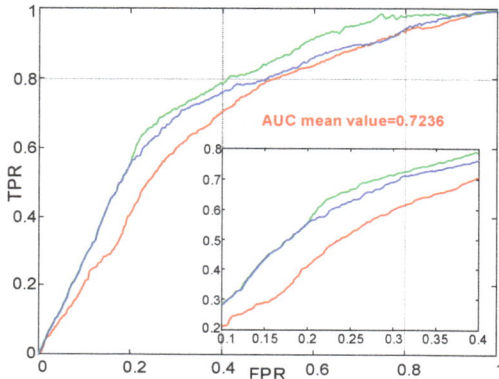

Figure 9. ROC curve detected by support vector machines with K-SMOTE.

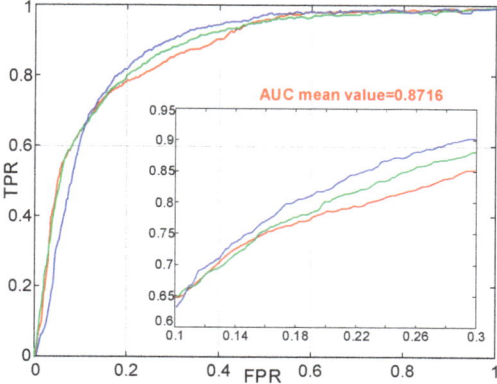

Figure 10. ROC curve detected by back-propagation neural network with K-SMOTE.

It can be concluded from the above test results that:

(1) Without K-SMOTE, the ACC value and AUC of RF detection method were relatively low. However, with K-SMOTE, the ACC and AUC value of three detection methods were obviously improved, which was increased about 10%. This indicates that unbalanced datasets would affect the accuracy of detection algorithm, and K-SMOTE plays an effective role in improving machine learning accuracy.

(2) The electricity user data processed by K-SMOTE were tested by BPN and SVM. The ACC mean values of SVM and BPN were 71.26% and 84.87%, respectively, and the mean values of AUC in SVM and BPN were 0.7236 and 0.8716, respectively. These indexes were lower than the ACC and AUC of RF, which were 94.53% and 0.9513, respectively. Thus, the performance of RF was superior to SVM and BPN.

5. Conclusions

In order to better adapt to the rapid development of the power grid, aiming at the unbalanced dataset on the user side and improving the efficiency and accuracy of electricity theft detection algorithms, this paper proposed a method based on K-SMOTE and RF classifier for detecting electricity theft. The main conclusions can be summarized as below:

(1) K-SMOTE was proposed to avoid the influence of unbalanced data on the accuracy of the classifier.
(2) The RF classifier, which was suitable for the nature of the user-side dataset, was used to detect electricity theft. The decision trees in RF classifier could work in parallel, which improved the detection efficiency and reduced the computational time.
(3) Compared with the conventional detection methods, the proposed method featured higher accuracy and stronger stability.

The method proposed in this paper can provide reliable targets for manual inspection, thereby reducing nontechnical losses in power systems and, hence, improving system reliability and security.

Author Contributions: Conceptualization, Z.Q.; Methodology, H.L.; Data Curation, Y.W.; Writing-Original Draft Preparation, J.Z.; Writing-Review & Editing, A.A.-S. Resources, Y.Y. All authors have read and agreed to the published version of the manuscript.

Funding: This work was supported by the National Key Research and Development Project of China [grant number 2016YFF0200105], and the National Natural Science Foundation of China [grant number 51777199].

Conflicts of Interest: The authors declare no conflict of interest.

Appendix A

Table A1. Nomenclature.

Acronyms	Full Name
SMOTE	synthetic minority oversampling technique
RF	random forest
PD	positive data
ND	negative data
OOB Error	out-of-bag data error
TL	Technical loss
NTL	non-technical losses
PCA	principal component analysis
FA	factor analysis
ELM	extreme learning machine
BC	balance cascade

Table A1. *Cont.*

Acronyms	Full Name
KNN	k-nearest neighbor
TN	true negative
TP	true positive
FP	false positive
FN	false negative
ACC	Accuracy
TPR	true positive rate
FPR	false positive rate
TNR	true negative rate
IE	information entropy
OOB	out of bag
SVM	support vector machines
BPN	back-propagation neural network

References

1. Lo, Y.L.; Huang, S.C.; Lu, C.N. Non-technical loss detection using smart distribution network measurement data. In Proceedings of the IEEE PES Innovative Smart Grid Technologies, Tianjin, China, 21–24 May 2012; pp. 1–5.
2. Smith, T.B. Electricity theft: A comparative analysis. *Energy Policy* **2004**, *32*, 2067–2076. [CrossRef]
3. Huang, S.C.; Lo, Y.L.C.N. Non-Technical Loss Detection Using State Estimation and Analysis of Variance. *IEEE Trans. Power Syst.* **2013**, *28*, 2959–2966. [CrossRef]
4. Bhavna, B.; Mohinder, G. Reforming the Power Sector, Controlling Electricity Theft and Improving Revenue. Public Policy for the Private Sector. 2004. Available online: http://rru.worldbank.org/PublicPolicyJourna (accessed on 16 December 2019).
5. Soma, S.S.R.D.; Wang, L.; Vijay, D.; Robert, C.G. High performance computing for detection of electricity theft. *Int. J. Electr. Power Energy Syst.* **2013**, *47*, 21–30.
6. Smart Meters Help Reduce Electricity Theft, Increase Safety. BCHydro. Available online: https://www.bchydro.com/news/conservation/2011/smart_meters_energy_theft.html (accessed on 16 December 2019).
7. Dzung, D.; Naedele, M.; Von Hoff, T.P.; Crevatin, M. Security for Industrial Communication Systems. *Proc. IEEE Secur. Ind. Commun. Syst.* **2005**, *93*, 1152–1177. [CrossRef]
8. Krebs, B. FBI: Smart Meter Hacks Likely to Spread. Available online: http://krebsonsecurity.com/2012/04/fbi-smart-meter-hacks-likely-to-spread (accessed on 16 December 2019).
9. Carlos, L.; Félix, B.; Iñigo, M.; Juan, I.; Guerrero, J.B.; Rocío, M. Integrated expert system applied to the analysis of non-technical losses in power utilities. *Expert Syst. Appl.* **2011**, *38*, 10274–10285.
10. Han, W.; Xiao, Y. NFD: A practical scheme to detect non-technical loss fraud in smart grid. In Proceedings of 2014 IEEE International Conference on Communications (ICC), Sydney, NSW, Australia, 10–14 June 2014; pp. 605–609.
11. Grochocki, D.; Huh, J.H.; Berthier, R. AMI threats, intrusion detection requirements and deployment recommendations. In Proceedings of the IEEE Third International Conference on Smart Grid Communications, Tainan, Taiwan, China, 5–8 November 2012; pp. 395–400.
12. Hao, R.; Ai, Q.; Xiao, F. Architecture based on multivariate big data platform for analyzing electricity consumption behavior. *Electr. Power Autom. Equip.* **2017**, *37*, 20–27.
13. Jiang, R.; Lu, R.; Wang, Y.; Luo, J.; Shen, C.; Shen, X. Energy-theft detection issues for advanced metering infrastructure in smart grid. *Tsinghua Sci. Technol.* **2014**, *19*, 105–120. [CrossRef]
14. Lo, C.-H.; Ansari, N. CONSUMER: A Novel Hybrid Intrusion Detection System for Distribution Networks in Smart Grid. *IEEE Trans. Emerg. Top. Comput.* **2013**, *1*, 33–44. [CrossRef]

15. McLaughlin, S.; Holbert, B.; Fawaz, A.; Berthier, R.; Zonouz, S. A multi-sensor energy theft detection framework for advanced metering infrastructures. *IEEE J. Sel. Areas Commun.* **2013**, *31*, 1319–1330. [CrossRef]
16. Leite, J.B.; Sanches mantovani, J.R. Detecting and locating non–technical losses in modern distribution networks. *IEEE Trans. Smart Grid* **2018**, *9*, 1023–1032. [CrossRef]
17. Tariq, M.; Poor, H. Electricity Theft Detection and Localization in Grid-Tied Microgrids. *IEEE Trans. Smart Grid* **2018**, *9*, 1920–1929. [CrossRef]
18. Salinas, S.; Li, P. Privacy-Preserving Energy Theft Detection in Microgrids: A State Estimation Approach. *IEEE Trans. Power Syst.* **2016**, *31*, 883–894. [CrossRef]
19. Cárdenas, A.A.; Amin, S.; Schwartz, G.; Dong, R.; Sastry, S. A game theory model for electricity theft detection and privacy-aware control in AMI systems. In Proceedings of the 50th Annual Allerton Conference on Communication, Control, and Computing (Allerton), Monticello, IL, USA, 1–5 October 2012; pp. 1830–1837.
20. O'Leary, D.E. Summary of Previous Papers in Expert Systems Review. *Intell. Syst. Account. Financ. Manag.* **2016**, *1*, 3–7. [CrossRef]
21. Punmiya, R.; Choe, S. Energy Theft Detection Using Gradient Boosting Theft Detector with Feature Engineering-Based Preprocessing. *IEEE Trans. Smart Grid* **2019**, *10*, 2326–2329. [CrossRef]
22. Coma-Puig, B.; Carmona, J. Bridging the Gap between Energy Consumption and Distribution through Non-Technical Loss Detection. *Energies* **2019**, *12*, 1748. [CrossRef]
23. Lu, X.; Zhou, Y.; Wang, Z.; Yi, Y.; Feng, L.; Wang, F. Knowledge Embedded Semi-Supervised Deep Learning for Detecting Non-Technical Losses in the Smart Grid. *Energies* **2019**, *12*, 3452. [CrossRef]
24. Nagi, J.; Yap, K.S.; Tiong, S.K.; Ahmed, S.K.; Mohamad, M. Nontechnical Loss Detection for Metered Customers in Power Utility Using Support Vector Machines. *IEEE Trans. Power Deliv.* **2009**, *25*, 1162–1171. [CrossRef]
25. Zhuang, C.J.; Zhang, B.; Hu, J.; Li, Q.; Zeng, R. Anomaly detection for power consumption patterns based on unsupervised learning. *Proc. CSEE* **2016**, *36*, 379–387. (In Chinese)
26. Ghasemi, A.A.; Gitizadeh, M. Detection of illegal consumers using pattern classification approach combined with Levenberg-Marquardt method in smart grid. *Int. J. Electr. Power Energy Syst.* **2018**, *99*, 363–375. [CrossRef]
27. Monedero, I.; Biscarri, F.; Leon, C.; Biscarri, J.; Millan, R. MIDAS: Detection of Non-technical Losses in Electrical Consumption Using Neural Networks and Statistical Techniques. In Proceedings of Computational Science and Its Applications-ICCSA 2006, International Conference, Glasgow, UK, 8–11 May 2006; Proceedings, Part V; Springer: Berlin/Heidelberg, Germany, 8 May 2006; pp. 725–734.
28. Nizar, A.H.; Dong, Z.Y.; Wang, Y. Power Utility Nontechnical Loss Analysis with Extreme Learning Machine Method. *IEEE Trans. Power Syst.* **2008**, *23*, 946–955. [CrossRef]
29. Muniz, C.; Figueiredo, K.; Vellasco, M.; Chavez, G.; Pacheco, M. Irregularity detection on low tension electric installations by neural network ensembles. In Proceedings of the International Joint Conference on Neural Networks, Atlanta, GA, USA, 14–19 June 2009; pp. 2176–2182.
30. Nagi, G.; Yap, K.S.; Tiong, S.K.; Ahmed, S.K.; Nagi, F. Improving SVM-Based Nontechnical Loss Detection in Power Utility Using the Fuzzy Inference System. *IEEE Trans. Power Deliv.* **2011**, *26*, 1284–1285. [CrossRef]
31. Viegas, J.L.; Esteves, P.R.; Vieira, S.M. Clustering-based novelty detection for identification of non-technical losses. *Int. J. Electr. Power Energy Syst.* **2018**, *101*, 301–310. [CrossRef]
32. Buzau, M.M.; Tejedor-Aguilera, J.; Cruz-Romero, P.; Gómez-Expósito, A. Detection of Non-Technical Losses Using Smart Meter Data and Supervised Learning. *IEEE Trans. Smart Grid* **2019**, *10*, 2661–2670. [CrossRef]
33. Messinis, G.M.; Hatziargyriou, N.D. Review of non-technical loss detection methods. *Electr. Power Syst. Res.* **2018**, *158*, 250–266. [CrossRef]
34. Jokar, P.; Arianpoo, N.; Leung, V.C.M. Electricity Theft Detection in AMI Using Customers' Consumption Patterns. *IEEE Trans. Smart Grid* **2015**, *7*, 216–226. [CrossRef]
35. Figueroa, G.; Chen, Y.; Avila, N.; Chu, C. Improved practices in machine learning algorithms for NTL detection with imbalanced data. In Proceedings of the 2017 IEEE Power & Energy Society General Meeting, Chicago, IL, USA, 16–20 July 2017; pp. 1–5.
36. Glauner, P.; Boechat, A.; Lautaro, D.; Radu, S.; Franck, B.; Yves, R.; Diogo, D. Large-scale detection of non-technical losses in unbalanced datasets. In Proceedings of the 2016 IEEE Power & Energy Society Innovative Smart Grid Technologies Conference (ISGT), Minneapolis, MN, USA, 6–9 September 2016; pp. 1–5.

37. Liu, X.; Wu, J.; Zhou, Z. Exploratory Under sampling for Class-Unbalance Learning. *IEEE Trans. Cybern.* **2009**, *39*, 539–550.
38. Victoria, L.; Alberto, F.; Salvador, G.; Palade, V.; Herrera, F. An insight into classification with unbalanced data: Empirical results and current trends on using data intrinsic characteristics. *Inf. Sci.* **2013**, *250*, 113–141.
39. Del Río, S.; López, V.; Benítez, J.M.; Herrera, F. On the use of MapReduce for unbalanced big data using Random Forest. *Inf. Sci.* **2014**, *285*, 112–137. [CrossRef]
40. Chawla, N.V.; Bowyer, K.W.; Hall, L.O.; Kegelmeyer, W.P. SMOTE: Synthetic minority over-sampling technique. *J. Artif. Intell. Res.* **2002**, *16*, 321–357. [CrossRef]
41. Feilong, C.; Yuanpeng, T.; Miaomiaom, C. Sparse algorithms of Random Weight Networks and applications. *Expert Syst. Appl.* **2014**, *41*, 2457–2462.
42. Song, X.; Guo, Z.; Guo, H.; Wu, S.; Wu, C. A new forecasting model based on forest for photovoltaic power generation. *Power Syst. Prot. Control* **2015**, *43*, 13–18.
43. Monedero, I.; Félix, B.; Carlos, L.; Guerrero, J.I.; Jesús, B.; Rocío, M. Detection of frauds and other non-technical losses in a power utility using Pearson coefficient, Bayesian networks and decision trees. *Int. J. Electr. Power Energy Syst.* **2012**, *34*, 90–98. [CrossRef]
44. Gao, D.; Zhang, Y.X.; Zhao, Y.H. Random forest algorithm for classification of multiwavelength data. *Res. Astron. Astrophys.* **2009**, *9*, 220–226. [CrossRef]
45. Fernandez, A.; García, S.; Herrera, F.; Chawla, N.V. SMOTE for Learning from Imbalanced Data: Progress and Challenges, Marking the 15-year Anniversary. *J. Artif. Intell. Res.* **2018**, *61*, 863–905. [CrossRef]
46. Bradley, R.; Fayyad, U. Refining Initial Points for K-Means Clustering. In Proceedings of the Fifteenth International Conference on Machine Learning; 1998; pp. 91–99. Available online: https://dl.acm.org/doi/10.5555/645527.657466 (accessed on 20 December 2019).
47. Breiman, L. Random forests. *Mach. Learn.* **2001**, *45*, 5–32. [CrossRef]
48. Breiman, L. Bagging predictors. *Mach. Learn.* **1996**, *24*, 123–140. [CrossRef]
49. Belgiu, M.; Drăguţ, L. Random forest in remote sensing: A review of applications and future directions. *ISPRS J. Photogramm. Remote Sens.* **2016**, *114*, 24–31. [CrossRef]
50. Alessia, S.; Antonio, C.; Aldo, Q. Random Forest Algorithm for the Classification of Neuroimaging Data in Alzheimer's Disease: A Systematic Review. *Front. Aging Neurosci.* **2017**, *9*, 329–336.
51. Quinlan, J.R. Decision trees and decision-making. *IEEE Trans. Syst. Man Cybern.* **1990**, *20*, 339–346. [CrossRef]
52. Wolpert, D.H.; Macready, W.G. An Efficient Method to Estimate Bagging's Generalization Error. *Mach. Learn.* **1999**, *35*, 41–55. [CrossRef]
53. Xu, G.; Tan, Y.P.; Dai, T.H. Sparse Random Forest Based Abnormal Behavior Pattern Detection of Electric Power User Side. *Power Syst. Technol.* **2017**, *41*, 1965–1973.
54. Rao, R.B.; Krishnan, S.; Niculescu, R.S. Data mining for improved cardiac care. *ACM SIGKDD Explor. Newsl.* **2006**, *1*, 3–10. [CrossRef]
55. Huang, Y.A.; You, Z.H.; Gao, X.; Wong, L.; Wang, L. Using Weighted Sparse Representation Model 561 Combined with Discrete Cosine Transformation to Predict Protein-Protein Interactions from Protein Sequence. *BioMed Res. Int.* **2015**, *2015*, 902198. [CrossRef] [PubMed]

© 2020 by the authors. Licensee MDPI, Basel, Switzerland. This article is an open access article distributed under the terms and conditions of the Creative Commons Attribution (CC BY) license (http://creativecommons.org/licenses/by/4.0/).

Article

A Novel Electricity Theft Detection Scheme Based on Text Convolutional Neural Networks

Xiaofeng Feng [1], Hengyu Hui [2,*], Ziyang Liang [2], Wenchong Guo [1], Huakun Que [1], Haoyang Feng [1], Yu Yao [2], Chengjin Ye [2] and Yi Ding [2]

1. Metrology Center of Guangdong Power Grid Corporation, Guangzhou 510080, China; ucihqtep@163.com (X.F.); wenchong1025@163.com (W.G.); quehuakun@126.com (H.Q.); fenghy111@163.com (H.F.)
2. College of Electrical Engineering, Zhejiang University, Hangzhou 310027, China; liangziyang@zju.edu.cn (Z.L.); zjuyaoyu@zju.edu.cn (Y.Y.); yechenjing@zju.edu.cn (C.Y.); yiding@zju.edu.cn (Y.D.)
* Correspondence: huihengyu@zju.edu.cn; Tel.: +86-136-2839-5130

Received: 30 September 2020; Accepted: 30 October 2020; Published: 3 November 2020

Abstract: Electricity theft decreases electricity revenues and brings risks to power usage's safety, which has been increasingly challenging nowadays. As the mainstream in the relevant studies, the state-of-the-art data-driven approaches mainly detect electricity theft events from the perspective of the correlations between different daily or weekly loads, which is relatively inadequate to extract features from hours or more of fine-grained temporal data. In view of the above deficiencies, we propose a novel electricity theft detection scheme based on text convolutional neural networks (TextCNN). Specifically, we convert electricity consumption measurements over a horizon of interest into a two-dimensional time-series containing the intraday electricity features. Based on the data structure, the proposed method can accurately capture various periodical features of electricity consumption. Moreover, a data augmentation method is proposed to cope with the imbalance of electricity theft data. Extensive experimental results based on realistic Chinese and Irish datasets indicate that the proposed model achieves a better performance compared with other existing methods.

Keywords: data-driven approaches; electricity theft detection; smart meters; text convolutional neural networks (TextCNN); time-series classification

1. Introduction

Electricity theft can be defined as the behavior of illegally altering an electric energy meter to avoid billing. This illegal behavior not only severely disrupts the normal utilization of electricity, but also causes huge economic losses to power systems. At the same time, the unauthorized modification of lines or meters easily leads to accidents such as power failures and fires, and poses a serious threat to the safety of the relevant power system [1,2]. According to the research released by an intelligence firm northeast group, llc in January 2017, electricity theft and other non-technical losses have rendered over $96 billion in losses per year globally [3]. State Grid Hunan Electric Power Company, China reported that nearly 40% of electrical fires and 28% of electric shock casualties are caused by electricity theft [4]. Therefore, it is necessary to develop effective techniques for electricity theft detection and ensure the security and economic operation of power system.

The electricity theft detection technologies can be divided into three categories: the network-oriented method, the data-oriented method and a hybrid-oriented method that mixes the previous two methods [5]. Network-oriented and hybrid-oriented approaches usually require the network topology [6,7] and even the installation of additional devices [8]. It is difficult to implement these approaches widely, because the network topology may be unattainable due to security concerns and the installation of addition devices is costly. Data-oriented approaches only focus on the data provided by smart meters and have no requirements of the network topology or additional devices, which helps with improving the cost-effectiveness for suspected electricity theft judgment and detection. Therefore, data-oriented approaches have been widely applied to electricity theft detection in recent years [9,10].

At present, there are two typical data-oriented methods to detect electricity theft: support vector machines (SVM) and neural networks. In [11], they proposed a SVM-based approach that uses customer consumption data to expose abnormal behavior and identify suspected thieves. The authors in [12,13] combined SVM and a fuzzy inference system to detect electricity theft. In [14], a comprehensive top-down scheme based on decision trees and SVM was proposed. The two-level data processing and analysis approach can detect and locate electricity theft at every level in power transmission and distribution. The authors in [15] proposed an ensemble approach combining the adaptive boosting algorithm and SVM.

More and more researchers are utilizing neural networks to detect electricity theft due to their effectiveness. In [16], a long short-term memory (LSTM) and bat-based random under-sampling boosting (RUSBoost) approach is proposed. The LSTM and bat-based RUSBoost are applied to detect abnormal patterns and optimize parameters, respectively. In [17], a method based on the wide and deep convolutional neural network (CNN) model is proposed. The deep CNN component can identify the periodicity of electricity consumption and the wide component can capture the global characteristics of electricity consumption data. The authors in [18] combined CNN and LSTM to detect electricity theft from the power consumption signature in time-series data. In [10], an end-to-end hybrid neural network is proposed, which can analyze daily energy consumption data and non-sequential data, such as geographic information.

The above methods have paved the way for building the structures of networks and dedicate to improving electricity theft detection's accuracy. However, they mainly focus on the daily or weekly electricity consumption patterns. As a result, if these methods are applied to hourly or more frequent electricity consumption data, their accuracy will decrease. This is because they fail to capture the intraday electricity consumption pattern, for example, the correlation between the electricity consumption at the same time on different days. In practice, some illegal users commit electricity theft for part of the day. Specifically, those who have precedent technology and large electricity demands (such as industrial electricity thieves) prefer to commit the crime during some specific hours after considering electricity prices, monitoring periods and the risk of being caught comprehensively. Meanwhile, new kinds of attacks such as interception communication and false data injection [19,20] make it easier to commit such crime. In a case of electricity theft caught by State Grid Shandong Electric Power Company, China, the illegal user had normal daily electricity consumption. However, he confused the metering time of his smart meter to avoid peak electricity tariffs [21]. In this way, his abnormal electricity consumption pattern can only be reflected in the intraday data. In paper [22], several possible attack models are proposed to confuse the metering time and commit electricity theft targeting time-based pricing. Therefore, it is necessary to construct an electricity theft detection scheme that can not only capture the daily features but the intraday features.

In order to better extract the periodical features from days and more frequent time periods, we utilize a two-dimensional grid structure for the raw input data in this paper. Based on the data structure, we propose a text convolutional neural network (TextCNN) to detect electricity theft. The main contributions of the proposed model include:

(1) We analyze the electricity data structure and transform it into a two-dimensional time-series. This structure carries the complete power consumption information of users, which means the consumption patterns of various time scales, such as the electricity consumption at the same period on different days and the daily consumption of different days.

(2) We propose a novel electricity theft detecting method based on TextCNN. The proposed method can extract features of different time scales from two-dimensional time-series. To improve the accuracy and efficiency of training and detection, we designed our detection network based on

TextCNN. To test the performance, we implemented extensive experiments on the residential and industrial datasets from a province in China and the public Irish residential dataset.

(3) We propose the data augmentation method to expand the training data in view of the shortage of electricity theft samples. Experimental analysis indicates that the method can improve the detection accuracy effectively with a proper augmentation process.

The remainder of this paper is organized as follows. The methodologies of data construction and CNN construction for electricity consumption data are described in Section 2. Section 3 proposes the neural network structure based on TextCNN and the data augmentation method. Then, the details of experimental datasets and methods used for comparison and metrics are given in Section 4. Section 5 presents the performance and superiority of the proposed model, analyzes the parameters and discuss the effectiveness of the data augmentation method. Finally, Section 6 concludes this paper.

2. Methodology

In this section, we introduce the characteristics of the electricity consumption data and then compare several data structures regarding their advantages and disadvantages. Finally, we introduce TextCNN and the reason why it is suitable for the two-dimensional time-series.

2.1. Data Structure Analysis

Smart meters can collect the electricity consumption data at a high frequency, such as once an hour. The datasets can be expressed as:

$$D_n = \left\{ x_{h_1}^{d_1}, x_{h_2}^{d_1}, \ldots, x_{h_{24}}^{d_1}, x_{h_1}^{d_2}, \ldots, x_{h_j}^{d_i} \right\} \tag{1}$$

where D_n represents the data of user n. $x_{h_j}^{d_i}$ is the value recorded by smart meters during time h_j on day d_i.

Most studies focus on periodical features of daily or weekly consumption patterns to detect electricity theft. Therefore, they merge the data of one day into one value and utilize the one-dimensional data structure or its variant, as shown in Figure 1. The datasets can be further expressed as:

$$D_n = \left\{ x_{d_1}, x_{d_2}, \ldots, x_{d_i} \right\} \tag{2}$$

where x_{d_i} represents the total amount of electricity consumption on day d_i.

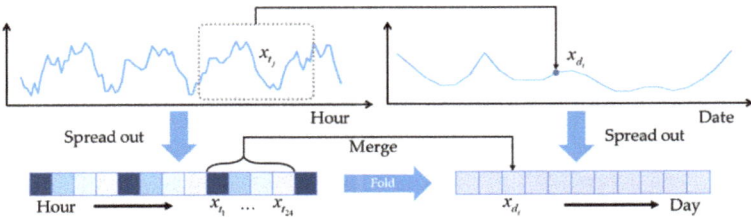

Figure 1. One-dimensional data structure of electricity consumption data.

In this way, they neglect the intraday electricity change and fail to capture the intraday features. In this paper, we construct the data into a two-dimensional grid, which is suitable for feature extraction from not only different days but different time periods. The two-dimensional grid can be expressed as:

$$D_n = \begin{bmatrix} x_{h_1}^{d_1} & x_{h_1}^{d_2} & \cdots & x_{h_1}^{d_i} \\ x_{h_2}^{d_1} & x_{h_2}^{d_2} & \cdots & x_{h_2}^{d_i} \\ \vdots & \vdots & \ddots & \vdots \\ x_{h_{24}}^{d_1} & x_{h_{24}}^{d_2} & \cdots & x_{h_{24}}^{d_i} \end{bmatrix} \quad (3)$$

The columns in Equation (3) represent the electricity consumption data of 24 h a day. In fact, smart meters may collect data more frequently, and may collect varieties of information, such as three-phase voltages and currents, power factors and so on. In this manner, in order to simplify the expression of Equation (3), we utilize the following column vector to represent the amount of data on day d_i:

$$\mathbf{x}_{d_i} = \{x_1, x_2, \ldots, x_j, \ldots, x_F\}^T \quad (4)$$

where F is the number of data from one day. So far, the dataset of one user can be expressed as:

$$D_n = \{\mathbf{x}_{d_1}, \mathbf{x}_{d_2}, \ldots, \mathbf{x}_{d_i}\} \quad (5)$$

Further, we use Figure 2 to explain the above two-dimensional structure. In the left figure, an individual curve represents the data x_j on different days and the cluster of curves demonstrates the daily electricity consumption. Then the expansion of the cluster is a grid, as shown in the right figure, which is also formulated as Equation (5). The height of the grid represents the number of data points from one day points. The length of the grid represents the number of days. In other words, the grid of electricity consumption data is a two-dimensional time-series.

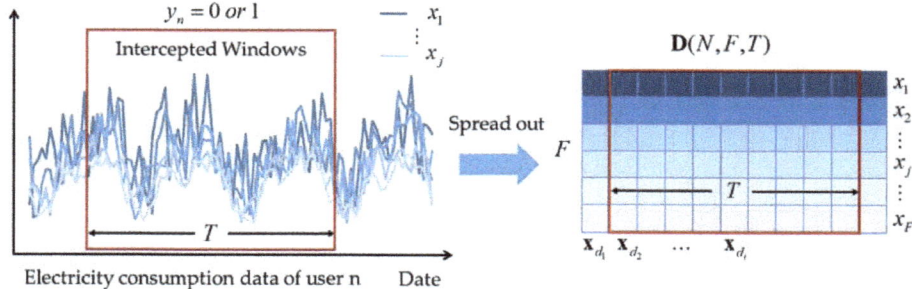

Figure 2. Two-dimensional data structure of electricity consumption data.

To extract the consumption patterns of different users, we utilize the same intercepted window to consecutively intercept different users' data. Then we can obtain a series of two-dimensional grids with the same length and height. For user n, we use y_n to label the intercepted window of the time-series to judge whether it is electricity theft or not, as shown in Figure 2. Then, we build a nonlinear map function from an input time-series to predict a class label y_n formula:

$$y_n = f(\mathbf{x}_{d_i}, T) \ d_i \in T \quad (6)$$

where T is the length of the intercepted window and $f(\cdot)$ is the key nonlinear function we aim to learn.

In order to conveniently express this data structure in CNN, we use $\mathbf{D}(N, F, T)$ to represent the intercepted segments, where N is the number of samples. For an individual data $\mathbf{D}(F, T)$ in the dataset $\mathbf{D}(N, F, T)$, the classification function $f(\cdot)$ needs learning. So far, we have constructed the two-dimensional structure that maintains the full information of the raw data and transforms the electricity theft detecting problem into the classification of time-series. Based on the two-dimensional time-series structure, we utilize TextCNN to learn the classification function.

2.2. CNN Structure Analysis

CNN specializes in processing data with a grid-like structure [23]. For different input data types, the structure of CNN should be selected further to achieve effectiveness—TextCNN, RCNN, etc. [24–26]. Considering the above two-dimensional time-series, we focus on TextCNN in this research. TextCNN is widely used in natural language processing (NLP) fields such as text classification, emotion analysis and sensitivity analysis for its simple structure and effectiveness [27,28].

2.2.1. Basic Introduction to CNN

The normal multilayered neural networks, which are also called deep neural networks (DNN), consist of input layers, hidden layers and output layers. CNN has an additional convolutional layer, as shown in Figure 3a. The discrete convolution is the key operation in convolutional layers. As shown in Figure 3b, we use a 2 × 2 kernel as an example to illustrate the discrete convolution. The input **I** has a value in each grid. Then, a two-dimensional kernel function $\mathbf{K} \in \mathbb{R}^{2\times 2}$ is used to extract features. The output **S** of the convolution is:

$$\mathbf{S}(i,j) = \sum_{k_i=0}^{1} \sum_{k_j=0}^{1} \mathbf{I}(i+k_i, j+k_j) \mathbf{K}(k_i, k_j) \tag{7}$$

Equation (7) and Figure 3b together illustrate that convolutional kernels map the neighboring information of the input into the output. Therefore, compared with DNN, CNN has an advantage of considering the information in the small neighborhoods, which is a crucial future in the classification of two-dimensional data, as the neighboring grids usually carry related information [29,30]. For example, if we regard **I** as a black and white picture, the kernels can efficiently extract features, such as edges, angles and shapes from neighboring pixels.

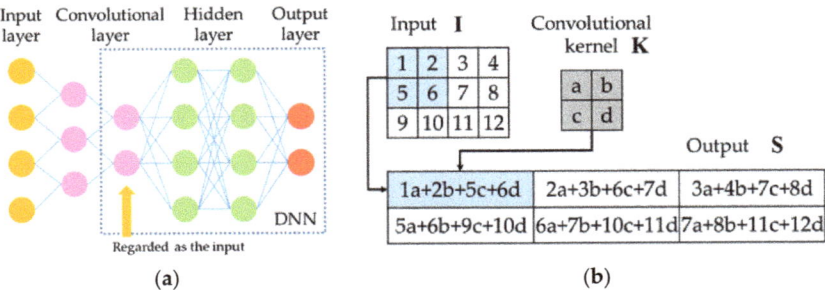

Figure 3. Diagrams of the CNN structure and the discrete convolution: (**a**) CNN structure; (**b**) discrete convolutions.

2.2.2. Differences between CNN and TextCNN

The kernel size is the main difference between CNN and TextCNN. As shown in Figure 4a, we use height and length to describe the size of a two-dimensional kernel. The commonly used kernel size in CNN is 3 × 3 [31,32], while in TextCNN the height of kernels is always equal to that of input data [27]. This is because for text classification, the most significant thing is to efficiently capture the internal features of an individual word and the correlations between multiple words. As shown in Figure 4a, the convolutional kernels are sliding windows with the same height as a single world. The kernel will only move in the length direction, so each time the kernel will slide over a complete word.

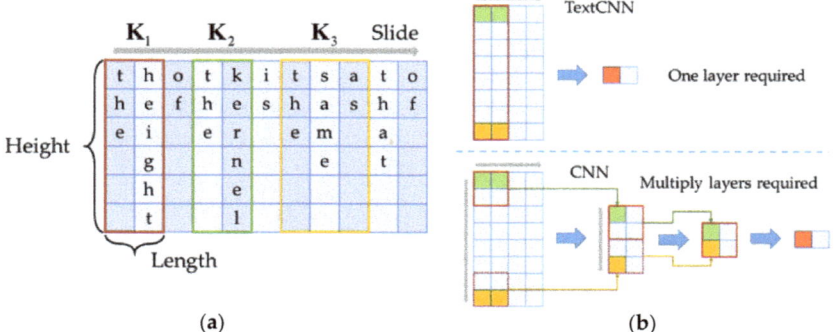

Figure 4. Characteristics of CNN: (**a**) feature schematic diagram of TextCNN's kernels; (**b**) differences between CNN and TextCNN.

The influences of different kernel sizes on the network are shown in Figure 4b. In order to capture the association between the green grids and the yellow grids, TextCNN requires only one convolutional layer, while CNN requires three convolutional layers. Therefore, TextCNN simplifies the structure of the neural network and reduces the parameters that require manual intervention. In this manner, the efficiency and effectiveness of capturing the internal features of a word and the correlations between multiple words are guaranteed.

In electricity theft detection, we aim to capture the features from the data correlations of weeks, days, hours and even more frequent time periods. Analogously, the intraday feature of electricity consumption is similar to the association between the green grids and the yellow grids in Figure 4b, and the multi-day correlations are extracted by different kernels, such as K_1, K_2 and K_3 in Figure 4a. Therefore, to efficiently extract features of electricity consumption data, we propose a neural network based on TextCNN for the classification of two-dimensional time-series.

3. Proposed Approach

In this section, we propose our electricity theft detection scheme. We introduce the data preprocess at first. Then, we propose a neural network structure based on TextCNN, which consists of convolutional layers, pooling layers and fully-connected layers. Finally, we propose the data augmentation method to increase the amount of electricity theft data for the balance of the training dataset. The total framework of the proposed electricity theft detection is demonstrated in Figure 5.

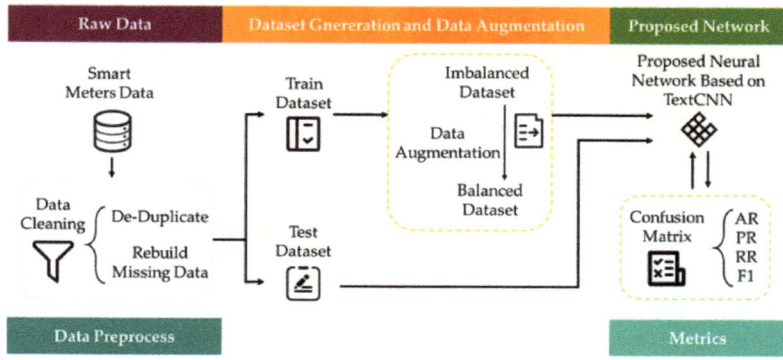

Figure 5. Proposed electricity detection scheme.

As shown in Figure 5, the raw data collected by smart meters gets through the data preprocess at first. Then, we divide it into the training dataset and the test dataset. If the training dataset is imbalanced, we utilize the proposed data augmentation method to balance it. Finally, we train the proposed network on the training dataset and validate the effect on the test dataset. The metrics used for training and testing are introduced in Section 4.3. It should be noted that the training process is supervised learning which requires labeled datasets.

3.1. Data Preprocess

During data collection, missing data, duplications and errors of electricity consumption data may occur. To avoid the adverse effects of faulty data on the electricity theft detection, reference [17] proposes an electricity data preprocessing method to recover the missing and erroneous data. Equation (8) represents the interpolation method to recover the mission data.

$$x_{d,t}^* = \begin{cases} \frac{x_{d,t-1}+x_{d,t+1}}{2} & x_{d,t} \in \text{NaN}, x_{d,t-1}, x_{d,t+1} \notin \text{NaN} \\ 0 & x_{d,t} \in \text{NaN}, x_{d,t-1} \text{ or } x_{d,t+1} \in \text{NaN} \\ x_{d,t} & x_{d,t} \notin \text{NaN} \end{cases} \qquad (8)$$

where $x_{d,t}$ is the electricity consumption data during time period t on day d. Additionally, NaN represents null and non-numeric character.

Moreover, the three-sigma rule of thumb [33] is used to recover the erroneous data as follows:

$$x_{d,t}^* = \begin{cases} \text{avg}(\mathbf{x}_d) + 2 \cdot \text{std}(\mathbf{x}_d) & \text{if } x_{d,t} > \text{avg}(\mathbf{x}_d) + 2 \cdot \text{std}(\mathbf{x}_d) \\ x_{d,t} & \text{otherwise} \end{cases} \qquad (9)$$

where $\mathbf{x}_d$ is a vector composed of $x_{d,t}$, $\text{avg}(\mathbf{x})$ and $\text{std}(\mathbf{x})$ stands for the average value and standard deviation value of vector $\mathbf{x}$.

3.2. Proposed Neural Network Structure Based on TextCNN

As shown in Figure 6, the proposed neural network structure based on TextCNN is mainly composed of convolutional layers, pooling layers and fully-connected layers. The features of each layer are demonstrated in detail below.

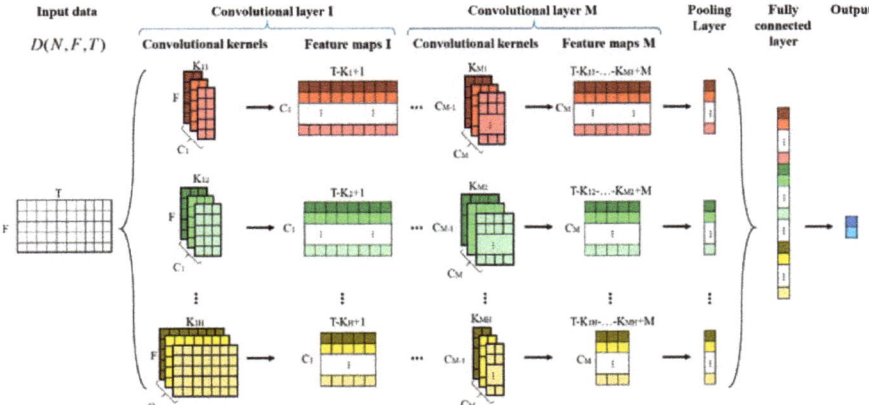

Figure 6. Proposed network structure for electricity theft detection.

3.2.1. Convolutional Layer

The proposed model has multiple convolutional layers. A convolutional layer has H different sizes of convolutional kernels. As mentioned in Section 2, in order to ensure the efficiency and effectiveness of the classification for two-dimensional time-series, the height of kernels is the same as the number of data points for one day. For convolution kernels of size H_i, $\mathbf{D}^u = (F, T)$ denotes the uth data sample. The corresponding kernel weight $\mathbf{w}_j^u \in \mathbb{R}^{F \times K}$ is used to extract features from the input data, where K is the kernel length. For example, the feature map $o_{j,i}^u$ is calculated by:

$$o_{j,i}^u = f_a\left(\mathbf{w}_j^u * \mathbf{D}^u + b_j^u\right) \tag{10}$$

where * means the convolutional operation. $b_j^u \in \mathbb{R}$ is a bias term and $f_a(\cdot)$ is a nonlinear activation function such as the rectified linear unit (ReLU) function. Without the activation function, the output of the next layer is a linear function of the input of the previous layer. Additionally, it is easy to prove that no matter how many convolutional layers there are, the output is a linear combination of inputs, which means the network has no hidden layer. Therefore, activation functions can improve the effectiveness of neural networks.

There are C kernels $\left\{\mathbf{w}_1^u, \mathbf{w}_2^u, \cdots, \mathbf{w}_j^u, \cdots \mathbf{w}_C^u\right\}$ of size H_i to produce C feature maps as follows:

$$\mathbf{o}_i^u = \left[o_{1,i}^u, o_{2,i}^u, \cdots, o_{j,i}^u, \cdots, o_{C,i}^u\right]^T \tag{11}$$

After first convolution, the feature maps of kernel size H_i are represented by $\mathbf{D}_i(N, C, T - K + 1)$.

In order to extract the time features and compress the amount of data, the feature maps of the first convolutional layer should be convoluted multiple times. Thus, there are multiple convolutional layers in the proposed neural network. It is worth noting that the kernel size of the previous layer is not necessarily equal to that of the next layer. For instance, the kernel size of $\mathbf{D}_i$ in the upper layer is H_{i1}, and in the next layer is H_{i2}. H_{i1} and H_{i2} are independent of each other. After passing through these convolutional layers, the feature maps of kernel size $\{H_{i1}, H_{i2}, \cdots, H_{iM}\}$ are expressed as:

$$\mathbf{D}_i(N, C, T - K_1 - K_2 - \cdots - K_M + M) \tag{12}$$

where K_M is the kernel length of convolutional layer M.

3.2.2. Pooling Layer

After multiple convolutional operations, the data come to the pooling layer. In this paper, a max pooling layer is adopted. In the max pooling layers, only the maxima of extracted feature values are retained and all others are discarded. The max pooling layer can extract the strongest feature and discard the weaker ones. After the max pooling operation, the output is described as $\mathbf{D}_i(N, C, 1)$.

3.2.3. Fully-Connected Layer

In the fully-connected layer, the input is the stack of the pooling layer's output. Then, we use a two-class classification, the softmax activation function, to calculate the classification result which consists of two probabilities. When the probability of committing electricity theft is greater than that of being normal, the input data are labeled as electricity theft. The final output of the entire model is expressed as:

$$f_{Softmax}\left[\mathbf{D}(N, \sum_i C, 1)\right] = \mathbf{D}(N, 2, 1) \tag{13}$$

3.2.4. Parameters of the Proposed Neural Network

The main parameters of the proposed neural network are as follows. We utilize two convolutional layers to extract the features. The two convolutional layers are selected to make a balance between the accuracy and computational time. In specific, the increase of convolutional layers may improve the accuracy, but the computation burden also increases significantly with the increase of convolutional layers. Therefore, we used two convolutional layers to balance the accuracy and the efficiency in the experiments in this paper. Moreover, more layers typically means a larger number of parameters, which makes the enlarged network more prone to overfitting [29].

Each convolutional layer has multiple kernels with different sizes. Considering the characteristics of the TextCNN, the height of the kernels is same as the number of data points from one day and the lengths of kernels are 2, 3, 5 and 7. Kernels with a length of 2 or 3 can capture features from adjacent days. Additionally, kernels with lengths of 5 and 7 can capture features from the periodicity of weekday and week, respectively. Besides, in order to reduce the risk of overfitting, the dropout rate of the proposed neural network is set at 0.4.

3.3. Data Augmentation

When using CNN to cope with the classification problem, it requires a large amount of data in various categories for training to obtain a more accurate classification result. Therefore, multiple methods are used to increase the image samples in the image classification problem [34,35]. In realistic datasets, since most users do not carry out electricity theft, there are less electricity theft data compared with the normal data. The imbalance of the datasets would affect the classification result easily, which could contribute to low accuracy or overfitting. Therefore, we propose a data augmentation method to address the imbalance problem.

The data augmentation is illustrated in Figure 7. Assuming the date of electricity theft is found on day D_T, then $[D_T - T, D_T]$ is an electricity theft sample. Due to the continuation of the theft behavior, electricity theft also occurs during the time $[D_T - T - 1, D_T - 1]$. Therefore, $[D_T - T - 1, D_T - 1]$ is also an electricity theft sample. If the intercepted window slides AG times, one electricity user datum can be transformed into $AG + 1$ samples. So far, the electricity theft samples can be increased effectively through this method. It is noted that the value of AG needs to be chosen appropriately. If AG is too large, it will classify the normal data into the theft samples and affect the classification result.

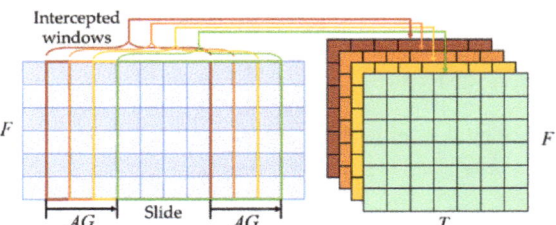

Figure 7. Diagram of the data augmentation method.

4. Experimental Settings

In this Section, we present the details of datasets. Then, we introduce several methods for comparison and the metrics to evaluate the accuracy of the classification model.

4.1. Datasets

Datasets (a) and (b) are realistic datasets from a certain province of China, containing electricity thieves and normal users. Dataset (c) is the public power data of Northern Ireland, which lacks the electricity theft data, so the electricity theft data in dataset (c) is artificially constructed.

(a) Residential user dataset

The amount of residential user data within 1277 days is quite large, so data filtering was required. There were 1063 electricity thieves; all of them were retained. The number of normal users is particularly large, reaching several million. In order to achieve a better classification effect, the ratio of electricity theft data to normal user data should be around 1:3. Therefore, the amount of normal user data finally obtained was 3564. The smart meters collected 5 data points per day, which means the number of data points from one day F was 5.

(b) Industrial user dataset

The industrial user dataset contains the electricity data of 8144 users within 1277 days, and smart meters also collected 5 data points per day, which means the number of data points from one day F was 5. Compared with the residential user dataset, the number of electricity thieves in the industrial user dataset is even smaller—only 92. The electricity thieves only occupy nearly 1% of all. Therefore, the proposed data augmentation method was used to increase the amount of the electricity theft data.

(c) Ireland residential user dataset

The dataset contains the electricity data of 5000 users within 535 days in Ireland, and smart meters collected 48 data per day (sampling every half hour), which means the number of data points from one day F was 48. These users were all normal users, so the dataset lacks electricity theft samples. As a result, we adopted a method introduced in [11] to produce electricity theft samples artificially. Since the electricity theft samples in this dataset are completely artificially generated, their number can be easily changed without using the data augmentation.

The details of each dataset are shown in Table 1.

Table 1. Datasets' information.

Datasets	(a)	(b)	(c)
Time	1 October 2015–31 March 2019	1 October 2015–31 March 2019	1 January 2009–31 December 2010
Total uses	4627	8144	5000
Normal uses	3564	8052	5000
Electricity thieves	1063	92	0

In dataset (a), the ratio of normal users to electricity thieves is 3.3:1, which satisfies the balance between positive data and negative data for training. However, in dataset (b) and dataset (c), electricity theft data needs to be augmented in order to satisfy the balance. In dataset (b), the proposed data augmentation method increases the electricity theft data in the training set to 520, and the ratio of normal data to electricity theft data in the training set is 2.2:1. In dataset (c), 1800 electricity theft samples are artificially generated, and the ratio is 2.8:1.

4.2. Baselines

Other than our proposed model, four classical models in machine learning are given for comparison. The basic parameters setting for baseline methods are summarized in Table 2.

- Logistic regression (LR). Logistic regression is a statistical model that models the probabilities for classification problems with the dependent variable being binary. It uses maximum likelihood estimation to estimate regression model coefficients that explain the relationship between input and output.
- Support vector machine (SVM). A support vector machine is a supervised learning model and can be used for classification. It uses a kernel trick to map the input into high-dimensional feature spaces implicitly. Then, SVMs construct hyperplane in high-dimensional space, and the hyperplane can be used for classification.

- Deep neural network (DNN). A deep neural network is a feedforward neural network with multilayered hidden layers. DNN can model complex non-linear relationships through the neurons in the hidden layer, which can be used for classification problem. Moreover, backpropagation algorithm is used to update the weight in DNN, because it can compute the gradient of the loss function with respect to the weights of the network efficiently.
- One-dimensional CNN (1D-CNN). The 1D-CNN is a classifier model which is similar to the proposed model. However, the user data are 1D electricity consumption data, and the dimensions of input data would be $D(N, 1, T)$. The structure of 1D CNN is the same as the proposed model mentioned in Section 3.2.

The general process of the experiments for all methods is as follows:

At first, we divide one dataset into two parts, one for training and the other for effect evaluation. The ratio of these two parts is called the training ratio. It is worth noting that positive samples and negative samples are divided separately, so the ratio of positive and negative samples in the training dataset is the same as that in the test dataset. At each training ratio, we implement ten experiments. The division of the training dataset and the test dataset is random and independent in each experiment. At last, we use the average result of these ten experiments to represent the final results.

Table 2. Parameter settings.

Baselines	Data Dimension	Parameters
LR	1-D	Penalty: L1 Solver: Liblinear Inverse of regularization strength: 1
SVM	1-D	Regularization parameter: 1.0 Kernel: RBF
DNN	1-D	Hidden layer: 3 Neurons in the hidden layer: 100, 60, 60
1D-CNN	1-D	Same as parameters of the proposed method
Proposed Method	2-D	Introduced in Section 3.2.4

4.3. Metrics

There are many ways to evaluate the classification accuracy. The evaluation metrics used in this paper are accuracy rate, precision rate, recall rate and F1.

The above four metrics were calculated based on the confusion matrix shown in Table 3.

Table 3. Confusion matrix.

Confusion Matrix		Actual	
		Negative (Normal)	Positive (Theft)
Classified	Negative (normal)	True Negative (TN)	False Negative (FN)
	Positive (theft)	False Positive (FP)	True Positive (TP)

In this paper, our purpose is to detect electricity theft. Therefore, we define electricity theft samples as positive samples, and normal samples as negative samples. Furthermore, metrics true positive (TP), true negative (TN), false positive (FP) and false negative (FN) can be obtained from the confusion matrix. TP and TN indicate that the actual attribute of the sample is the same as the classified one, which means the classification result is accurate. FP indicates that the sample is actually negative, but the classified result is positive. FN indicates that the sample is actually positive, while the classified result is negative. The contrast between actual and classified results reflects the inaccuracy of the classification model.

Accuracy rate (AR) is the proportion of correctly classified samples in all samples. It is the most intuitive and commonly used criterion to measure the classification effect of the model. The formula is as follows:

$$AR = \frac{TP + TN}{TP + TN + FP + FN} \times 100\% \quad (14)$$

However, most samples in the training set are normal, and only a few of them committed electricity theft, which means that there are far more actual negative samples than actual positive samples. If the model classifies all actual positive samples into negative, the accuracy rate of the model will still be very high. Therefore, only using the AR criterion to evaluate the accuracy is not comprehensive.

Precision rate (PR) refers to the proportion of actual positive results in the classified positive samples, which indicates the classification accuracy in the classified positive samples. The formula is as follows:

$$PR = \frac{TP}{TP + FP} \times 100\% \quad (15)$$

Recall rate (RR) is defined as the proportion of classified positive results in the actual positive samples, which means the classification accuracy in the actual positive samples. The formula is as follows:

$$RR = \frac{TP}{TP + FN} \times 100\% \quad (16)$$

F_score is the harmonic mean of the precision rate and the recall rate, so it is more comprehensive to evaluate the accuracy. The formula is as follows:

$$F\_score = \frac{(\alpha^2 + 1) \times PR \times RR}{\alpha^2 \times (PR + RR)} \times 100\% \quad (17)$$

where α is a parameter greater than 0. In particular, when α equals one, the F_score is expressed as F1, which is the most representative criterion in common use. The formula is as follows:

$$F1 = \frac{2 \times PR \times RR}{PR + RR} \times 100\% \quad (18)$$

All in all, we construct a confusion matrix and four indicators AR, PR, RR and F1 to comprehensively consider the accuracy of the classification model. In the next section, we will analyze different models in different datasets based on the proposed metrics.

5. Results and Analysis

In this Section, we present the experimental results and analysis. We compare the performances of the proposed model with those of other methods first. Then, we study the influences of the parameters on the results. Last, we discuss the effectiveness of the proposed data augmentation method.

5.1. Performance Comparison

The performance comparison between the proposed model and other models in three datasets is demonstrated in Table 4.

The proposed model performs better than other models in different training ratios, as shown in Table 4. Take the result of a 70% training ratio as an example. The proposed model has the highest PR and RR for each dataset. However, other models had better ARs in some dataset. For example, the AR of the 1D-CNN model was the highest for dataset (a)—4.3% higher than that of the proposed model. However, F1 (which is the most comprehensive indicator of the classification performance) of the proposed model was the highest in each dataset and reached 0.757, 0.850 and 0.904 in dataset (a), dataset (b) and dataset (c), which is 20.1%, 15.2% and 8.9% higher than the second-place model respectively. Meanwhile, the proposed model performed better with the increase in the training ratio.

For example, in dataset (c), the F1 increased from 0.759 to 0. 896 as the training ratio increased from 50 to 80%.

Table 4. Results on different datasets with different models.

Model	Training = 50%											
	Dataset (a)				Dataset (b)				Dataset (c)			
	AR	PR	RR	F1	AR	PR	RR	F1	AR	PR	RR	F1
LR	0.851	0.623	0.488	0.547	0.577	0.487	0.442	0.463	0.867	0.706	0.827	0.762
SVM	0.694	0.733	0.022	0.042	0.523	1.000	0.023	0.045	0.793	0.886	0.223	0.356
DNN	0.843	0.596	0.429	0.487	0.514	0.384	0.349	0.366	0.844	0.655	0.543	0.575
1D-CNN	0.871	0.689	0.439	0.536	0.714	0.913	0.488	0.636	0.843	0.682	0.677	0.787
Proposed CNN	0.830	0.956	0.601	0.738	0.795	0.945	0.634	0.759	0.870	0.719	0.803	0.835

Model	Training = 60%											
	Dataset (a)				Dataset (b)				Dataset (c)			
	AR	PR	RR	F1	AR	PR	RR	F1	AR	PR	RR	F1
LR	0.851	0.614	0.519	0.562	0.654	0.586	0.531	0.557	0.898	0.748	0.888	0.812
SVM	0.700	0.917	0.027	0.052	0.676	0.727	0.094	0.167	0.810	0.838	0.290	0.431
DNN	0.851	0.618	0.429	0.497	0.732	0.375	0.562	0.450	0.863	0.683	0.600	0.635
1D-CNN	0.890	0.730	0.500	0.594	0.719	0.944	0.531	0.680	0.812	0.610	0.615	0.744
Proposed CNN	0.846	0.931	0.669	0.779	0.834	0.952	0.720	0.819	0.919	0.825	0.876	0.897

Model	Training = 70%											
	Dataset (a)				Dataset (b)				Dataset (c)			
	AR	PR	RR	F1	AR	PR	RR	F1	AR	PR	RR	F1
LR	0.852	0.633	0.496	0.556	0.678	0.714	0.400	0.513	0.907	0.725	0.930	0.815
SVM	0.700	0.833	0.030	0.058	0.660	0.500	0.094	0.158	0.833	0.793	0.324	0.460
DNN	0.855	0.636	0.411	0.494	0.833	0.692	0.360	0.474	0.856	0.660	0.624	0.636
1D-CNN	0.875	0.778	0.398	0.527	0.735	0.833	0.600	0.698	0.840	0.750	0.500	0.750
Proposed CNN	0.893	0.839	0.690	0.757	0.844	0.952	0.756	0.850	0.920	0.785	0.966	0.904

Model	Training = 80%											
	Dataset (a)				Dataset (b)				Dataset (c)			
	AR	PR	RR	F1	AR	PR	RR	F1	AR	PR	RR	F1
LR	0.837	0.610	0.472	0.532	0.652	0.529	0.529	0.529	0.917	0.712	0.977	0.824
SVM	0.695	0.778	0.035	0.066	0.660	0.615	0.094	0.163	0.833	0.684	0.302	0.419
DNN	0.856	0.630	0.434	0.511	0.784	0.522	0.706	0.600	0.857	0.653	0.630	0.635
1D-CNN	0.859	0.800	0.359	0.496	0.714	0.909	0.588	0.741	0.833	0.705	0.574	0.762
Proposed CNN	0.723	0.908	0.742	0.816	0.901	0.958	0.841	0.896	0.958	0.857	1.000	0.947

It is also worth noting that the proposed model had better universality and performance in the realistic dataset. Comparing dataset (c) with the realistic dataset (a) and dataset (b), the F1 of the proposed model with dataset (c) reached about 0.95, but 0.816 for dataset (a) and 0.896 for dataset (b) when the training ratio was 80%. This is mainly because that the electricity theft data in dataset (c) were artificially generated, of which the data features can be identified and extracted easily by machine learning models. However, realistic electricity theft data are more complicated and lack regularity. Therefore, the results in realistic datasets are relatively worse than those in dataset (c). However, compared with other models, the performance of the proposed model was still the highest.

The comparing results show that the proposed model has better performance overall, which implies that the proposed model has higher accuracy in electricity theft detection.

5.2. Parameter Study

To study the effect of the length of the intercepted window T on the proposed model, we conducted an experiment on dataset (a) and dataset (b) by changing the value of T from 10 to 500 with a step size of 10. Figure 8a,b shows the experiment results of dataset (a) and dataset (b), respectively.

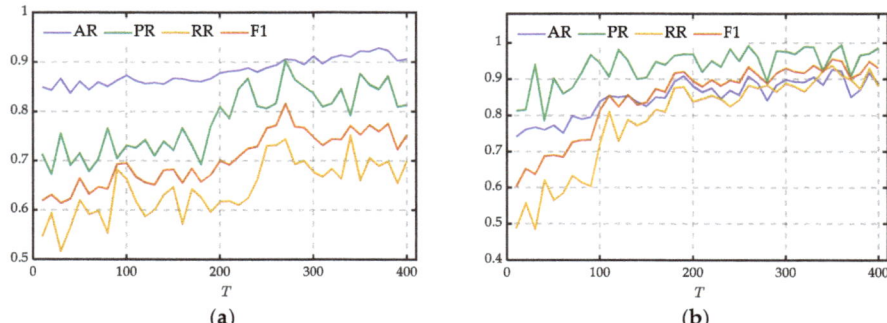

Figure 8. Performances of different T. (**a**) Trends of indicators in dataset (a); (**b**) trends of indicators in dataset (b).

The length of the intercepted window has an important impact on the performance, as shown in Figure 8. Four indicators all increased with the increasing of T at first. Then they no longer increased and began to fluctuate when T exceeded a certain value and continued to increase. In Figure 8a, PR, RR and F1 achieved their maxima when T was about 270. In Figure 8b, four indicators achieved their maxima when T was about 200. This is mainly because that the features of electricity theft data are easier to be extracted with more electricity consumption information when T increases, which leads to the improvement of the performance.

Therefore, to achieve the best performance of the proposed model, it is necessary to investigate an appropriate length of the intercepted window T for electricity theft detection.

5.3. Data Augmentation Analysis

The proposed data augmentation method was used to augment the electricity theft data in dataset (b). To study the effectiveness of the proposed data augmentation method, we varied the value of AG, which represents the repeated times, from 0 to 20 with a step of 1. At the same time, other parameters were fixed.

The comparison results of different AG in training ratios of 50% and 80% are given in Figure 9a,b respectively. The four indicators all increased at first as AG increased, while the classification accuracy decreased after AG exceeded a value. The indicators had a positive relationship with AG in the early stage because the increase in the amount of electricity theft data during the training was of great help to the classification accuracy, which can effectively increase the number of TP (true positives) in the classification result. Therefore, all four indicators had an upward trend.

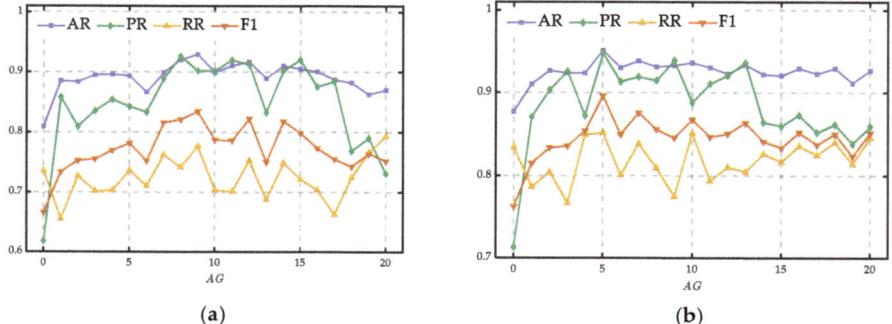

Figure 9. Performances of different AG. (**a**) Trends of indicators with a 50% training ratio; (**b**) trends of indicators with an 80% training ratio.

When *AG* continued to increase, AR, RR and F1 dropped, while PR fluctuated. The main reason is that the normal data were labeled as electricity theft during the data preprocessing when the repeated time was too large, so the training model tended to classify the normal users into abnormal users. As a result, the numbers of FP (false positives) and FN (false negatives) increased, while the numbers of TP and TN (true negative) declined in the classification result. Therefore, the classification accuracy dropped and most indicators decreased, especially the most important indicator F1.

All in all, the data augmentation which increases the number of electricity theft data points for CNN training can improve the classification accuracy effectively. It is also important to choose an appropriate *AG* to achieve better classification results, because the indicators for accuracy may fluctuate with inappropriate *AG*.

6. Conclusions

In this paper, we propose a novel electricity theft detection scheme based on TextCNN. We innovatively formulated the electricity data into two-dimensional time-series in order to capture the intraday and daily correlations of electricity consumption data. Then, we discussed the relationship between DNN, CNN and TextCNN, and explained why TextCNN is the most suitable classifier for our purposes, considering both the efficiency and effectiveness. Additionally, in order to balance the electricity consumption dataset, we proposed a data augmentation method. We conducted extensive experiments on different realistic datasets to prove the effectiveness of the proposed scheme, including the residential and industrial datasets from a province in China and the public Irish residential dataset. The experimental results show that the proposed method outperforms other methods, such as LR, SVM, DNN and 1D CNN. At the same time, we analyzed the importance and effectiveness of data augmentation.

Author Contributions: Conceptualization and methodology, X.F. and H.H.; software, Z.L. and Y.Y.; validation, W.G., H.Q. and H.F.; writing, H.H., Z.L., Y.Y. and C.Y.; supervision and project administration, X.F., H.Q., H.F. and Y.D. All authors have read and agreed to the published version of the manuscript.

Funding: This research was funded by Project Supported by the China Southern Power Grid Corporation, grant number GDKJXM20185800.

Conflicts of Interest: The authors declare no conflict of interest.

References

1. Depuru, S.S.S.R.; Wang, L.; Devabhaktuni, V. Electricity theft: Overview, issues, prevention and a smart meter based approach to control theft. *Energy Policy* **2011**, *39*, 1007–1015. [CrossRef]
2. Venkatachary, S.K.; Prasad, J.; Samikannu, R. Overview, issues and prevention of energy theft in smart grids and virtual power plants in Indian context. *Energy Policy* **2017**, *110*, 365–374. [CrossRef]
3. Northeast Group, LLC. Electricity Theft & Non-Technical Losses: Global Markets, Solutions, and Vendors. Available online: http://www.northeast-group.com/reports/Brochure-Electricity%20Theft%20&%20Non-Technical%20Losses%20-%20Northeast%20Group.pdf (accessed on 20 September 2020).
4. Liu, Z. Over 110 MWh in 35 Years, Electricity Theft Arrested in Shaoyang. Available online: http://epaper.voc.com.cn/sxdsb/html/2018-08/02/content_1329743.htm?div=-1 (accessed on 20 September 2020).
5. Messinis, G.M.; Hatziargyriou, N.D. Review of non-technical loss detection methods. *Electr. Power Syst. Res.* **2018**, *158*, 250–266. [CrossRef]
6. Short, T.A. Advanced Metering for Phase Identification, Transformer Identification, and Secondary Modeling. *IEEE Trans. Smart Grid* **2013**, *4*, 651–658. [CrossRef]
7. Leite, J.B.; Mantovani, J.R.S. Detecting and Locating Non-Technical Losses in Modern Distribution Networks. *IEEE Trans. Smart Grid* **2018**, *9*, 1023–1032. [CrossRef]
8. Jiang, R.; Lu, R.; Wang, Y.; Luo, J.; Shen, C.; Shen, X. Energy-theft detection issues for advanced metering infrastructure in smart grid. *Tsinghua Sci. Technol.* **2014**, *19*, 105–120. [CrossRef]

9. Glauner, P.; Dahringer, N.; Puhachov, O.; Meira, J.A.; Valtchev, P.; State, R.; Duarte, D. Identifying Irregular Power Usage by Turning Predictions into Holographic Spatial Visualizations. In Proceedings of the 2017 IEEE International Conference on Data Mining Workshops (ICDMW), New Orleans, LA, USA, 18–21 November 2017; pp. 258–265.
10. Buzau, M.-M.; Tejedor-Aguilera, J.; Cruz-Romero, P.; Gomez-Exposito, A. Hybrid Deep Neural Networks for Detection of Non-Technical Losses in Electricity Smart Meters. *IEEE Trans. Power Syst.* **2020**, *35*, 1254–1263. [CrossRef]
11. Jokar, P.; Arianpoo, N.; Leung, V.C.M. Electricity Theft Detection in AMI Using Customers' Consumption Patterns. *IEEE Trans. Smart Grid* **2016**, *7*, 216–226. [CrossRef]
12. Nagi, J.; Yap, K.S.; Tiong, S.K.; Ahmed, S.K.; Mohamad, M. Nontechnical Loss Detection for Metered Customers in Power Utility Using Support Vector Machines. *IEEE Trans. Power Deliv.* **2010**, *25*, 1162–1171. [CrossRef]
13. Nagi, J.; Yap, K.S.; Tiong, S.K.; Ahmed, S.K.; Nagi, F. Improving SVM-Based Nontechnical Loss Detection in Power Utility Using the Fuzzy Inference System. *IEEE Trans. Power Deliv.* **2011**, *26*, 1284–1285. [CrossRef]
14. Jindal, A.; Dua, A.; Kaur, K.; Singh, M.; Kumar, N.; Mishra, S. Decision Tree and SVM-Based Data Analytics for Theft Detection in Smart Grid. *IEEE Trans. Ind. Inf.* **2016**, *12*, 1005–1016. [CrossRef]
15. Wu, R.; Wang, L.; Hu, T. AdaBoost-SVM for Electrical Theft Detection and GRNN for Stealing Time Periods Identification. In Proceedings of the IECON 2018—44th Annual Conference of the IEEE Industrial Electronics Society, Washington, DC, USA, 21–23 October 2018; pp. 3073–3078.
16. Adil, M.; Javaid, N.; Qasim, U.; Ullah, I.; Shafiq, M.; Choi, J.-G. LSTM and Bat-Based RUSBoost Approach for Electricity Theft Detection. *Appl. Sci.* **2020**, *10*, 4378. [CrossRef]
17. Zheng, Z.; Yang, Y.; Niu, X.; Dai, H.-N.; Zhou, Y. Wide and Deep Convolutional Neural Networks for Electricity-Theft Detection to Secure Smart Grids. *IEEE Trans. Ind. Inf.* **2018**, *14*, 1606–1615. [CrossRef]
18. Hasan, M.N.; Toma, R.N.; Nahid, A.-A.; Islam, M.M.M.; Kim, J.-M. Electricity Theft Detection in Smart Grid Systems: A CNN-LSTM Based Approach. *Energies* **2019**, *12*, 3310. [CrossRef]
19. Kim, T.T.; Poor, H.V. Strategic Protection Against Data Injection Attacks on Power Grids. *IEEE Trans. Smart Grid* **2011**, *2*, 326–333. [CrossRef]
20. Zanetti, M.; Jamhour, E.; Pellenz, M.; Penna, M.; Zambenedetti, V.; Chueiri, I. A Tunable Fraud Detection System for Advanced Metering Infrastructure Using Short-Lived Patterns. *IEEE Trans. Smart Grid* **2019**, *10*, 830–840. [CrossRef]
21. Wang, X. Analysis of Typical Electricity Theft Cases—Adjust the Metering Time of Meters to Avoid the Peak Period Tariffs. Available online: https://www.zhangqiaokeyan.com/academic-conference-cn_meeting-7953_thesis/020222030513.html (accessed on 22 September 2020).
22. Han, W.; Xiao, Y. Combating TNTL: Non-Technical Loss Fraud Targeting Time-Based Pricing in Smart Grid. In Proceedings of the Cloud Computing and Security, Nanjing, China, 29–31 July 2016; Volume 10040, pp. 48–57.
23. Dhillon, A.; Verma, G.K. Convolutional neural network: A review of models, methodologies and applications to object detection. *Prog. Artif. Intell.* **2020**, *9*, 85–112. [CrossRef]
24. Jiao, J.; Zhao, M.; Lin, J.; Liang, K. A comprehensive review on convolutional neural network in machine fault diagnosis. *Neurocomputing* **2020**, *417*, 36–63. [CrossRef]
25. Girshick, R.; Donahue, J.; Darrell, T.; Malik, J. Rich Feature Hierarchies for Accurate Object Detection and Semantic Segmentation. In Proceedings of the 2014 IEEE Conference on Computer Vision and Pattern Recognition, Columbus, OH, USA, 23–28 June 2014; pp. 580–587.
26. Girshick, R. Fast R-CNN. In Proceedings of the 2015 IEEE International Conference on Computer Vision (ICCV), Santiago, Chile, 7–13 December 2015; pp. 1440–1448.
27. Kim, Y. Convolutional Neural Networks for Sentence Classification. In Proceedings of the 2014 Conference on Empirical Methods in Natural Language Processing (EMNLP), Doha, Qatar, 25–29 October 2014; pp. 1746–1751.
28. Zhang, Y.; Wallace, B. A Sensitivity Analysis of (and Practitioners' Guide to) Convolutional Neural Networks for Sentence Classification. In Proceedings of the Eighth International Joint Conference on Natural Language Processing, Taipei, Taiwan, 27 November–1 December 2017; Volume 1, pp. 253–263.

29. Szegedy, C.; Wei, L.; Yangqing, J.; Sermanet, P.; Reed, S.; Anguelov, D.; Erhan, D.; Vanhoucke, V.; Rabinovich, A. Going deeper with convolutions. In Proceedings of the 2015 IEEE Conference on Computer Vision and Pattern Recognition (CVPR), Boston, MA, USA, 7–12 June 2015; pp. 1–9.
30. Krizhevsky, A.; Sutskever, I.; Hinton, G.E. ImageNet classification with deep convolutional neural networks. *Commun. ACM* **2017**, *60*, 84–90. [CrossRef]
31. Simonyan, K.; Zisserman, A. Very Deep Convolutional Networks for Large-Scale Image Recognition. In Proceedings of the 3rd International Conference on Learning Representations, ICLR 2015, San Diego, CA, USA, 7–9 May 2015.
32. Szegedy, C.; Vanhoucke, V.; Ioffe, S.; Shlens, J.; Wojna, Z. Rethinking the Inception Architecture for Computer Vision. In Proceedings of the 2016 IEEE Conference on Computer Vision and Pattern Recognition (CVPR), Las Vegas, NV, USA, 27–30 June 2016; pp. 2818–2826.
33. Chandola, V.; Banerjee, A.; Kumar, V. Anomaly detection: A survey. *ACM Comput. Surv.* **2009**, *41*, 15:1–15:58. [CrossRef]
34. Haixiang, G.; Yijing, L.; Shang, J.; Mingyun, G.; Yuanyue, H.; Bing, G. Learning from class-imbalanced data: Review of methods and applications. *Expert Syst. Appl.* **2017**, *73*, 220–239. [CrossRef]
35. Zhou, L. Performance of corporate bankruptcy prediction models on imbalanced dataset: The effect of sampling methods. *Knowl. Based Syst.* **2013**, *41*, 16–25. [CrossRef]

Publisher's Note: MDPI stays neutral with regard to jurisdictional claims in published maps and institutional affiliations.

© 2020 by the authors. Licensee MDPI, Basel, Switzerland. This article is an open access article distributed under the terms and conditions of the Creative Commons Attribution (CC BY) license (http://creativecommons.org/licenses/by/4.0/).

Identification of the State of Electrical Appliances with the Use of a Pulse Signal Generator

Augustyn Wójcik [1,*], Piotr Bilski [1,*], Robert Łukaszewski [1,*], Krzysztof Dowalla [1,*] and Ryszard Kowalik [2,*]

1 Institute of Radioelectronics and Multimedia Technologies, Warsaw University of Technology, Nowowiejska 15/19, 00-665 Warsaw, Poland
2 Institute of Electrical Power Engineering, Warsaw University of Technology, Koszykowa 75, 00-662 Warsaw, Poland
* Correspondence: a.wojcik@ire.pw.edu.pl (A.W.); p.bilski@ire.pw.edu.pl (P.B.); r.lukaszewski@ire.pw.edu.pl (R.Ł.); k.dowalla@ire.pw.edu.pl (K.D.); ryszard.kowalik@ee.pw.edu.pl or r.kowalik@ien.pw.edu.pl (R.K.)

Abstract: The paper presents the novel HF-GEN method for determining the characteristics of Electrical Appliance (EA) operating in the end-user environment. The method includes a measurement system that uses a pulse signal generator to improve the quality of EA identification. Its structure and the principles of operation are presented. A method for determining the characteristics of the current signals' transients using the cross-correlation is described. Its result is the appliance signature with a set of features characterizing its state of operation. The quality of the obtained signature is evaluated in the standard classification task with the aim of identifying the particular appliance's state based on the analysis of features by three independent algorithms. Experimental results for 15 EAs categories show the usefulness of the proposed approach.

Keywords: NILM; signature; load disaggregation; transients; pulse generator

1. Introduction

The Non-Intrusive Appliance Load Monitoring (NIALM or NILM) [1] is a solution for the problem of collecting electrical energy consumption data more accurately than using only typical electricity meters. The methodology (also known as energy disaggregation [2]) is used for power systems analysis, in which demand for energy continuously increases. The purpose of the appliances' load identification is to provide information about the energy consumption of individual devices. This may lead to a decrease in electricity consumption and suppressing environmental pollution [3]. According to [4] the application of NIALM approaches might lead to a reduction of household energy consumption by at least 12%. Another potential application is the diagnostics of electrical appliances [5], like monitoring device degradation or detecting supply network's state in the presence of external disturbances, like voltage spikes, insulation decrease, etc. In the NILM architecture, measurements are done close to the energy meter, in contrast to intrusive systems where every socket or device is equipped with a suitable sensor [6]. When new appliances are plugged into such systems, the measurement hardware is not expanded. Acquired values are typically aggregated currents and voltages [7]. Characteristic features allowing for the identification of a particular Electrical Appliance (EA) are obtained individually during training in the specific deployment location.

Over the past 20 years, the topic was widely explored [7–11]. Public databases were prepared to allow for the verification of new approaches [12–14]. The main achievements are summarized, for instance, in [15].

The taxonomy of NILM methods considers multiple criteria. Firstly, they can be classified based on the frequency of the measured signals [7,8]. In [16] four types of frequency-based methods were identified: LF (Low Frequency), MF (Medium Frequency),

HF (High Frequency), and EHF (Extra-High Frequency). The first one exploits the RMS of the current and voltage waveform or the amplitude of its first harmonic collected with the sampling frequency from below one to several Hz. The MF approach processes signal samples collected with a frequency from 1 kHz to dozens of kHz. The HF method operates on transients collected with a sampling frequency from dozens of kHz to several dozens of MHz. Finally, EHF methods operate on sampling frequencies above a dozen of MHz. The latter group was not investigated so far.

The second taxonomy criterion [10,11], is the moment of the appliance analysis. These can be steady-state (SS) or transient state (TS), depending on whether the waveforms are sampled during the state change or after it is already done and no transients are present in the signals. The SS-LF methods are currently the most popular because of the low cost of sensors [17] and the simple mathematical apparatus required to process data with low computational power requirements [18].

The frequency of the measured signals determines features characterizing the analyzed devices. For SS-LF methods the most commonly used parameters are the average power [19–24], reactive power [25,26], and power factor [19]. The SS-MF methods work with the amplitude [27] and phase [28] of the subsequent current harmonics. Power characteristics may be considered as well [3]. The real and imaginary parts of the current odd harmonics were used in [29]. In SS-HF approaches, disturbances not being harmonics of the fundamental component of the power grid (i.e., 50 Hz or 60 Hz) are analyzed. For example, it can be EMI noise specific for different electronic appliances [30].

Some electrical devices, when turned on, generate a short-term current pulse with an amplitude significantly exceeding their nominal supply current [31]. These transients can be used to identify the EA state [32]. Approaches using such features belong to the TS-HF group. In [32], the fundamental voltage component was filtered, from which the frequency components of the disturbances appearing at the moment of switching on the device were extracted. In [33], voltage harmonics were eliminated using Notch filters, and then the signal was analyzed using Wavelet Transform (WT). In [34] the current of device in the transient states was recorded with the 100 kHz sampling frequency. The model of the transient state was created on the basis of the currents collected when turning the EA on. The energy spectra calculated using WT were used in [34]. Next, the energy distribution in the subbands was used to identify appliances.

The areas of HF and EHF methods, although more challenging, especially from the data acquisition point of view, rely on phenomena that cannot be observed at lower sampling rates. These give possibilities of distinguishing between different appliances, as is shown in this paper.

During our previous research [35], it was discovered that parameters of electrical signals related to the operation of a particular EA depend on other devices operating in the network at the same time. As a result, the set of devices working in the background determines features extracted from the waveforms. Such a phenomenon can be used to identify states of individual EAs.

To verify the practical application of the presented phenomenon, the HF-GEN method was developed. It includes a measurement system acquiring the supply current signal complemented by the known impulse signal generator. To detect changes in the impulse parameters it is necessary to use the time-frequency analysis, which allows for comparing impulse signals generated when different loads are switched on. The paper also presents the data processing method used for EAs identification, based on cross-correlation of the reference signal with the analyzed signal.

The outline of the paper is as follows. Section 2 presents the method of determining the characteristics of EA. Section 3 contains the quality assessment method. Section 4 covers experimental results, while Section 5 holds conclusions and future prospects.

2. HF-GEN Method for Determining the Characteristics of EA

Known methods exploiting the analysis of the transient currents and voltages during the appliance's state change are still a minority. Most EA identification algorithms rely on the characteristic features determined in steady states of operation. The transient signal is a result of changing the state of the device (for instance, by turning it on). Electrical signals recorded at the moment of the transient state change must be analyzed. The key to detecting the EA state change is to find proper features of the impulse signal. They should clearly distinguish impulse signals appearing as a result of changes in the states of various EAs. Two problems emerge that significantly limit the applicability of such approaches. First, EAs are switched on with a random voltage phase, so the transient states of the examined EAs also have the random voltage phase. Secondly, EAs in the background influence parameters of the transient signals. Both factors affect the shape of the analyzed impulse and make assigning the specific transient to a device difficult.

In the HF-GEN method, the generated current pulse signal is introduced into the tested power network circuit. The analyzed impulse is therefore the effect of a deliberately created transition state, not related to any EA. The pulse is generated many times at regular intervals. When the EA state changes, the pulse shape also changes, because it is characteristic of the particular EA. Detection of the EA state change consists of observing corresponding changes in the pulse features, which form the EA signature. The latter should unequivocally identify the specific EA. The principle of the HF-GEN method is illustrated in Figure 1. The pulse shape changes between the appliance's "on" and "off" states.

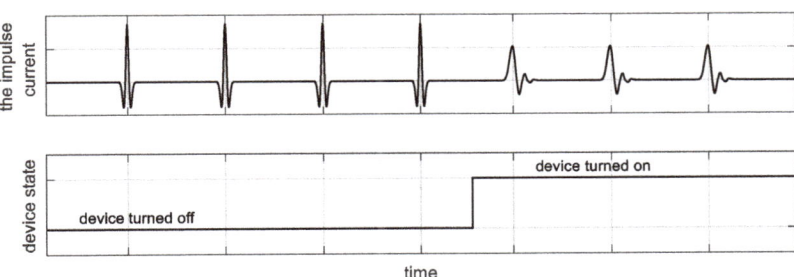

Figure 1. Illustration of the HF-GEN method.

The following were the experiments' assumptions:

- the analysis covers impulse signals generated by the same source,
- the source of transients is a pulse signal generator connected to the tested power network,
- the pulse signal appears in a specific phase of the supply voltage,
- a pulse signal is produced at regular intervals,
- the maximum frequency of measured signals is 15 MHz.

The purpose is to find changes in the pulse signal caused by the load change in the supply circuit. The load on the power circuit depends on the set of EAs connected to it. Characteristics of the impulse signal are related to the specific EA, therefore enabling identification of the moment when the particular device is turned on. The block diagram of the HF-GEN method is shown in Figure 2. The first step is the generation of the impulse signal. The generator detects the supply voltage phase and then inputs the pulse signal to the LV (Low-Voltage) circuit. In the second step, the pulse current is measured with the sampling frequency of 30 MS/s. The acquired samples are processed to select their subset acquired during 4 ms after the pulse detection. Next, cross-correlations between the samples' vector and transients patterns stored in the dictionary are calculated. A signature characterizing the pulse signal is then prepared. Finally, the signature quality is determine Subsequent steps are presented in detail in the sections below.

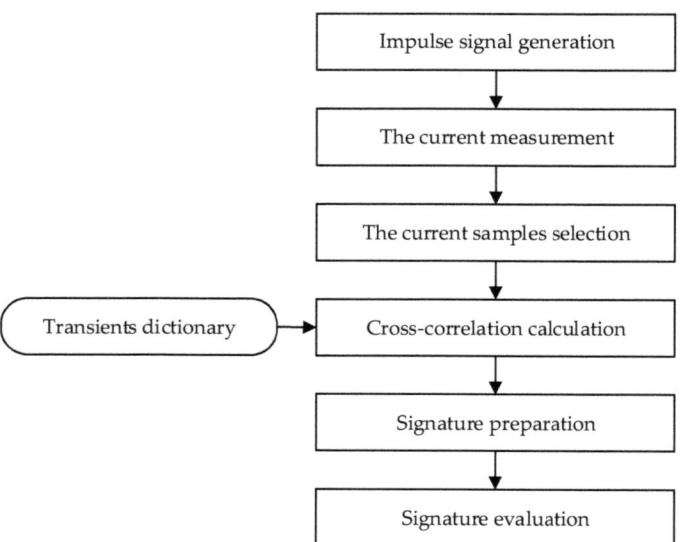

Figure 2. Block diagram of the HF-GEN method.

2.1. Pulse Signal Generation

The block diagram of the pulse signal generator with two connectors/ports is show in Figure 3. The first one, the input and output port (I/O), is connected to the tested circu of the power network with a voltage of 230 V and a frequency of 50 Hz. This port is marke as input (I) because it is used to supply the generator with voltage. It is also treated as t output port (O), because of providing the impulse current signal to the power network. T O-SYN port is used to get the synchronization signal outside the generator. It determin time instances of the pulse signal generation. The synchronization output is used to contr the acquisition system. When designing the generator, the following parameters we assumed:

- the maximum value of the current pulse signal is 10 A,
- the rise time of the pulse is 60μs,
- total pulse duration is less than 1 ms,
- interval between successive impulse triggers is less than 1 s.

The pulse signal generator consists of a matching circuit (MC-GEN), an Analog-t Digital converter (AD-GEN), a Digital Output (DO), a Relay (RE), and an Attached Loa (AL). The measurement and generation system is connected to a computer (PC-GEN) o which the Control Software (CS) is running.

The pulse amplitude depends primarily on the voltage phase in which the Attache Load (AL) is connected to the power grid. The pulse amplitude is proportional to t voltage value at the moment of turning the AL on. Setting the constant voltage phas (the same each time) is the biggest challenge. The time instant must be synchronize with the phase of the supply voltage. The process is as follows: the main voltage $u($ is applied to the MC-GEN, which converts the voltage $u(t)$ into the voltage $u_{AD-GEN}($ whose amplitude matches the dynamic range of the AD-GEN input. The AD-GEN conver voltage $u_{AD-GEN}(t)$ to samples u_n with a speed of 250 kS/s. Based on the voltage samples u the CS detects the supply voltage phase. As a result of its operation, the logic signal o_n given to the input of DO, assuming a high value when the impulse signal is generated. D converts the logic signal o_n to the voltage $u_{SYN}(t)$. The O-SYN synchronization output triggered at the right moment by a high voltage level. The main function of the RE is to apply the supply voltage to AL when a high voltage level appears at $u_{SYN}(t)$.

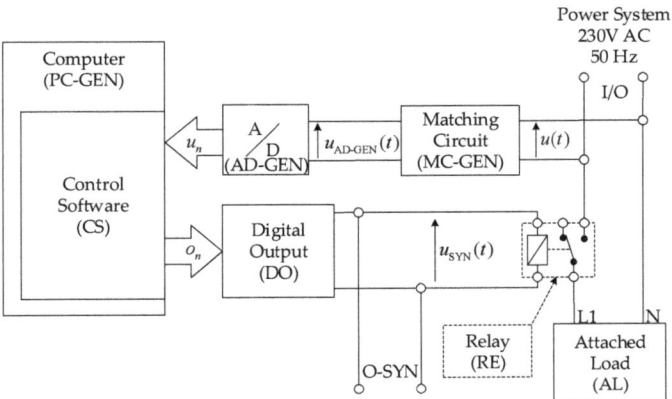

Figure 3. Block diagram of the pulse signal generator.

The pulse shape parameters are determined by the AL. The rise time of the pulse and its total duration depends on the AL impedance. Contrary to the tested EA, AL is a known load with a specific transmittance, temporarily connected to the supply network to change the parameters of the network. In practice, any appliance approved for use in the LV grid, for example, an energy-saving light bulb, can be used as AL. In such a situation, the pulse generator is no different than other appliances in the network. It is a typical load, connected to the network at specified intervals (e.g., 1 s) for a specific time (e.g., 40 ms).

2.2. Measurement Method

In the HF-GEN method, the measured signal is the impulse in the current introduced to the tested circuit by the signal generator. The parameters of the signal change with the load of the tested network after introducing the specific EA. This fact is used to detect the change in the EA state.

The measurement system from Figure 4 consists of a transient generator (GEN), an electrical appliance energy receiver (EA), a Current-Voltage Converter (CVC), an Acquisition Card (AC), a computer (PC), software (SW), and memory (MM). The tested EA and GEN are powered from the network with an RMS voltage of 230 V and frequency of 50 Hz.

The supply network voltage $u(t)$ is provided to GEN through the I/O terminals connected to the phase conductor L1 and the neutral conductor N. The synchronization voltage $u_{SYN}(t)$ is supplied from the synchronization output O-SYN of GEN to the synchronization input of the Analog-to-Digital Converter. High levels of $u_{SYN}(t)$ determine time instants for pulse generation. The current $i(t)$ is converted by the CVC into $u_{AD}(t)$ voltage with a level adjusted to the dynamic range of the analog input of the acquisition card (AC) converter, providing samples i_n.

The voltage $u_{SYN}(t)$ also triggers the acquisition of current samples when a pulse is generated. The SW running on PC controls the AC operation and collects the current samples i_n storing them in MM for further analysis. Due to triggering the AC converter acquisition, the amount of data for processing is significantly reduced.

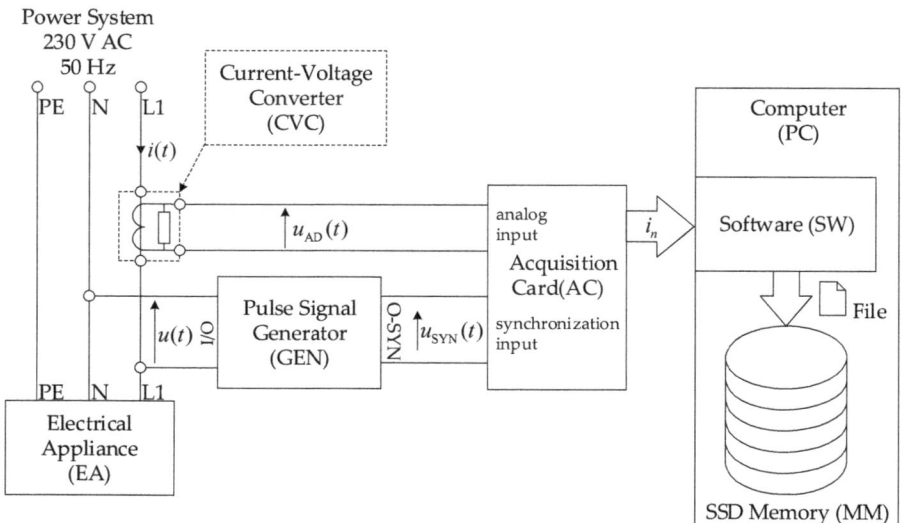

Figure 4. Diagram of the measuring system of the HF-GEN method.

2.3. Selection of Current Samples

The result of data acquisition is the current vector $i = [i_1 \ldots i_N]$ (see Figure 5). contains current samples recorded around (before and after) the pulse manifestation.

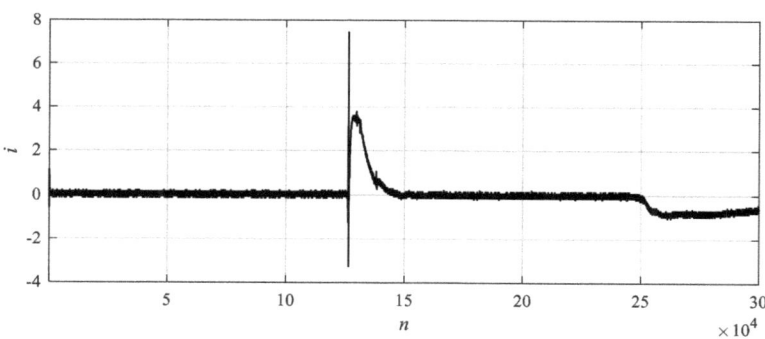

Figure 5. The sampled current vector i.

Due to the effectiveness of further calculations, only a selected fragment of the current vector is analyzed. This is because some fragments of the obtained current data do not contain useful information. Specifically, the current vector contains data measured prior to generating the current pulse (e.g., current vector samples from 1 to 125,000 in Figure 5. The data in this fragment of the current vector bear no information characteristic for the tested EA.

The most relevant is the fragment of the current vector near the largest pulse peak. Therefore only part of the original vector (i.e., i_{SEL}) is extracted for analysis. The vector i is filtered by the high-pass filter with a cut-off frequency of 1 kHz, which enables effective suppression of the 50 Hz component and its harmonics (100 Hz, 150 Hz, and so on. Then, the maximum of the high-frequency components (i.e., above 10 kHz) is found. The vector i_{SEL} contains 2700 selected samples around the maximum of the high-frequency components. Figure 6 shows example of the i_{SEL} vector.

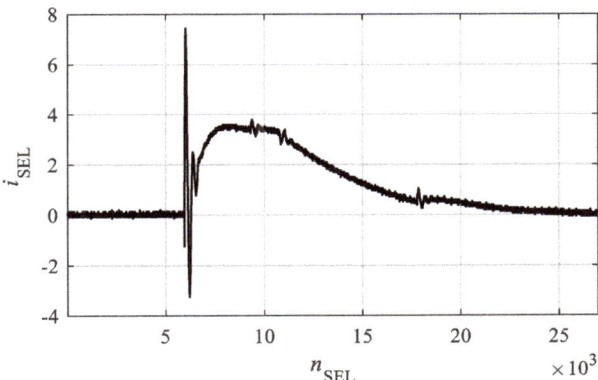

Figure 6. Current vector i_{SEL} for sample measurement data.

2.4. Preparation of a Dictionary of Transients

The dictionary of transients D is a set of selected fragments of the current vectors containing pulses for various appliances:

$$D = \left\{ i_{DIC}^{(1)}, i_{DIC}^{(2)}, \ldots, i_{DIC}^{(l_D)}, \ldots i_{DIC}^{(LDIC)} \right\}, \quad (1)$$

where $i_{DIC}^{(l_D)}$ are the most interesting fragments of vectors i describing the pulse and $LDIC$ is the number of examples. Figure 7 shows the method of preparing the dictionary. Samples from vector i are selected as in Section 2.3. Then, the initial and terminal indexes of the transition are marked, leading to the structure presented below.

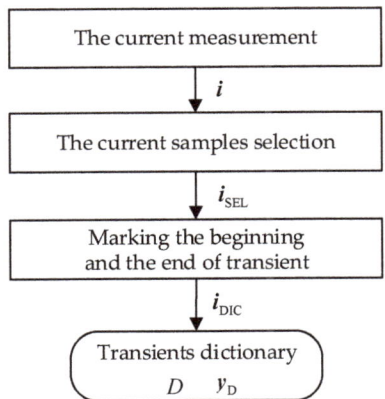

Figure 7. Preparation of dictionary of transients.

The marking process is performed by specifying the initial n_{START} and terminal n_{STOP} indexes. The fragment i_{DIC} is then extracted as follows:

$$i_{DIC} = [i_{DIC,1} \ldots i_{DIC,N_{DIC}}] = \{i_{SEL,n_{START}}, i_{SEL,n_{START}+1}, \ldots, i_{SEL,n_{STOP}-1}, i_{SEL,n_{STOP}}\}, \quad (2)$$

where $N_{DIC} = n_{STOP} - n_{START} + 1$ denotes the number of samples in i_{DIC}.

Figure 8 shows the example of i_{SEL} with the marked indices n_{START} and n_{STOP} (**a**) and the extracted i_{DIC} (**b**).

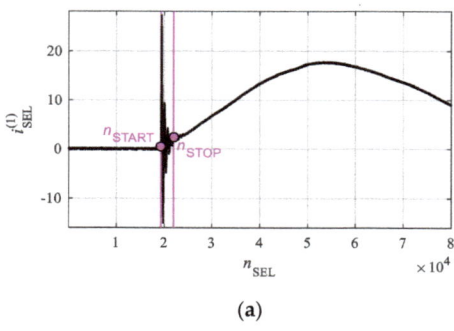 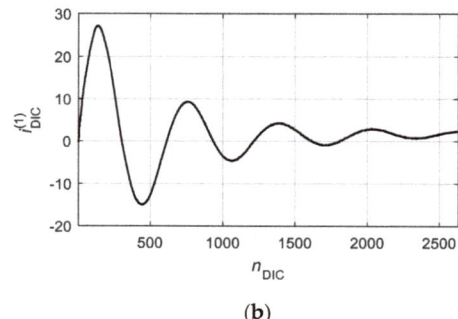

Figure 8. Current vectors for the sample measurement data: i_{SEL} (**a**) and i_{DIC} (**b**).

For each considered EA, 10 examples of transition states were added to the dictionary. They differ in amplitude and shape. The selected number is the compromise between the variety of stored data and the computational effort required to obtain examples. An example is a current vector and corresponding category from the set D_{CAT} (which cardinality determines the number of identified appliances N_{EA}). Therefore, the number of vectors i_D in the dictionary is $LDIC = 10 \cdot N_{EA}$.

2.5. Determining the Cross-Correlation

In this stage, the maximum correlation between the measured signal i_{SEL} and subsequent dictionary entries i_{DIC} is found. The vector i_{SEL} is longer than the current vector from the dictionary i_{DIC}, so the correlation is calculated for all possible shifts between i_D and i_{SEL}.

The vector i_{SEL} has $N_{SEL} = 120,000$ samples (representing the duration of 4 ms for sampling frequency $f_S = 30$ MHz). The correlation will be determined many times for each transition state. Therefore, the method of determining the cross-correlation should be computationally efficient. The determination of the cross-correlation without normalization was considered due to the simplicity and efficiency of calculations. In the discussed problem, the cross-correlation without normalization cannot be used, because the elements of current vectors mainly contain a fundamental component of the current signal with frequency of 50 Hz. On the other hand, pattern vectors only contain components with frequencies at least 200 times greater than the fundamental component. The 50 Hz frequency component significantly changes the average value of the current vector, and as a result significantly affects the value of cross-correlation without the normalization. The measure of similarity between sample vectors based on the Pearson correlation coefficient was used. The mean and standard deviation for each fragment of the vector i_{SEL} was calculated, which requires significant computational effort. Therefore, the optimized calculation method [30] was used.

As a result, vectors of correlations **r** and shifts **c** were obtained. Figures 9–11 illustrate the procedure.

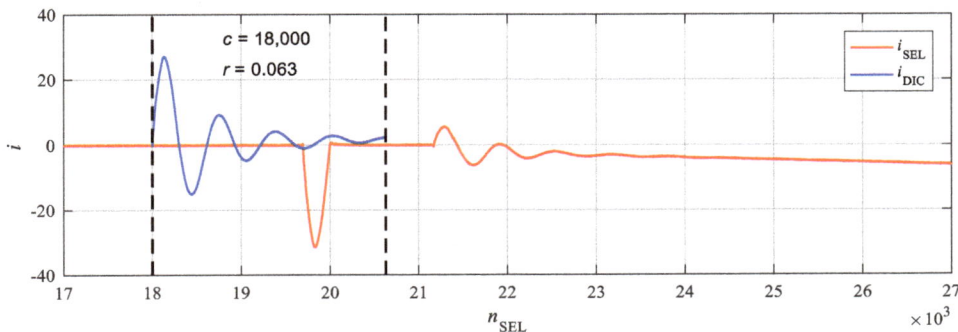

Figure 9. Determining the correlation for sample measurement data, delay c = 18,000.

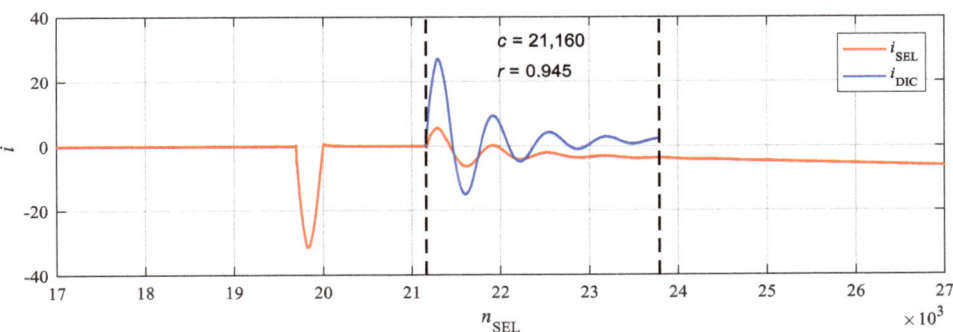

Figure 10. Determining the correlation for sample measurement data, delay c = 21,160.

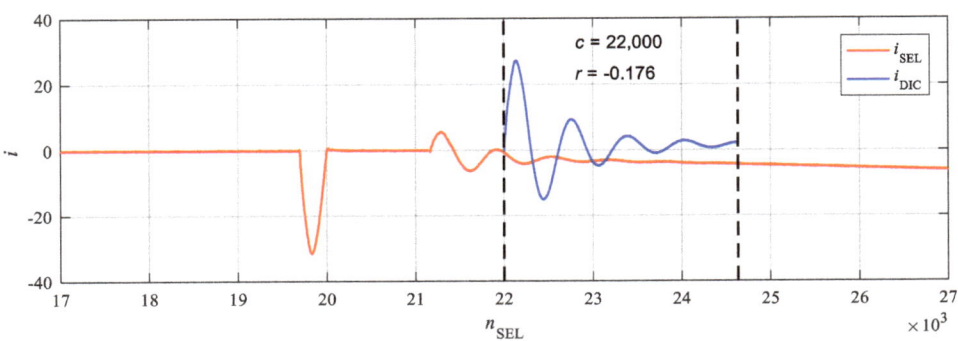

Figure 11. Determining the correlation for sample measurement data, delay c = 22,000.

Figure 12a shows the cross-correlation vector r as a function of delay c for the example of measurement data. Figure 12b shows the same relationship for the vector fragment r with the highest correlation values.

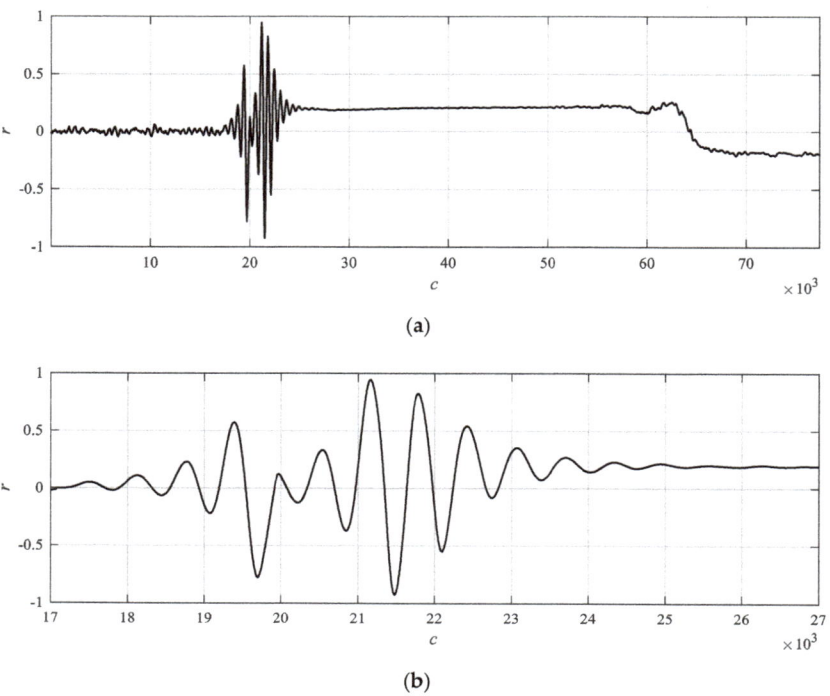

Figure 12. Cross-correlation as a function of delay for an example of the measurement data; the whole correlation vector (**a**); the fragment of the vector r that contains the highest correlation values (**b**).

2.6. Signature Calculation

The signature parameters are the maximum cross-correlation determined between the current vector i_{SEL} and all current vectors $i_{DIC}^{(I_D)}$ from the dictionary of transients. Correlation vectors for successive current vectors $i_{DIC}^{(I_D)}$ are denoted as r_{I_D}. The set of categories D_{CA} from the dictionary of transitions is used to name successive signature features. The idea is presented in Figure 13.

The EA signature contains maximum values of the cross-correlation between the analyzed current vector and the individual dictionary elements. Signature features are determined as the maximum absolute value of the cross-correlation r_{I_D} between the analyzed current vector i_{SEL} and the stored current vector $i_{DIC}^{(I_D)}$:

$$COR\_x\_y = \max \left| r_{I_D} \right|, \qquad (3)$$

where $x \in \{1, \ldots, N_{EA}\}$, $y \in \{A, B, C, D, E, F, G, H, I, J\}$.

The computed cross-correlation with the marked maximum value for sample measurement data are presented in Figure 14.

A signature s_l consists of $P_{HF-COR} = 10 \cdot N_{EA}$ features, arranged in a specific order. Names of features and their acronyms are listed in Table 1.

$$\mathbf{s}_l = \begin{bmatrix} s_{l,1} \ldots s_{l,p} \ldots s_{l,P_{HF-COR}} \end{bmatrix}^T. \qquad (4)$$

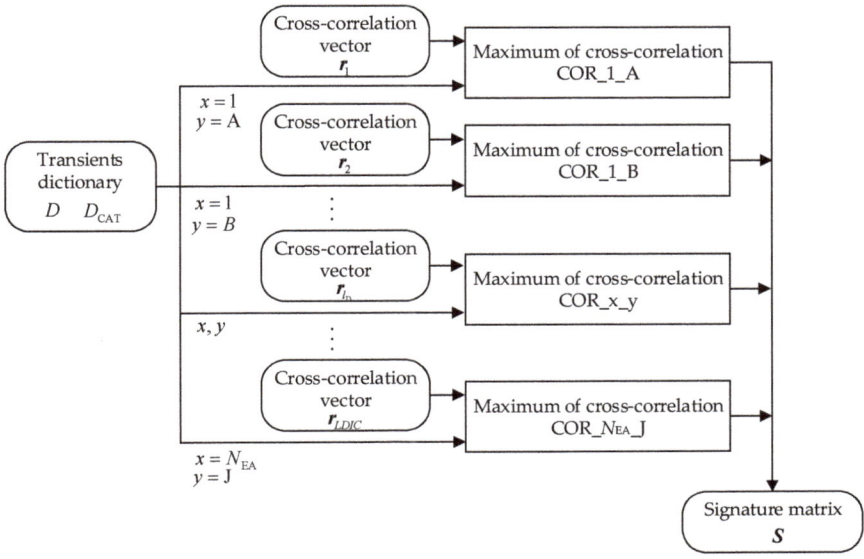

Figure 13. Signature determination algorithm for the HF-COR method.

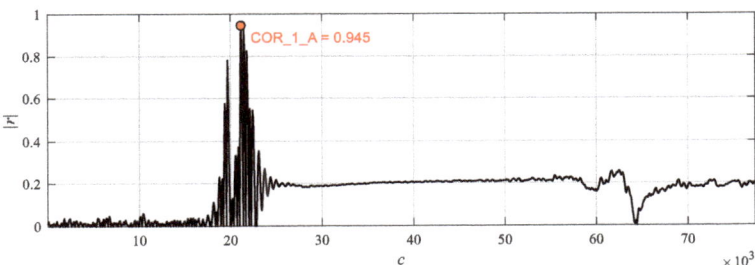

Figure 14. The absolute value of cross-correlation with the maximum value marked for example measurement data.

Table 1. Signature features in the HF-COR method.

p	Full Name	Acronym
1	The highest value of the correlation module with example A of the EA dictionary with the category number 1	COR_1_A
2	The highest value of the correlation module with example B of the EA dictionary with the category number 1	COR_1_B
3	The highest value of the correlation module with example C of the EA dictionary with the category number 1	COR_1_C
⋮	⋮	⋮
p_{HF-COR}	The highest value of the correlation module with the example J of the EA dictionary with the category number N_{EA}	COR_N_{EA}_J

Signature vectors for individual transient states constitute successive columns of the signature array S:

$$S = [s_1 \ldots s_l \ldots s_{LSP}], \qquad (5)$$

where LSP is the total number of transients processed.

3. Signature Quality Assessment Method

The signature well describes devices if its features allow for distinguishing between them. Feature vectors for the same appliance should be similar to each other. The purpose of the signatures quality assessment is to verify if they can be used to identify appliances.

The process is presented in Figure 15. Division of available data into training and testing sets is important. The K-fold Cross-Validation (CV) with $K = 10$ was used here. The data set is split K times into training and testing subsets (with the ratio of 9:1) in such a way that each EA is represented by the single signature in the testing set. The training sets were used to extract knowledge for the intelligent classifier, while the testing ones were applied to verify their generalization abilities. The classification accuracy was averaged of all trials.

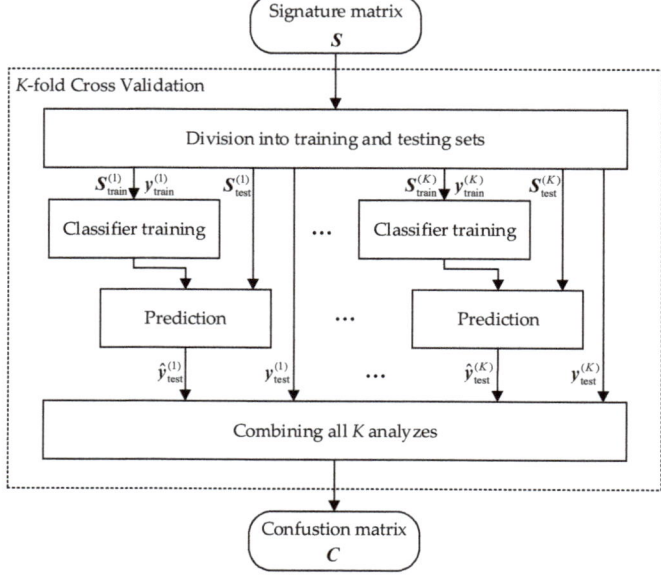

Figure 15. Block diagram of the signature quality assessment method.

The classifier processes signatures $S_{test}^{(k)}$ to predict appliance identifiers (represented by category estimates $\hat{y}_{test}^{(k)}$). The latter are compared to actual categories $y_{test}^{(k)}$ so sample errors can be calculated. Among many types of candidates for classifiers, the following were selected:

- Decision Tree (DT) representing rule-based decision systems,
- Artificial Neural Network (ANN) representing numerical decision-making systems,
- the k-Nearest Neighbors (kNN) classifier representing distance-based systems.

For each round of the CV, each classifier is trained and tested separately (see Figure 16). This way all approaches can be compared. Also, their fusion may be applied if necessary. Each algorithm has specific advantages and hyperparameters. For instance, DT during training selects features based on which rules are constructed. This is the problem for kNN, where the subset of signature values must be manually selected or weighted. Also, the number of neighbors influences diagnostic accuracy. One CV round produces four vectors:

- actual appliances identifiers in the testing set—$y_{test}^{(k)}$,
- category estimates based on ANN prediction—$\hat{y}_{NN}^{(k)}$,
- category estimates based on DT prediction—$\hat{y}_{DT}^{(k)}$,
- category estimates based on kNN prediction—$\hat{y}_{kNN}^{(k)}$.

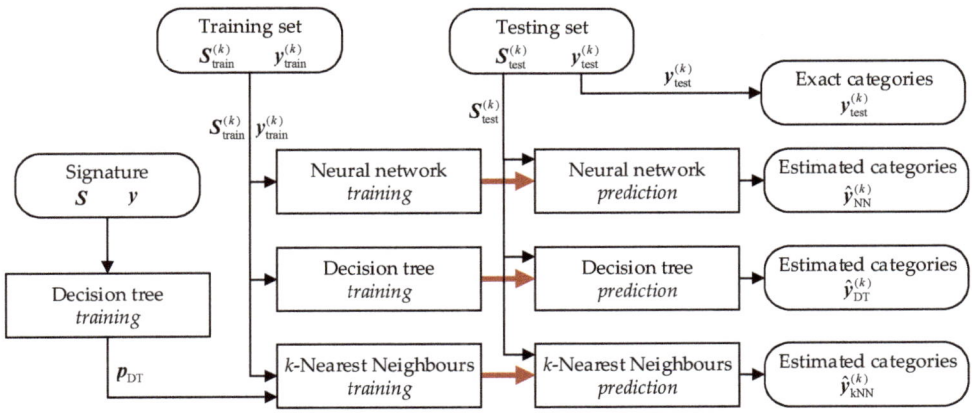

Figure 16. Detailed diagram of one cross-validation attempt.

3.1. Decision Tree

The DT is a tool storing knowledge in the form of a tree (Figure 17). Nodes indicated by circles represent tests on the selected feature (in our case, one of the signature parameters) and its threshold value (like $x_1 > 15$). The result of the test redirects the analyzed vector of features to the node one level below until the terminal node (leaf) is reached. The leaves (rectangles) represent appliance categories. Classification of the example is then based on exploring the tree from the root (yellow node) to one of the leaves. Tests performed at each node indicate which way to take next. Generation of the DT is done using one of the machine learning algorithms like C4.5 or CART, which differ in the method of selecting tests for nodes.

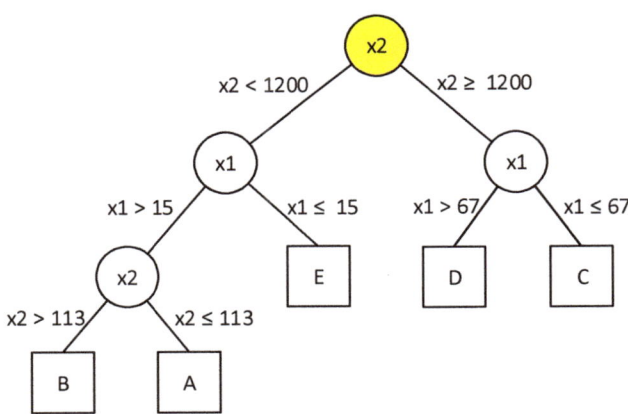

Figure 17. Decision tree example.

3.2. Neural Network

The ANN is widely used in classification. The feed-forward structures, like multilayered perceptrons or RBF networks, are the most popular. Their hyperparameters include the number of hidden layers or the number of neurons in them s_{HL}. Also, the output layer category coding is important, depending on the activation functions (like sigmoidal ones or softmax). The optimal structure of ANN is then found to maximize the classification accuracy for the minimum number of neurons. Knowledge extraction is performed using gradient-based algorithms.

3.3. K-Nearest Neighbors

The kNN classifier is one of the simplest non-parametric classification methods. Using the distance measure, k examples from the dictionary closest to the classified feature vector are found. The analyzed example is assigned to the categories supported by the majority of k voting vectors. The hyperparameters include the value of k, voting strategy and the distance measure selection. This is the only one of the applied classifiers not extracting knowledge from data during the machine learning process. The problem here is determining the significance of available features, for example, by using the information capacity or correlation methods. In the presented research the DT was used to preselect them for the Euclidean measure calculation between each pair of examples l_1 and l_2:

$$d_{\text{EUCLID},l_1,l_2} = \sqrt{\left(\mathbf{S}_{l_1,\mathbf{p}_{\text{DT}}} - \mathbf{S}_{l_2,\mathbf{p}_{\text{DT}}}\right)\left(\mathbf{S}_{l_1,\mathbf{p}_{\text{DT}}} - \mathbf{S}_{l_2,\mathbf{p}_{\text{DT}}}\right)^T},$$

where $\mathbf{S}$ denotes the signature array and $\mathbf{p}_{\text{DT}}$ is the number of the signature features selected by DT.

3.4. Classification Accuracy

The standard method of evaluating the classifier in the multi-category identification problem is the confusion matrix. To determine the overall quality, the accuracy should be calculated as the number of correctly identified examples from the testing set. This can be done for each category η_{EA} separately:

$$\eta_{n_{\text{EA}}} = \frac{|LSP_{\text{EA}} : y = \hat{y}|}{|LSP_{\text{EA}}|}$$

or on the whole set (of N_{EA} categories):

$$\eta_{\text{ALL}} = \frac{1}{N_{\text{EA}}} \cdot \frac{|LSP : y = \hat{y}|}{|LSP|}$$

4. Experimental Results

The following section discusses details of experiments, including the laboratory test stand, collected data, and classification results.

The HF-GEN method was tested in the laboratory conditions on a fixed set of 1 appliances. For each of them, 150 current pulses were recorded. From the vector i in the transient state l a signature vector s_l was obtained. The signature set S contains all LS signature vectors.

The used EAs included a vacuum cleaner, a slow juicer, an "Osram" light bulb, the "Philips" light bulb, an "Omega" light bulb, a "Lexman" lamp with four bulbs, a laptop, irons, sharpeners, grinders, kettle, jigsaw, coffee machine, air conditioner and planer.

4.1. Laboratory Test Stand

The measurement system consists of the single analyzed electrical appliance (EA), current-voltage converter of the SCT-013-020 (CVC) type, the Advantech PCIE-1744 data acquisition card (AC), signal generator (GEN), and computer (PC) with the LabVIEW-based virtual instrument (SW) installed. The EA was connected to the power network. The CVC was installed on the L1 cable supplying EA through the resistor $R = 47\,\Omega$. The GEN input-output (I/O) connectors were connected to L1 and N power cables. The AC was configured in such a way that the high level of the sync voltage $u_{\text{SYN}}(t)$ applied to the synchronization input would trigger the acquisition of the signal $u_{\text{AD}}(t)$ fed to an analog input. The signal $u_{\text{AC}}(t)$ was recorded for 10 ms since the occurrence of the high level of synchronization voltage $u_{\text{SYN}}(t)$. The AC sampling rate was 30 MS/s. The data stream containing the samples was captured by a SW running on a PC and saved in the *.tdms file format.

The pulse signal generator consisted of AL, i.e., a lamp with an "Osram" LED bulb, (type AB30526) and a "Relpol" relay (type RM699V-3011-85-1005-RE), voltage transformer (MC-GEN), Advantech PCIE-1816H acquisition card containing an analog-to-digital converter (AD-GEN) and digital output (DO) and a computer (PC-GEN) running the virtual instrument (CS). AL was connected to the supply network via the RE relay, while MC-GEN—to the supply network via the L1 and N conductors. The CVC of the type SCT-013-020 type converts voltage $u(t)$ to $u_{AD-GEN}(t)$. Its measuring range is about 120A. Laboratory tests proved that SCT-013-020 allows for accurate measurements of signals with frequency up to 400 kHz which is enough for the presented HF-GEN method. The voltage $u_{AD-GEN}(t)$ was fed to the analog input no 0 (AD-GEN) of the acquisition card.

AD-GEN samples voltage $u_{AD-GEN}(t)$ at 250 kS/s. Based on them, CS detects the voltage phase by actuating a logic signal o_n. The AL is switched on when the voltage $u(t)$ reaches the value of 300 V. DO converts logic signal o_n to voltage $u_{SYN}(t)$. The RE becomes closed when the high state appears on $u_{SYN}(t)$.

4.2. Measurement Procedure

During experiments, the following measurement procedure was implemented:
1. Connecting to the power grid and switching on EA under test;
2. Setting CS so that the GEN generates a pulse signal 150 times (in the case of its acquisition for quality evaluation) or at least 10 times (for the transients' dictionary);
3. Setting the SW to acquire all current pulses;
4. Starting the impulse generation and acquisition process;
5. Switching off the tested device.

These steps are performed for each tested EA. A separate series of measurements is carried out with no EA connected (only steps 2–4 are then taken).

4.3. Analysis of Measured Current Vectors

As a result of measurements for 16 categories (15 types of EA and no-EA), $150 \times 16 = 2400$ vectors of current samples were collected. Details of the recorded vectors are in Table 2. Each current vector i has 300,000 samples (representing duration of 10 ms).

Table 2. Information on recorded current vectors.

Category Number n_{EA}	EA Name	Type	Nominal Power	Indexes of Current Vectors l
0	no EA	-	-	1 ... 150
1	vacuum cleaner	Zelmer ZVC425HT	1000 W	151 ... 300
2	slow juicer	Eldom PJ400	400 W	301 ... 450
3	lamp with LED bulb "Osram"	Osram AB42758	11.5 W	451 ... 600
4	lamp with LED bulb "Philips"	Philips 9290012345C	13 W	601 ... 750
5	lamp with LED bulb "Omega"	no data	11 W	751 ... 900
6	wall lamp with four LED bulbs "Lexman"	no data	5 W	901 ... 1050
7	laptop	Dell PP36L	100–240 V ~1.6 A	1051 ... 1200
8	iron	Philips GC 4410	2000–2400 W	1201 ... 1350
9	sharpener	SilverCrest SEAS 20 B1	20 W	1351 ... 1500
10	grinder	Makita GB801	550 W	1501 ... 1650
11	kettle	Zelmer CK1004	1850–2200 W	1651 ... 1800
12	jigsaw	Bosch GST 90 BE	650 W	1801 ... 1950
13	coffee machine	Saeco HD8917	1850 W	1951 ... 2100
14	air conditioner	Cooper&Hunter CH-S09RX4	2700–2850 W	2101 ... 2250
15	planer	Makita 2012NB	1650 W	2251 ... 2400

The current vectors were selected according to Section 2.3. The result were current vectors i_{SEL} which length of N_{SEL} = 120,000 (duration of 4 ms). The vectors $i_{SEL}^{(l)}$ for selected examples l are in Figures 18 and 19.

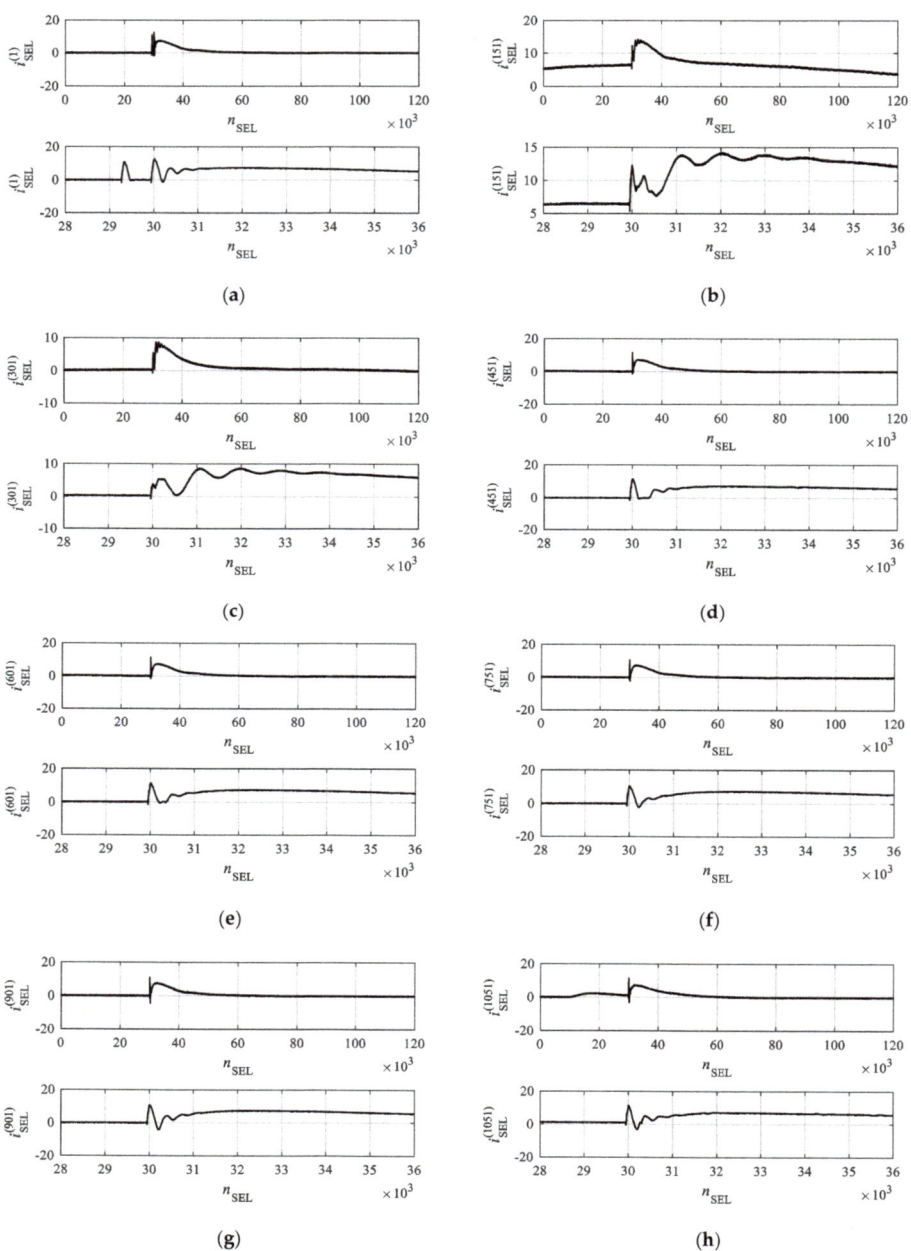

Figure 18. Current vector $i_{SEL}^{(l)}$ for selected examples l belonging to categories 0–7. $l = 1$: No EA (**a**), $l = 151$: vacuum cleaner (**b**), $l = 301$: slow juicer (**c**), $l = 451$: lamp with bulb "Osram" (**d**), $l = 601$: lamp with bulb "Philips" (**e**), $l = 751$: lamp with bulb "Omega" (**f**), $l = 901$: wall lamp with four bulbs "Lexman" (**g**), $l = 1051$: laptop (**h**).

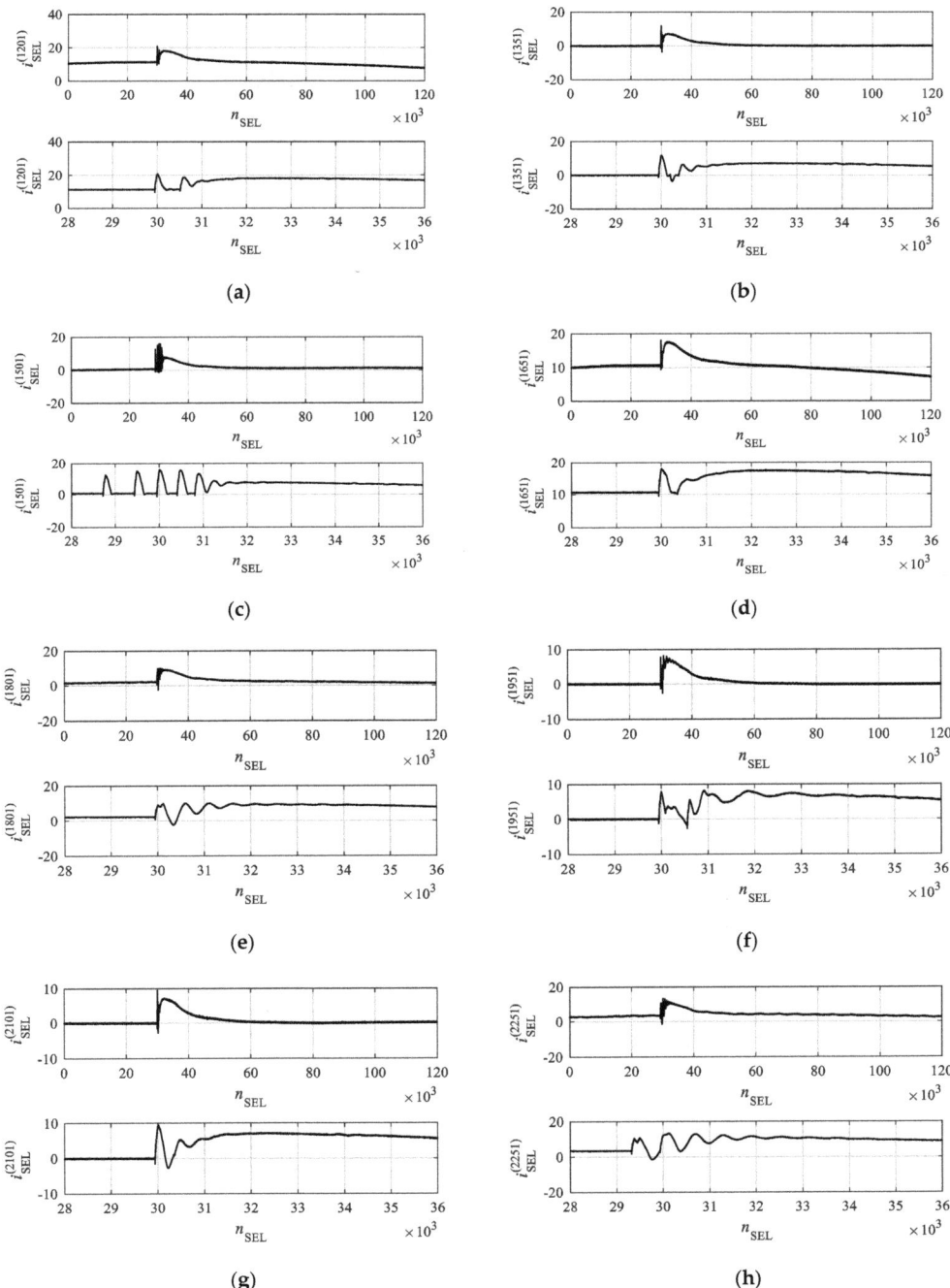

Figure 19. Current vector $i_{SEL}^{(l)}$ for selected examples l belonging to categories 8–15. $l = 1201$: iron (**a**), $l = 1351$: sharpener (**b**), $l = 1501$: grinder (**c**), $l = 1651$: kettle (**d**), $l = 1801$: jigsaw (**e**), $l = 1951$: coffee machine (**f**), $l = 2101$: air conditioner (**g**), $l = 2251$: planer (**h**).

The average current value $i_{SEL}^{(l)}$ depends on the analyzed EA. For instance, examples $l \in \{151, 1201, 1651, 2251\}$ representing vacuum cleaner, iron, kettle, and planer are characterized by relatively high power. The pulses are generated for the voltage $u = 300$ V when the instantaneous current levels of EAs are close to the maximum value. Therefore, a high average current value is observed here.

The direction of the pulse current is always the same. It results from forcing the voltage phase at the moment of generating the pulse.

All waveforms presented in Figures 18 and 19 are characterized by a rise in the average current value starting approx. at $n_{SEL} = 30,000$. In turn, for the $n_{SEL} = 34,000 \ldots 50,000$ current values drop until reaching the level as before the pulse appearance.

For examples $l \in \{1, 1201, 1501\}$, the multiple contact of the RE is visible in the form of many similar oscillations which quickly converge to the average current value $i_{SEL}^{(l)}$. For category 0 (no EA), this oscillation is visible for the $n_{SEL} = 29,400$, while for category (iron) it is for $n_{SEL} = 30,000$, and for category 10 (grinder), 4 such oscillations are visible for $n_{SEL} \in \{28,800, 29,500, 30,000, 30,500\}$.

The starting point for further analysis is the current vector obtained for category 0 when no EA is connected. The shape of the pulse for the example $l = 1$ (Figure 18) is the impulse response of AL after switching on the supply voltage. All other current vectors are the impulse response of the system in which two electricity receivers are simultaneously connected to the power supply: AL and the tested EA. The change in the shape of the current waveform $i_{SEL}^{(l)}$ is proportional to the influence of the tested EA on the total impedance of these two parallelly connected loads in the supply network.

Examples $l \in \{751, 901, 1051\}$, i.e., lamp with bulb "Omega", wall lamp with four bulbs "Lexman" and kettle are similar to the example $l = 1$. Examples $l = 451$ (lamp with bulb "Osram") and $l = 601$ (lamp with bulb "Philips") are distinguished by the lack of the minimum of the 2A-amplitude instantaneous current for $n_{SEL} = 30,300$.

The example $l = 1951$ recorded for the coffee machine has a characteristic shape especially in the area $n_{SEL} = 30,000 \ldots 31,000$, where rapid changes in the instantaneous current values are visible, and the characteristic for many other examples of quasi-periodic oscillations cannot be found.

A vacuum cleaner ($l = 151$), slow juicer ($l = 301$), jigsaw ($l = 1801$), and planer ($l = 2251$) reduce the frequency of current oscillations, and increase the number of visible oscillations, which is unique for each EA. Specifically, for example $l = 151$, five oscillations have period of approximately 940 samples corresponding to a frequency of 31.9 kHz. For the example $l = 301$ (slow juicer), four periods exist (923 samples each) which corresponds to a frequency of 32.5 kHz. For the example $l = 1801$ (jigsaw), five periods of 500 samples correspond to a frequency of 60 kHz. For the example $l = 2251$ (planer), there are six periods, each 610 samples long, which corresponds to a frequency of 49.2 kHz. All these categories have motors, which may shape the current pulse.

4.4. Dictionary of Transients

The measurement data for the transient dictionary does not coincide with the measurement data used to train and test the classification algorithms. The set of measurement data used in the transient state dictionary was prepared independently of the data set described in Section 4.3. Selection of current vectors for the dictionary does not disturb the obtained classification results.

To prepare the dictionary of transients, the procedure presented in Section 2.4 was used. For each of sixteen categories, 10 examples of transient current i were collected, from which current vectors i_{SEL} were obtained. The resulting dictionary of transients is presented in Table 3. The most important fragments of current vectors $i_{DIC}^{(l_D)}$ for selected categories are in Figures 20–23.

Table 3. Information about the transient dictionary.

Category Number x	Device Name	Indexes of Current Vectors l_D	Signature Features Related to the Dictionary Examples
0	no EA	1 ... 10	COR_0_A ... COR_0_J
1	vacuum cleaner	11 ... 20	COR_1_A ... COR_1_J
2	slow juicer	21 ... 30	COR_2_A ... COR_2_J
3	lamp with bulb "Osram"	31 ... 40	COR_3_A ... COR_3_J
4	lamp with bulb "Philips"	41 ... 50	COR_4_A ... COR_4_J
5	lamp with bulb "Omega"	51 ... 60	COR_5_A ... COR_5_J
6	wall lamp with five "Lexman" bulbs	61 ... 70	COR_6_A ... COR_6_J
7	Laptop	71 ... 80	COR_7_A ... COR_7_J
8	Iron	81 ... 90	COR_8_A ... COR_8_J
9	Sharpener	91 ... 100	COR_9_A ... COR_9_J
10	Grinder	101 ... 110	COR_10_A ... COR_10_J
11	Kettle	111 ... 120	COR_11_A ... COR_11_J
12	Jigsaw	121 ... 130	COR_12_A ... COR_12_J
13	coffee machine	131 ... 140	COR_13_A ... COR_13_J
14	air conditioner	141 ... 150	COR_14_A ... COR_14_J
15	Planer	151 ... 160	COR_15_A ... COR_15_J

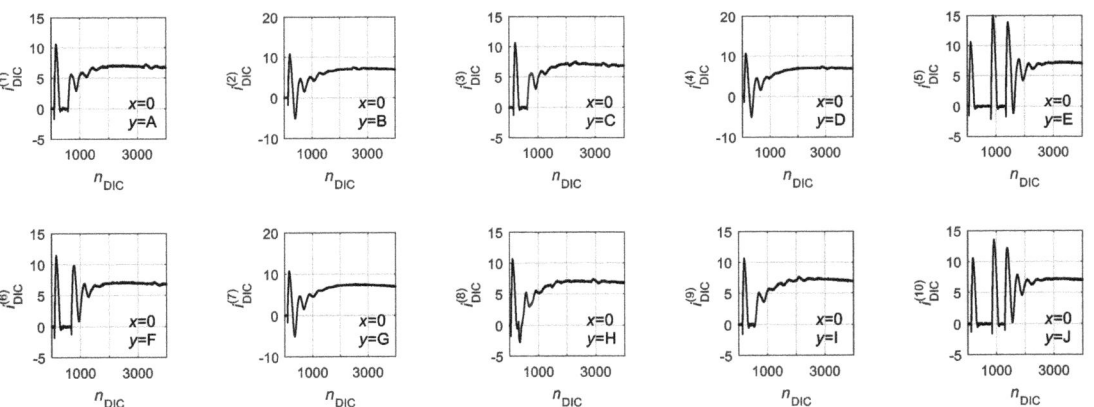

Figure 20. Current vectors i_{DIC} for dictionary elements for the category no EA ($x = 0$).

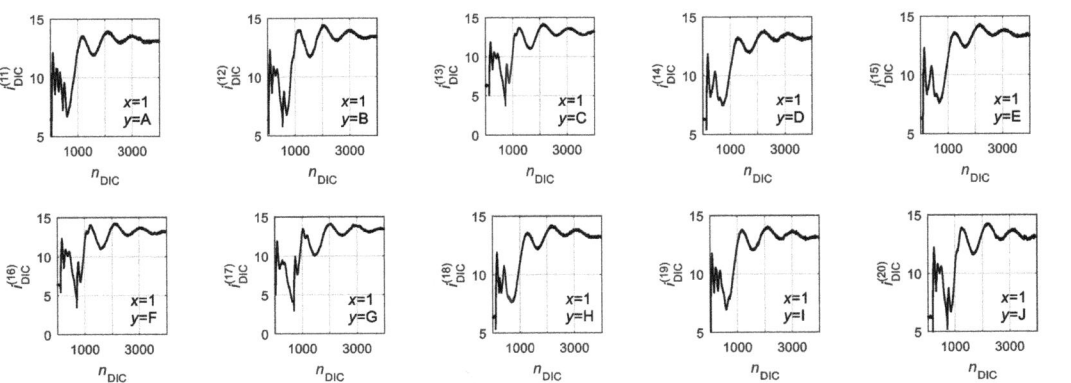

Figure 21. Current vectors i_{DIC} for the vacuum cleaner ($x = 1$).

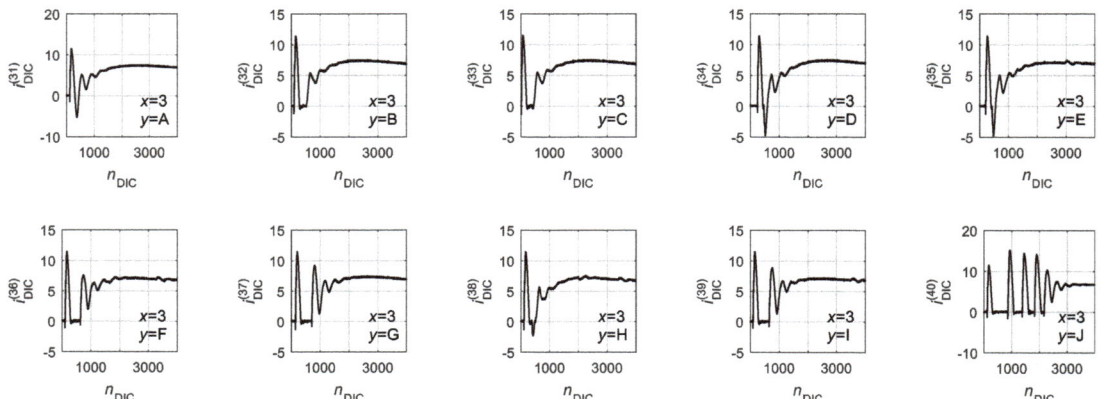

Figure 22. Current vectors i_{DIC} for the dictionary elements for the category lamp with a bulb "Osram" ($x = 3$).

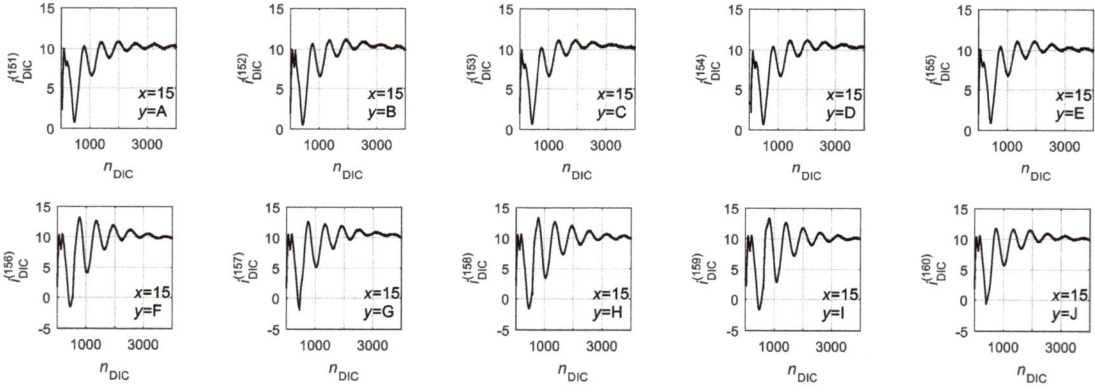

Figure 23. Current vectors i_{DIC} for dictionary elements for the planer category ($x = 15$).

Despite ensuring similar conditions for generating impulses, the current vectors i_D for the no-EA category (Figure 20) differ from each other. The first difference between examples is the multiple contact phenomenon described in Section 4.3, visible for the examples $l_D \in \{5, 6, 10\}$.

In all waveforms, a quasi-periodic oscillation is present, disappearing after about three periods. A characteristic of these vectors is a rising edge on which the oscillation located. In the example $l_D = 7$, the slope is visible for $n_{DIC} = 1 \ldots 2000$. The rising edge is characteristic feature of applied AL.

For dictionary examples representing the vacuum cleaner (Figure 21), three types of waveforms can be distinguished. They differ mainly in the shape of the initial part of the vector i_{DIC} ($n_{DIC} = 1 \ldots 1000$). The first type is present in examples $l_D = 11$ and $l_D = 19$. The second type is visible in examples $l_D \in \{14, 15, 18\}$, while the third one—in examples $l_D \{12, 13, 16, 17, 20\}$. The oscillation frequency in the second part of the vector i_{DIC} is lower than in the no EA case (category 0 in Figure 20). Duration of the oscillation between samples $n_{DIC} = 1000$ and $n_{DIC} = 4000$ is the same for all vectors in this category with period of about 940 samples, which corresponds to a frequency of 31.9 kHz.

Waveforms in Figure 22 represent lamps with the "Osram" bulb. Here, the multiple-contact phenomenon of the relay is visible, especially for the example $l_D = 40$, where four similar oscillations are present in the first part of the current vector.

Vectors for the planer (Figure 23) are different from other appliances, and at the same time, they are similar to each other. Their distinguishing feature is the shape of the first part of the vector $i_{DIC}(n_{DIC} = 1 \ldots 300)$. Here examples $l_D \in \{151, 153, 154, 155\}$ have one maximum above the slope of the oscillation. It is present around the sample $n_{DIC} = 100$. Examples $l_D \in \{152, 156, 157, 158, 159, 160\}$ have two visible maxima, one for $n_{DIC} = 90$ and the other one at $n_{DIC} = 190$. In all examples for the planer, at least five periods of oscillation with a period of about 610 samples are present, corresponding to a frequency of about 49.2 kHz.

4.5. Signature Parameters

Based on the determined correlation vectors for each pair of the transition (for $l = 1 \ldots 2400$) and the dictionary example $l_D = 1 \ldots 160$, the values of features in the signature S were determined. For each example l, the signature contains 160 features. Values of two characteristic features for each category are presented in Figures 24 and 25.

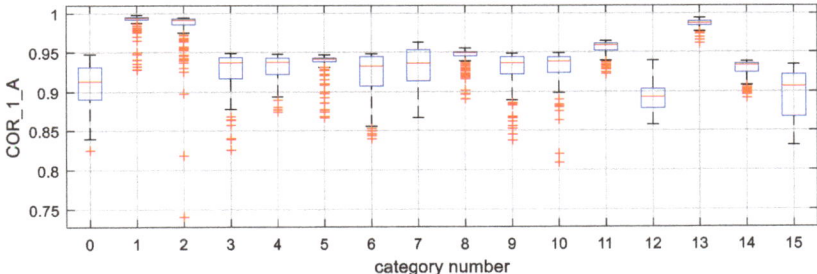

Figure 24. Box plot of feature COR_1_A.

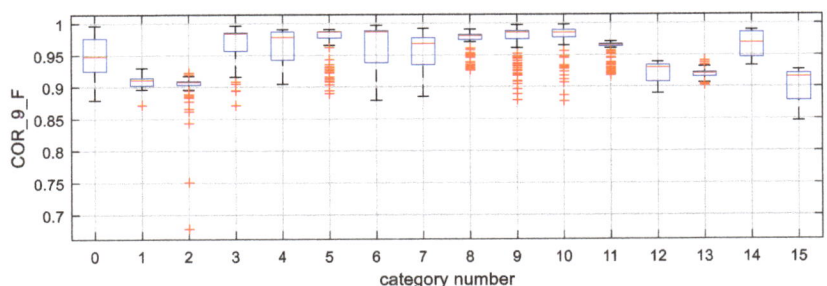

Figure 25. Box plot of feature COR_9_F.

The COR_1_A feature (Figure 24) can be used to distinguish between three categories: 1 (vacuum cleaner), 2 (slow juicer), and 13 (coffee machine). A majority of examples belonging to these categories have the value of the COR_1_A in the range between 0.97 and 1.0. Almost all observations of the remaining categories assume values of this feature in the range 0.85–0.97, so it is not suitable for distinguishing between them.

The COR_9_F (Figure 25) feature is characterized by high values for almost all examples. Only three vectors from category 3 (lamp with the "Osram" bulb) have COR_9_F values below 0.85. Observations for all other categories assume values of this feature in the range 0.85–1. Even though the COR_9_F feature was determined for the dictionary examples belonging to the grinder category, the value distribution of this feature is similar for each category. Therefore it is not useful in most cases.

4.6. Classification Results

This section presents results of three classifiers' operation for the available data partitioned using the K-fold cross validation (where K = 10).

4.6.1. Neural Network

The application of ANN required selecting the optimal number of neurons in the hidden layer. For that purpose, the network was trained many times, with the number of neurons in the hidden layer ranging from 1 to 24. The classification error as a function of the number of neurons in the hidden layer is in Figure 26.

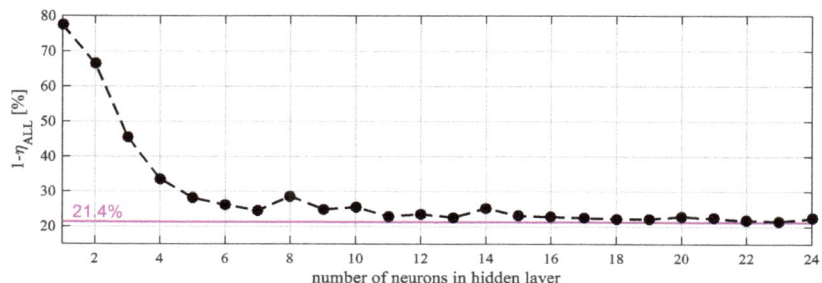

Figure 26. Classification error as a function of the number of neurons in the hidden layer.

The total classification error is minimal for 15 neurons in the hidden layer. Further increase of this parameter does not significantly affect the classification error.

The confusion matrix C_{NN} and the classification accuracy $\eta_{n_{EA}, NN}$ for 16 examined categories in the optimal ANN structure are presented in Table 4.

The overall classification accuracy was as follows:

$$\eta_{ALL, NN} = \frac{1}{16} \sum_{n_{EA}=0}^{15} \eta_{n_{EA},NN} = 78.6\%.$$

The accuracy of at least 95% was obtained for nine appliances: 1 (vacuum cleaner), 2 (slow juicer), 7 (laptop), 8 (iron), 11 (kettle), 12 (jigsaw), 13 (coffee machine), 14 (air conditioner), and 15 (planer). Signatures for EAs 11–15 significantly differ from all others, which makes them easily identifiable (see columns 11–15). The lowest classification accuracy (24%) was achieved for category 9 (sharpener) and for category 3 (lamp with the "Osram" bulb) at the level of 28%. Total 56 (37%) examples of category 9 were incorrectly assigned to category 10. More than a half (51%) of examples of category 4 (lamp with the "Philips" bulb) were incorrectly assigned to category 5 (lamp with the "Omega" bulb). This is because categories 4 and 5 have similar signatures.

4.6.2. Decision Tree

None of the categories was faultlessly identified by DT (see Table 5). The best score of 97% was obtained for appliances 12 (jigsaw), 13 (coffee machine), 14 (air conditioner), and 15 (planer). For category 1 (vacuum cleaner), the accuracy was 95%. Categories 12 and 13 have very few examples from other incorrectly assigned categories. The worst classification result (29%) was obtained for category 3 (lamp with "Osram" bulb). There are also two groups of indistinguishable appliances. The first one consists of categories 0 (no EA), 3 (lamp with the "Osram" bulb), 6 (wall lamp with four "Lexman" bulbs), 9 (sharpener), and

10 (grinder). The second group consists of category 4 (lamp with bulb "Philips") and 5 (lamp with bulb "Omega"). The overall classification efficiency was as follows:

$$\eta_{\text{ALL, DT}} = \frac{1}{16} \sum_{n_{\text{EA}}=0}^{15} \eta_{n_{\text{EA}}, \text{DT}} = 76.2\%. \tag{10}$$

Table 4. Confusion matrix and the classification accuracy of the neural network for 15 neurons in the hidden layer.

	\multicolumn{16}{c}{Assigned Category}	$\eta_{n_{\text{EA}}, \text{NN}}$
	0	1	2	3	4	5	6	7	8	9	10	11	12	13	14	15	
0	89	0	0	14	0	0	15	0	2	20	9	0	0	0	1	0	59%
1	0	147	3	0	0	0	0	0	0	0	0	0	0	0	0	0	98%
2	0	3	146	0	0	0	0	0	0	0	0	0	0	0	0	1	97%
3	40	0	0	42	0	0	6	1	0	30	31	0	0	0	0	0	28%
4	0	0	0	0	73	76	0	0	1	0	0	0	0	0	0	0	49%
5	0	0	0	0	32	117	0	0	0	0	0	1	0	0	0	0	78%
6	31	0	0	8	0	1	106	0	0	3	1	0	0	0	0	0	71%
7	0	0	0	0	0	1	0	148	1	0	0	0	0	0	0	0	99%
8	0	0	0	0	4	2	0	0	142	1	1	0	0	0	0	0	95%
9	21	0	0	30	0	0	5	1	1	36	56	0	0	0	0	0	24%
10	11	0	0	13	0	0	2	0	0	31	93	0	0	0	0	0	62%
11	0	0	0	0	0	1	0	0	0	0	0	149	0	0	0	0	99%
12	0	0	0	0	0	0	0	0	0	0	0	0	150	0	0	0	100%
13	0	0	0	0	0	0	0	0	0	0	0	0	0	150	0	0	100%
14	0	0	0	0	0	0	0	0	0	0	0	0	0	0	150	0	100%
15	0	0	0	0	0	0	0	0	0	0	0	1	0	0	149		99%

Table 5. Confusion matrix and classification accuracy for the decision tree algorithm.

	\multicolumn{16}{c}{Assigned Category}	$\eta_{n_{\text{EA}}, \text{DT}}$
	0	1	2	3	4	5	6	7	8	9	10	11	12	13	14	15	
0	60	0	0	25	3	1	28	1	3	25	4	0	0	0	0	0	40%
1	0	143	3	0	0	0	0	0	0	0	0	0	0	4	0	0	95%
2	0	5	140	0	0	0	0	3	0	0	0	1	0	1	0	0	93%
3	29	0	0	44	0	0	14	1	0	36	26	0	0	0	0	0	29%
4	0	0	0	0	111	33	2	2	1	0	0	0	0	0	1	0	74%
5	1	0	0	0	39	105	2	0	1	0	0	0	0	0	2	0	70%
6	23	0	0	11	2	1	100	2	0	9	2	0	0	0	0	0	67%
7	2	0	2	0	0	1	0	136	3	2	3	0	0	1	0	0	91%
8	2	0	2	1	4	1	0	0	134	1	0	3	0	0	2	0	89%
9	25	0	0	33	1	1	7	2	2	57	22	0	0	0	0	0	38%
10	8	0	0	30	0	0	6	0	0	28	78	0	0	0	0	0	52%
11	0	0	2	0	0	0	0	3	5	0	0	139	1	0	0	0	93%
12	0	0	0	0	0	1	0	0	1	0	0	0	145	0	2	1	97%
13	0	3	0	0	0	0	0	1	0	0	0	0	0	146	0	0	97%
14	1	0	0	0	0	0	0	0	3	1	0	0	0	0	145	0	97%
15	0	0	0	0	1	2	0	0	0	0	0	1	0	0	0	146	97%

4.6.3. The kNN Algorithm

To identify the most significant predictors, the DT was first trained for all 2400 examples of transients. In the DT training process, 82 signature features were selected as the most significant ones. Using selected features, a classification was made for each of the cross-validation attempts. The confusion matrix C_{kNN} for all cross-validation trials and the classification accuracy $\eta_{n_{EA}, kNN}$ is presented in Table 6.

Table 6. Confusion matrix and classification accuracy for the k-nearest neighbors algorithm.

		0	1	2	3	4	5	6	7	8	9	10	11	12	13	14	15	$\eta_{n_{EA}, kNN}$
	0	70	0	0	38	0	1	14	2	1	17	7	0	0	0	0	0	47%
	1	0	149	1	0	0	0	0	0	0	0	0	0	0	0	0	0	99%
	2	0	3	147	0	0	0	0	0	0	0	0	0	0	0	0	0	98%
	3	38	0	0	65	0	0	3	0	1	21	22	0	0	0	0	0	43%
	4	1	0	0	0	120	29	0	0	0	0	0	0	0	0	0	0	80%
	5	3	0	0	0	31	113	2	0	0	0	0	0	0	0	1	0	75%
True Category	6	19	0	0	11	1	0	107	1	0	7	4	0	0	0	0	0	71%
	7	1	0	0	1	0	0	0	145	1	2	0	0	0	0	0	0	97%
	8	0	0	0	1	0	2	0	0	145	1	0	0	0	0	1	0	97%
	9	25	0	0	26	0	0	6	0	3	59	31	0	0	0	0	0	39%
	10	10	0	0	29	0	0	1	0	0	25	85	0	0	0	0	0	57%
	11	0	0	0	0	0	1	0	0	0	0	0	149	0	0	0	0	99%
	12	0	0	0	0	0	0	0	0	0	0	0	0	150	0	0	0	100%
	13	0	0	0	0	0	0	0	0	0	0	0	0	0	150	0	0	100%
	14	0	0	0	0	0	0	0	0	0	0	0	0	0	0	150	0	100%
	15	0	0	0	0	0	0	0	0	0	0	0	0	0	0	0	150	100%

The k-nearest neighbors algorithm classified nine categories with an accuracy of least 95%. These are: 1 (vacuum cleaner), 2 (slow juicer), 7 (laptop), 8 (iron), 11 (grinder), 12 (jigsaw), 13 (coffee machine), 14 (air conditioner), and 15 (planer). Categories 11, 12, 13 and 15 have been identified flawlessly. The worst classification result (39%) was obtained for category 10 (grinder). There are two groups of similar categories: (0, 3, 6, 9, 10) and (4, 5). The overall classification efficiency is as follows:

$$\eta_{ALL, kNN} = \frac{1}{16} \sum_{n_{EA}=0}^{15} \eta_{n_{EA}, kNN} = 81.4\%. \qquad (1)$$

5. Summary

The paper presented the system for NILM task based on the transient features of the generated pulse analysis. It exploits mutual influence between devices operating in the same power circuit to identify the moment of introducing the new one.

The pulse signal generator was designed. Its task is to generate current pulses in predetermined time intervals in a fixed phase of the supply voltage. The measurement system acquires current pulses and stores them as sample vectors. A method for processing current signals was designed to determine their characteristic features based on the cross correlation calculated for each pair of EAs. The method uses information about the phase and amplitude of all (periodic and non-periodic) components of the current pulse appearing in the transient state of the device turned on. The processing result is a signature with features characterizing the EA. The signature quality was verified using three different classifiers.

The presented experiments show that devices connected to one circuit of the supply network influence each other. A significant impact of a background device in the steady state on the current pulse on another device being turned on was observed. When a known load is switched on under repeated conditions, the change in the shape of this pulse may be characteristic for a device operating in the background. Results of the classification show that in the best case, 9 out of 15 EAs are recognizable with an accuracy of at least 97%. Satisfying results were obtained for majority of tested EAs. There are types of EA for which this method fails. Therefore multiple different identification methods should be implemented simultaneously.

In practical application of NILM system the changing set of devices operating at the same time must be considered. This makes the task difficult, as the change of the pulse shape will be a certain superposition of all working EA. Therefore additional research are required to approach this challenge. Results of presented experiments show that in the highly controlled environment (especially when only a single appliance is operating) the proposed approach provides high identification accuracy Its applicability should be further investigated in the future.

Author Contributions: Conceptualization, R.K. and R.Ł.; Data curation, A.W.; Formal analysis, R.K.; Funding acquisition, R.Ł. and P.B.; Investigation, A.W. and P.B.; Methodology, R.K. and P.B.; Resources, A.W. and R.Ł.; Visualization, A.W. and K.D.; Writing—original draft, A.W.; Writing—review & editing, A.W., R.K., P.B. and K.D. All authors have read and agreed to the published version of the manuscript.

Funding: This research received no external funding.

Data Availability Statement: The data presented in this study are available on request from the corresponding author.

Conflicts of Interest: The authors declare no conflict of interest.

References

Hart, G.W. Nonintrusive Appliance Load Monitoring. *Proc. IEEE* **1992**, *80*, 1870–1891. [CrossRef]

Tang, G.; Wu, K.; Lei, J. A Distributed and Scalable Approach to Semi-Intrusive Load Monitoring. *IEEE Trans. Parallel Distrib. Syst.* **2016**, *27*, 1553–1565. [CrossRef]

Agyeman, K.A.; Han, S.; Han, S. Real-time recognition non-intrusive electrical appliance monitoring algorithm for a residential building energy management system. *Energies* **2015**, *8*, 9029–9048. [CrossRef]

Carrie Armel, K.; Gupta, A.; Shrimali, G.; Albert, A. Is disaggregation the holy grail of energy efficiency? The case of electricity. *Energy Policy* **2013**, *52*, 213–234. [CrossRef]

Nguyen, T.K.; Azarkh, I.; Nicolle, B.; Jacquemod, G.; Dekneuvel, E. Applying NIALM technology to predictive maintenance for industrial machines. In Proceedings of the IEEE International Conference on Industrial Technology, Lyon, France, 20–22 February 2018; pp. 341–345.

He, D.; Lin, W.; Liu, N.; Harley, R.G.; Habetler, T.G. Incorporating non-intrusive load monitoring into building level demand response. *IEEE Trans. Smart Grid* **2013**, *4*, 1870–1877.

Zeifman, M.; Roth, K. Nonintrusive appliance load monitoring: Review and outlook. *IEEE Trans. Consum. Electron.* **2011**, *57*, 76–84. [CrossRef]

Esa, N.F.; Abdullah, M.P.; Hassan, M.Y. A review disaggregation method in Non-intrusive Appliance Load Monitoring. *Renew. Sustain. Energy Rev.* **2016**, *66*, 163–173. [CrossRef]

Liszewski, K.; Łukaszewski, R.; Kowalik, R.; Łukasz, N.; Winiecki, W. Different appliance identification methods in Non-Intrusive Appliance Load Monitoring. In *Advanced Data Acquisition and Intelligent Data Processing*; Haasz, V., Madani, K., Eds.; River Publishers: Aalborg, Denmark, 2014; pp. 31–58.

Wong, Y.F.; Ahmet Sekercioglu, Y.; Drummond, T.; Wong, V.S. Recent approaches to non-intrusive load monitoring techniques in residential settings. In Proceedings of the IEEE Symposium on Computational Intelligence Applications in Smart Grid (CIASG), Singapore, 16–19 April 2013; pp. 73–79.

Abubakar, I.; Khalis, S.N.; Mustafa, M.W.; Shareef, H.; Mustapha, M. An Overview of Non-Intrusive Load Monitoring Methodologies. In Proceedings of the IEEE Conference on Energy Conversion (CENCON), Johor Bahru, Malaysia, 9–20 October 2015; pp. 54–59.

Gao, J.; Giri, S.; Kara, E.C.; Bergés, M. PLAID: A public dataset of high-resolution electrical appliance measurements for load identification research. In *Proceedings of the BuildSys 2014—Proceedings of the 1st ACM Conference on Embedded Systems for Energy-Efficient Buildings, Memphis, Tennessee, 4–6 November 2014*; Association for Computing Machinery: New York, NY, USA, 2014; pp. 198–199.

13. Kelly, J.; Knottenbelt, W. The UK-DALE dataset, domestic appliance-level electricity demand and whole-house demand from five UK homes. *Sci. Data* **2015**, *2*, 1–14. [CrossRef]
14. Renaux, D.P.B.; Pottker, F.; Ancelmo, H.C.; Lazzaretti, A.E.; Lima, C.R.E.; Linhares, R.R.; Oroski, E.; da Silva Nolasco, L.; Lim, L.T.; Mulinari, B.M.; et al. A dataset for non-intrusive load monitoring: Design and implementation. *Energies* **2020**, *13*, 5371. [CrossRef]
15. Pereira, L.; Nunes, N. Performance evaluation in non-intrusive load monitoring: Datasets, metrics, and tools—A review. *Wiley Interdiscip. Rev. Data Min. Knowl. Discov.* **2018**, *8*, 1–17. [CrossRef]
16. Wójcik, A.; Łukaszewski, R.; Kowalik, R.; Winiecki, W. Nonintrusive appliance load monitoring: An overview, laboratory test results and research directions. *Sensors* **2019**, *19*, 3621. [CrossRef] [PubMed]
17. Mauch, L.; Yang, B. A new approach for supervised power disaggregation by using a deep recurrent LSTM network. In Proceedings of the 2015 IEEE Global Conference on Signal and Information Processing, Orlando, FL, USA, 14–16 December 2015; pp. 63–67.
18. Koziy, K.; Gou, B.; Aslakson, J. A low-cost power-quality meter with series arc-fault detection capability for smart grid. *IEEE Trans. Power Deliv.* **2013**, *28*, 1584–1591. [CrossRef]
19. Wang, Z.; Zheng, G. Residential appliances identification and monitoring by a nonintrusive method. *IEEE Trans. Smart Grid* **2012**, *3*, 80–92. [CrossRef]
20. Zhao, B.; Stankovic, L.; Stankovic, V. On a Training-Less Solution for Non-Intrusive Appliance Load Monitoring Using Graph Signal Processing. *IEEE Access* **2016**, *4*, 1784–1799. [CrossRef]
21. Li, D.; Dick, S. Whole-house Non-Intrusive Appliance Load Monitoring via multi-label classification. In Proceedings of the International Joint Conference on Neural Networks, Vancouver, BC, Canada, 24–29 July 2016; pp. 2749–2755.
22. Henao, N.; Agbossou, K.; Kelouwani, S.; Dube, Y.; Fournier, M. Approach in Nonintrusive Type I Load Monitoring Using Subtractive Clustering. *IEEE Trans. Smart Grid* **2017**, *8*, 812–821. [CrossRef]
23. Liu, B.; Luan, W.; Yu, Y. Dynamic time warping based non-intrusive load transient identification. *Appl. Energy* **2017**, *195*, 634–645. [CrossRef]
24. Kong, W.; Dong, Z.; Xu, Y.; Hill, D. An Enhanced Bootstrap Filtering Method for Non- Intrusive Load Monitoring. In Proceedings of the 2016 IEEE Power and Energy Society General Meeting (PESGM), Boston, MA, USA, 17–21 July 2016; pp. 1–5.
25. Ducange, P.; Marcelloni, F.; Antonelli, M. A novel approach based on finite-state machines with fuzzy transitions for nonintrusive home appliance monitoring. *IEEE Trans. Ind. Inform.* **2014**, *10*, 1185–1197. [CrossRef]
26. Sultanem, F. Using appliance signatures for monitoring residential loads at meter panel level. *IEEE Trans. Power Deliv.* **1991**, 1380–1385. [CrossRef]
27. Reinhardt, A.; Burkhardt, D.; Zaheer, M.; Steinmetz, R. Electric appliance classification based on distributed high resolution current sensing. In Proceedings of the 37th Annual IEEE Conference on Local Computer Networks Workshops, Clearwater, FL, USA, 22–25 October 2012; pp. 999–1005.
28. Yoshimoto, K.; Nakano, Y.; Amano, Y.; Kermanshahi, B. Non-intrusive appliances load monitoring system using neural networks. In Proceedings of the Proceedings ACEEE Summer Study on Energy Efficiency in Buildings; 2000; Volume 7, pp. 2–5. Available online: https://www.eceee.org/static/media/uploads/site-2/library/conference_proceedings/ACEEE_buildings/2000/Panel_7/p6_17/paper.pdf (accessed on 28 January 2021).
29. Srinivasan, D.; Ng, W.S.; Liew, A.C. Neural-network-based signature recognition for harmonic source identification. *IEEE Trans. Power Deliv.* **2006**, *21*, 398–405. [CrossRef]
30. Gupta, S.; Reynolds, M.S.; Patel, S.N. ElectriSense: Single-point sensing using EMI for electrical event detection and classification in the home. In Proceedings of the 12th ACM International Conference on Ubiquitous Computing, Copenhagen, Denmark, 26–29 September 2010; pp. 139–148.
31. Torquato, R.; Acharya, J.R.; Xu, W. A Method to Determine Stray Voltage Sources—Part II: Verifications and Applications. *IEEE Trans. Power Deliv.* **2015**, *30*, 720–727. [CrossRef]
32. Duarte, C.; Delmar, P.; Goossen, K.W.; Barner, K.; Gomez-Luna, E. Non-intrusive load monitoring based on switching voltage transients and wavelet transforms. In Proceedings of the 2012 Future of Instrumentation International Workshop, Gatlinburg, TN, USA, 8–9 October 2012; pp. 101–104.
33. Duarte, C. *Non-Intrusive Monitoring of Electrical Loads Based on Switching Transient Voltage Analysis: Signal Acquisition and Feature Extraction*; Univeristy of Delaware: Newark, DE, USA, 2013.
34. Chang, H.H.; Lian, K.L.; Su, Y.C.; Lee, W.J. Power-spectrum-based wavelet transform for nonintrusive demand monitoring and load identification. *IEEE Trans. Ind. Appl.* **2014**, *50*, 2081–2089. [CrossRef]
35. Wójcik, A.; Winiecki, W.; Łukaszewski, R.; Bilski, P. Analysis of Transient State Signatures in Electrical Household Appliances. In Proceedings of the 10th IEEE International Conference on Intelligent Data Acquisition and Advanced Computing Systems: Technology and Applications (IDAACS), Metz, France, 18–21 September 2019; pp. 639–644.
36. Lewis, J.P. Fast Normalized Cross-Correlation. *Vis. Interface* **2010**, 120–123.

Article

Satisfaction-Based Energy Allocation with Energy Constraint Applying Cooperative Game Theory

Samira Ortiz [1], Mandoye Ndoye [2,*] and Marcel Castro-Sitiriche [1,3]

1. Department of Electrical and Computer Engineering, University of Puerto Rico-Mayagüez Campus, Mayagüez, PR 00682, USA; samira.ortiz@upr.edu (S.O.); marcel.castro@upr.edu (M.C.-S.)
2. Department of Electrical and Computer Engineering, Tuskegee University, Tuskegee, AL 36088, USA
3. Nelson Mandela African Institution of Science and Technology, Arusha, Tanzania
* Correspondence: mndoye@tuskegee.edu; Tel.: +1-(334)-727-8623

Abstract: There has been an effort for a few decades to keep energy consumption at a minimum or at least within a low-level range. This effort is more meaningful and complex by including a customer's satisfaction variable to ensure that customers can achieve the best quality of life that could be derived from how energy is used by different devices. We use the concept of Shapley Value from cooperative game theory to solve the multi-objective optimization problem (MOO) to responsibly fulfill user's satisfaction by maximizing satisfaction while minimizing the power consumption, with energy constrains since highly limited resources scenarios are studied. The novel method uses the concept of a quantifiable user satisfaction, along the concepts of power satisfaction (PS) and energy satisfaction (ES). The model is being validated by representing a single house (with a small PV system) that is connected to the utility grid. The main objectives are to (i) present the innovative energy satisfaction model based on responsible wellbeing, (ii) demonstrate its implementation using a Shapley-value-based algorithm, and (iii) include the impact of a solar photovoltaic (PV) system in the energy satisfaction model. The proposed technique determines in which hours the energy should be allocated to maximize the ES for each scenario, and then it is compared to cases in which devices are usually operated. Through the proposed technique, the energy consumption was reduced 75% and the ES increased 40% under the energy constraints.

Keywords: electrical energy; load scheduling; satisfaction; Shapley Value; smart meter; solar photovoltaics

1. Introduction

Energy is the backbone of modern society. It provides the means to support everyday infrastructure, such as hospitals, schools, and homes. In the case of homes, for most of the 90 percent of the global population with access, it is difficult to imagine living without electric energy [1]; it powers our essential needs, such as water pumping, lighting, cooling, and very often also cooking, among others. Additionally, electric energy provides comfort and entertainment, and although these are not essential needs, they can also improve quality of life.

Throughout the years, with an increasing energy demand, managing energy consumption has become important. Demand Side Management (DSM) encapsulates those strategies that change the main power consumption to better match the power supply. Through DSM methodologies, one of the purposes is to create an energy demand scheduling to benefit a household.

Many DSM studies focus on minimizing energy cost, achieving utility stability, and shifting peak demand. Wu et al. [2] proposed a mixed-integer linear programming (MILP) model for the energy system optimization to reduce the annual cost in a building distributed heating network. Similarly, Tang et al. [3] proposed a game theoretic method to maximize net profit and reduced demand fluctuation using real data of building on a campus in

Hong Kong. Conversely, Lokeshgupta et al. [4] proposed a mathematical model of intelligent multi-objective home energy management (HEM) to simultaneously minimi the consumer's bill and system peak demand.

Some authors have aimed at the time of device usage instead of the costs while e suring a level of satisfaction. Yang et al. [5] implemented a Nash-based game theoret approach to optimize time-of-use (ToU) pricing strategies considering the costs of flu tuating demands to the utility company and the satisfactions costs of user. Additonal Marzband et al. [6] introduced a satisfaction function in a bi-level model to maximi and allocate the profit. The authors included a satisfaction function as part of the payo function. The function is calculated at the end of each time slot and depends on the amou of energy generation.

Among the studies that include satisfaction as part of the objectives, it is defined terms on how much their expectations are met [7,8]. The aforementioned methodologi tackle the comfort/satisfaction/welfare part as an indirect measure. It is derived fro another variable taking this into account. Ogunjuyigbe et al. [8], on the other han developed a cost per unit satisfaction index. Their model considered individual devices each time of the day. The index was maximized by using a genetic algorithm.

Game theory approaches have become one of the tools adopted for modeling an analyzing energy consumption, due to its effectiveness to capture complex interactio between multiple players. The Stackelberg game is one of the most used game strategi for demand response problems [3,9–12]. Another largely used non-cooperative approac as a solution for DSM, is the Nash equilibrium strategy [3,13–16].

Summarizing the aforementioned studies, even though research efforts are starting emerge in the point of confluence of analyzing energy systems while considering quality life, they are still less prevalent. Two knowledge gaps in these studies are that satisfactio is not computed directly, neither is satisfaction considered in most cases from the point time dependent. This study provides a motivation for such a granular level of smart met data with distinct energy uses. The proposed study explores more ways to harness sma meters data to improve people's wellbeing. Detailed and granular information on energ consumption is expected to be broadly available in most households in the near futu This study is centered on the specific power usages at the household level. The benefi they bring to the household, as captured by human satisfaction, is also studied. The us satisfaction is not studied as a posteriori parameter to test the model but as a key part the problem.

DSM requires the processing of a high amount of data to coherently use consumptio patterns and manage demand. Smart metering infrastructure (SMI) provides the mea to gather this high amount of electrical consumption information. However, it is still challenge to consider people´s wellbeing while using smart meter data. Buchanan et al. [1 studied how smart meters can affect consumer's wellbeing. Under the 'five ways to we being framework' [18], they explored other areas that may be found with the consum acceptance and engagement with smart meter enabled services (SMES). To address th gap, this work is also contributing to a new platform to insert smart meter research direct into the exploration of wellbeing and the human impact of energy socio-technical system The present research offers a Shapley Value (SV) game-theory approach to solve the mu objective optimization problem (MOO) to optimize energy consumption. The hours of t day for which energy should be allocated are found. Quantifiable user satisfaction metr is used through the concepts of power satisfaction (PS) and energy satisfaction (ES). I and ES were recently developed by the authors [19]. PS and ES were computed hour and incorporated the detrimental impact that excess consumption can have in the quali of life. Although the state of art may offer other traditional [20] and metaheuristics [2 multi-objective based approaches, the present novel SV-based game-theoretic approach, seen in mentioned research, offers a simpler and more intuitive way to tackle the proble

Chambers [22] proposed responsible wellbeing to combine the concept of wellbei with personal responsibility. Castro-Sitiriche and Ozik [23] delved into the matter whe considering responsible wellbeing in terms of energy consumption. The energy threshold hypothesis is defined. It was previously presented by Max-Neef [24] in terms of economic growth and quality of life. The proposed MOO consists of responsibly fulfilling user's needs by maximizing satisfaction while minimizing the power consumption. A novel model is proposed to include customer's satisfaction in an optimization problem to minimize the energy consumption. To summarize, the contributions of this paper can be highlighted as follows:

- Energy satisfaction (ES) is proposed to capture the benefit of energy uses and to model the optimization problem.
- The Shapley Value algorithm is implemented to maximize ES and minimize energy consumption.
- The proposed model also integrated solar-based renewable-energy resources (RESs). Real data from [25] was used for validation.

2. Cooperative Game Theory

2.1. Overview of Game Theory

Game theory provides a series of analytical tools, which allows us to understand what is observed in decision-making interactions. The foundation of the theory is formed by two basic assumptions: decision-makers are rational, and reason strategically by considering the expectations of other decision-makers' behaviors. Real-life situations are modeled by game theory through highly abstract representations, thus, allowing their use to study problems in many fields [26].

2.2. Types of Games

There are noncooperative games and cooperative (or coalitional) games. In the former, each action is taken by a single player in response to the other players [27]. In cooperative games, the model consists of the set of joint actions that each group of players (or coalition) can take in response to the other players. Cooperative games are concerned with the interactions among players, the value of each coalition and how the value can be distributed to the participating players.

2.3. Shapley Value

The Shapley Value [28] is a solution concept for coalitional games along with the core, nucleolus and Pareto optimal, among others. Given a coalitional game $(\mathcal{N}, v)$, there is a unique feasible payoff division $x(v) = \varphi(\mathcal{N}, v)$ that divides the full payoff of the grand coalition. The Shapley Value can be defined as [29],

$$\varphi_i(\mathcal{N}, v) = \frac{1}{N!} \sum_{R=1}^{N!} [v(P_i(R) \cup i) - v(P_i(R))] \tag{1}$$

where R is the set of all $N!$ orderings of $\mathcal{N}$, $P_i(R)$ is the set of players preceding i in the ordering R and $\varphi_i(\mathcal{N}, v)$ is the expected marginal contribution over all orders of player i to the set of players who are preceding it [26].

The Shapley Value also satisfies the following axioms [30]:

- Symmetry: The symmetry axiom states that interchangeable players i and j should receive the same payments: $\varphi_i(\mathcal{N}, v) = \varphi_j(\mathcal{N}, v)$. Two agents are interchangeable if they contribute the same amount to every coalition with the other agents.
- Dummy Axiom: The dummy player i contributes to any coalition the same amount that i can achieve alone. Thus, for any v, if i is a dummy player, then $\varphi_i(\mathcal{N}, v) = v(i)$.
- Additivity: The additivity axiom states that for any two coalitional game problems, defined by v_1 and v_2, we have for any player i that $\varphi_i(\mathcal{N}, v_1 + v_2) = \varphi_i(\mathcal{N}, v_1) + \varphi_i(\mathcal{N}, v_2)$, where the game $\varphi_i(\mathcal{N}, v_1 + v_2)$ is defined by $(v_1 + v_2)(S) = v_1(S) + v_2(S)$ for every coalition S.

- Efficiency: The efficiency axioms states that the entire payoff is divided among the players, so $\sum_1^N \varphi_i = v(\mathcal{N})$, where φ_i is the Shapley Value of player i.

3. Satisfaction Model

3.1. Satisfaction Concept

The human experience is complex to describe since it depends on the reality of each person. Satisfaction, on the other hand, is a more general concept and describes the user perception and how its expectation can be fulfilled [7]. This research proposes a model to quantify satisfaction in the context of a household with different devices. The use of each device provides different levels of satisfaction at different times of the day. Daily moments are captured and summed up to build the energy satisfaction concept. As part of this model, we also consider excessive and poor consumption.

3.2. User Input Satisfaction

The proposed algorithm will be finding the hours of the day for which energy should be allocated to achieve the maximum satisfaction at a minimum energy consumption. For the user satisfaction, it is assumed:

- Satisfaction among devices can be compared at two levels: time-based, satisfaction Ω, and device-based satisfaction Δ, as defined by Ogunjuyigbe [8]. The former implies that if there is a device no. 1, then the satisfaction it is providing at time t_1, ($\Omega_1(t_1)$) can be compared with the satisfaction the same device is providing at different time t ($\Omega_1(t_2)$). For the latter, if the two devices need to be used at the same time, then there is a satisfaction derived from using device no. 1 ($\Delta_1(t_1)$), which can be compared with the satisfaction derived from using device no. 2 ($\Delta_2(t_1)$) at that hour.
- Time-based satisfaction Ω has an integer numerical value from zero (0) to six (6), where six (6) means completely satisfied, three (3) means neutral, and small satisfaction values, such zero (0), one (1) and two (2) will denote dissatisfaction.
- Three (3) levels of time-based satisfaction are identified according with these seven (7) scores, see Table 1.

Table 1. Levels of Satisfaction.

Level	Respondent's Answer
Satisfied	4, 5, 6
Neutral	3
Unsatisfied	0, 1, 2

3.3. Power Satisfaction

The power satisfaction will not only depend on the user input satisfaction set by the household's head. PS also depends on the energy consumption patterns at each hour of the day. Thus, PS depends on the number of hours of continuous usage (CLoU) and the total length of use in a day (LoU). To find the PS at time t, we need to analyze the previous 24 h, i.e., from time $t - 24$ until time $t - 1$. It is suggested that PS cannot be tested based on how satisfied a person is at t but it will be affected by its perception of the last 24 h.

3.4. Equation of Power Satisfaction

Power satisfaction at a given time/hour can be expressed as the function in Equation (2) below:

$$PS_i[t] = (\alpha_i[t] - \beta_i[t]|\widehat{t}_{u_i}[t] - t_{u_i}| - \gamma[t]|\widehat{t}_{t_i}[t] - t_{t_i}|) u_i[t] \qquad (2)$$

where,

$$\beta_i[t] = \gamma_i[t] = \frac{\alpha_i[t] + 3}{48 - t_{u_i} - t_{t_i}} \qquad (3)$$

where, α is related to the customer's answer, β is related to CLoU and γ is the related to LoU. The main idea is to penalize the initial satisfaction, α according to an excess or poor consumption. Variables β and γ depends on α and convert $\widehat{f_u}$ and $\widehat{f_t}$, respectively, into a value that can be deducted from α and,

$$\alpha_i[t] = \frac{\Omega_i[t]}{t_{t_i}}, \qquad (4)$$

where Ω_i is the time-based input satisfaction for device i and it is divided by the responsible LoU, t_{t_i}, hence α will reach its maximum if it used the expected time t_{t_i} hours during a 24-h period. To find PS, 24 h of experiment is needed. For each device i, the array $u_i = \{u_n : n = 1, 2, \ldots 24\}$ where u_n is 0 or 1. It is the input vector for each device with the operational status of the devices, it will be one (1) when 'ON' and (0) if it is 'OFF'. For further details, see previous work reference [19].

3.5. Energy Satisfaction

Each user has a subset $\{i : i = 1, 2, \ldots, N\}$ of N participant devices. Following the well-known concept of electric energy, to find the energy satisfaction, ES, in time t, the previous 24 values of PS are required, i.e., $\{PS_i[t-24], \ldots, PS_i[t-1]\}$. To compute the first PS, $PS_i[t-24]$, the past 24 h before this time are also required. Hence, to compute ES, 48 h of experiment are required. ES is defined as in Equation (5).

$$ES[t] = \frac{1}{N} \sum_{i \in N} \sum_{n=k} PS_{i,n}, \qquad (5)$$

where N is the number of participant devices, k is the initial time of the experiment, $t = k + 23$ and $\Delta_{i,t}$ is the device-based satisfaction.

Equation (5) is modified to include the concept of device-based satisfaction Δ. ES becomes a weighted summation and reflects the specific needs at that current time t. Subsequently, the average is computed to obtain the ES value at time t,

$$ES'[t] = \frac{1}{N} \left(\sum_{i \in N} \Delta_{i,t} PS_{i,t} + \left(\sum_{n=k}^{t-1} PS_{i,n} \right) \right), \qquad (6)$$

4. Problem Formulation

4.1. Electric Energy Function

The energy consumption in an hour from time t until time $t + 1$ is defined as in Equation (7),

$$L[t] = \sum_{i \in N} e_{i,t} u_i[t], \qquad (7)$$

where $e_{i,t}$ represents an usual energy consumption of device i in one hour and u_i is the input vector for each device i with its operational status. It will be one (1) when 'ON' and zero (0) if it is 'OFF'.

4.2. Optimization Problem

The problem of jointly maximizing the satisfaction while minimizing the power consumption $f_2(u)$ can be formulated as a multi-objective (MOO) problem:

$$\min[f_1(u,t), -f_2(u,t)], \qquad (8)$$

$$s.t.\ L[t] \leq E_{con}, \qquad (9)$$

where,

$$f_1(t,u) = L[t], \qquad (10)$$

$$f_2(t,u) = ES'[t], \qquad (11)$$

$$t \in \{k, k+1, \ldots, k+23\} \tag{1}$$

4.3. Constraint

The algorithm is subjected to the constraint that the total energy consumption of the user is less or equal to a pre-defined energy budget, E_{con},

$$s.t.\ L[t] \leq E_{con}, \tag{1}$$

4.4. Cooperative Game Model Implementation

A mathematical model based on cooperative game theory is developed to capture the complex interactions among the different devices. The Shapley Value algorithm based on the cooperative game implementation is applied to all participating devices to obtain the u_i vector. Such that energy can be allocated to simultaneously maximize the ES' and minimize the energy consumption L for a corresponding energy reference. The flow chart for the proposed algorithm is shown in Figure 1. Time-based and device-based satisfaction tables and power consumption for each device are established. The proposed approach to design a consumption scheduling for a selected number of hours n. Additionally, we should decide time t to start the experiment and the number N of participating devices. Considering the experiment should collect energy data for the 48 h before this time t, then the status matrix is of size $48 \times N$. The number of possible action profiles A, is four (2^N). Since each possible action profile contains the possible devices' status, each of the possible profiles are multiplied for the correspondent Device-based satisfaction Δ according to the time of the day and the device. PS and ES are computed using Equations (2) and (at each time slot, respectively. Next, the worth of the coalition $v(S_k)$ can be found for each time slot for each profile. Lastly, the group of actions with maximum SV can be selected.

To implement this algorithm the following considerations are followed:

- Players:

  Devices-agents $i = 1, 2, \ldots, N$ are considered the players.

- Actions:

  Devices can be ON or OFF, so the actions would be turning ON or OFF the participating devices (players.) The u_i array will be zeros or one depending on the device status.

- Payoffs:

  Consumption and ES are combined and modeled as optimization function using the characteristic function.

- Value of coalition:

  Agents form coalitions and every coalition and corresponding actions have different value. For $K \subseteq N$, S_K is the group of K devices that agreed to make a coalition to maximize the total expected payoff $v(S_K)$ that they can achieve together. The value of a coalition for this cooperative game is,

$$v(S_K)_t = \begin{cases} 0, & |L[t] > E_{ref} \cap ES'[t] < 4 \\ \left[1 - f\left|\frac{L[t] - P_{PV}[t]}{max(L[t], P_{PV}[t])}\right|\right]\left[\frac{ES[t]}{6}\right], & P_{PV}[t] < L[t] < E_{ref} \\ \left[1 - \left|\frac{L[t] - P_{PV}[t]}{max(L[t], P_{PV}[t])}\right|\right]\left[\frac{ES[t]}{6}\right] & \text{otherwise} \end{cases} \tag{1}$$

where f is a factor to penalize level of consumption among the energy reference and solar energy less than cases when less then solar energy and $P_{PV}[t]$ is the minimum among the set of power output of generation($P_{pv}[t]$) and the power inverter ($P_{inv}[t]$), i.e., $P_{PV}[t] = min(P_{pv}[t], P_{inv}[t])$, for the cases with just utility and $P_{PV}[t] = min(P_{pv}[t] + P_{bat}, P_{inv}[t])$ for the cases with batteries and utility, where P_{bat} the set of power output from batteries.

There are three levels of consumption: $L \leq P_{PV}$, $P_{PV} < L \leq E_{ref}$ and $L > E_{ref}$. Where E_{ref} would be a consumption that is acceptable, there will be penalty if it is less than P_{PV} or greater than E_{ref}.

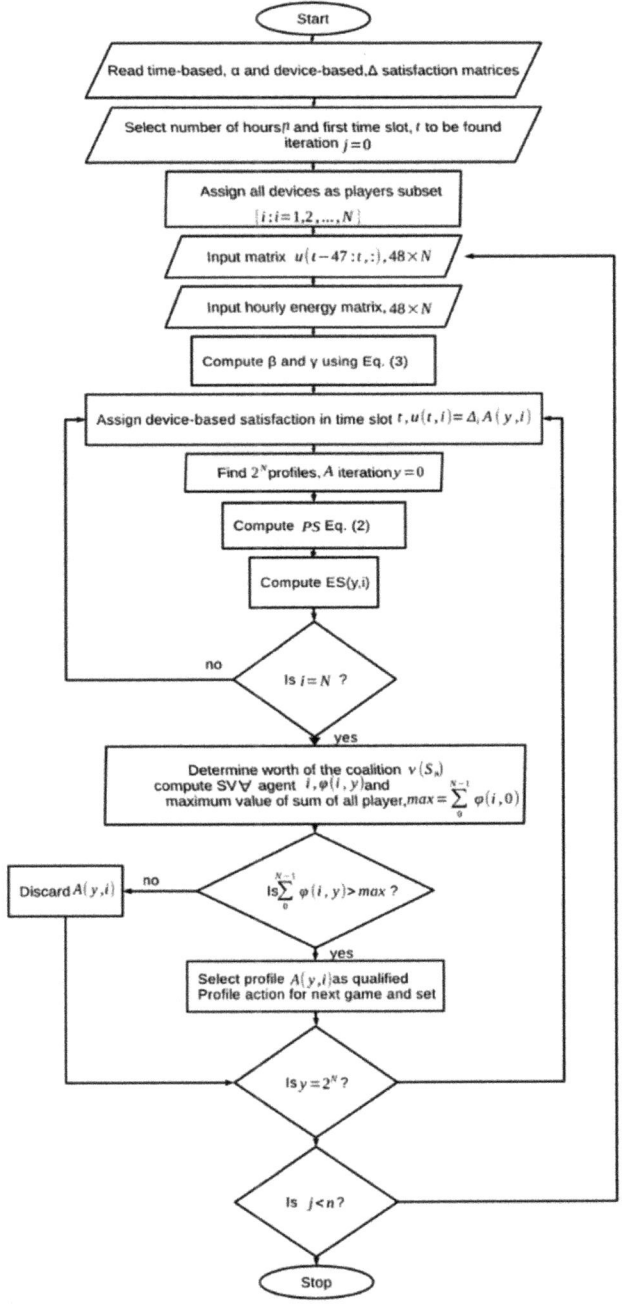

Figure 1. Flowchart for cooperative game implementation.

5. Experimental Results and Discussion

5.1. Case Study: House with Photovolatic System (PV) and Utility or House with PV, Batteries and Utility

The proposed model is being validated by representing a single rural residential house that may contain a small photovoltaic (PV) system that is composed of one or more solar panels combined, a DC/AC inverter, and it is grid-connected. Figure 2 describes a generalized version of the system. Energy data to simulate the case study is obtained in two ways:

- Reference Energy Disaggregation Data Set (REDD) [31] for power ratings and status vector (u_i) for refrigerator, microwave, lighting, stove, and water heater.
- Power ratings for TV, AC, Radio and phone, are obtained from Ogunjuyigbe [8]. Status vectors for these four devices are randomly generated.

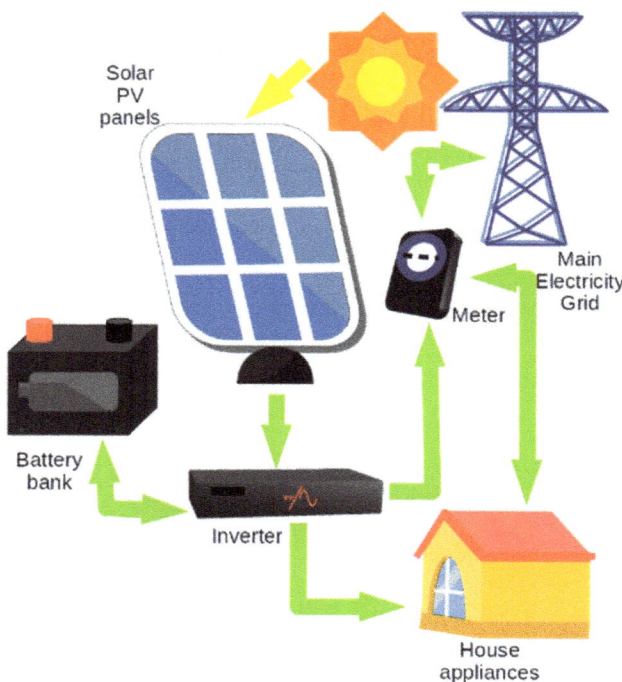

Figure 2. Photovoltaic (PV) system.

5.2. Data Characterization for the Algorithm Calibration

5.2.1. Household's Load

For the load data, a single house's devices (four of them) are analyzed, the REDD data [31] is used to represent a part of house's loads. Data that is shown in Table 2 contains average power reading for the individual circuits of the house. The REDD data was sampled every three seconds. Hence, there are 20 data points per minute. The u_i array is an hourly vector. For this vector, it needs to be decided how much time within an hour a certain device must be ON to assign a one (1) in that time slot.

Table 2. Load Description.

Device	Rating (W)	t_u	t_t
Lightings	240	4	8
Refrigerator	170	24	24
Stove	2	3	9
Microwave (Oven)	1200	1	4

5.2.2. House Head's Satisfaction

Satisfaction matrices were generated according to author's personal experience and they are described in Tables 3 and 4. This first stage of results is simply used as a sanity check. In the Section 5.3, actual load description (see Table 5) is used for each device based on REDD database and Ogunjuyigbe et al.'s [8] work. Besides, actual values of satisfaction were used to generate Tables 6 and 7. The authors´ personal experience reflects and summarizes how devices were prioritized by most of the population in extreme conditions such as those experienced after hurricane Maria in 2017 in Puerto Rico. However, future work will include field data to build the time-based satisfaction and device-based satisfaction tables.

5.2.3. Results

Figure 3 depicts the amount of energy used by each of the four devices under the proposed SV allocation algorithm and the actual energy usage according to the REDD database [31]. The proposed optimization problem has been constrained. Thus, showing how with the same amount or less energy than in the actual REDD scenario. ES (see Figure 4) is higher at each time slot.

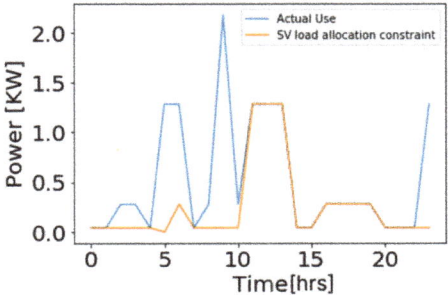

Figure 3. Power usage with the proposed Shapley Value (SV) load allocation algorithm vs. real usage according with the REDD [31] database.

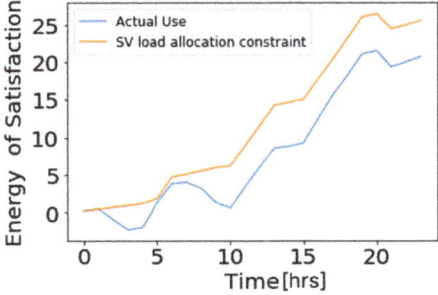

Figure 4. Energy satisfaction (ES) with the proposed SV load allocation using the same or less energy than real REDD [31] database vs. ES satisfaction using REDD [31] database.

Table 3. Time-based satisfaction.

S/N	Equipment	Hours																							
		1	2	3	4	5	6	7	8	9	10	11	12	13	14	15	16	17	18	19	20	21	22	23	24
1	Lightings	-3	-3	-3	-3	-3	6	6	6	-2	-2	-2	-2	-2	-2	-2	-2	6	6	6	6	6	-3	-3	-3
2	Refrigerator	4	4	4	4	4	6	6	6	6	6	6	6	6	6	6	6	6	6	6	6	6	4	4	4
3	Stove	-2	-2	-2	-2	-2	6	6	6	-2	-2	-2	5	6	6	5	-2	-2	-2	-2	6	6	-2	-2	-2
4	Microwave	4	4	0	0	0	6	6	6	6	0	0	6	6	6	0	0	0	0	6	6	6	0	0	0

Table 4. Device-based satisfaction.

S/N	Equipment	Hours																							
		1	2	3	4	5	6	7	8	9	10	11	12	13	14	15	16	17	18	19	20	21	22	23	24
1	Lightings	4	4	4	4	4	4	4	4	4	4	4	4	4	4	4	4	4	4	4	4	4	4	4	4
2	Refrigerator	1	1	1	1	1	1	1	1	1	1	1	1	1	1	1	1	1	1	1	1	1	1	1	1
3	Stove	-2	-2	-2	-2	-2	-2	-2	-2	-2	-2	-2	-2	-2	-2	-2	-2	-2	-2	-2	-2	-2	-2	-2	-2
4	Microwave	4	4	4	4	4	4	4	4	4	4	4	4	4	4	4	4	4	4	4	4	4	4	4	4

Table 5. Load description.

Device	Rating (W)	t_u	t_t
Lightings	135	4	8
Microwave (Oven)	1200	1	4
TV	200	2	4
AC	800	6	8
Radio	50	5	12
Water Heater	2000	1	2
Laptop	100	8	12
Phone	10	1	3

Figure 5 depicts how power is used in both scenarios. Figure 6 shows the resulting ES from that energy usage. Lighting is suggested to be on early in the morning and in the afternoon when the satisfaction derived from them is the highest. Therefore, the refrigerator needs to be off for one hour because it is a priority (see Table 4 to have lighting on since its contribution to the total ES is higher. However, analyzing ES brought by refrigerator in Figure 6 is slightly lower for the proposed algorithm since it was turned off. According to the REDD database [31], the stove is in on status during one hour but it is on at a time that is not bringing any satisfaction, hence the proposed algorithm recommends to not turn it on under the energy restrictions. For the microwave, in the REDD database [31], it is on at different times of the day however it is not on when the satisfaction is the highest. For example, in the morning it is preferable to have lighting on instead of the microwave, since the energy usage of lighting is less than the energy usage of the microwave.

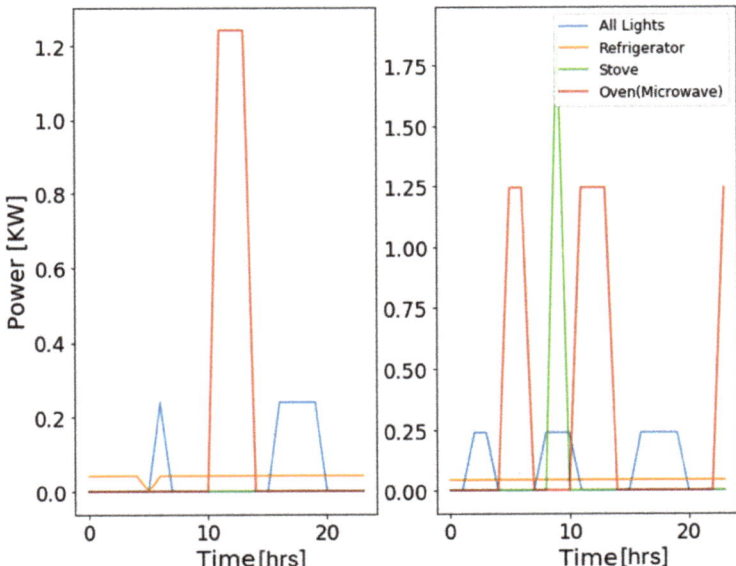

Figure 5. Power usage by each device with the proposed SV load allocations vs. real usage according with the REDD [31] database.

Table 6. Time-based satisfaction.

S/N	Equipment	Hours																							
		1	2	3	4	5	6	7	8	9	10	11	12	13	14	15	16	17	18	19	20	21	22	23	24
1	TV	0.0	0.0	0.0	0.0	0.0	0.0	−2.4	−2.4	0.0	0.0	0.0	0.0	0.0	0.0	−2.4	−2.4	−2.4	−1.8	4.4	6.0	6.0	5.2	4.4	−0.6
2	Lighting	0.0	0.0	0.0	0.0	0.0	4.8	4.0	−0.6	−2.4	0.0	0.0	0.0	0.0	0.0	0.0	0.0	−1.2	4.0	4.4	6.0	6.0	5.2	4.4	−1.2
3	AC	0.0	0.0	0.0	0.0	−2.4	−0.6	−1.8	−2.4	0.0	−2.4	0.0	0.0	0.0	0.0	0.0	0.0	−2.4	−1.2	−1.2	4.4	6.0	6.0	−1.8	−2.4
4	Radio	0.0	0.0	0.0	0.0	−2.4	−1.8	6.0	−1.8	−1.8	0.0	0.0	0.0	0.0	0.0	0.0	0.0	0.0	0.0	−2.4	−1.8	−1.8	0.0	0.0	0.0
5	Water Heater	0.0	0.0	0.0	−2.4	−1.2	6.0	4.0	−1.8	−2.4	0.0	0.0	0.0	0.0	0.0	0.0	0.0	0.0	0.0	0.0	0.0	0.0	0.0	0.0	0.0
6	Lighting	0.0	0.0	0.0	0.0	−1.2	6.0	−1.2	−1.8	−2.4	0.0	0.0	0.0	0.0	0.0	0.0	0.0	0.0	−1.8	−2.4	−1.8	6.0	4.0	−1.2	−1.8
7	Microwave Oven	0.0	0.0	0.0	0.0	−2.4	6.0	6.0	−2.4	0.0	0.0	0.0	0.0	0.0	0.0	0.0	0.0	−2.4	−2.4	4.8	6.0	−1.2	−2.4	0.0	0.0
8	Lighting	0.0	0.0	0.0	0.0	−1.2	6.0	4.0	−1.2	−2.4	0.0	0.0	0.0	0.0	0.0	0.0	0.0	0.0	−2.4	4.8	6.0	−0.6	−1.8	−2.4	0.0
9	Lighting	6.0	6.0	5.6	5.2	4.8	4.4	−1.2	−2.4	0.0	0.0	0.0	0.0	0.0	0.0	0.0	0.0	−2.4	−2.4	−1.8	4.0	4.8	5.2	5.6	6.0
10	Lighting	0.0	0.0	0.0	0.0	−1.8	6.0	4.8	−1.2	−2.4	0.0	0.0	0.0	0.0	0.0	0.0	0.0	0.0	−2.4	−1.8	−0.6	4.0	4.8	6.0	−0.6
11	Laptop	0.0	0.0	0.0	0.0	−2.4	4.0	−1.2	−2.4	−2.4	0.0	0.0	0.0	0.0	0.0	0.0	0.0	0.0	0.0	−2.4	−2.4	−1.2	4.0	6.0	5.2
12	Phone	6.0	4.8	4.4	−0.6	−1.2	−1.2	−1.8	−2.4	0.0	0.0	0.0	0.0	0.0	0.0	0.0	0.0	0.0	0.0	−2.4	−2.4	−2.4	−1.2	4.0	6.0

Table 7. Device-based satisfaction.

S/N	Equipment	Hours																							
		1	2	3	4	5	6	7	8	9	10	11	12	13	14	15	16	17	18	19	20	21	22	23	24
1	TV	0.0	0.0	0.0	0.0	0.0	0.0	−2.4	−2.4	0.0	0.0	0.0	0.0	0.0	0.0	−2.4	−2.4	−2.4	−1.8	4.4	6.0	6.0	5.2	4.4	−0.6
2	Lighting	0.0	0.0	0.0	0.0	0.0	4.8	4.0	−0.6	−2.4	0.0	0.0	0.0	0.0	0.0	0.0	0.0	−1.2	4.0	4.4	6.0	6.0	5.2	4.4	−1.2
3	AC	0.0	0.0	0.0	0.0	−2.4	−0.6	−1.8	−2.4	0.0	−2.4	0.0	0.0	0.0	0.0	0.0	0.0	−2.4	−1.2	−1.2	4.4	6.0	6.0	−1.8	−2.4
4	Radio	0.0	0.0	0.0	0.0	−2.4	−1.8	6.0	−1.8	−1.8	0.0	0.0	0.0	0.0	0.0	0.0	0.0	0.0	0.0	−2.4	−1.8	−1.8	0.0	0.0	0.0
5	Water Heater	0.0	0.0	0.0	−2.4	−1.2	6.0	4.0	−1.8	−2.4	0.0	0.0	0.0	0.0	0.0	0.0	0.0	0.0	0.0	0.0	0.0	0.0	0.0	0.0	0.0
6	Lighting	0.0	0.0	0.0	0.0	−1.2	6.0	−1.2	−1.8	−2.4	0.0	0.0	0.0	0.0	0.0	0.0	0.0	0.0	−1.8	−2.4	−1.8	6.0	4.0	−1.2	−1.8
7	Microwave Oven	0.0	0.0	0.0	0.0	−2.4	6.0	6.0	−2.4	0.0	0.0	0.0	0.0	0.0	0.0	0.0	0.0	−2.4	−2.4	4.8	6.0	−1.2	−2.4	0.0	0.0
8	Lighting	0.0	0.0	0.0	0.0	−1.2	6.0	4.0	−1.2	−2.4	0.0	0.0	0.0	0.0	0.0	0.0	0.0	0.0	−2.4	4.8	6.0	−0.6	−1.8	−2.4	0.0
9	Lighting	6.0	6.0	5.6	5.2	4.8	4.4	−1.2	−2.4	0.0	0.0	0.0	0.0	0.0	0.0	0.0	0.0	−2.4	−2.4	−1.8	4.0	4.8	5.2	5.6	6.0
10	Lighting	0.0	0.0	0.0	0.0	−1.8	6.0	4.8	−1.2	−2.4	0.0	0.0	0.0	0.0	0.0	0.0	0.0	0.0	−2.4	−1.8	−0.6	4.0	4.8	6.0	−0.6
11	Laptop	0.0	0.0	0.0	0.0	−2.4	4.0	−1.2	−2.4	−2.4	0.0	0.0	0.0	0.0	0.0	0.0	0.0	0.0	0.0	−2.4	−2.4	−1.2	4.0	6.0	5.2
12	Phone	6.0	4.8	4.4	−0.6	−1.2	−1.2	−1.8	−2.4	0.0	0.0	0.0	0.0	0.0	0.0	0.0	0.0	0.0	0.0	−2.4	−2.4	−2.4	−1.2	4.0	6.0

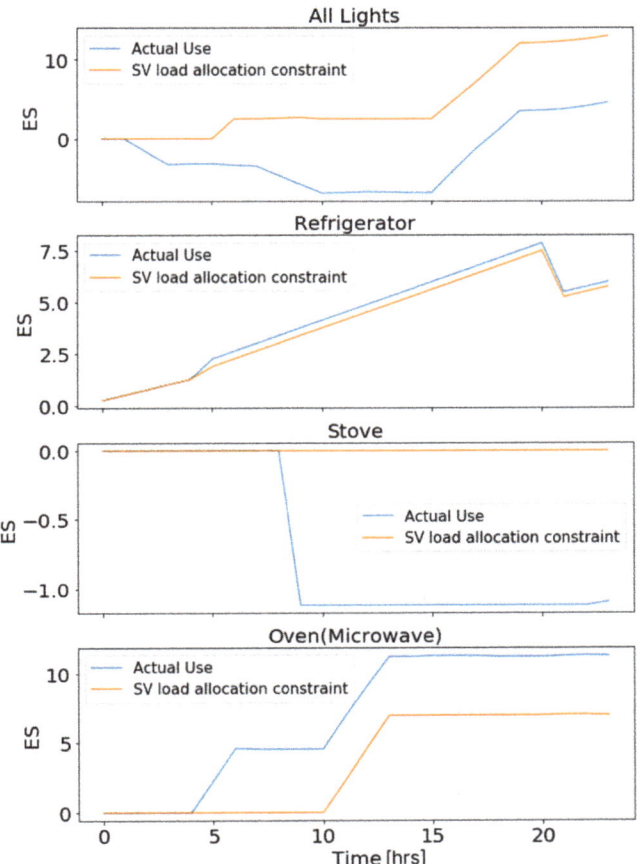

Figure 6. ES derived from each device with the proposed SV load allocation algorithm vs. real usage according to the REDD [31] database.

For the calibration part, we made sure the algorithm was suggesting those profiles where the Energy Satisfaction was highest at a minimum energy usage. In the following section, we will be testing the algorithm for a different set of devices.

5.3. Data Characterization for the Algorithm Testing.

Eight devices were selected according to the ones used in Ogunjuyigbe et al. [8] for testing purposes. For the algorithm simplicity, all lights were considered as a single appliance. Ogunjuyigbe et al. [8] used some loads for which there are not data available in the REDD database [31], such as TV, AC, Radio, and phone. Hence, for these ones, Unit Wattage data from [8] was used. Additionally, u_i was randomly generated for those devices, such that a comparison can be made with the proposed algorithm's u_i output. Table 5 describes the electrical appliances used by a user, their rating, their optimum CLoU, t_u and their optimum LoU, t_t for a responsible consumption.

5.3.1. House Head's Satisfaction

The algorithm also needs practical input data for satisfaction to create the model. Data from Ogunjuyigbe et al. [8] (σ^t and σ^d) are being used for this purpose. Data from time-based satisfaction was mapped into satisfaction levels described in Table 1 in the following fashion,

$$\Omega_t = \begin{cases} 0, & if\ \sigma^t = 0.5 \\ 6 - \frac{2(1-\sigma^t[t])}{0.5}, & if\ \sigma^t \geq 0.5 \\ 6\sigma^t[t] - 3, & if\ \sigma^t < 0.5 \end{cases} \tag{1}$$

Dissatisfaction and 'indifference' values (0, 1, 2 and 3) are mapped into negative values and zero (−3, −2, −1 and 0, respectively). The proposed model introduced negative values when low satisfaction, thus making it preferable to have them 'OFF', representing by itself before the optimization problem, a more accurate satisfaction model, which allows making decisions not only for energy and economic savings but also responsibly fulfilling the customer's satisfaction. Table 6 shows complete resulting time-based satisfaction table and Table 7 shows device-based satisfaction after mapping d-domain satisfaction found Ogunjuyigbe et al. [8] (σ^d) by using Equation (16).

$$\Delta[t] = 10\sigma^d[t], \tag{1}$$

5.3.2. Results

One of the main results to report is that the implementation of the SV optimization provided a consumption pattern that represents an energy consumption less or equal than the initial actual use and a higher energy satisfaction for almost all hours in all devices Figure 7 provides the graphical comparison of the power consumption at each hour for the actual based case and the case with the SV optimization. Figure 8 presents the energy satisfaction at each hour showing how the SV optimization outperforms the base case particularly increasing its advantage in the early morning hours and the late evening hours

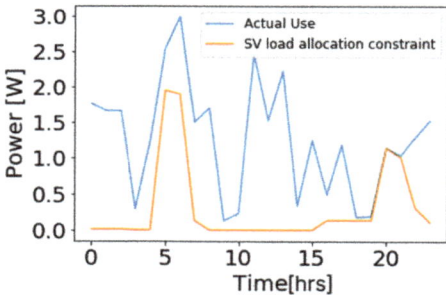

Figure 7. Power usage by each device with the proposed SV load allocations vs. real usage according to the REDD [31] database.

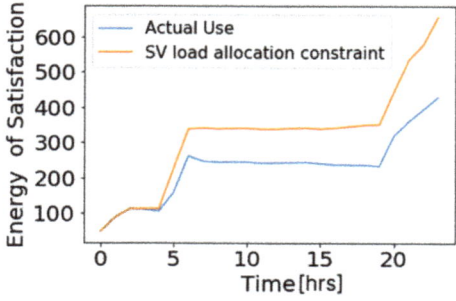

Figure 8. ES Satisfaction with the proposed SV load allocation using the same or less energy than real REDD [31] database vs. real usage according to the REDD [31] database.

Figure 9 depicts a comparison of the hourly power consumption of all devices between the proposed SV allocation algorithm and the actual power consumption. In Figure 9 (left side), the algorithm attempts to meet desired time-based and device-based satisfaction tables (See Tables 6 and 7) while consuming equal or less hourly power than the one shown in 9 (right side). The energy reduction was of approximately 75%, from 32.6 KWh to 7.35 KWh. Equally important is the energy satisfaction increase of 40% with the SV algorithm, from 5500 to 7825. A consumption plan is scheduled for the user by managing devices based on the SV game theory approach. Next, a reliability signal or economic signal will be sent to a Human-Machine Interface (HMI). A reliability signal will ensure that the electric system keeps operating when a house is not connected to the grid, while an economic signal ensures the same purpose when connected to the grid. This way, the customer is aware of the situation and can make a final decision based on the available information.

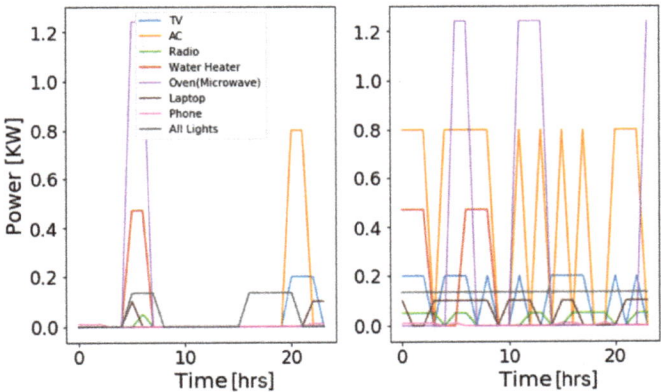

Figure 9. Power usage by each device with the proposed SV load allocations (**left** side) vs. real usage according to the REDD (**right** side) [31] database.

Figure 10 includes the ES for each device and it seems that the better SV performance is due mainly to the microwave use in the morning and the TV at night. Since operational status vectors, u for TV, AC, Radio, and phone, were randomly generated, Figure 9 shows an atypical consumption pattern. Figure 10 shows a comparison between the hourly ES obtained through the SV allocation algorithm and the ES obtained in the actual case representation, for each of the devices. When attempting to meet the energy constraints imposed by the actual case scenario, with the proposed SV algorithm, the ES obtained at each hour is equal or higher for almost each of the hours for every device. This is one of the most important results of this study. Only in the case of the laptop, the ES is higher at the last hour of the day in the actual case scenario.

Similarly, Ogunjuyigbe et al. [8] presented their results in 24 h plots for three different daily budget constraints to provide a maximum satisfaction at those predefined budgets. On the other hand, the present research used energy constraints rather than budget, and thus including a key component of the research when penalizing excessive and low consumption, because of its detrimental impact in the quality of life. He compared a 'desired satisfaction' with an 'achieved satisfaction'. The 'achieved satisfaction' was the output of their load-satisfaction algorithm, which is analogous to the present 'SV load allocation' algorithm. We did not choose to compare the output results with the 'desired satisfaction' (as Seen in Tables 6 and 7). Instead, we compare it with an 'actual' scenario represented by using the REDD database. Ogunjuyigbe et al. [8] implemented a genetic algorithm (GA) approach. While the GA approach may have a good convergence speed and good efficiency, the present approach does not have to deal with convergence times and offers a more intuitive optimization framework.

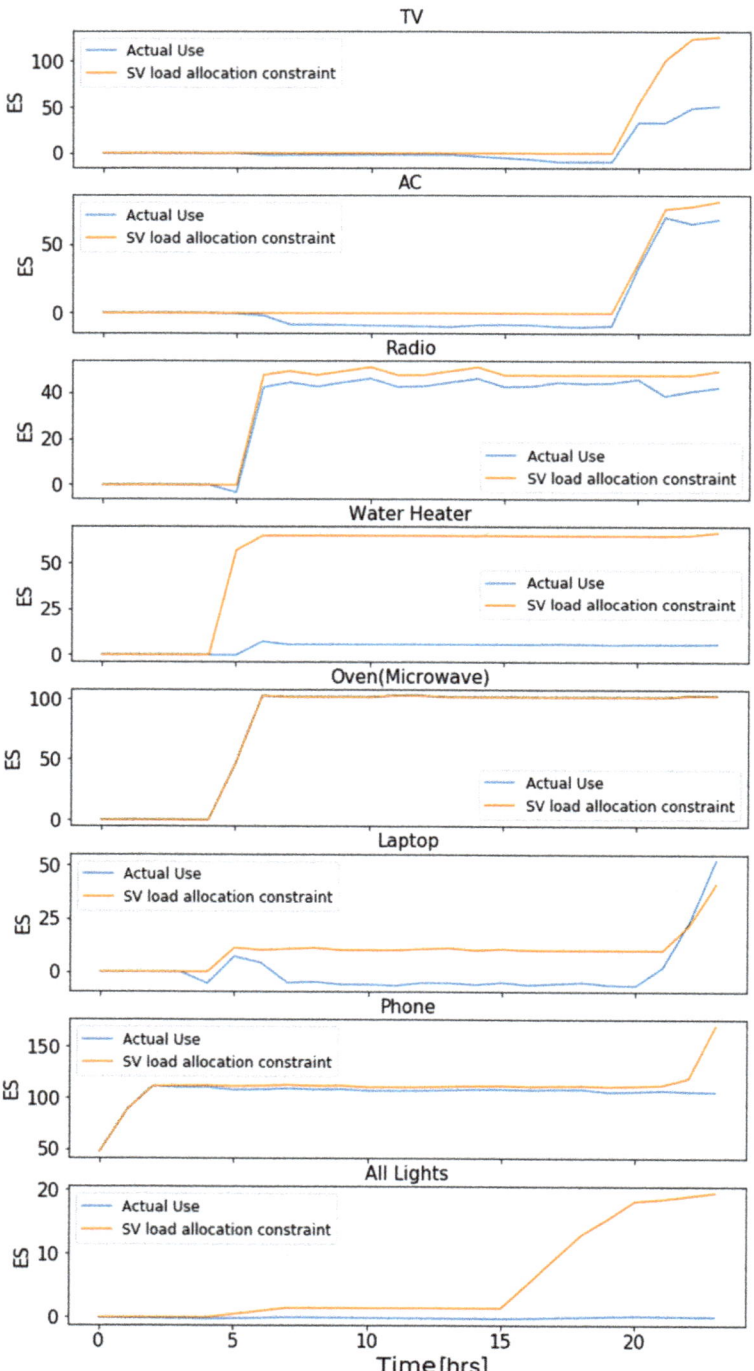

Figure 10. ES derived from each device with the proposed SV load allocation algorithm vs. real usage according to the REDD [31] database.

6. Conclusions

A novel model to include customer's satisfaction in an optimization problem was introduced. A quantifiable user satisfaction was developed. The satisfaction concept through the novel concepts of power satisfaction (PS) and energy satisfaction (ES) included the detrimental impact that excess consumption could have in the quality of life. The algorithm provided the hours of the day for which the energy should be allocated to achieve maximum satisfaction at an energy constraint imposed by the energy consumption in actual case scenarios. Actual case scenarios were represented by using the REDD database. The Shapley Value (SV) concept from the game theory framework was implemented to obtain a recommendation on how energy should be allocated. The results showed how the algorithm maximized user´s ES at a minimum energy consumption. The proposed approach reduced energy consumption 75%, while increasing ES 40%.

SV-based optimization successfully achieved to maximize satisfaction at a minimum energy consumption although it also has high computational complexity. Further work can be done to decrease computational complexity and thus the required reduction processing times. This could be achieved by using more powerful computers or through the derivation recursive and/or parallel implementation of the proposed algorithm. The proposed methodology is validated by simulating a rural single house with limited resources connected to the grid. It should be pointed out that the satisfaction model can be readily applied in a real case scenario of rural communities.

Author Contributions: Conceptualization, S.O., and M.C.-S.; data curation, S.O.; formal analysis, S.O.; funding acquisition, M.C.-S.; investigation, S.O.; methodology, S.O. and M.N.; software, S.O.; validation, S.O.; visualization, S.O.; writing—original draft, S.O.; writing–review and editing, S.O., M.N. and M.C.-S. All authors have read and agreed to the published version of the manuscript.

Funding: This document is the result of the research project partially funded by the National Science Foundation through the project titled "Cultivating Responsible Wellbeing in STEM: Social Engagement through Personal Ethics", award #1449489, at the University of Puerto Rico in Mayagüez.

Data Availability Statement: Public REDD dataset can be found at http://redd.csail.mit.edu/.

Conflicts of Interest: The authors declare no conflict of interest.

References

Access to Energy | Sustainable Energy for All. Available online: https://www.seforall.org/goal-7-targets/access (accessed on 24 February 2021).

Wu, Q.; Ren, H.; Gao, W.; Ren, J. Benefit allocation for distributed energy network participants applying game theory based solutions. *Energy* **2017**, *119*, 384–391. [CrossRef]

Tang, R.; Wang, S.; Wang, H. Optimal power demand management for cluster-level commercial buildings using the game theoretic method. *Energy Procedia* **2019**, *159*, 186–191. [CrossRef]

Lokeshgupta, B.; Sivasubramani, S. Cooperative game theory approach for multi-objective home energy management with renewable energy integration. *IET Smart Grid* **2019**, *2*, 34–41. [CrossRef]

Yang, P.; Tang, G.; Nehorai, A. A game-theoretic approach for optimal time-of-use electricity pricing. *IEEE Trans. Power Syst.* **2012**, *28*, 884–892. [CrossRef]

Marzband, M.; Ardeshiri, R.R.; Moafi, M.; Uppal, H. Distributed generation for economic benefit maximization through coalition formation–based game theory concept. *Int. Trans. Electr. Energy Syst.* **2017**, *27*, e2313. [CrossRef]

Shahzad, S.; Brennan, J.; Theodossopoulos, D.; Hughes, B.; Calautit, J.K. Energy and comfort in contemporary open plan and traditional personal offices. *Appl. Energy* **2017**, *185*, 1542–1555. [CrossRef]

Ogunjuyigbe, A.S.O.; Ayodele, T.R.; Akinola, O.A. User satisfaction-induced demand side load management in residential buildings with user budget constraint. *Appl. Energy* **2017**, *187*, 352–366. [CrossRef]

Yoon, S.G.; Choi, Y.J.; Park, J.K.; Bahk, S. Stackelberg-Game-Based Demand Response for At-Home Electric Vehicle Charging. *IEEE Trans. Veh. Technol.* **2016**, *65*, 4172–4184. [CrossRef]

Yu, M.; Hong, S.H. A Real-Time Demand-Response Algorithm for Smart Grids: A Stackelberg Game Approach. *IEEE Trans. Smart Grid* **2016**, *7*, 879–888. [CrossRef]

Azimian, B.; Fijani, R.F.; Ghotbi, E.; Wang, X. Stackelberg game approach on modeling of supply demand behavior considering BEV uncertainty. In Proceedings of the 2018 IEEE International Conference on Probabilistic Methods Applied to Power Systems (PMAPS), Boise, ID, USA, 24–28 June 2018. [CrossRef]

12. Jia, L.; Tong, L. Dynamic Pricing and Distributed Energy Management for Demand Response. *IEEE Trans. Smart Grid* **20**, *7*, 1128–1136. [CrossRef]
13. Ma, L.; Liu, N.; Wang, L.; Zhang, J.; Lei, J.; Zeng, Z.; Wang, C.; Cheng, M. Multi-party energy management for smart building cluster with PV systems using automatic demand response. *Energy Build.* **2016**, *121*, 11–21. [CrossRef]
14. Agyeman, K.; Han, S.; Han, S. Real-Time Recognition Non-Intrusive Electrical Appliance Monitoring Algorithm for a Residential Building Energy Management System. *Energies* **2015**, *8*, 9029–9048. [CrossRef]
15. Pal, S.; Thakur, S.; Kumar, R.; Panigrahi, B.K. A strategical game theoretic based demand response model for residential consumers in a fair environment. *Int. J. Electr. Power Energy Syst.* **2018**, *97*, 201–210. [CrossRef]
16. Wang, Z.; Paranjape, R. Optimal Residential Demand Response for Multiple Heterogeneous Homes with Real-Time Price Prediction in a Multiagent Framework. *IEEE Trans. Smart Grid* **2017**, *8*, 1173–1184. [CrossRef]
17. Buchanan, K.; Banks, N.; Preston, I.; Russo, R. Corrigendum to The British public's perception of the UK smart metering initiative: Threats and opportunities. *Energy Policy* **2016**, *91*, 87–97. [CrossRef]
18. Aked, J.; Cordon, C.; Marks, N. Five Ways to Wellbeing: A Report Presented to the Foresight Project on Communicating the Evidence Base for Improving People's Well-Being | Repository for Arts and Health Resources. Available online: https://www.artshealthresources.org.uk/docs/five-ways-to-wellbeing-a-report-presented-to-the-foresight-project-on-communicating-the-evidence-base-for-improving-peoples-well-being/ (accessed on 5 February 2021).
19. Ortiz, S.; Ndoye, M.; Castro-Sitiriche, M. Toward an Energy Threshold Hypothesis Model for Social Benefit and Household Energy Usage. In Proceedings of the International Conference on Appropriate Technology 2020, Pretoria, South Africa, 23– November 2020; pp. 346–355.
20. Sarker, E.; Halder, P.; Seyedmahmoudian, M.; Jamei, E.; Horan, B.; Mekhilef, S.; Stojcevski, A. Progress on the demand side management in smart grid and optimization approaches. *Int. J. Energy Res.* **2021**, *45*, 36–64. [CrossRef]
21. Tzanetos, A.; Dounias, G. Nature inspired optimization algorithms or simply variations of metaheuristics? *Artif. Intell. Rev.* **20**. [CrossRef]
22. Robert, C. Editorial: Responsible well-being—A personal agenda for development. *World Dev.* **1997**, *25*, 1743–1754. Available online: https://ideas.repec.org/a/eee/wdevel/v25y1997i11p1743-1754.html (accessed on 5 February 2021).
23. Castro-Sitiriche, M.J.; Ndoye, M. On the Links between Sustainable Wellbeing and Electric Energy Consumption. *Afr. J. S Technol. Innov. Dev.* **2013**, *5*, 327–335. [CrossRef]
24. Max-Neef, M. Economic growth and quality of life: A threshold hypothesis. *Ecol. Econ.* **1995**, *15*, 115–118. [CrossRef]
25. NREL. Solar Resource Data. Jurutungo Community, Puerto Rico. Available online: https://pvwatts.nrel.gov/pvwatts.p (accessed on 11 November 2020).
26. Osborne, M.J.; Rubinstein, A. *A Course in Game Theory*; The MIT Press: Cambridge, MA, USA, 2011.
27. O'Brien, G.; El Gamal, A.; Rajagopal, R. Shapley Value Estimation for Compensation of Participants in Demand Response Programs. *IEEE Trans. Smart Grid* **2015**, *6*, 2837–2844. [CrossRef]
28. Jackson, M.O.; Leyton-Brown, K.; Yoav, S. Game Theory Online. Available online: http://game-theory-class.org/game-theory.html (accessed on 24 February 2021).
29. Arif, M.A.; Ndoye, M.; Murphy, G.V.; Aganah, K. A cooperative game theory algorithm for distributed reactive power reserve optimization and voltage profile improvement. In Proceedings of the 2017 North American Power Symposium (NAP Morgantown, WV, USA, 17–19 September 2017. [CrossRef]
30. Leyton-Brown, K.; Shoham, Y. Essentials of Game Theory: A Concise Multidisciplinary Introduction. *Synth. Lect. Artif. Intel Mach. Learn.* **2008**, *2*, 1–88. [CrossRef]
31. Kolter, J.Z.; Batra, S.; Ng, A.Y. Energy disaggregation via discriminative sparse coding. In Proceedings of the Advances in Neural Information Processing Systems 23: 24th Annual Conference on Neural Information Processing Systems 2010, Vancouver, B Canada, 6–9 December 2010.

Article

The Impact of Ambient Sensing on the Recognition of Electrical Appliances

Jana Huchtkoetter, Marcel Alwin Tepe and Andreas Reinhardt *

Department of Informatics, TU Clausthal, 38678 Clausthal-Zellerfeld, Germany; jana.huchtkoetter@tu-clausthal.de (J.H.); mat14@tu-clausthal.de (M.A.T.)
* Correspondence: andreas.reinhardt@tu-clausthal.de

Abstract: Smart spaces are characterized by their ability to capture a holistic picture of their contextual situation. This often includes the detection of the operative states of electrical appliances, which in turn allows for the recognition of user activities and intentions. For electrical appliances with largely different power consumption characteristics, their types and operational times can be easily inferred from data collected at a single metering point (typically, a smart meter). However, a disambiguation between consumers of the same type and model, yet located in different areas of a smart building, is not possible this way. Likewise, small consumers (e.g., wall chargers) are often indiscernible from measurement noise and spurious power consumption events of other appliances. As a consequence thereof, we investigate how additional sensing modalities, i.e., data beyond electrical signals, can be leveraged to improve the appliance detection accuracy. Through a set of practical experiments, recording ambient influences in eight dimensions and testing their effects on 21 appliance types, we evaluate the importance of such added features in the context of appliance recognition. Our results show that electrical power measurements already yield a high appliance recognition accuracy, yet further accuracy improvements are possible when considering ambient parameters as well.

Keywords: appliance load signatures; ambient influences; device classification accuracy

1. Introduction

The number of different electrical appliances in households keeps rising. As such, it is becoming increasingly important to recognize the potentials for energy savings and demand side management, i.e., the possibility to defer power consumption in order to improve the stability of the power grid. For these purposes, it is not only vital to have complete knowledge about the appliances present in a building, but also their individual energy demands and operational times. Monthly electricity bills only provide insufficient information to accomplish this task, however, as the required information can only be estimated from the household total. Even current-generation smart meters are incapable of providing a detailed and unambiguous itemization of energy consumption, so more fine-grained means for load monitoring in a home are needed to provide enhanced consumption feedback and accomplish energy savings (which were documented to reach up to 12% in [1]).

Two fundamentally different approaches exist for the collection of the required data at appliance level. One the one hand, Intrusive Load Monitoring relies on the installation of power sensing devices for each appliance under consideration (or at least every electrical circuit in the home). The advantages of being able to attribute power consumption to individual devices, however, come at a high cost for instrumenting the environment with sensors, and maintaining their operability during their lifetime. On the other hand, Non-Intrusive Load Monitoring (NILM) methods collect the electrical information of a whole building or apartment and use algorithms to disaggregate the total power demand into the contributions of individual devices. The non-intrusive approach is often preferred due to lower costs and installation efforts. However, current NILM methods do not always succeed in accurately and unambiguously disaggregating power data from households [

Most current methods for appliance recognition and load disaggregation consider electrical consumption data (e.g., active power) [2,3]. When data are available at a sufficient resolution, however, more complex features like spectral components can be computed and used supplementally. Using more, especially more complex, features has been shown to improve the rate of correct device recognitions [4]. Confusion may still exist between certain appliance types, and some appliances have been reported to be "hard to disaggregate" [5, based on their electricity consumption alone. One potential candidate to alleviate the current limitations of NILM is the additional use of contextual information, as documented in [7–9]. For example, the distributions of On- and Off-durations as well as dependencies between device usages are modeled into a Factorial Hidden Markov Model (FHMM) in [The resulting performance shows a marked improvement, even for a larger number of active appliances, and reaches improvements of up to 25%.

A similar approach based on user presence and time constraints was presented in [The application of time constraints alone was shown to achieve a small improvement of about 3%. However, as soon as indicators of the user presence were included in combination with time constraints, improvements of about 14% were reported.

But there are more parameters besides the aforementioned attributes. It is well-known that many electrical devices generate acoustic, magnetic, or optical emissions, or dissipate the consumed energy as heat during their activity. The potentials of using such information in the appliance recognition task have been investigated in [5,10,11] and further studies presented in Section 2. We, however, believe that our work is first to present a holistic and comprehensive study that determines the information gain of a range of additional sensing modalities. A thorough understanding of the importance of ambient sensing features is vital to optimally support appliance recognition (e.g., by lowering the number of candidate devices for the classification task). We strongly expect monitoring systems to profit from deeper understanding of the features that characterize the operation of electrical appliances. System operators could then decide to specifically collect data based on the importance of certain sensor types, i.e., their usefulness, to better evaluate if costs outweigh possible benefits. Costs typically arise from hardware purchases and the device deployment in the optimum locations (e.g., luminosity sensors need to be mounted next to the light-emitting parts). However, non-monetary costs may also play a role, e.g., when sensors have the potential to compromise on user security or privacy. Understanding the importance of ambient sensing features will thus ease the considerations which sensor types to deploy. Accordingly, our work seeks to establish the foundation to enable further work in this context by providing an answer to the following question: Which ambient sensors can lead to improved appliance recognition results, and what are the most useful sensor types to facilitate the categorization of electrical consumers by their types?

In order to answer this question, we design a study to be conducted in two sequential steps. The first step, the design and implementation of a comprehensive data acquisition setup, is essential due to the unavailability of publicly released data that contains information beyond electrical power consumption data. In our second step, data analysis, we assess the contribution of each sensing modality to the overall appliance recognition task. Beyond the recognition of specific appliances, we also determine the set of features to facilitate the detection of appliance classes (i.e., the distinction between devices of different categories Ultimately, the analyses presented in this work allow us to derive recommendations for future data collection campaigns, similar to the set of guidelines for electrical datasets presented in [12].

Our manuscript is organized as follows. In Section 2 we provide an overview of further studies which considered additional ambient sensing and illustrate how these works were considered for the design of our study. In Section 3 we present our system design to collect data from eight ambient parameters, both during appliance operation and inactivity. The concept for our subsequent evaluation and its parameter choices are detailed in Section 4. We evaluate to what extent devices could be recognized and what ambient sensors carried the most information in Section 5, and we summarize the insights gained in our study in Section 6.

2. Related Work

A number of studies have considered further data besides electrical information sources to improve the device identification and accordingly improve the disaggregation process: Acoustic sensors, light, temperature, vibration, electro-magnetic fields, or acceleration data [5,10,11,13–16]. Remarkably, however, the aforementioned works have considered these parameters largely in isolation, as shown in Table 1. Opposed to this, we present a comprehensive study that relies on all sensor types in this work.

Table 1. Types of Ambient Sensors Used in Related Work.

Reference	Sound	Light	Temperature	Electro-Magnetic Fields	Acceleration	Vibration
[5]	✓					
[15]	✓	✓		✓		
[16]	✓	✓	✓		✓	✓
[10]	✓	✓	✓		✓	
[14]	✓					
this paper	✓	✓	✓	✓	✓	✓

The sensor deployment methodologies differ as well. The authors of [5,11,13,14] use a small number of sensors, which are not fixed to the appliances under consideration, but rather monitor the ambient conditions in general. As such, they can collect information from appliances that are operating simultaneously. In contrast to this, the collection setups presented in [10,15,16] use separate sensors for each appliance, thus recorded values can be unambiguously attributed. These sensor placements are mostly related to the different concepts of the respective studies. The authors of [11,13,17] present different sensing platforms and concepts, but only briefly evaluate the possibility to identify electrical devices. In [14] the authors introduce a system which solely includes ambient audio information, collected on a per-room granularity. The system is considered as a load monitoring system and uses different collected audio features as a first disaggregation layer. Only if the audio features do not allow for disaggregation, the electrical features are evaluated and can overwrite the decision. The authors of [18] follow a similar approach, albeit they implement a smartphone-based system to detect household activities. Through the annotation of activities with corresponding energy consumption data, the authors enable basic load monitoring based on audio information. Lastly, the authors of [10] present the appliance-agnostic usage of "multi-modal signatures" through the common evaluation of all sensor data and their changes that allow for device identification. They recognize the potential of closely correlated environmental data as trusted sources of appliance activations, allowing NILM systems to validate or re-train themselves. While the general usefulness of multimodal signatures is proven in [10], the used sensor types are not evaluated concerning their individual usefulness. Aligned with these insights, the authors of [19] have also remarked the potential of environmental sensors in disaggregation tasks.

Bearing these related findings in mind, we present our data acquisition concept and its evaluation in the following sections. As the related studies did use approaches with and without consideration of electrical data for the recognition process, our study will accordingly include evaluations for both approaches.

3. Data Acquisition Concept

A number of datasets are widely used in related research on energy data analytics so far (e.g., [20–24]). Collecting such datasets, however, is often motivated by the desire to capture a large continuous stream of electrical energy consumption readings for data

processing tasks like pattern recognition or forecasting. Ambient features or user-speci
details (e.g., presence) are not part of most datasets, and there is no dataset available th
comprises electrical signals as well as the full set of ambient conditions we consider in th
work. As a result of this shortcoming of published datasets, it was necessary to run o
own data collection campaign. We have decided to design a collection system for bo
electrical and ambient sensor data and use it to collect a dataset for the data analysis v
conduct in Section 4. We describe our rationales behind the design of the system as well
the data preprocessing steps we apply in the following subsections.

3.1. Selection of Appliances

As we aim for a generalizable evaluation of device types and their emissions of no
electrical signals, the first decision to make is the selection of the set of appliances und
consideration. Our goal is to determine a representative set of electrical devices, th
will be operated in a controlled environment in order to collect the input dataset for a
further analyses. To make an informed choice of devices, we have consulted studies of
electrical appliance ownership worldwide [25–27]. Through considering the applian
types reportedly owned by at least $2/3$ of the households, we have been able to identify
set of 13 appliances that are present in many households in developed countries. Min
household devices are typically not part of the aforementioned surveys due to their larg
diversity and the negligible contributions to the monthly energy bill. Still, several u
cases for their recognition in load data are conceivable, e.g., the identification of us
activities that are tightly bound to the use of these devices. Accordingly, we have chose
eight additional devices related to cooking (e.g., a mini oven), personal hygiene (such
a hair dryer), and office activities (e.g., a printer). The full set of all 21 appliances und
consideration is provided in Section 4.

3.2. Selection of Monitored Parameters

Having selected the appliances under consideration, their (expected) ambient infl
ences need to be determined, in order to derive the sensors required to capture the
parameters. For the evaluation of possible ambient influences we have extracted possib
emissions from the devices' data sheets, the general construction of devices (thus implicit
considering the laws of physics), and moreover inspected the devices under test manua
during their operation. The complete list of all eight captured sensor parameters is give
in Table 2.

3.3. Data Collection System Design

Based on the derived set of requirements pertaining to the parameters to monitor,
collection system was prototypically designed and implemented. As it was our intentic
to collect the sensor measurements as close to the Device under Test (DuT) as possib
most of the ambient sensors were wired up to an embedded microcontroller system, base
on the PJRC Teensy 3.2 board. Its compact size offered the possibility to be mounted ve
close to the DuT. Sensors for the parameters of interest were interfaced to the board eith
via a digital two-wire (I^2C) interface, or through analog signals that were converted in
the digital domain by the microcontroller's integrated 16-bit Analog-to-Digital Convert
(ADC). The microcontroller system was programmed to use a periodic sampling schedu
and capture the considered parameters as synchronously as possible. Retrieved senso
values (e.g., digitized temperature readings) are scaled in order to report their data in
units (e.g., °C). The data sampling rates are shown alongside the sensor types in Table
They were selected such that the microcontroller system could perform data processir
and transmit them (across its USB-serial connection) in real-time. The choice of 400 Hz fe
vibration measurements aligns well with typical rotational speeds of the evaluated intern
motors (e.g., internal motors of DVD and CD players typically spin at 200 rpm to 570 rpm

Table 2. Summary of the Used Sensing Devices and the Properties of the Data They Collect.

Symbol	Sensing Device	Interface	Collected Information [Unit]	Sampling Frequency
T	temperature sensor	I²C	ambient temperature [°C]	1/5 Hz
H	humidity sensor	I²C	relative humidity [%]	1/5 Hz
UV	UV radiation sensor	I²C	UV steps from the sensor	400 Hz
IR	IR radiation sensor	ADC	ADC steps	400 Hz
LDR	visible light sensor	ADC	ADC steps	400 Hz
B	magnetic flux density sensor	ADC	magnetic flux density [mT]	400 Hz
Vib	vibration sensor	ADC	ADC steps	400 Hz
Aud	microphone	USB	dominant audio frequency [Hz]	44.1 kHz
U	oscilloscope	USB	mains voltage [V]	10 kHz
I	oscilloscope	USB	appliance current [A]	10 kHz
P	oscilloscope	USB	appliance power [W]	10 kHz

The sensor platform was connected to a personal computer in charge of centrally collecting all sampled data, to which two more sensors were attached. First, electrical signals for both voltage and current are collected at 10 kHz through a USB-interfaced PicoScope 4444 oscilloscope, equipped with a Hall effect current probe and a passive differential measurement voltage probe. Second, the collection of audio information and the determination of the most dominant frequency was accomplished through a connected USB microphone, sampling audio at 44.1 kHz. Temporal synchronization between the data recorded from the heterogeneous sensing modalities is ensured through inter-process signaling on the data collecting system. The raw data is collected into a file containing electrical information, an audio file, and a CSV file containing the ambient measurements.

3.4. Measurement Environment

With the exception of large and immobile appliances (washing machine, dryer, refrigerator), measurements were collected in the same ambient conditions of an office environment. The remaining measurements were collected in-situ, i.e., a kitchen (for the fridge) and the laundry room (for washing machine and dryer). The ambient sensors were placed directly on the DuT, oriented according to the expected maximum emission strength for each captured feature. As such, e.g., light sensors were placed in front of light emitting devices and magnetic flux density sensors were placed close to motors wherever possible. While such a placement may not be realistic for real-world deployments, note that the intention of our approach is to determine importance of such features in the first place, for which as detailed information as possible are required. An example of the sensor placement for the mini oven is shown in Figure 1. Unless practically impossible, measurements were collected for full working cycles of devices. Each measurement was succeeded by a phase of appliance inactivity, in order to allow for sensor offset calibration. Data from devices with continuously variable power demand or without deterministic operation durations (e.g., computer monitors, lamps) were collected for two to five minutes. The only device measured for shorter duration was the food hand mixer, as it could only be operated for up to one minute before requiring a cool-off period, according to its user manual.

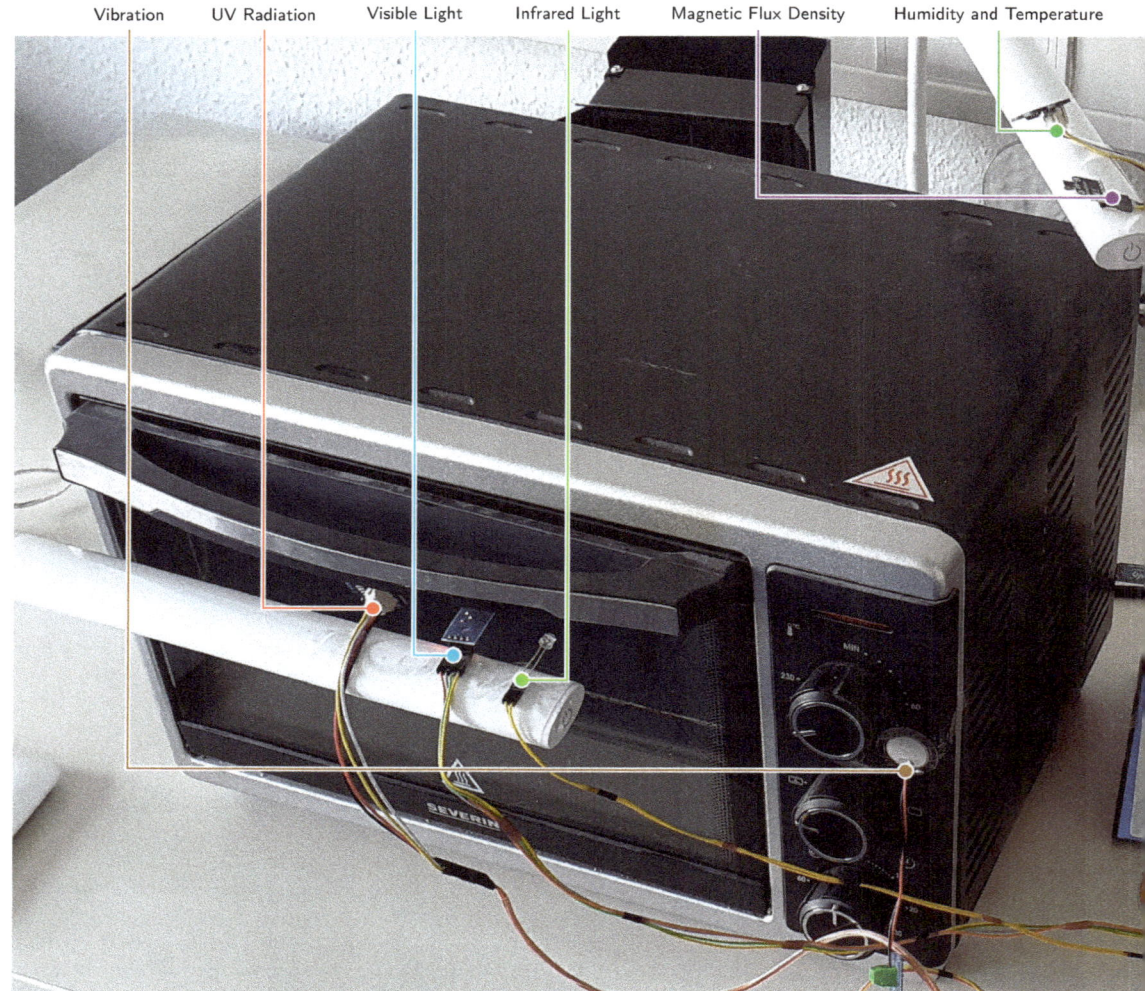

Figure 1. Practical Setup of the Data Collection Test Bench. The Microphone is not Pictured, due to its Positioning Outside of the Image Boundaries.

At least two operational cycles were recorded for each DuT in order to permit cross checking the recorded sensor data. All measurements were manually checked for correctness before storage, in order to ensure consistency regarding the collected data. In the rare occasion of obviously inconsistent data, the data collection was repeated, and the faulty data trace discarded. For devices that could be operated in different states (such as the hair dryer or fan), measurements were collected for each of the states individually, and treated as the same device during the evaluations.

3.5. Data Postprocessing and Dataset Creation

For the feature evaluation presented in Section 4, we only consider a simplified feature subset, consisting of either the maximum changes or a binary activation indication for the ambient features under evaluation. These simplified features were chosen because they could easily be determined locally on the low-power sensing device. Besides their fast computation, omitting raw data from collection also caters to user privacy protection [28

Additionally, findings based on such simple features are also reproducible on systems using higher sampling rates. All three points were vital to allow the results of this study to be used as a guideline for a wide range of practical systems, some of which are expected to only provide low data resolutions.

To compute these features, each data collection period T_{coll} was succeeded by an offset calibration phase T_{cal}. Both were designed to have approximately the same duration (i.e., 2–5 min, cf. Section 3.4). A clear delineation between both phases was easily possible due to the corresponding changes in electrical current consumption. Changes to the ambient humidity and temperature values were detected using additional temperature and humidity values, collected through a secondary measurement device in the room. During the manual evaluation of each trace, the steady-state value of each ambient sensor readings was determined for T_{cal}, and used as the baseline value for the uninfluenced ambient readings. For each of the used sensor types, the difference between both values in T_{coll} and the baseline is used as a feature in our analysis. In addition to the use of absolute values by which each parameter has changed, we also consider them in a binary form, according to the following rule set:

- T, H, Ultraviolet (UV): A change was reported when the values recorded during T_{coll} differed by at least 1 (°C, %, or UV step) from the values recorded during T_{cal}.
- Infrared (IR): We observed the IR sensor's readings to oscillate by a maximum of ± 5 ADC steps during T_{cal}. A change of the IR signal was thus logged when the average of the IR samples recorded during T_{coll} differed by at least 1 step from the highest non-singular value in T_{cal}.
- Visible light (LDR): Readings collected from the Light Dependent Resistor (LDR) were determined to fluctuate by ± 15 ADC steps. A change was thus logged when collected measurements differed by at least 50 ADC steps from the average value during T_{cal}.
- Magnetic flux density (B): To account for naturally occurring variations, a change in the magnetic flux density was detected if the mean values during the steady phase of T_{cal} and T_{coll} differed by at least 0.2 mT.
- Vibration (Vib): To minimize the impact of spurious vibration signals, the mean value seen during T_{coll} was required to differ by at least ± 15 ADC steps from readings seen during T_{cal}.
- Audio information (Aud): The recorded audio files were checked manually to ensure that no undesired ambient noise would affect the evaluations. During this check it was furthermore verified that the device actually produced sound emissions. This approach was chosen because an algorithmic threshold solution would not have accounted for possible undesired sound collections from the collection environment.

We would like to note that supplementally collected electrical features (voltage, current, power) were not translated into a binary form, given that voltage readings remained constant and current samples always showed variations during a device's operation. Instead, the electrical features used in this paper were chosen such that they represent electrical information already used during load monitoring, at a complexity similar to the considered ambient features. This enables a comparison between use cases only using ambient data and use cases combining ambient data with the already present electrical data. All electrical features are calculated during a 40 ms (i.e., two mains periods) long section of the measurement, selected such that the appliance's current consumption is maximal, and requiring that its value is identical in both successive mains periods. Based on the data from this excerpt, the RMS voltage (U) and current (I) as well as the active power (P) are calculated.

Based on the collected and post-processed data, we have generated four variations of the dataset to serve as the foundation for our evaluations. All datasets contain a total of 144 traces. Their details are given as follows.

- **Ambient Parameters; binarized (A_{bin}):** Changes in the sensor measurements were evaluated in a binary way in order to evaluate if the corresponding ambient characteristic is influenced by the device's activation. The resulting data consisted of a binary value for each sensor, which we use in our evaluation as indicators whether the DuT operation had an impact on the corresponding characteristic.

- **Ambient and Electrical Parameters; binarized (AE_{bin})** is an extension to A_{bin} that makes the (numerical) electrical readings (U, I, P) available to the evaluation system in addition to the (binary) indicators of changes in the ambient data.
- **Ambient Parameters; differences (A_Δ)**: In contrast to the previously described A dataset, the maximum change is used now, i.e., the difference between the maximum sensor value during T_{coll} and the average of the values collected during T_{cal}. While most sensor values could be evaluated on the raw data, the vibration and audio measurements where evaluated in the frequency spectrum. As such the measured time sequence was transformed into the frequency spectrum and the strongest frequency was chosen.
- **Ambient and Electrical Parameters; differences (AE_Δ)**: Analogous to AE_{bin}, the AE_Δ dataset is comprised of the (numerical) values from A_Δ in conjunction with the (numerical) values for the three electrical features.

Let us consider an example of the data collection and processing sequence for the mini oven appliance as follows. The mini oven was equipped with the sensors according Figure 1. Sensor data was collected during the mini oven's operation twice, with sufficient time between measurements to allow for a cooling down. Raw environment sensor data from one of the measurement run are plotted in Figure 2. The average sensor values during T_{cal} and from ambient measurements are then postprocessed to create the four aforementioned variants of the dataset. In the figure, the first 20 min of the sample, during which an electrical current flow was recorded, constitute T_{coll}. The remaining about 20 min of the collected trace constitute T_{cal}. The postprocessing is applied as described, a visualization of the process is included in Figure 3. The resulting entries for the four datasets introduced above are computed. They are shown for reference in Table 3.

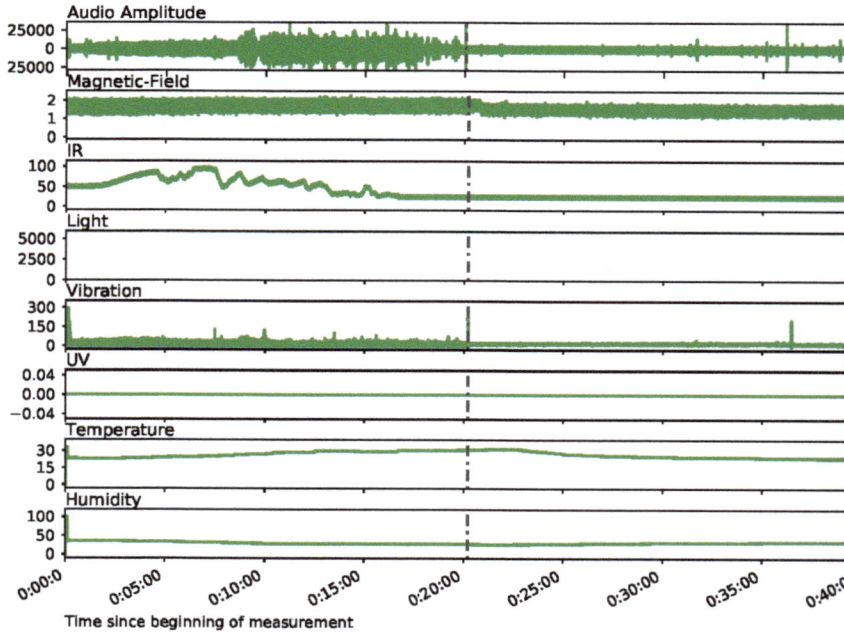

Figure 2. Data Collection Example, Displaying the Environmental Sensor Traces. Units According Table 2, the End of T_{coll} is Marked as a Dotted Grey Line.

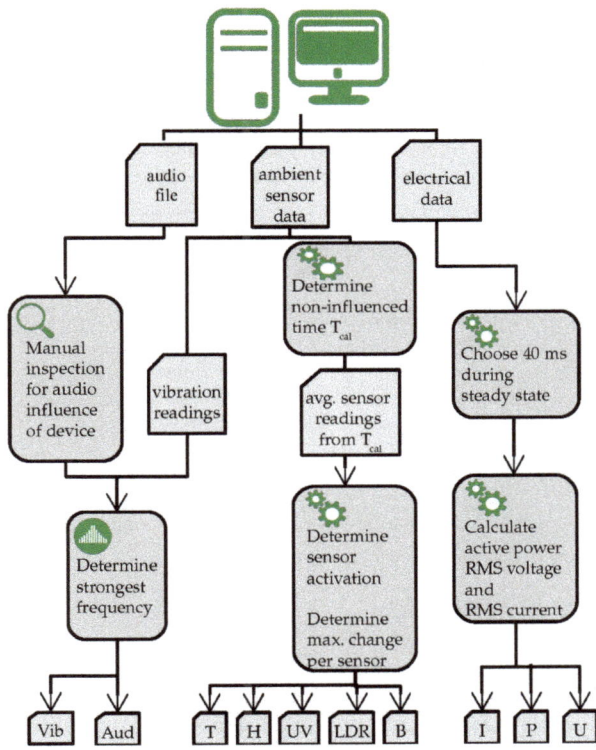

Figure 3. Feature Extraction Processing Flow, Applied to Collected Raw Data.

Table 3. Sample Data Collected from a Mini Oven, Following the Processing Flow Depicted in Figure 3.

Dataset	T	H	UV	IR	LDR	B	Vib	Aud	I	P	U
A_{bin}	TRUE	TRUE	FALSE	TRUE	FALSE	TRUE	TRUE	TRUE			
A_Δ	+9 °C	−8 %	0	+74	0	+0.59 mT	37.2 Hz	7838 Hz			
AE_{bin}	TRUE	TRUE	FALSE	TRUE	FALSE	TRUE	TRUE	TRUE	4.39 A	995 W	226.8 V
AE_Δ	+9 °C	−8 %	0	+74	0	+0.59 mT	37.2 Hz	7838 Hz	4.39 A	995 W	226.8 V

4. Data Evaluation Concept

Our research objective is to assess how knowledge of the ambient conditions in an appliance's environment can support the recognition of electrical appliances. A total of eleven attributes are available for analysis (cf. Section 3): Eight ambient sensor attributes and three electrical quantities (voltage, current, and power). Instrumenting residential environments with sensing devices to capture all of these parameters, however, has several drawbacks. Besides the monetary costs for purchasing and installing sensors as well as ensuring their continuous operability, the continuous collection of data may be perceived as an intrusion into user privacy.

We hence conduct a methodological evaluation how each of the sensed attributes impact the appliance recognition rate, in order to determine the most information-rich subset of features.

We begin our evaluations with a determination of the importance of the contributions of each of the collected features when used to distinguish between the 21 appliance types listed in Table 4. In subsequent evaluations, however, we also present three evaluations considering the categorization of appliances into classes, as well as appliance recognition results when the appliance class is known a priori. Through this set of evaluation guidelines on the best-suited feature subsets for different appliance recognition scenarios are derived.

Table 4. List of Considered Appliances and Associated Categories.

Device Type	Classification By		
	Operation State	Energy Consumption	Load Curve
DVD player	Single-State	small	inductive
Fan	Multi-State	small	ohmic
Coffee machine	Single-State	large	ohmic
Fridge	Single-State	large	inductive
Lamps	Single-State	small	ohmic/SMPS
Laptop computer	Infinite-State	small	SMPS
Personal computer	Infinite-State	small	SMPS
Microwave	Single-State	large	inductive
Monitor	Infinite-State	small	SMPS
Vacuum cleaner	Single-State	large	inductive
Washing machine	Multi-State	large	ohmic
Dryer	Multi-State	large	ohmic
Printer	Infinite-State	small	ohmic/SMPS
Hand mixer	Multi-State	small	inductive
Mini oven	Single-State	large	ohmic
Toaster	Single-State	large	ohmic
Mouth wash	Multi-State	small	inductive
Hair dryer	Multi-State	large	ohmic
Clothes iron	Single-State	large	ohmic
FM radio	Infinite-State	small	inductive
Kettle	Single-State	large	ohmic

4.1. Determining the Feature Importance for Appliance Recognition and Classification

Determining the usefulness of features for classification purposes is a task that occurs across many research domains [29]. Considering the appliance recognition and classification case of this study, the usefulness of features is considered to allow cost-efficient data collection through the exclusion of features that carry little or no information. Additionally, feature selection methods allow for the comparison of the usefulness of features or subsets of features for different use cases. Note that the usefulness of a feature is highly specific to a given use case. For our contribution, we have chosen appliance recognition as a use case, i.e., the classification of appliances by their types, depending on the values of the available feature set. Appliance recognition is a typical classification use case from the field of energy data analysis: Based on a set of features, the single most likely appliance type should be returned. As follows, we assess the importance of the features we have described in Section 3.5 for the task at hand. Instead of conducting a single study on the general feature importance, however, we proceed in a more fine-grained fashion by considering several subsets of appliances (cf. Table 4). This way, we seek to provide a more detailed picture of the feature relevance for different use cases.

4.2. Methodology for Determining the Distinctiveness of Features

For the evaluations we conduct below, two pieces of information are of primary interest:

1. A score to describe the importance of each feature, and
2. the most expressive subset of features, referred to as the optimal subset.

While the individual determination of a feature's importance helps in assessing to what extent each feature can reduce the chance of misclassification, it generally cannot identify the feature combination that leads to the best classification result overall. In order to find such combinations, the determination of an optimal subset is required. This feature subset considers which features work best together, indicating an ideal set of features to be used for the considered use cases [29]. When combined, both methods (individual feature relevance and best feature subset) allow for the development of better appliance monitoring systems.

4.2.1. Feature Importance

The usefulness of each feature is determined through the usage of a Random Forest of Trees. A decision tree is a structure which continuously divides the whole input data into subsets, such that the new subsets become more pure, i.e., features that lead to different output values become divisive elements [30]. In other words, the features that enable the cleanest division of input data into categories are considered the most important. In contrast to a simple decision tree, the Random Forest of Trees generates multiple trees for randomly selected subspaces of the total feature space. Only a subset of the input sets of the feature values is evaluated in each tree, and the resulting trees are then combined by averaging the determined probabilities. This ultimately allows for greater classification accuracy improvements as compared to a singular decision tree [30].

The Gini Impurity is defined as the rate of misclassification when an additional decision element is added to an existing decision tree [31]. It is widely used for the feature selection in Random Forests of Trees in order to annotate each division of input data into new subsets with an importance score. Only the decision that yields the greatest reduction of the Gini impurity is maintained, which corresponds to a decrease in the probability of misclassification. The averaged Gini impurity scores are used as feature importance scores in our present study.

4.2.2. Optimal Feature Subset

To confirm that attained results can be generalized and allow to determine an optimal feature subset, we rely on the Recursive Feature Elimination [29]. The algorithm starts with the full set of features and greedily excludes the least informative feature after each evaluation iteration. A ranking criterion is calculated for all features, and the feature with the lowest ranking criterion is eliminated. This process is repeated until the desired size of the feature subset is reached [32]. If the size of the optimal subset is unknown in advance, a performance rating for the trained classifier results can be introduced. For this study, the accuracy was chosen as a performance rating for the trained classifier, such that the subset with the greatest overall accuracy result is chosen as the optimal feature subset.

4.2.3. Avoiding Overfitting

Small subsets of data can be prone to overfitting their input data, i.e., adapting to their characteristics too well. Cross-validation is a methodology which allows to counteract overfitting, as the available data of input features and known correct output classifications is not simply divided into disjunct subsets of training and testing data, but broken down into multiple smaller sets, so-called folds. The classifier is then trained in a leave-one-out manner: Each fold is once used as the test subset, while all other folds are used for training. The results of these training phases are averaged and given as the Cross-Validation Score (CVS) [29,32].

4.2.4. Implementation

All aforementioned feature selection methods were implemented in Python using the `scikit-learn` library [33]. The Recursive Feature Elimination is implemented using a Support Vector Classification and uses the accuracy as its performance rating. We specify the average accuracy in percent, with an accuracy of 100 % indicating that all input samples could be correctly categorized. As the collected datasets are rather small, only a 2-fold cross-validation is conducted. The 2-fold validation is stratified to ensure that the class distributions between the test and training sets remain comparable to the full set of data.

4.3. Device Categorization

Let us next introduce the device categorization used during our evaluation study. A complete overview of the appliances under consideration and their corresponding classes according to the three categorization approaches are given in Table 4. Categorizing appliances by their classes allows us to not only run analyses on the entire dataset, but also on subsets of the data in which all devices share a commonality (e.g., an inductive load curve). For example, the device of type mini oven is a member of the Single-State devices class, belongs to the large consumers (classified by its power consumption), and is an ohmic appliance (due to the load type of its heating rods).

4.3.1. Categorization by Number of Operational States

Classifications by an appliance's number of operational states can be found in works considering load monitoring [2,34,35]. The typical device classes listed in the context of Non-Intrusive Load Monitoring (NILM), based on the complexity of their consumption patterns, are given as follows.

- Single-State Devices: Also known as On/Off appliances, such devices exhibit one steady electrical power consumption value during their activity.
- Multi-State Devices: These devices have multiple states of usually different power consumption levels, that occur in defined (usually sequential) patterns.
- Infinite-State and/or Continuously Variable Devices: This class of devices exhibit constantly variable power consumption values.

We included this categorization for two reasons: First, different works in load monitoring have reported that Single-State appliances are often easier to disaggregate [2]. Second, we expect greater differences in the importance of ambient features when appliances exhibit multiple operational states.

4.3.2. Categorization by Power and Energy Consumption

We wish to note at this point that NILM is not the only use case for energy data analysis. For example, applications like demand side management favor the availability of loads with a large energy consumption. As large energy consumers typically emit more excess heat and generate stronger magnetic fields, differences in the feature importance are likely to occur. Finding the optimal feature subset for such application cases is thus a prerequisite for the realization of such services. Hence, we classify the appliances under consideration by their power consumption, which we define as follows:

- Large Consumer: Devices surpassing 1 kW of consumption or 10 kWh per year consumption, which was determined based on the data sheet information concerning expected annual consumption.
- Small Consumer: All electrical devices that do not count as large consumers.

4.3.3. Categorization by Load Type

Finally, we consider device classification according to their load type, according to the classification scheme proposed in [34]. The load type is determined by the phase shift between mains voltage and an appliance's current consumption as well as the presence of non-linear loads within devices. To allow to determine the load type for the devices in our evaluation, we followed the definition and examples given in [34], and evaluated the devices' data sheets.

- Ohmic Consumers: Devices that have neither recognizable reactive power nor exhibit harmonics beyond the fundamental mains frequency
- Inductive Consumers: Devices whose inductive component dominates their current intake (e.g., transformers).

- Switched-Mode Power Supplies (SMPS): Devices with a large amount of harmonic content, whose characteristics may moreover vary depending on the currently exhibited power demand.

It needs to be noted that we excluded the "composite load" class proposed in [34], as for our test only the washing machine and dryer could have fit the corresponding definition. Furthermore, the proposed "Capacitive Consumers" presented in [34] do not have a representation in our set of devices, as they are not typically found in residential spaces.

4.4. Conducted Evaluations

We evaluate the ambient features with respect to three general use cases:

1. First and foremost, we assess how well the feature set allows for correctly determining the type of the appliance. This appliance recognition test is conducted for each of the datasets A_{bin}, A_Δ, AE_{bin}, and AE_Δ. The available features are provided at the input, and the system trained to correctly categorized them into the corresponding (known) output, i.e., the device type, reporting an overall accuracy score eventually.
2. We assess the correct recognition of an appliance's device class. This evaluation group considers the usage of features calculated from ambient sensor data for the classification of devices according to their operation state, energy consumption, and load curve (as introduced in Table 4).
3. We assess how well the system can distinguish between device type that are confirmed to belong to the same class. To this end, we train the system with data from all appliances of a single class only, and verify which features allow to distinguish between the remaining electrical devices best. Using the evaluations in this scenario, we seek to find out if and to which degree ambient information is informative if a device class is already known, but the devices inside the class are supposed to be distinguished.

5. Evaluation of the Feature Importance

As follows, we present the results of the different evaluation settings introduced in Section 4.4. Unless noted otherwise, we present the Cross-Validation Score (CVS) and optimal feature subset for each of the four sets of input data (cf. Section 3.5). The symbol notations for the sensed modalities introduced in Table 2 are used in the results.

5.1. Evaluation Results for the Distinction between Devices on the Full Dataset

Our first evaluation was conducted on the whole generated datasets and considered the device types as output. The results for the Random Forest of Trees method are displayed in Figure 4, whereas the results for the Recursive Feature Elimination are documented in Table 5. The bar graphs indicate the calculated feature importances, wherein the y-axis denominates the feature, while the x-axis denominates the corresponding feature importance in percent.

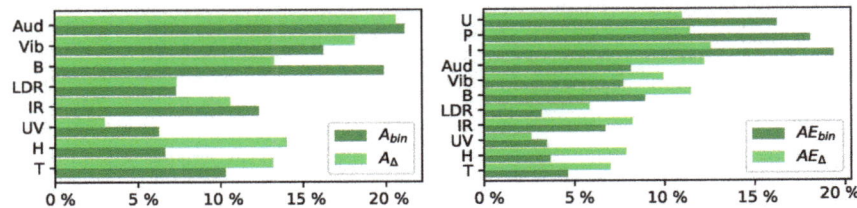

Figure 4. Feature Importances for the Appliance Recognition Across the Whole Dataset.

Table 5. Cross-Validation Score (CVS) Values for the Evaluated Appliance Recognition Scenarios.

	CVS [%]	Features in the Best Feature Subset										
		T	H	UV	IR	LDR	B	Vib	Aud	U	I	P
A_{bin}	68.8	✓	✓		✓	✓	✓	✓	✓	n/a	n/a	n/a
AE_{bin}	70.8	✓	✓		✓		✓	✓	✓			✓
A_{Δ}	70.8	✓	✓		✓		✓	✓	✓	n/a	n/a	n/a
AE_{Δ}	76.4	✓	✓		✓		✓	✓		✓		✓

The Random Forest of Trees method indicates a high usefulness for the vibration and sound measurements. For the case of using binary change indicators, the magnet flux density is deemed similarly important. The inclusion of electrical parameters slight equalizes the importance levels, yet the addition of electrical parameters only leads to slig improvements of the results when compared to the use of ambient sensors alone. This resu is confirmed through the Feature Elimination, which shows that the inclusion of electric parameters allows for slightly improved Cross-Validation Scores, but the best feature subs does still require nearly all features. The UV and visible light emissions are only useful fo a very small number of devices; the UV readings exclusively for lamps. While this indicate that the sensor type is not of interest for most evaluations, it furthermore shows that there potential to improve lamp recognition by means of only a single sensor type. Furthermor audio frequency features can be considered to be highly informative, as they are not on present for most devices, but distinctly different between them. The maximally observe sensor changes, as collected in A_{Δ} and AE_{Δ}, enable more fine-grained assessments of th impact of observed signal changes. While the subsequent measurements on a device di generally not exhibit strong variations, variations did exist between different devices of th same type. The differences in ambient influences between two devices of the same typ can, however, be high. For example measurements from two different microwaves resulte in consistent dominant audio frequencies for each of the devices across measuremen Nonetheless, one microwave exhibited a dominant frequency of 172 Hz, whereas the othe had a dominant frequency of 344 Hz. This indicates that setups evaluating ambient da need to consider the similarity of devices of the same type and if multiple devices of th same type are present. Such findings must be considered during training, as they indica that training done on one household may not be transferable to other environments.

5.2. Evaluation Results for Device Categories

The following evaluations were conducted such that, based on the ambient me surements, each DuT was classified into its corresponding class for each categorizatio presented in Section 4.3. Each presented evaluation provides the feature importanc considering the Random Forest of Trees method and the optimal feature subset dete mined through the Recursive Feature Elimination, as well as the CVS achieved with th optimal subset.

5.2.1. Classification According to the Number of States

In this evaluation each measurement was classified as belonging to either a Singl State, Multi-State or Infinite-State appliance. The results are listed in Figure 5. Bot evaluated algorithms show a high importance of audio and vibration features for the class fication into Single-State, Multi-State and Infinite-State Appliances. The binary evaluatio achieves a better Cross-Validation Score, indicating that the simple presence of sound o vibrations could provide relevant contextual information for this classification. Howeve both CVS values are low, such that a classification based on the ambient features alone ma not be fruitful.

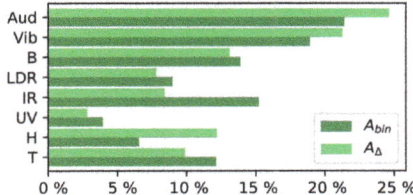

Figure 5. Feature Importances [%] and Best Feature Subset for the Classification into Single-State, Multi-State and Infinite-State Devices—Considering Only Ambient Features.

5.2.2. Classification According to Electrical Power Consumption

This evaluation considers the feature usefulness to determine if a set of ambient measurements belongs to a large consumer or small consumer appliance. Its results are given in Figure 6. The evaluation according to consumption ranks the temperature and humidity sensors as most relevant for this distinction. This is unsurprising, as appliances designed to significantly heat or cool the environment generally exhibit a high power consumption. However, the Feature Elimination furthermore reveals that the UV radiation is a relevant feature for this distinction, which is reasonable considering that only the light installations under evaluation emitted UV radiation, all of which belong to the class of small consumers.

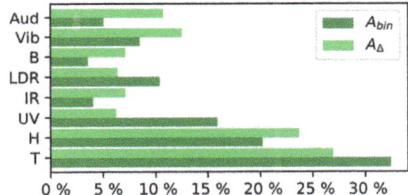

Figure 6. Feature Importances (%) and Best Feature Subset for the Classification into Small Consumers and Large Consumes—Considering Only Ambient Features.

5.2.3. Classification According to Load Curves

The evaluation considering load types evaluates feature usefulness to determine if measurements were taken from an ohmic, inductive or SMPS appliance. The results are contained in Figure 7. To classify devices according to their load curves, the Feature Elimination indicates that maximum changes are more effective. However, the results for the two applied methods differ considerably. This indicates that the combination of features is a lot more informative than singular features. The similar usefulness of most features determined through the Random Forest of Trees matches this finding. However, it needs to be remarked that the CVS for the binary evaluation is very low, indicating that this evaluation is not well-suited for a classification into load types.

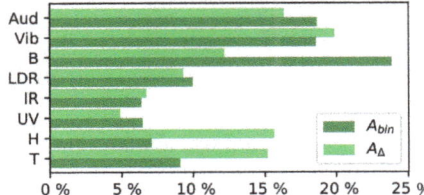

Figure 7. Feature Importances (%) and Best Feature Subset for the Classification into Ohmic, Inductive, and Switched-Mode Power Supplies Devices—Considering Only Ambient Features.

5.3. Device Type Distinction within Different Device Catgetories

The following evaluations consider if and to which degree ambient information informative if a device class is already known, but the instances of the device type with the given class shall be distinguished. Such an evaluation allows us to identify the device classes whose instances can be distinguished algorithmically. In contrast to the classification task considered in the previous section, the evaluations were not run on the full dataset with a target output from the set of device classes. Instead, we have run them on subsets the full data, divided such that only one class of devices is part of the subset. The results are recorded in Tables 6 and 7 and presented in the following subsections.

Table 6. CVS Values for the Evaluated Appliance Destinction Scenarios—Binary Features.

	A_{bin}	Best Feature Subset								AE_{bin}	Best Feature Subset								
	CVS [%]	T	H	UV	IR	LDR	B	Vib	Aud	CVS [%]	T	H	UV	IR	LDR	B	Vib	Aud	U	I
Single-State	83.3	✓	✓	✓	✓	✓	✓	✓	✓	97.2		✓								
Multi-State	53.6	✓	✓				✓		✓	60.7	✓		✓		✓			✓		
Infinite-State	100	✓	✓				✓	✓	✓	100										
Large	42.0				✓	✓	✓	✓		75.3		✓			✓	✓			✓	✓
Small	61.9	✓	✓			✓	✓	✓		71.6	✓					✓	✓			
Ohmic	48.8				✓	✓	✓	✓		78.8	✓			✓	✓	✓		✓		
Inductive	57.4	✓	✓		✓	✓	✓	✓	✓	59.5				✓						
SMPS	100		✓		✓		✓		✓	100			✓							

Table 7. CVS Values for the Evaluated Appliance Destinction Scenarios—Maximum Feature Changes.

	A_{Δ}	Best Feature Subset								AE_{Δ}	Best Feature Subset								
	CVS [%]	T	H	UV	IR	LDR	B	Vib	Aud	CVS [%]	T	H	UV	IR	LDR	B	Vib	Aud	U	I
Single-state	88.9	✓	✓		✓	✓		✓	✓	90.3										
Multi-state	62.5	✓	✓		✓		✓	✓	✓	75	✓	✓		✓			✓	✓	✓	
Infinite-state	100	✓	✓				✓			100										
Large	77.8	✓	✓		✓			✓	✓	69.2	✓	✓		✓		✓				
Small	90.4	✓	✓			✓		✓	✓	84.2	✓	✓			✓		✓	✓		
Ohmic	76.3	✓	✓		✓	✓	✓	✓	✓	77.5	✓	✓		✓	✓		✓	✓		
Inductive	87.1	✓			✓			✓	✓	87.2	✓						✓			
SMPS	100					✓			✓	100					✓					

5.3.1. Distinguishing between Devices Sharing the Same Number of States

Recall that the device classes categorization used in this paper divides the set considered appliances into Single-State, Multi-State and Infinite-State appliances. Again we have computed the best feature subset within each of these categories, as well as ranking the feature importances. The Feature Elimination results can be found in Tables 6 and in the first to third row, while the results for the Random Forest of Trees are depicted Figure 8.

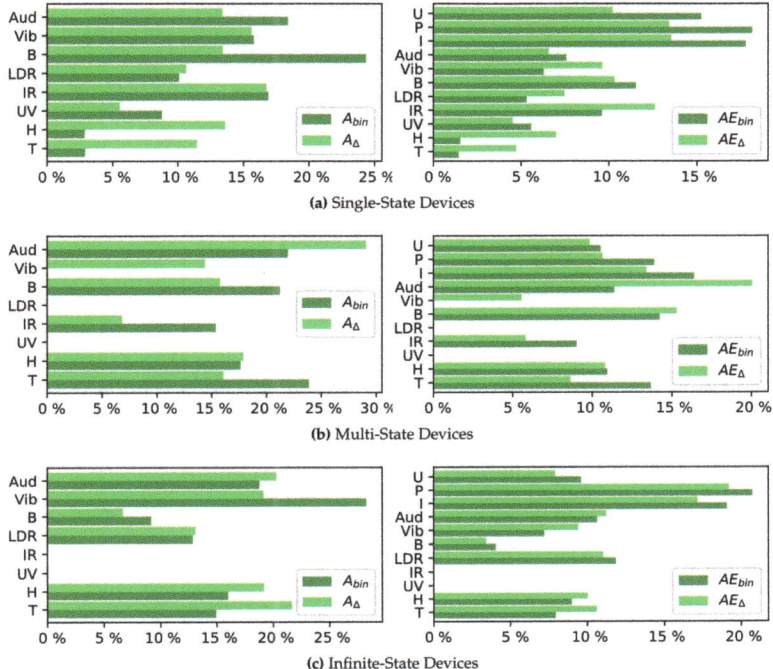

Figure 8. Feature Importances (%) for Recognition of Appliances of a Single Class—Appliance Categorization By: Number of State.

Concerning the recognition of device types from measurements of either only Single-State or only Multi-State Appliances, a rather large set of ambient measurements is required for best results. None of the features are particularly indicative of a specific appliance when considered on its own. However, the visible light and UV radiation are not present in Multi-State appliances, as none of the considered Multi-State devices emitted light. The distinction of Infinite-State devices differs as their distinction reaches the maximal Cross-Validation Score, in the case of maximum change evaluations, with a small dataset which only contains changes in temperature, humidity, and vibration. Sound emissions were identified as similarly important through the Random Forest of Trees method. The inclusion of electrical features was found to be useful for all three intra-category distinctions (as visible in the columns on the right-hand side of the tables).

5.3.2. Distinguishing between Devices of the Same Consumption Class

For the following evaluations, the datasets of ambient features were divided, such as to create datasets that only contain measurements from either large or small consumers. The resulting datasets were evaluated to assess to what extent the device type of a measurement can be determined based on the data. The Feature Elimination results are documented in Tables 6 and 7 in the fourth and fifth row, and the results for the Random Forest of Trees are visualized in Figure 9.

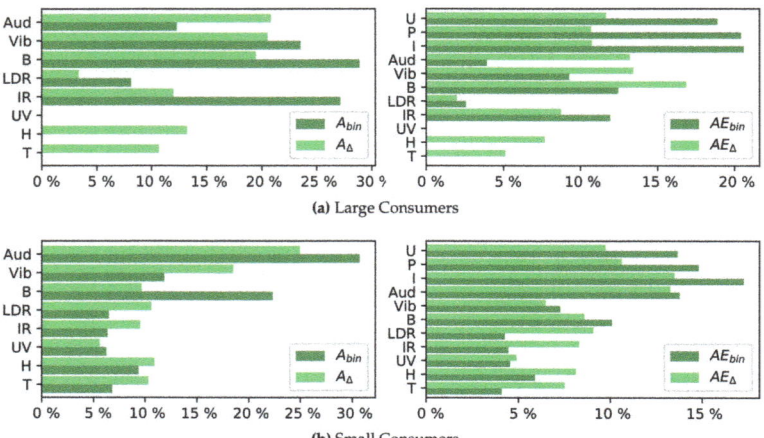

Figure 9. Feature Importances (%) for Recognition of Appliances of a Single Class—Appliance Categorization By: Power Consumption.

Considering the large consumers (see Table 4), most appliances include a heating or cooling element, and multiple of them a motor, all of which can be expected to generate magnetic fields, vibrations, and sounds. Accordingly, both feature selection methods identify the magnetic flux density, vibrations, and audio features as important. The presence of IR radiation is furthermore shown to be distinctive. Small consumers do show low Cross-Validation Scores when only considering the presence of emissions, however maximum change evaluation scores indicate adequate results. Two findings are of special interest: First, the maximum change evaluation reaches better Cross-Validation Scores for feature subsets only including ambient features. Additionally, the feature subsets are quite large, but the majority of sensors is present for the recognition of only small and for the recognition of only large consumers.

5.3.3. Distinguishing between Devices of the Same Load Type

To allow to evaluate the usefulness of ambient feature information for appliance recognition for appliances belonging to a certain load type, the whole datasets were split to only contain measurements of devices belonging to one load type and than evaluated such that for each measurement the device type should be distinguished. The Feature Elimination results are documented in Tables 6 and 7 in the sixth to ninth row, whereas the results for the Random Forest of Trees are depicted in Figure 10.

While the intra-class distinction for inductive and SMPS devices generates small optimal feature subsets for AE_Δ with three or less features, the distinction of ohmic appliances always requires bigger subsets. While acknowledge that our results for SMPS might be potentially biased, given that many of these devices were emitting light and thus the great importance of the LDR is not surprising. Still, the differences between ohmic and inductive appliances indicate that systems using features best-suited for ohmic devices could gain additional information to ease the distinction of inductive devices with low additional data requirements.

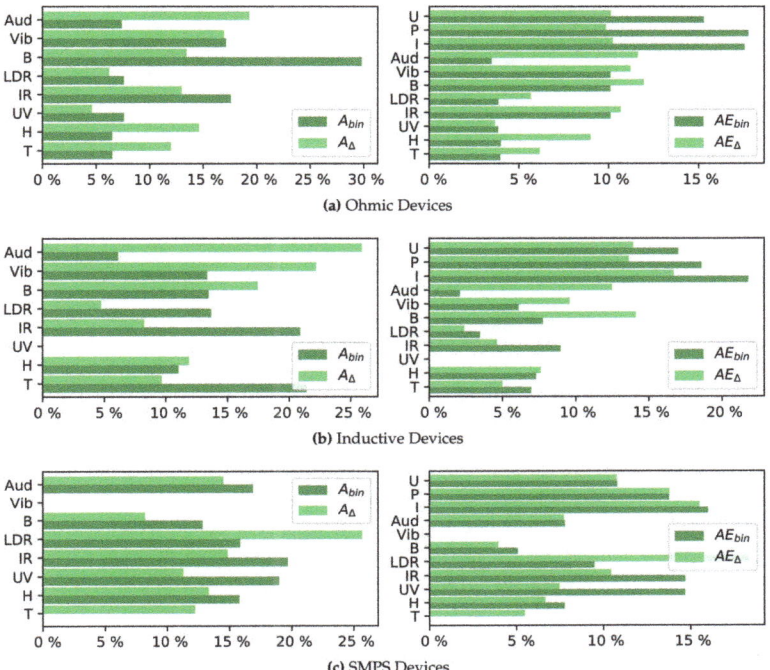

Figure 10. Feature Importances (%) for Recognition of Appliances of a Single Class—Appliance Categorization By: Load Curve.

5.4. Interpretation and Discussion

We evaluated a set of features calculated from ambient sensor readings considering their usefulness and importances for different decision scenarios. All selected and evaluated features were shown to be relevant for at least one of the considered evaluations. However, the UV radiation was found to ease the distinction of different lamps, but could not be detected for any other appliance. Accordingly, its usefulness is restricted to scenarios involving such appliances.

To enable further consideration of the usefulness of individual features, we have accumulated how often each feature was part of an optimal feature subset. The sums are displayed in Table 8 and reconfirm the restricted usage potentials for the UV measurements. It can be furthermore seen that the audio and vibration data have an overall high importance, and were marked especially useful if only binary ambient information is available (i.e., if operating devices leads to the presence of acoustic signals or vibrations). A comparably high usefulness can be attributed to the temperature, magnetic flux density, and humidity measurements; they are included in more than half of the optimal feature subsets. Their importance is even higher when only ambient features are available for evaluation. This evaluation of emissions and feature selection methods illustrates that multiple ambient features might be necessary to properly distinguish electrical devices.

Table 8. Number of Times each Feature was included in the Optimal Feature Subset.

Dataset	# of Evaluations	Best Feature Subset										
		T	H	UV	IR	LDR	B	Vib	Aud	U	I	P
A_{bin}	12	8	8	3	5	8	9	10	11	n/a	n/a	n/a
A_{Δ}	12	10	10	1	8	5	9	9	9	n/a	n/a	n/a
AE_{bin}	9	1	3	0	5	1	5	4	4	1	2	3
AE_{Δ}	9	5	6	0	6	3	3	5	3	1	0	1

Considering the usage of electrical and ambient features, our results show that the inclusion of electrical features allows to achieve distinction with less ambient sensors and generally achieves higher Cross-Validation Scores. This indicates that systems integrating ambient information in the decision process should consider electrical and ambient features at the same time and not within different decision mechanisms.

During the evaluation of the usefulness for different scenarios, this work also considered the possibility to use ambient sensor data to assign device classes. While the classifications specific to the appliance's internal workings, based on the number of appliance states (Single-State, Multi-State, or Infinite-State appliances) or based on the load curves (ohmic, inductive, or Switching Power Supplies), achieved low Cross-Validation Scores, the distinction between smaller and larger consumers could be shown to be feasible based on the easy-to-calculate features evaluated in this study from a small set of ambient sensors. Considering the classification according to load types or number of states, the use of binary features (indicating the change of an ambient parameter) is outperformed by the use of amplitude information from the audio and vibration sensors. As other studies have shown the feasible usage of more complex audio features (e.g., [5,14]), the additional information gain through such features could be investigated in future work to improve the results for the usage of ambient data for appliance classification even further.

Lastly, we conducted evaluations on two kinds of features: A binary change evaluation and the evaluation of the maximum changes of sensor values. Considering the results, we have observed that binary features of ambient influences typically yield lower Cross Validation Scores and often contain more features in the resulting optimal feature subset. The usage of maximum change information resulted in a rise of 15.3 percentage points in the CVS, with a maximum improvement of nearly 30 percentage points for appliance recognition scenarios using only ambient information. A further 5.1 percentage points increase in CVS could be observed when including electrical information in appliance recognition scenarios.

We expect the results of this work to be a useful guideline for the creation of future energy data collection systems. Our careful analysis of the information gains of sensor types beyond the traditionally used electrical signals showcases the great potentials of using ambient information in conjunction with electrical data. We are convinced that the analyzed scenarios regarding certain classes of devices and the usefulness of ambient sensor types can help developers to advance and improve existing systems and algorithms based on our findings.

6. Conclusions and Outlook

In this work, we have conducted an evaluation of ambient sensor data and its possible uses in the context of electrical appliance recognition and load signature analysis in general. Based on two feature selection and ranking methods, we have demonstrated that sensor data for temperature, humidity, audio, vibrations, the magnetic flux density and IR radiation are useful for nearly any load monitoring use case. It could additionally be shown that a single measure for the change of ambient features during an appliance's operation is often sufficient to reach decent Cross-Validation Scores, indicating that such a data collection with low resource usage can already improve load monitoring systems. By considering different evaluation scenarios, we were able to show that the combination of both electrical and ambient sensor data has been proven to provide the best benefits. We would like to reiterate at this point that our work is not primarily contributing to the field of NILM, i.e., the disaggregation of aggregate load data. Rather than that, we have methodologically determined the most information-rich sensor parameters that can improve energy data analysis methods (such as NILM) in a more general and algorithm-independent way.

Author Contributions: Conceptualization, J.H., M.A.T., A.R.; Data curation, M.A.T., J.H.; Formal analysis, J.H., M.A.T.; Funding acquisition, A.R.; Investigation, M.A.T., J.H.; Methodology, M.A.T., A.R., J.H.; Software, M.A.T.; Supervision, A.R., J.H.; Validation, M.A.T., J.H.; Visualization, J.H., A.R.; Writing—original draft, M.A.T.; Writing—review & editing, J.H., A.R. All authors have read and agreed to the publication of this version of the manuscript.

Funding: This work was supported by Deutsche Forschungsgemeinschaft grant no. RE 3857/2-1.

Institutional Review Board Statement: Not applicable.

Informed Consent Statement: Not applicable.

Conflicts of Interest: The authors declare no conflict of interest.

Abbreviations

The following abbreviations are used in this manuscript:

A_{bin}	Ambient Parameters; binarized
AE_{bin}	Ambient and Electrical Parameters; binarized
A_Δ	Ambient Parameters; differences
AE_Δ	Ambient and Electrical Parameters; differences
ADC	Analog-to-Digital Converter
CSV	Comma-Separated Value
CVS	Cross-Validation Score
DuT	Device under Test
FHMM	Factorial Hidden Markov Model
IR	Infrared
LDR	Light Dependent Resistor
NILM	Non-Intrusive Load Monitoring
SMPS	Switched-Mode Power Supplies
UV	Ultraviolet

References

Armel, K.C.; Gupta, A.; Shrimali, G.; Albert, A. Is Disaggregation the Holy Grail of Energy Efficiency? The Case of Electricity. *Energy Policy* **2013**, *52*, 213–234. [CrossRef]

Zoha, A.; Gluhak, A.; Imran, M.A.; Rajasegarar, S. Non-Intrusive Load Monitoring Approaches for Disaggregated Energy Sensing: A Survey. *Sensors* **2012**, *12*, 16838–16866. [CrossRef] [PubMed]

Paradiso, F.; Paganelli, F.; Giuli, D.; Capobianco, S. Context-Based Energy Disaggregation in Smart Homes. *Future Internet* **2016**, *8*, 4. [CrossRef]

Gupta, S.; Reynolds, M.S.; Patel, S.N. ElectriSense: Single-point Sensing Using EMI for Electrical Event Detection and Classification in the Home. In Proceedings of the 12th ACM International Conference on Ubiquitous Computing (UbiComp), Copenhagen, Denmark, 26–29 September 2010; pp. 139–148.

Guvensan, M.A.; Taysi, Z.C.; Melodia, T. Energy Monitoring in Residential Spaces With Audio Sensor Nodes: TinyEARS. *Ad Hoc Netw.* **2013**, *11*, 1539–1555. [CrossRef]

Reinhardt, A.; Klemenjak, C. How does Load Disaggregation Performance Depend on Data Characteristics? Insights from a Benchmarking Study. In Proceedings of the 11th ACM International Conference on Future Energy Systems (e-Energy), Melbourne, Australia, 22–26 June 2020; pp. 167–177.

Kim, H.; Marwah, M.; Arlitt, M.; Lyon, G.; Han, J. Unsupervised Disaggregation of Low Frequency Power Measurements. In *Proceedings of the 2011 SIAM International Conference on Data Mining*; Society for Industrial and Applied Mathematics: Philadelphia, PA, USA, 2011; pp. 747–758.

Ruzzelli, A.; Nicolas, C.; Schoofs, A.; O'Hare, G. Real-Time Recognition and Profiling of Appliances through a Single Electricity Sensor. In Proceedings of the 7th Annual IEEE Communications Society Conference on Sensor Mesh and Ad Hoc Communications and Networks (SECON), Boston, MA, USA, 21–25 June 2010; pp. 1–9.

9. Reinhardt, A.; Baumann, P.; Burgstahler, D.; Hollick, M.; Chonov, H.; Werner, M.; Steinmetz, R. On the Accuracy of Appliance Identification Based on Distributed Load Metering Data. In Proceedings of the 2nd IFIP Conference on Sustainable Internet and ICT for Sustainability (SustainIT), Pisa, Italy, 4–5 October 2012; pp. 1–9.
10. Bergés, M.; Soibelman, L.; Matthews, H.S. Leveraging Data From Environmental Sensors to Enhance Electrical Load Disaggregation Algorithms. In Proceedings of the 13th International Conference on Computing in Civil and Building Engineering (ICCBE), Nottingham, UK, 30 June–2 July 2010.
11. Brunelli, D.; Minakov, I.; Passerone, R.; Rossi, M. POVOMON: An Ad-Hoc Wireless Sensor Network for Indoor Environment Monitoring. In Proceedings of the 2014 IEEE Workshop on Environmental, Energy, and Structural Monitoring Systems (EESMS), Naples, Italy, 17–18 September 2014; pp. 1–6.
12. Klemenjak, C.; Reinhardt, A.; Pereira, L.; Berges, M.; Makonin, S.; Elmenreich, W. Electricity Consumption Data Sets: Pitfalls and Opportunities. In Proceedings of the 6th ACM International Conference on Systems for Energy-Efficient Buildings, Cities, and Transportation (BuildSys), New York, NY, USA, 13–14 November 2019; pp. 159–162.
13. Lifton, J.; Feldmeier, M.; Ono, Y.; Lewis, C.; Paradiso, J.A. A Platform for Ubiquitous Sensor Deployment in Occupational and Domestic Environments. In Proceedings of the 6th International Symposium on Information Processing in Sensor Networks (IPSN), Cambridge, MA, USA, 25–27 April 2007; pp. 119–127.
14. Taysi, Z.C.; Guvensan, M.A.; Melodia, T. TinyEARS: Spying on House Appliances with Audio Sensor Nodes. In Proceedings of the 2nd ACM Workshop on Embedded Sensing Systems for Energy-Efficiency in Building (BuildSys), Zurich, Switzerland, 3–5 November 2010; pp. 31–36.
15. Kim, Y.; Schmid, T.; Charbiwala, Z.M.; Srivastava, M.B. ViridiScope: Design and Implementation of a Fine Grained Power Monitoring System for Homes. In Proceedings of the 11th International Conference on Ubiquitous Computing (UbiComp), Orlando, FL, USA, 30 September–3 October 2009; pp. 245–254.
16. Schoofs, A.; Guerrieri, A.; Delaney, D.T.; O'Hare, G.M.P.; Ruzzelli, A.G. ANNOT: Automated Electricity Data Annotation Using Wireless Sensor Networks. In Proceedings of the 7th Annual IEEE Communications Society Conference on Sensor, Mesh and Ad Hoc Communications and Networks (SECON), Boston, MA, USA, 21–25 June 2010; pp. 1–9.
17. Klemenjak, C.; Elmenreich, W. YaY—An Open-hardware Energy Measurement System for Feedback and Appliance Detection Based on the Arduino Platform. In Proceedings of the 13th Workshop on Intelligent Solutions in Embedded Systems (WISES), Hamburg, Germany, 12–13 June 2017; pp. 1–8.
18. Englert, F.; Schmitt, T.; Kößler, S.; Reinhardt, A.; Steinmetz, R. How to Auto-Configure Your Smart Home? High-Resolution Power Measurements to the Rescue. In Proceedings of the 4th ACM International Conference on Future Energy Systems (e-Energy), Berkeley, CA, USA, 22–24 May 2013; pp. 215–224.
19. Jiang, X.; Van Ly, M.; Taneja, J.; Dutta, P.; Culler, D. Experiences with a High-fidelity Wireless Building Energy Auditing Network. In Proceedings of the 7th ACM Conference on Embedded Networked Sensor Systems (SenSys), Berkeley, CA, USA, 3–6 November 2009; pp. 113–126.
20. Iqbal, H.K.; Malik, F.H.; Muhammad, A.; Qureshi, M.A.; Abbasi, M.N.; Chishti, A.R. A Critical Review of State-of-the-art Non-intrusive Load Monitoring Datasets. *Electr. Power Syst. Res.* **2020**. [CrossRef]
21. Kolter, J.Z.; Johnson, M.J. REDD: A Public Data Set for Energy Disaggregation Research. In Proceedings of the Workshop on Data Mining Applications in Sustainability (SIGKDD), San Diego, CA, USA, 21 August 2011; Volume 25, pp. 59–62.
22. Anderson, K.; Ocneanu, A.; Carlson, D.R.; Rowe, A.; Bergés, M. BLUED: A Fully Labeled Public Dataset for Event-Based Non-Intrusive Load Monitoring Research. In Proceedings of the 2nd KDD Workshop on Data Mining Applications in Sustainability (SustKDD), Beijing, China, 12 August 2012; pp. 1–5.
23. Barker, S.; Mishra, A.; Irwin, D.; Cecchet, E.; Shenoy, P.; Albrecht, J. Smart*: An Open Data Set and Tools for Enabling Research in Sustainable Homes. In Proceedings of the 2nd KDD Workshop on Data Mining Applications in Sustainability (SustKDD), Beijing, China, 12 August 2012; pp. 1–6.
24. Beckel, C.; Kleiminger, W.; Cicchetti, R.; Staake, T.; Santini, S. The ECO Data Set and the Performance of Non-intrusive Load Monitoring Algorithms. In Proceedings of the 1st ACM Conference on Embedded Systems for Energy-Efficient Buildings (BuildSys), Memphis, TN, USA, 3–6 November 2014; pp. 80–89.
25. Cabeza, L.F.; Ürge Vorsatz, D.; Palacios, A.; Ürge, D.; Serrano, S.; Barreneche, C. Trends in Penetration and Ownership of Household Appliances. *Renew. Sust. Energy Rev.* **2018**, *82*, 4044–4059. [CrossRef]
26. Jones, R.V.; Lomas, K.J. Determinants of High Electrical Energy Demand in UK Homes: Appliance Ownership and Use. *Energy Build.* **2016**, *117*, 71–82. [CrossRef]
27. Federal Statistical Office of Germany (Destatis) Equipment with Consumer Durables. 2018. Available online: http://www.destatis.de/EN/Themes/Society-Environment/Income-Consumption-Living-Conditions/Equipment-Consumer-Durables/_node.html (accessed on 10 December 2020).
28. Laput, G.; Zhang, Y.; Harrison, C. Synthetic Sensors: Towards General-Purpose Sensing. In Proceedings of the CHI Conference on Human Factors in Computing Systems (CHI), Denver, CO, USA, 6–11 May 2017; pp. 3986–3999.
29. Guyon, I.; Elisseeff, A. An Introduction to Variable and Feature Selection. *J. Mach. Learn. Res.* **2003**, *3*, 1157–1182.
30. Ho, T.K. Random Decision Forests. In Proceedings of the 3rd International Conference on Document Analysis and Recognition (ICDAR), Montreal, QC, Canada, 14–16 August 1995; Volume 1, pp. 278–282.
31. Loh, W.Y. Fifty Years of Classification and Regression Trees. *Int. Stat. Rev.* **2014**, *82*, 329–348. [CrossRef]

Guyon, I.; Weston, J.; Barnhill, S.; Vapnik, V. Gene Selection for Cancer Classification Using Support Vector Machines. *Mach. Learn.* **2002**, *46*, 389–422. [CrossRef]

Pedregosa, F.; Varoquaux, G.; Gramfort, A.; Michel, V.; Thirion, B.; Grisel, O.; Blondel, M.; Prettenhofer, P.; Weiss, R.; Dubourg, V.; et al. Scikit-Learn: Machine Learning in Python. *J. Mach. Learn. Res.* **2011**, *12*, 2825–2830.

Najafi, B.; Moaveninejad, S.; Rinaldi, F. Chapter 17—Data Analytics for Energy Disaggregation: Methods and Applications. In *Big Data Application in Power Systems*; Arghandeh, R., Zhou, Y., Eds.; Elsevier: Amsterdam, The Netherlands, 2018; pp. 377–408.

Kelly, J.; Knottenbelt, W. Neural NILM: Deep Neural Networks Applied to Energy Disaggregation. In Proceedings of the 2nd ACM International Conference on Embedded Systems for Energy-Efficient Built Environments (BuildSys), Seoul, Korea, 4–5 November 2015; pp. 55–64.

Article

A Framework to Generate and Label Datasets for Non-Intrusive Load Monitoring

Benjamin Völker *, Marc Pfeifer, Philipp M. Scholl and Bernd Becker

Chair for Computer Architecture, University of Freiburg, 79110 Freiburg, Germany; pfeiferm@informatik.uni-freiburg.de (M.P.); pscholl@informatik.uni-freiburg.de (P.M.S.); becker@informatik.uni-freiburg.de (B.B.)
* Correspondence: voelkerb@informatik.uni-freiburg.de

Abstract: In order to reduce the electricity consumption in our homes, a first step is to make the user aware of it. Raising such awareness, however, demands to pinpoint users of specific appliances that unnecessarily consume electricity. A retrofittable and scalable way to provide appliance-specific consumption is provided by Non-Intrusive Load Monitoring methods. These methods use a single electricity meter to record the aggregated consumption of all appliances and disaggregate it into the consumption of each individual appliance using advanced algorithms usually utilizing machine-learning approaches. Since these approaches are often supervised, labelled ground-truth data need to be collected in advance. Labeling on-phases of devices is already a tedious process, but, if further information about internal device states is required (e.g., intensity of an HVAC), manual post-processing quickly becomes infeasible. We propose a novel data collection and labeling framework for Non-Intrusive Load Monitoring. The framework is comprised of the hardware and software required to record and (semi-automatically) label the data. The hardware setup includes a smart-meter device to record aggregated consumption data and multiple socket meters to record appliance level data. Labeling is performed in a semi-automatic post-processing step guided by a graphical user interface, which reduced the labeling effort by 72% compared to a manual approach. We evaluated our framework and present the FIRED dataset. The dataset features uninterrupted, time synced aggregated, and individual device voltage and current waveforms with distinct state transition labels for a total of 101 days.

Keywords: data annotation; non-intrusive load monitoring; semi-automatic labeling; smart meter

1. Introduction

The United Nations has outlined 17 Sustainable Development Goals [1] for 2030. Related to the production and consumption of electric energy are three of them: *stop global warming* by *clean energy* in *sustainable cities*.

One important step to achieve these goals is to reduce the electricity consumption in our homes. In the residential domain, energy monitoring and 'eco-feedback' techniques have proven to help by raising the awareness of an unnecessary electricity consumption of a particular device. In addition, these techniques can be combined with demand-side flexibility to schedule their usage, so that mostly renewable energy is used. Ehrhardt-Martinez et al. [2] found that per device consumption feedback can achieve high energy savings when provided frequently. More specifically, according to this meta-study, real-time aggregated electricity consumption feedback can preserve around 8.6 % of electricity on average. If the feedback is provided appliance-wise, they spotted that the savings are up to an average of 13.7 %. These savings are achieved by simply raising the user awareness. The actual savings might even be increased, if the feedback system is combined with a smart home agent. Smart home agents can learn user behavior, adapt knowledge of other agents, and can either directly control smart appliances to save electricity or recommend specific energy saving strategies to the user.

There are mainly two possibilities to obtain the device specific electricity consumption of certain devices in a home. (1) Each device is equipped with a dedicated electricity meter known as Intrusive Load Monitoring (ILM). (2) A single electricity meter is installed that measures the composite load of all appliances. Specially designed and often individually trained algorithms disaggregate this aggregated load into the load of each individual consumer. This approach is known as Non-Intrusive Load Monitoring (NILM) and is said to be more feasible (compared to ILM) as it only requires a single smart electricity meter.

In many countries, the standard electricity meter (Ferrari meter) has already been exchanged by a smart electricity meter. For instance, the roll-out of smart meters in Germany began with heavy consumers (>6000 kW h per year) by the beginning of 2020 [3]. Smart meters are promoted to bring features like device level electricity feedback to our homes by using NILM.

NILM research already started in 1992 when a first NILM prototype was introduced by G.W. Hart [4,5]. Recent promotion of smart meters, associated research funding (e.g., SINTEG [6]), and emerging machine learning algorithms accelerated research in this field. Even if the concept is already known for 35 years, Armel et al. [7] stated that "disaggregation may be the lynch-pin to realizing large-scale, cost-effective energy savings in residential and commercial buildings." Over the last three decades, various NILM algorithms have been developed by researchers. These can be roughly categorized into (1) event-based algorithms which relate signal state changes to appliance state changes (such as [4] or [8]) and (2) event-less algorithms which estimate an overall system state using techniques such as Factorial Hidden Markov Models [9,10].

To train, evaluate, and compare these algorithms, public available datasets are used. Even though a lot of datasets have been published such as REDD [9], UK-Dale [11], BLOND [12], and many more (see [10] for a comprehensive overview), they can only hardly be used to compare different disaggregation techniques because of a low sampling frequency which does not allow to test event-based approaches or because of missing or incorrect ground truth information. Besides these, no datasets—except BLUED [13] to some extent—includes labels of internal appliance state changes (e.g., changing the channel of television). Unfortunately, such information is of particular interest for event-based NILM approaches and electricity-based human activity recognition systems such as [14,15].

Retrospectively generating fine grain labels is not possible for datasets that have already been recorded years ago and hardly possible while generating new datasets. It would require the residents to manually log every action in the home (e.g., every key-press of the television remote) with precise timestamps. In other domains such as activity recognition, the problem of generating ground truth data is typically addressed by recording video contemporaneous to e.g., accelerometer signals. The labeling step is then performed manually afterwards by going through the video on a frame by frame basis. This technique could hardly be applied to electricity datasets as: (1) electricity datasets typically cover long time period of several weeks or months, (2) privacy concerns if all rooms or residents are equipped with a camera, and (3) internal or automatic state changes of appliances (like the cooling cycle of the fridge) can not be identified via video.

Therefore, we propose a hardware and software framework to generate and label data for NILM that feature fine-grained labels based on intrusive meters, additional sensors and a smart labeling tool. The system allows to record time synced data of a home electrical input (aggregated data) and nearly all individual consumers. Furthermore smart appliances are incorporated to log their states. For devices that do not expose their states (e.g., old TVs), custom logging devices are used such as infrared sniffers. A post conducted, semi-automatic algorithm identifies appliance steady states and state changes in the individual appliance data and applies preliminary labels to the data. Therewith the overall labeling effort is reduced significantly to a human supervision step.

This report summarizes two publications [16,17] and extends them by (1) including more related work, (2) an in depth explanation of the used hardware and software components bundled into the proposed expandable framework, and (3) a deeper evaluation of the FIRED dataset that has also been extended to include more recording days and high quality event labels.

The remainder of the paper is structured as follows: Section 1.1 describes how others have recorded and labeled NILM datasets. In Section 2, we identify remaining challenges to record NILM datasets and describe the hardware and software of our proposed framework. We successfully utilized the framework to generate and label the FIRED [17] dataset which we present in Section 3. Finally, a discussion of the FIRED dataset and our observations with our framework concludes the paper in Sections 4 and 5.

1.1. Related Work

As interest in evaluating and comparing electricity related algorithms has increased over the recent decade, several datasets and the hardware used to record them have been published. Some of these datasets which have been recorded either in residential or industrial environments are briefly discussed in the following.

The Reference Energy Disaggregation Dataset (REDD) was introduced by Kolter et al. in 2011 [9]. The authors used a custom-built meter to record the whole house electricity consumption of six different homes in the US. *NI-9239* (National Instruments) analog to digital converters were used to measure the mains' voltage and *SCT-013* (YHDC) split core current transformers to measure current in a secure and non-intrusive way. The readings were acquired by a recording laptop at 16.5 kHz with an ADC resolution of 24 bit. High frequency mains' data of the complete recording duration is, however, only available as compressed files generated with a custom lossy compression. Socket and sub-circuit-level data are only available as unevenly sampled low frequency data of approximately 1/3 Hz. Furthermore, these data show gaps of several days.

The UK Domestic Appliance-Level Electricity dataset (UK-DALE) introduced by Kelly et al. in 2015 [11] covers the whole house electricity demand of five homes in the UK. In particular, three of the houses (1, 2, and 5) have been recorded at a sampling rate of 16 kHz. The aggregated power consumption was recorded with off-the-shelf USB sound cards with stereo line input. AC-AC transformers were used to scale down the voltage, while split core current transformers were used to measure current. The recording duration of house 1 was up to 1629 days resulting in the longest whole house recording known to us. Appliance level data was sampled using off-the-shelf 433 MHz electricity meter plugs (Eco Manager Transmitter Plugs developed by Current Cost) paired with a custom self-developed base station. Devices directly connected to the mains are metered using current clamp meters (Current Cost transmitter) that are sampled using the same custom base station. The data of these devices were sampled with a low sampling rate of around 1/6 Hz and contains several gaps too.

The Electricity Consumption and Occupancy (ECO) dataset was introduced by Beckel et al. in 2014 [18]. The authors leverage the communication interface of an off-the-shelf smart electricity meter (*E750 from Landis + Gyr*) to read out the aggregated consumption of six homes in Switzerland. The consumption data include different electricity related metrics such as active power, RMS voltage, and current as well as the phase shifts of all three supply legs. Furthermore, 6–10 *Plugwise* smart plugs have been deployed per house to record individual appliance active power measurements of selected appliances at around 1 Hz (the actual sampling rate varied due to a sequential readout, but the data have been resampled to 1 Hz). Home occupancy information is also available recorded by tablet computers and passive infrared sensors. The low sampling rate, dropouts, and low individual appliance coverage makes it difficult to use the dataset to evaluate event-based NILM and activity recognition approaches, as multiple events may happen between two samples.

The Almanac of Minutely Power dataset (AMPds) was introduced by Makonin et al. [19] in 2013. It features electricity, water, and gas readings at one minute resolution of a residential building in Canada. They used an off-the-shelf *Powerscout18* meter (DENT Instruments) to record the whole house consumption and the consumption of individual circuits over a time period of two years. Data from the same house is also available at 1 Hz resolution the Rainforest Automation Energy (RAE) Dataset [20] introduced by the same authors 2018. The RAE dataset covers 72 days of electricity data without any power events marke

In the non-residential domain, Kriechbaumer et al. proposed the Building-Level Offi eNvironment Dataset (BLOND) in 2018 [12]. They recorded aggregated and device lev data of an office building in Germany over a time period of around 260 days. The autho used custom-built hardware for both aggregated and individual appliance readings. At aggregated level, they used Hall effect current transformers and AC-AC transforme to record the 3-phase power grid with up to 250 kHz. Individual appliances have be recorded using the same principle (AC-AC transformer + Hall effect current transforme embedded into off the shelf power strips with up to 50 kHz. Their dataset is split into tv measurement series. BLOND-50 features 50 kHz aggregated and 6.4 kHz device level da over 213 days and BLOND-250 features 250 kHz aggregated and 50 kHz device level da over 50 days. For both sets, the 1 Hz apparent power has been derived from the volta; and current waveforms. However, downloading the dataset requires storing ≈40 TB data. Moreover, the authors have not used their recording system to generate a residenti dataset yet.

These datasets have successfully been used to evaluate different event-less NIL algorithms (e.g., in [10,18]). Event-based NILM methods, however, require informatic about all appliance events in order to evaluate the detection and classification of the events. Such information is not available in the presented datasets. The lack of datase for event-based NILM algorithms has already been explored by Pareira et al. in [21]. The proposed a post-conducted labelling approach which can be applied to the individu device data of a dataset. Their method uses an automatic event detector based on the l likelihood ratio to recognize events in the power signal. They evaluated the detector usir the REDD [9] and AMPds [19] datasets. It achieved F_1 scores of 84.52% for REDD ar 94.87% for AMPds. The detector results highly depend on the quality of the data and s parameters. Therefore, a supervision is still required.

The Building-Level fUlly-labeled dataset for Electricity Disaggregation (BLUED) i troduced by Anderson et al. [13] in 2012 was specifically recorded with event-detection mind. The authors recorded voltage and current measurements with a resolution of 16 b and a sampling rate of 12 kHz. Significant appliance transients were labeled manual and using additional sensors and switchable sockets. However, no individual applian electricity measurements are available in this dataset.

We identified different shortcomings of existing datasets: (1) Larger time periods which no data or only a part of the data are available (REDD, ECO, UK-DALE). (2) Rel tive low sampling rate for appliance level data (REDD, UK-DALE, ECO, AMPds) or r appliance level data at all (BLUED). (3) Missing information about the time and type appliance events (REDD, ECO, UK-DALE, BLOND, AMPds). (4) Unknown number ar type of devices which are not monitored individually (ECO, UK-DALE, BLUED, AMPd (5) No standard procedure to load the data or explore the dataset.

2. Materials and Methods

The shortcomings of existing datasets have been expressed in Section 1.1. In ord to overcome these shortcomings, we define a set of challenges that need to be addresse when recording datasets to evaluate load monitoring or other electricity related algorithn in the residential domain. In particular, event-based NILM algorithms and event detectic algorithms cannot be evaluated using the existing datasets, as they lack ground tru information (time and type) of events. The challenges have been summarized in Table 1

Table 1. Challenges that need to be addressed when recording datasets for Non-Intrusive Load Monitoring.

ID	Challenge
C1	**Simultaneous recordings** of a home's aggregated electricity consumption and the consumption of the individual appliances. The individual data can be used to validate the appliance estimates of NILM algorithms. Furthermore, it can be explored how semi-supervised hybrid NILM algorithms such as [22] can benefit from individual appliance data.
C2	**High sampling rates** of the aggregated and individual appliance data. This allows for the extraction of high frequency features from the individual waveforms which might further improve traditional NILM algorithms. Kriechbaumer et al. [12] focused on recording a dataset with a very high sampling rate but, therefore, require to download ≈75 GB of data per day. To not sacrifice usability, a trade-off between high sampling rates and file size needs to be examined.
C3	**Continuous data recording** for multiple days is crucial to understand and explore different consumption behavior based on the *time-of-day* or *day-of-week*.
C4	**High quality dataset labels** to evaluate event-based NILM and event detection algorithms. These labels should consist of a timestamp describing when the event occurred, the device responsible for the event and a textual description of the event.
C5	**High temporal accuracy** of the data and its labels is required. Labels should always reflect the associated change in the signal. This requires that the data streams are in sync and do not drift apart.
C6	**Usability** is one of the most underrated factors of a dataset. However, researchers should be able to explore and utilize a dataset in a quick and easy way.

Based on the stated challenges, we have developed a framework to record and label NILM datasets. The overall flow of this framework is shown in Figure 1. It consists of an aggregated electricity meter (Smart Meter) which records high frequency voltage and current waveforms at the aggregated level, and multiple distributed meters which record voltage and current waveforms of individual appliances. Further sensors can be added to measure other quantities (e.g., temperature or movement). The current and voltage waveforms as well as the sensor data are collected by a recording PC and stored in multimedia containers. Other electricity related metrics such as active and reactive power are derived from the raw current and voltage waveforms. These power data is stored with different sampling rates and is used to generate data labels semi-automatically. A post-processing step extracts events and assigns labels to these events. Both events and labels are refined by a human using a graphic user interface (GUI) resulting in a final set of label files. Each part of the framework is explained in more detail in the remainder of this section.

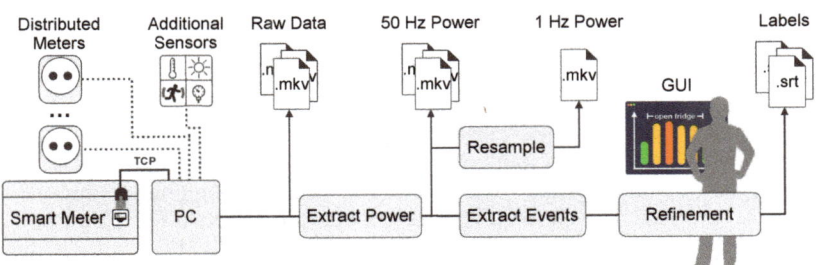

Figure 1. Overall flow of the presented framework to record advanced NILM datasets.

2.1. Smart Meter

Aggregated data are recorded using a custom-built measurement system referenced as the *SmartMeter* from now on. The system was introduced by Völker et al. in [23,24]. A schematic wiring diagram of the smart meter can be seen in Figure 2. It shows the required connections to the power grid. As the analog to digital converter (ADC) requires input voltage levels of 2 V maximum, we use a voltage dividers with a ratio of 1:1000 to scale down the mains voltage levels. Likewise, we use current transformers (*YHDC SCT-013* with a ratio of 1:2000) to convert the home's current consumption into a voltage signal that can be measured by the ADC. Using the split-core variant of the current transformer allows us to measure current in the most non-intrusive way. The home's scaled voltage levels and current consumptions are sampled at up to 32 kHz using the *ADE9000* ADC from *Analog Devices* [25]. The ADC can handle seven input signals at a resolution of 24 bit and a signal-to-noise ratio of 96 dB. It further has an internal Digital Signal Processor (DSP) to calculate attributes like active or apparent power as well as electrical energy. The sampled data is retrieved by an *ESP32* microcontroller (Espressif Systems) over an isolated SPI interface. The microcontroller converts the raw fixed point data to 32 bit float values representing the actual voltage and current measurements (in *Volt* and *Milliampere*, respectively). The data can be sent to a sink via either a *USB Serial*, a *TCP* or a *UDP* connection. An external flash memory allows for buffering the data on short network disconnections. We used 8 MB in the installed system which can hold up to ≈ 41 s of data at a sampling rate of 8 kHz. Furthermore, a Real Time Clock (RTC) is used to sync the sampling rate of the installed ADC (see Section 3.3.4). An Ethernet connection adds a reliable cable connection to the measurement system using the *LAN8720* chip (Microchip Technology) with an ordinary RJ 45 connector. It is also possible to use WiFi communication, as the ESP32 comes with WiFi on-board. However, in our findings, Ethernet is more reliable in a fuse box environment and should be preferred.

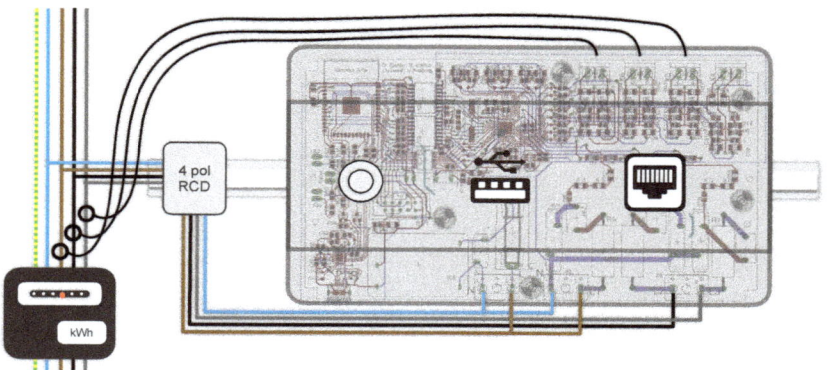

Figure 2. Schematic of the SmartMeter wiring for a three-phase power supply inside the fuse box.

The measurement system is encapsulated in a fire proof DIN housing. This allows the system to be installed at a DIN rail inside the fuse box (as shown on the left side of Figure 6).

2.2. Distributed Meters

Individual appliances can be recorded using a set of *PowerMeters* (see right side of Figure 6). PowerMeters are custom-built smart plugs designed to measure current and voltage waveforms of individual appliances. The plugs were introduced by Völker et al. in [22,23]. Their general system architecture is nearly identical to the architecture of the SmartMeter. A PowerMeter scales down the power outlet's voltage by a factor of 1:58 using a voltage divider and the current drawn by the connected appliance using a 3 mΩ shunt resistor. The analog signals are sampled using a *STPM32* ADC from *STMicroelectronics* [26]. The ADC can sample at up to 7.875 kHz with 24 bit resolution and has an internal DSP. The DSP again allows for calculating other electricity metrics directly inside the smart plug. Data from the ADC is collected by the same ESP32 microcontroller used inside the SmartMeter. 4 MB external flash storage allows for being resilient against ≈ 250 s network dropouts at a sampling rate of 2 kHz. Each distributed meter also includes an RTC for clock synchronization. Data is sent over WiFi, as a wireless solution should be preferred over a wired interface in such a distributed setup. The cost of a single PowerMeter is comparatively low with approximately €35.

The power consumption of each PowerMeter itself is quite low (0.56 W) and compared to the smart plugs used to record other NILM datasets e.g., *Plugwise* as used in [11] (0.5 W

2.3. Additional Sensors

The framework allows for recording arbitrary sensor values or additional appliance information together with the electricity measurements using an MQTT-API. MQTT [27] provides a standardized publish–subscribe messaging system and has emerged to one of the standard protocols in the world of IoT. If an instance wants to share some information, it can send a message for a given *topic*. If other instances are interested in this information, they can *subscribe* to the specific topic. Each new message *published* under a certain topic is relayed by the MQTT broker to all instances subscribed to this topic. MQTT builds on top of the TCP network protocol which guarantees the successful transmission of data.

The recording PC (see Figure 1) hosts a central MQTT broker. A small *Python* script listens for incoming messages under a general topic *recording* and will handle storing the incoming data into *CSV* files. If a sensor should be added to the recording infrastructure, it simply needs to connect to the broker and send its data on a unique sub-topic (e.g., *recording/livingroom_temp*). Data must follow the *JSON* format. Each JSON key corresponds to a header entry in the resulting CSV file. A timestamp is added by the Python script for each entry if the key "ts" is not present in the data. An example for a valid message of a temperature sensor is *recording/livingroom_temp {"value": 20.5}*.

We further highlight three examples of how additional appliance states or sensors can be added:

- **Smart lighting:**
  Many light bulbs are nowadays substituted with smart light bulbs. Most of these can be controlled via a *ZigBee* gateway. Such a gateway can be incorporated to pass information if a light bulb changes its state, dimm state, or color. We have implemented a Python script which interfaces with such a gateway to log the state changes of all light bulbs connected to the gateway using our MQTT-API. This allows for deriving power consumption estimates without intrusively metering each light individually and provides further room occupancy information.

- **Sensors:**
  We show an example flow of how a custom sensor can be developed using the provided MQTT-API in Figure 3. The ESP32 has WiFi built-in and provides certain inter-system interfaces such as *SPI, I2C*, or *UART*. This allows for rapid prototyping different sensors like temperature or occupancy.

- **Bridges:**
  The same system overview as shown in Figure 3 can be used to develop different gateways. As an example, we developed a 433 MHz gateway that logs state changes of switchable sockets, wall switches, or remote button presses of devices that are equipped with 433 MHz. We further implemented an infrared sniffer that logs all commands received from off-the-shelf remotes to MQTT. This helps to capture interactions with televisions, HiFi systems, and air conditioners.

Figure 3. Example of extending the recording system by logging additional sensor data using the MQTT-API.

The Python script further publishes recording information each 600 s under the topic *recordingInfo*. The information includes the current recording state, the number of meters active, and the average power of each meter since the last message.

2.4. Extracting Events

Evaluating event-based NILM algorithms requires having ground truth data for events in the dataset. The authors of the UK-Dale [11] dataset therefore recorded appliance turn on/off events for house 1 using switchable sockets. If the user pressed the button of such a switchable socket. The current timestamp, device, and state of the socket is logged. We particularly see three drawbacks of such an approach: (1) Devices that are hardwired to the mains like the stove or lighting cannot be equipped with such a socket. (2) Only on/off events can be logged. Most of our household appliances are multi state devices that have more than just a binary state *on* or *off*. (3) Devices that change their state without user interactions can not be labeled (e.g., a kettle turns off automatically if the water is boiling).

We build on the idea of a post-conducted labelling approach introduced by Pereira et al. in [21] and developed a semi-automatic labeling algorithm that consists of three steps event detection, unique event identification, and high variance filtering.

2.4.1. Event Detection

The event detector utilizes the Log-Likelihood Ratio (LLR) test introduced by Pereira et al. in [21]. It has been enhanced by adaptive thresholding by Völker et al. in [23]. The detector calculates the likelihood ($L(i)$) that an event has happened at sample i by using a detection window over the power signal ($S(i)$). The detection window splits into two sub-windows, the *pre-event* window $[a, i[$, and the *post-event* window $[i, b]$. L calculates as

$$L(i) = \ln\left(\frac{\sigma_{[a,i[}}{\sigma_{[i,b]}}\right) + \frac{\left(S(i) - \mu_{[a,i[}\right)^2}{2 \cdot \sigma_{[a,i[}^2} - \frac{\left(S(i) - \mu_{[i,b]}\right)^2}{2 \cdot \sigma_{[i,b]}^2},$$

where $\sigma_{[a,i[}$, $\sigma_{[i,b]}$, $\mu_{[a,i[}$ and $\mu_{[i,b]}$ are the standard deviations and means of the pre-event and post-event window, respectively. This signal is cleaned using an adaptive threshold ($thres_i$). If the change of the mean value between the pre- and post-event drops below the threshold, $L(i)$ is forced to zero using

$$L(i) = \begin{cases} L(i), & \text{if } |\mu_{[a,i[} - \mu_{[i,b]}| > thres_i \\ 0, & \text{otherwise} \end{cases}.$$

$thres_i$ is defined as

$$thres_i = thres_{min} + m \cdot \mu_{[a,i[},$$

with $thres_{min}$ being the minimum power change of interest and m a linear coefficient.

This coefficient causes a linear increase of $thres_i$ with the current power drawn (power of the pre-event window). Typically, the variance in the power signal is proportional to the amount of power drawn. This effect is caused by increasing noise in the appliance of the analog frontend of the electricity meter. If a fixed small threshold $thres_i$ is set, a large number of false events may occur at regions where more power is drawn. If a fixed high threshold is set, low power events may be missed. Pereira et al. used a relative large threshold of 30 W [28] which does not allow for detecting state changes of low power devices such as battery chargers or lights. We use the linearly increasing threshold as it adapts to possible larger fluctuations, preventing false events and missed events.

If an event is detected at sample i, the likelihood will also be non-zero around that sample as a mean change is still observable in close proximity to the event depending on pre-event and post-event window sizes. The exact sample at which the event occurred is identified using a *voting window*. This window is applied to the signal L. Inside the window only the maximum of the absolute value of L is kept. We further restrict the minimum distance between two events with an additional parameter l.

This algorithm has six adjustable parameters: the duration of pre-event, post-event, and voting window, the minimum detection threshold $thres_{min}$, the linear coefficient m, and the minimum distance between two events l. A user should specifically adjust the parameters $thres_{min}$ and l according to prior knowledge of the data: a low threshold $thres_{min}$ is required if events with small mean changes are expected, and a short l should be chosen if events can happen close in time. Values that seem to work quite well across different devices are: pre-event window = 1 s, post-event window = 1.5 s, voting window = 2 s, $thres_{min} = 3$ W, $m = 0.005$ and $l = 1$ s.

2.4.2. Unique Event Identification

To further simplify the labeling effort, we try to identify similar events of the appliance to label them accordingly. We therefore utilize the fact that most of our home appliances draw different but constant power before and after an event (e.g., the kettle after switched on) which represent constant states of the home appliance (e.g., *off* and *on* for the kettle). Depending on its complexity, an appliance can easily have more than ten unique states (e.g., a dishwasher).

The data is split at each event and the mean power demand between these splits is calculated. Unique mean values (representing unique appliance states) are then identified using hierarchical clustering with a distance threshold determined by $thres_{min}$. Each cluster is given a textual ID which is used to assign labels to each event (S0, S1, ..., as shown in Figure 11). As some appliances show a higher rush-in power followed by a power settling due to moving parts in the appliance, we remove the 10 % of the highest and lowest values before calculating the mean value.

2.4.3. High Variance Filtering

Appliances such as PCs or televisions draw variable power depending on the current context (i.e., calculations of the PC, content of a television). This causes a large number of false events using the LLR test. To filter these false events, we first identify regions in the signal that show such high variance and afterwards remove all events found in those regions. We therefore calculate the mean ($\mu(i)$) and variance ($\sigma(i)$) of a sliding window. If $\sigma(i)$ is larger than $n \cdot \mu(i)$, the window is marked. If the length of consecutively marked windows exceeds a certain length (w), all events in these windows are removed. The parameters n and w can be adjusted. Values which show good results were found empirically as $w = 4$ s and $n = 0.005$.

By using the event extraction algorithm, an appliance power signal can be pre-labeled. Each found event is marked and a unique label is assigned. The extensive task of labeling can therewith be reduced to supervision and inspection: Remaining falsely classified events (FP) need to be removed, events not found (FN) need to be added, and each unique state label should be changed to a meaningful label representing the state of the appliance.

2.5. Human Supervision

To combine the presented automatic event labeling (Section 2.4) with a simple graphic based human supervision, we have developed the Annoticity inspection and labeling tool. The tool is introduced by Völker et al. in [16]. Annoticity is implemented as an interactive web application. Its workflow is depicted in Figure 4. The tool allows for uploading your own data in the *Matroska* multimedia (MKV) [29] or *CSV* format. A user can further select data form several existing datasets such as REDD [9], ECO [18], BLOND [12], UK-DALE [11], or FIRED [17].

Annoticity is split into a server backend and client frontend. The backend loads the data and prepares it for visualisation. Data is down-sampled to a reasonable sampling rate according to the current time-span selected by the user. Furthermore, the automatic labeling algorithm presented in Section 2.4 can be performed, and file downloads (labels or data) are provided.

The graphical user interface of the client frontend is shown in Figure 5. After either uploading a file or selecting a device of an available dataset, the user can visually inspect the data. All available measures (e.g., active and reactive power) can be selected. Zooming into the data reveals more information as it leads to a data download at a higher sampling rate. The user can further execute the automatic labeling algorithm resulting in an initial set of labels. Each label consists of a start time and a textual description representing the event or the state after the event. The initial set can be adjusted by the user. Labels can be added, removed, or its text can be modified. If the user is only interested in the events' timestamps, all text can be removed. Furthermore, it is possible to modify all labels with the same text in one step. The frontend also allows for adjusting the parameters of the automatic labeling algorithm explained in Section 2.4. The final set of labels can be stored either as plain *CSV*, *ASS*, or *SRT* files or embedded into a *MKV* container with the original data.

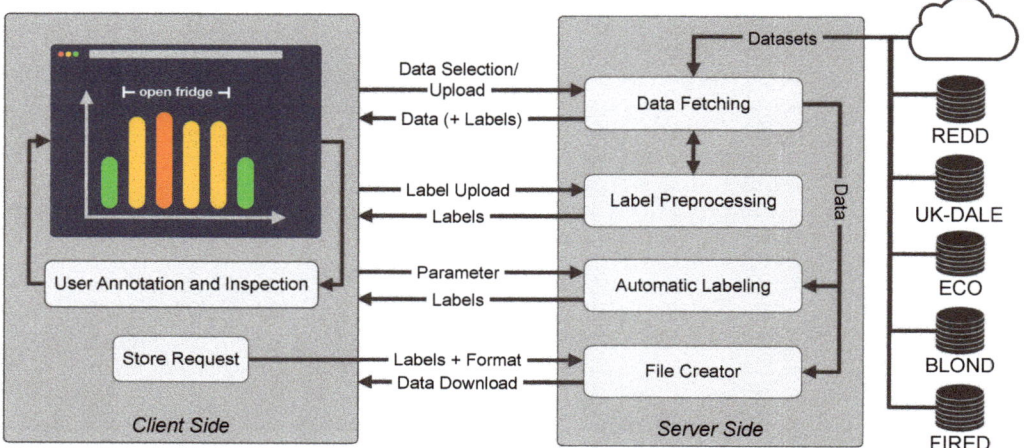

Figure 4. Flow of the Annoticity labeling tool. Data fetching, automatic labeling, and file creation are performed on the server side, while labeling and user interaction is performed on the client side (modified from [16]).

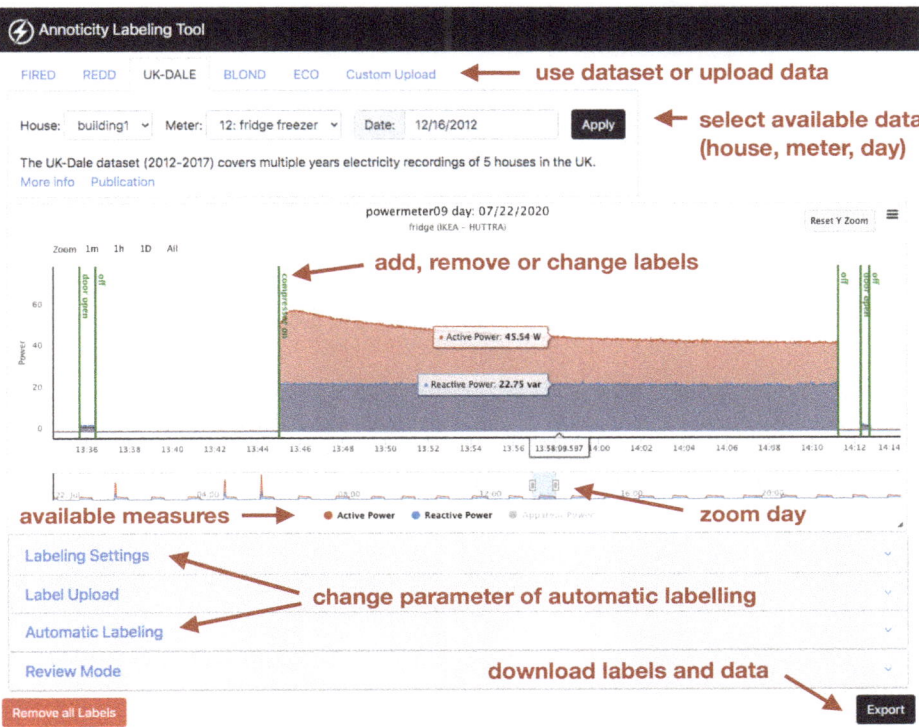

Figure 5. The graphical user interface of the Annoticity inspection and labeling tool. The fridge events were generated and clustered automatically. Each text description (*off*, *compressor on*, and *door open*) was only set once by the user. All other occurrences were labeled accordingly (modified from [16]).

3. Results

We utilized the framework explained in Section 2 to record and label the Fully-labeled hIgh-fRequency Electricity Disaggregation (FIRED) dataset which was first introduced by Völker et al. in [17]. Version 2 of the FIRED dataset extends its time period to 101 days. The data includes aggregated three phase current and voltage measurements sampled at 8 kHz as well as 21 time synced individual appliance measurements sampled at 2 kHz from a residential apartment in Germany. Furthermore, it includes sensor readings such as room temperatures and additional state information of certain appliances and each light bulb in the apartment. Annoticity has been used to fully label all state changes of the sub-metered appliances over a time period of two weeks.

The FIRED dataset was collected at an apartment building constructed in 2017 with seven apartments on four floors. Data was collected from a three-room apartment with 79 m^2 of space (open combined kitchen and living room, bedroom, child's room partly used as office, hallway, bathroom, and storage room). The apartment is inhabited by two adults and one child. The building is heated via a district heating and most rooms are equipped with air filters with built-in recuperators. According to the building's energy certificate, it requires a primary energy consumption of 12 kWh/(m^{2}a). The apartment's power grid is a three-phase 50 Hz system consisting of L_1, L_2, L_3, and neutral (N) wires. L_1–L_3 has a phase shift of 120°. Access to the apartment's electrical system is given through a fuse box located in the hallway. All lights installed in the apartment are off-the-shelf smart light bulbs with a built-in *ZigBee* module. This allows the lights to be turned on or off via a smartphone application, voice assistant, or regular wall-light-switch. It further allowed to log all state changes during the recording of the dataset as explained in Section 2.3. The washing machine, dryer, and freezer are located in the basement of the building and are not part of the recording.

A SmartMeter (see Section 2.1) was installed in the apartment's fuse box. Split core current transformers were attached to the three incoming supply legs. For voltage measurements, L_1, L_2, L_3, and N were connected in parallel. The meter is supplied with power by an additional L_1 leg that is secured by a separate 16 A fuse. The final installation is shown in Figure 6 (left).

We further deployed 21 PowerMeters (see Section 2.2) in the apartment and connected them to WiFi. We further checked that the WiFi signal quality (*RSSI*) of each PowerMeter exceeded −60 dBm to be certain that data could be sent flawlessly. Some devices like the oven and the exhaust hood were directly connected to the mains. To measure those appliances, we connected a special version of our PowerMeters with screw terminals. Figure 6 (right) shows two PowerMeters connected to the espresso machine and coffee grinder.

Modern households can easily include more than 40 appliances (68 in the FIRED household). Many of these devices are only plugged in occasionally and sometimes in a different socket than before. Therefore, connecting a continuously sensing meter to each appliance is infeasible. Instead, we connected devices of the same category (e.g., routers) or devices which are only used simultaneously (e.g., monitor and PC) to the same PowerMeter. Devices which are only plugged in occasionally and typically not at the same time (e.g., mixer or vacuum cleaner) were connected to a dedicated PowerMeter (*pm1*). If an appliance was connected or disconnected, a corresponding entry was manually added to a log file. This means that, even if the socket was continuously sampling data, the appliance connected to it changed.

Moreover, temperature and humidity sensors were installed in most of the rooms, a ZigBee logger was set up, and both a 433 MHz and an infrared bridge were installed (as described in Section 2.3).

To properly connect all individual sensors to a central recording PC, the apartment was equipped with four WiFi access points. The power consumption of the recording PC and access points were recorded individually and also contribute to the apartment's aggregated consumption. The recording PC gathered the data of all electricity meters and sensors, stored these into files frequently, and pushed the files to a cloud server for persistent storage.

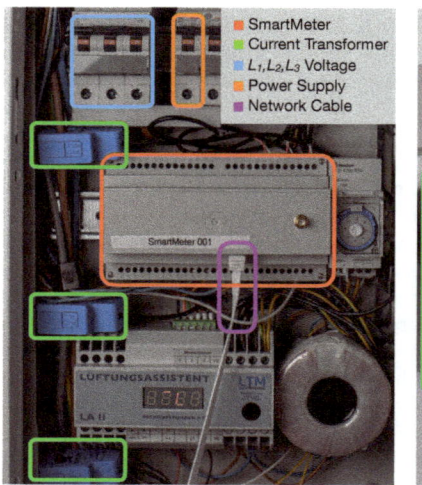 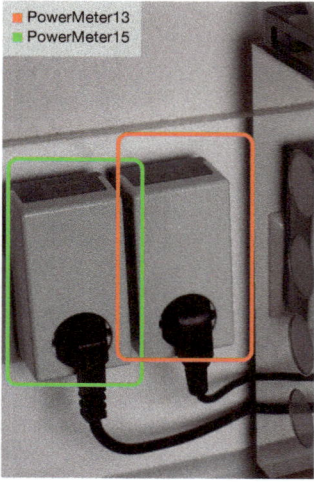

Figure 6. (**Left**) SmartMeter installed in the apartment's fuse box. (**Right**) PowerMeters with ID and 15 connected to the coffee grinder and the espresso machine.

3.1. Data Records

The provided data include voltage and current measurements at high sampling rates taken from the aggregated mains' signal and 21 individual sockets. Furthermore, FIRED contains per day and device summary files with derived power measurements to get a quick insight into the data. The root directory of the dataset contains folders with the *raw* and *summary* data. The data are stored as multiple *MKV* container into sub-folders named *powermeter<ID>* and *smartmeter001*, respectively. We used multimedia containers to store the data, as we previously explored their benefits in [30]. While being optimized for audio or video streams, these containers allow for storing regularly sampled sensor data as time synced audio streams. Text based labels can further be stored as subtitle streams in the same file. The FIRED data of each metering device are stored as a single *WavPack* [31] encoded audio stream inside the multimedia container. Each stream has multiple channels for the voltage and current signals. As stated in [30], WavPack allows for a lossless compression while maintaining high compression rates for time series data. In particular, we achieved a compression ratio of 1.46 for the voltage and current measurements using WavPack (only 1.42 could be achieved with *hdf5* [32], which has been used e.g., for UK-DALE [11] and BLOND [12]). We also store different Metadata for each of the streams including the start timestamp (with microseconds resolution), the particular meter used, the sampling frequency, codec information, the name of the measured attributes, and the stream's duration. Therewith, each file is self descriptive and can be used without prior knowledge. File size and, therewith, file loading times are kept reasonable by splitting all files at regular time intervals. The local time of the first sample is appended to each filename as *<year>_<month>_<day>__<hour>_<min>_<sec>*.

Table 2 shows the mapping of the recorded appliance to the used PowerMeter (ID). For more information about the appliance, its brand and model are shown with its power rating (P) according to the device manufacturer as well as the average $\overline{P}$ and maximum power P_{max} observed during recording. Φ shows the live wire (L_1, L_2 or L_3) to which the PowerMeter was connected. A complete list of all appliances in the apartment is part of the dataset. It contains additional information and website links for each appliance.

3.1.1. Voltage and Current Data

All PowerMeters sampled the current and voltage waveforms at a rate of 2 kHz. While in theory data can be sampled and sent at up to ≈8 kHz using our PowerMeters, the available WiFi bandwidth limits the amount of data that can be sent simultaneously by all meters. Therefore, we chose a sampling rate of 2 kHz as a trade-off between reliability and temporal data resolution (see Section 3.3.3 for more information). The SmartMeter recorded voltage and current waveforms of L_1, L_2, and L_3 with a sampling rate of 8 kHz. The ADC installed in the SmartMeter allows for sampling these waveforms with a rate of up to 32 kHz, but, again, we preferred a higher reliability over a better time resolution. This is in line with Armel et al. [7] who stated that sampling rates between 15 kHz to 40 kHz will not improve the performance of NILM algorithms as higher frequency signal components are distracted by noise in real buildings.

Each file contains 600 s of data stored into a single audio stream inside the multimedia container. For the aggregated data, each audio stream has six channels (*v_l1*, *i_l1*, *v_l2*, *i_l2*, *v_l3*, *i_l3*) representing the current and voltage waveforms for the three supply legs. The audio streams for the individual appliance data contain two channels (*v*,*i*). The number of samples in each file should match the time distance to the next file. If this is not the case, no data is available for this particular meter during this time period. This occurred during a reliability reset each day at midnight and rarely for single meters due to occasional data loss as depicted in Section 3.3.3.

Data pre-processing is typically not required, as both the SmartMeter and all PowerMeters calculate the physical quantities (*Volt* for voltage and *Milliampere* for current measurements) from the raw ADC samples.

Table 2. Appliances recorded via PowerMeters. *ID* represents the identifier of the PowerMeter used for recording. For *PowerMeter11*, the connected appliance changed during recording. P is the power according to the device manufacturer, Φ is the *Live Wire* the device is connected to (L_1, L_2 or L_3), P_m is the maximum average power drawn for the duration of one second and $\overline{P}$ is the per day average power seen during the recording. The unit of all power measurements is Watts.

ID	Connected Appliance	Brand	Model	P	Φ	P_{max}	$\overline{P}$
08	Baby Heat Lamp	Reer	FeelWell	600	2	611.93	0.30
09	Fridge	IKEA	HUTTRA	1000	3	1138.79	18.02
10	Smartphone Charger #1	-	2 Port USB	10	3	12.74	1.68
11	Changing Device				3	1898.71	3.10
12	Smartphone Charger #2	-	4 Port USB	25	1	27.86	2.77
13	Coffee Grinder	Graef	Cm800	128	3	206.89	0.17
14	Smart Speaker	Apple	HomePod	15	3	3.60	0.23
15	Espresso Machine	Rocket	Appartamento	1200	3	1230.62	29.82
16	Kettle	Aigostar	Adam 30GOM	2200	3	1958.76	2.89
17	Hairdryer	Remington	D3190	2200	1	1934.85	1.00
18	Router #1	Apple	Airport Extreme A1521	10.3	1	27.97	19.34
18	Router #2	Telekom	Speedport Smart 1	10	1	27.97	19.34
18	Telephone	Gigaset	A400	1	1	27.97	19.34
19	Printer	EPSON	Stylus SX435W	15	1	21.45	0.16
20	Office PC	Apple	Mac Mini A1993	85	2	236.08	59.49
20	27" Display	Apple	Thunderbolt display	200	2	236.08	59.49
20	Speaker	Logitech	Z2300	240	2	236.08	59.49
20	Smartphone Charger #3	Apple	MD813ZM/A	5	2	236.08	59.49
20	Access Point #2	Apple	Airport Express A1264	8	2	236.08	59.49
21	Media PC	Apple	Mac Mini A1347	85	3	45.38	12.98
22	HiFi System	Onkyo	TX-SR507	160	3	85.86	15.27
22	Subwoofer	Onkyo	SKW-501E	105	3	85.86	15.27
23	Television	Samsung	UE48JU6450	64	3	150.96	12.31
24	Light+Driver	IKEA	-	40	3	36.56	1.78
25	Oven	IKEA	MIRAKULÖS	3480	3	2491.03	9.69
26	Access Point #3	Apple	Airport Express A1392	2.2	3	2.67	2.21
27	Router #3	Netgear	R6250	30	1	56.91	15.74
27	Recording PC	Intel	NUC8v5PNK	60	1	56.91	15.74
28	Fume Extractor	IKEA	WINDIG	250	3	249.26	1.11

The provided voltage and current data without any pre-processing can be seen in Figure 7. Plots 1–3 show data of the SmartMeter while plots 4 and 5 show the simultaneous measurements of two additional PowerMeters. The figure does not only highlight the high temporal resolution of the data but also the achieved clock synchronization. The rush-in current shown in the PowerMeter data (Figure 7 plot 4) matches the rush-in current seen in L_3 of the SmartMeter (Figure 7 plot 3). A time shift between the measurement devices of around 10 ms can be observed. Even after 16 h of continuous recording, the offset between the SmartMeter and PowerMeters is below one mains cycle, highlighting the effectiveness of the realized clock synchronization (see Section 3.3.4).

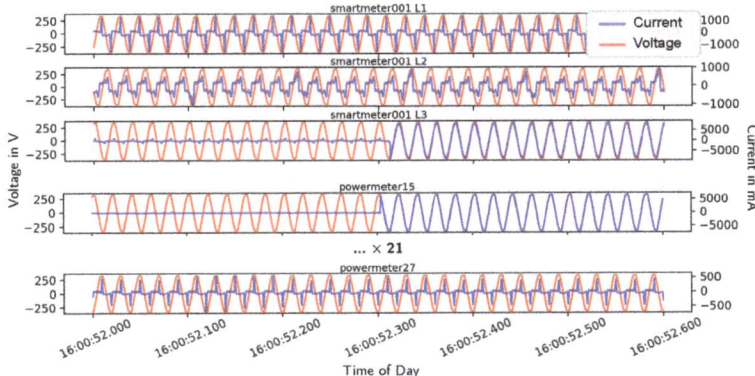

Figure 7. Voltage (red) and current (blue) waveforms of the SmartMeter, powermeter15 and powermeter27. The recording was taken on 9 June 2020 at around 4:00 p.m. The same appliance switch-on event of the espresso machine is visible in the recording of L3 of the smartmeter and of powermeter15.

3.1.2. Power Data

The power data was derived from the voltage (V) and current (I) waveforms. We calculated active, reactive, and apparent power from the raw voltage and current data of all recording devices. The data is stored as a single file for each day of the recording. The formulas used to calculate the individual powers based on the mains frequency $f_l = 50\,\text{Hz}$ are shown in (4), (5), and (6), respectively:

$$P(n) = \frac{1}{N} \cdot \sum_{i=0}^{N-1} V(i) \cdot I(i) \tag{4}$$

$$S(n) = I_{RMS}(n) \cdot V_{RMS}(n) \tag{5}$$

$$Q(n) = \sqrt{S(n)^2 - P(n)^2} \tag{6}$$

P, Q, and S are calculated for each non-overlapping window n. The length of the window is $N = \frac{f_s}{f_l}$. f_s is the sampling rate of the voltage and current measurements. For the 8 kHz, SmartMeter data N is $\frac{8000\,\text{Hz}}{50\,\text{Hz}} = 160$. I_{RMS} and V_{RMS} are calculated as follows:

$$I_{RMS}(n) = \sqrt{\frac{1}{N} \cdot \sum_{i=0}^{N-1} I(i)^2} \tag{7}$$

$$V_{RMS}(n) = \sqrt{\frac{1}{N} \cdot \sum_{i=0}^{N-1} V(i)^2} \tag{8}$$

Since data of commercial smart meters have a sampling frequency of 1 Hz to 0.01 Hz, an additional 1 Hz version of the power data is provided.

The 50 Hz and 1 Hz power are stored for each meter individually and contain one day of data. Times for which the power could not be calculated as no voltage and current data being available are marked with a power of constant zero to maintain an equidistant time period between samples.

Figure 8 shows the single day apparent power consumption of the apartment. The contribution of the six appliances which consumed the most power on this day is shown as individual colored blocks. All other appliances are summed and plotted as *Others*. The aggregated power consumption of the SmartMeter is shown as *mains*. Ideally, the superposition of the apparent power of all individual meters should match the aggregated apparent power. Nevertheless, a small margin can be observed in Figure 8. This gap is caused by hard-wired appliances such as lights and the ventilation system, which are not individually monitored (see Section 3.3.2 for more information).

3.1.3. Logs

The *annotation* folder contains 33 tab-separated *CSV* files. The first column of each f represents the timestamp associated with the event or sensor reading. We divide these fil into three categories:

- **Smart lights:**
State changes of each light in the apartment is logged. The filenames of these lo have the form:
$$light\_\_<room>\_\_<deviceName>\_\_<deviceModel>.csv$$
room is the room the light is installed in, *deviceName* represents how this light is use (e.g., ceiling light) and *deviceModel* matches the light model name. The file's secor column represents the state of the light (*on* or *off*), the third column is the ligh intensity (0 % to 100 %) while the last column represents the light's *hex* color. If settir different colors is not supported by the light, the column only shows *None* values. As individual measurements of the apartment's lighting have shown, the install smart lights consume constant apparent power linearly increasing with the ligh intensity (dimm setting). As information of the lights' state and dimm setting available for the complete recording duration, this information can be used to estima the power consumption of each light.
The smart light logs of two days are shown in Figure 9. The *hallway ceiling lig* consists of three light bulbs and is triggered by a passive infrared sensor. Hence, three light bulbs are turned on if a resident walks through the hallway, which can l seen in Figure 9 throughout the days.

- **Sensor readings:**
The readings of temperature and humidity sensors are stored in files following t name scheme:
$$sensor\_\_<room>\_\_<sensorType>.csv$$
sensorType is either *hum* or *temp* for humidity or temperature readings. Each file second column represents the sensor reading. Temperature readings are stored degrees Celsius and humidity readings in %, respectively. All values have floatir point precision. Samples are not acquired equidistantly, as the sensors only send ne values if they have changed.

- **Device info:**
Certain smart devices or bridges as explained in Section 2.3 allow for capturing even of devices in the apartment, e.g., pressing a certain key of the television remote. The events are logged in files with the following name format:
$$device\_\_<room>\_\_<deviceName>\_\_<deviceModel>.csv$$
Each file's second column gives information about the current device state or ha pened event. The file of the HiFi system, for instance, includes key-presses such *power* or *vol_up*. Figure 10 shows the logs for the television, the HiFi system, and t espresso machine.

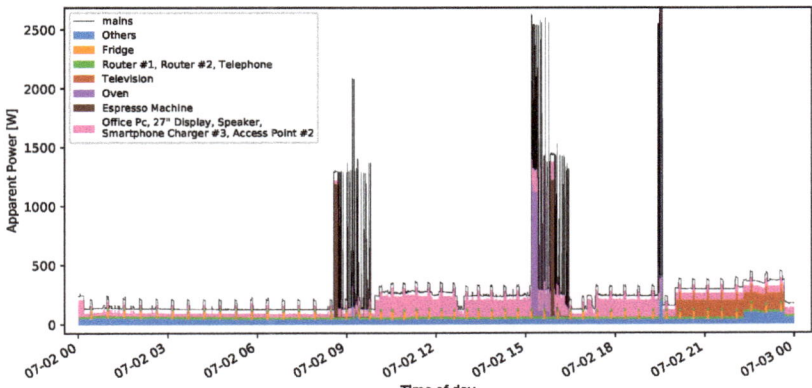

Figure 8. The power consumption of the apartment over one full day (2nd July 2020). The power is down-sampled to one sample every 3 s. The black line indicates the power consumption recorded by the SmartMeter. The contribution of the six top-most consumers is shown as stacked colored blocks. The consumption of the remaining individually metered appliances are aggregated and shown as the blue block *Others*. A slight offset between the SmartMeter and the accumulated power of all PowerMeters can be seen.

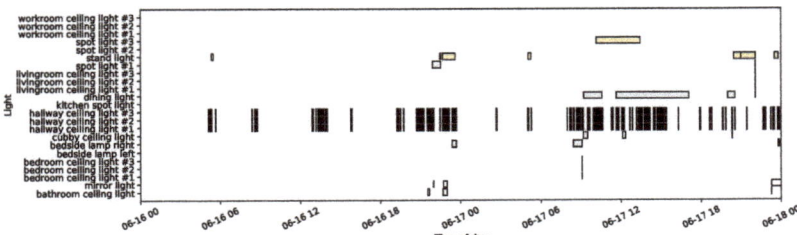

Figure 9. Two days of light usage information (16th–17th June 2020). The time of day is shown on the *x*-axis while the particular light is shown on the *y*-axis. If a light was used, a black box is shown during this period. The box is filled with the set color value and dimm setting.

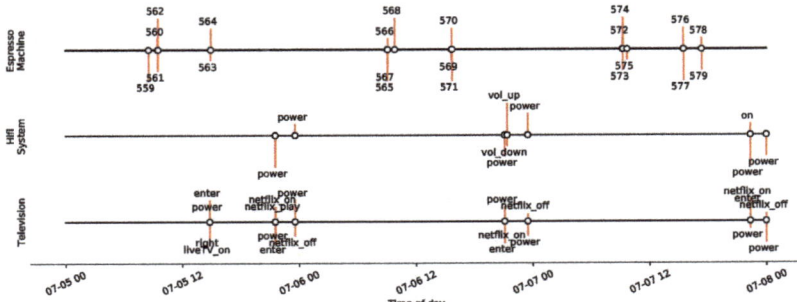

Figure 10. Three days of device logs (19th–22nd June 2020). The data of the espresso machine shows the numbers of espressos made, while the data of the HiFi system and television show key-presses on the remote. The television content is provided by a connected media PC.

3.1.4. Data Labeling

We used the Annoticity labeling tool to fully label all events that happened within two weeks of the FIRED data (22nd July 2020–4th August 2020). The automatic labeling tool was used to generate an initial set of labels. This set was modified by visually inspecting the data. We removed false events, added missing events, and assigned a distinct and descriptive label to each appliance state. The labels were stored as *CSV* files and are part of the dataset. During labeling, we also stored the initial set of labels obtained by the automatic labeling algorithm to evaluate the algorithm's performance. Figure 11 shows both the initial set of labels and the final labeled data of the 'espresso machine'.

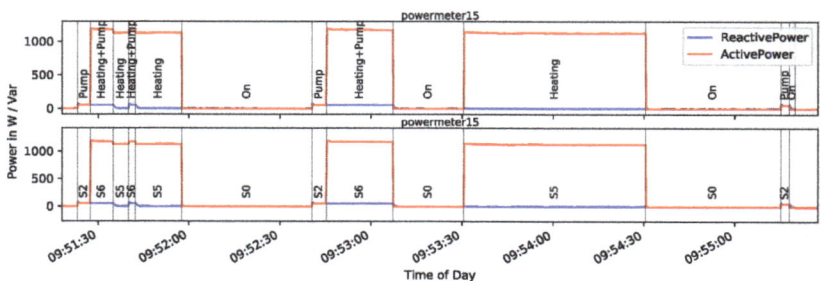

Figure 11. The fully labeled data of the espresso machine (30th of July at 9:50 a.m.). The bottom plot shows the initial labeling of the automatic labeling algorithm, while the top plot shows the final labeling after human supervision. (The rightmost event has been missed by the algorithm.)

To evaluate the performance of the labeling tool (see Section 2.4), we compared its result with the final set of manually labeled events. Evaluation was done in terms of True Positives (TP) i.e., true events, False Positives (FP), i.e., falsely classified events and False Negatives (FN), i.e., events not found by the tool. A TP is defined as a classified event which is reflected within 2 s in the manually labeled data. According to that, a FN is an event in the manually labeled data not found within 2 s during detection. An FP is a classified event which is not present in the manually labeled data. The F_1 score was used to summarize these metrics into a single score. We fixed the algorithm's parameters for all appliances to simplify the evaluation: pre-event window length = 1 s, post-event window length = 1.5 s, voting window length = 2 s, $thres_{min} = 3$ W, $m = 0.005$, and $l = 2$ s.

We experienced that the labeling algorithm is performing fairly well for devices which show distinct states in the power signal (such as the oven, kettle, or the espresso machine shown in Figure 11). For devices that draw variable power in between states (such as a PC or a coffee grinder), a large number of false events were triggered.

To put this into perspective, Section 3.1.4 shows the evaluation results split into two groups: #1 represents appliances that show distinct states and #2 represents appliances which draw variable power. Appliances for which no distinct events were labeled manually (e.g., network equipment) are omitted.

Section 3.1.4 indicates that most of the appliances present in residential homes (group #1) can be labeled in a semi-automatic way. The *Coffee Grinder* and the *HiFi System* show a comparable low performance with a high number of FP. This is due to higher variance if the motor is active or music is playing and could have been avoided by using a higher linear factor m or a higher threshold $thres_{min}$.

To get an overall estimate of how much the labeling effort can be reduced by using the automatic labeling, we compared the raw number of clicks required to label the data from scratch with the number of clicks required to supervise and modify the pre-labeled set generated by the automatic labeling algorithm for group #1. In total, 4379 events were labeled manually. If we omit the task of applying textual labels, labeling events would still have required at least 4379 clicks. As shown in Table 3, for devices in group #1, the labeling algorithm automatically placed 3232 events at the correct position. With 159 false classified events, 377 missing events, and the 770 missing labels of group #2 (which would require manual labeling), 1306 clicks were required to remove false events and add missing events. Therewith, the sheer amount of clicks was already reduced by 70.18% not accounting for the support provided by the Annoticity tool while applying textual labels:

$$Reduction = 1 - \frac{t_{add} \cdot MissedEvents + t_{del} \cdot FalseEvents}{t_{Add} \cdot AllEvents} \quad (9)$$

If we also accommodate the fact that it typically takes less time to remove a falsely placed label (t_{del}) compared to manually adding a label from scratch (t_{add}), we can apply Equation (9). Using $t_{add} = 10$ s and $t_{del} = 5$ s as a reasonable guess of this difference, the reduction in labeling effort is actually 71.99% compared to a fully manual approach. Considering that the parameters could have been manually adjusted and optimized for each appliance, the actual reduction might be even higher.

Table 3. Results for the automatic labeling algorithm split into two appliance groups. In #1, appliances are grouped which show distinct states in the power signal in which nearly constant power is drawn. #2 groups appliances that draw variable power. *Events* marks the number of ground truth events labeled manually.

Group	Appliance	Events	TP	FP	FN	F1
#1	Baby Heat Lamp	6	6	0	0	100.00
	Fridge	1006	863	2	143	92.25
	Coffee Grinder	348	250	114	98	70.22
	Espresso Machine	1880	1760	0	120	96.70
	Kettle	30	30	0	0	100.00
	Hairdryer	18	17	0	1	97.14
	Hifi System, Subwoofer	45	44	37	1	69.84
	Television	79	65	4	14	87.84
	Kitchen Spot Light	12	12	0	0	100.00
	Oven	138	138	1	0	99.64
	Fume Extractor	47	47	1	0	98.95
	Sum	3609	3232	159	377	92.34
#2	Smartphone Charger #1	96	83	1491	13	9.94
	Smartphone Charger #2	63	45	7999	18	1.11
	Office Pc	583	410	85,367	173	0.95
	Media Pc	28	10	26,632	18	0.07
	Sum	770	548	121,489	222	0.89

3.2. Data Statistics

Overall, we collected 53,328 h of raw current and voltage waveforms using the proposed framework. Figure 7 highlights the richness of the captured data for the SmartMeter and the PowerMeter units. Figure 8 shows the apparent power extracted for each individual appliance. It further emphasizes the contribution of each appliance to the total power consumption on this day.

According to [33], the average consumption of a comparable three person household in Germany is 7.12 kW h per day. Evaluating the SmartMeter data of the FIRED dataset results in an average electricity consumption of 6.06 kW h per day, which is slightly lower compared to the average, but this is expected as the data does not contain the electricity consumed by the washing machine, dryer, and freezer.

Figure 12a shows the consumption of six appliances at the hour of day averaged over the whole recording duration. This delivers a good indication for usage patterns. For example, the 'espresso machine' shows two distinct peaks, one in the morning around 9:00 a.m. (morning coffee) and one in the afternoon at 3:00 p.m. (coffee break). In comparison, the 'router' does not show any significant peak. It can also be seen that the 'office PC' has a high standby consumption of around 35 W and is used mainly between 9:00 a.m. and 6:00 p.m. Figure 12b shows the distribution of power demand for the same appliances. Some state information can already be derived from these plots. The 'hairdryer' shows two distinct states representing two different temperature settings. The 'office PC' shows three peaks. The peak around 35 W represents the already mentioned standby consumption, the peak around 50 W represents the PC in its *On* state, and the 140 W peak includes the *On* state of the 27-inch monitor connected to the same meter. The 'espresso machine' consumes a huge amount of power (1200 W) during its heating cycles but mostly idle (5 W) in between.

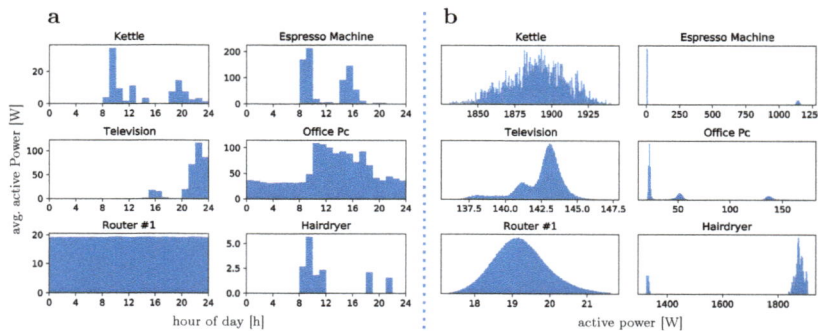

Figure 12. Appliance usage over the complete recording duration. (**a**) shows the daily usage pattern of the devices as the average power for each hour-of-day; (**b**) shows the histogram of the power demands. A 2 W threshold was set to omit data in which no power is drawn.

3.3. Technical Validation

Measurements of the FIRED dataset are provided without applying any pre-processing or filtering rather than the conversion to physical quantities. The recording framework is equipped with different mechanisms to cope with real world effects such as network dropouts or clock drifts, and the data integrity was analyzed in detail.

3.3.1. Calibration

We calibrated each meter in advance using an *ENERGY-LOGGER 4000* [34] (with stated accuracy of 1%) as a reference. We used ten loads with different power consumption ranging from 5 W to 2000 W and applied a linear calibration. The calibration parameters of each meter were stored permanently in their non-volatile memory. We repeated the calibration after the recording duration to see if aging affects have already occurred. Such effects could not be identified.

3.3.2. Residual Power

The sum of all PowerMeters' apparent power matches the apparent power recorded by the SmartMeter with a slight offset (residual power). The residual power is the portion of the total consumed power which is not metered by an individual meter, i.e., the portion for which no ground truth data is available. Our goal was to minimize this portion in order to provide reliable ground truth data that can be used by supervised machine learning algorithms.

The residual power observed in the FIRED dataset (see Figure 8) is mainly due to non-monitored, hard-wired devices in the apartment such as the lighting and the ventilation system but also due to the power consumption of the distributed meters. The individual consumption of each light-bulb can be estimated with the log files which we provide with our dataset. To show that this is feasible, we generated apparent power estimates using these log files and the provided individual light recordings. The consumption of the remaining unmonitored appliances (including the consumption of 21 PowerMeters) is the base power consumption of the apartment. It can be estimated at times when lights are turned off and the majority of appliances do not consume any power which is typically during the night or in the case of owner absence. We calculated the base power P_{baseLx} for each individual supply leg $x \in [1, 2, 3]$ as

$$PM_{Lx} := \{ pm \in PM \mid \text{phase of } pm \text{ is } x \}, \tag{10}$$

$$L_{Lx} := \{ l \in Lights \mid \text{phase of } l \text{ is } x \}, \tag{11}$$

$$\mathbf{P}_{baseLx} = \mathbf{P}(SM_{Lx}) - \sum_{pm \in PM_{Lx}} \mathbf{P}(pm) - \sum_{l \in L_{Lx}} \mathbf{P}(l). \tag{12}$$

SM_{Lx} is the SmartMeter data of live wire Lx, PM is the set of all PowerMeters, PM_{Lx} is the set of PowerMeters that are connected to live wire Lx, $Lights$ is the set of all lights, L_{Lx} is the set of lights connected to Lx and $\mathbf{P}(X)$ is the extracted power trace of a meter or light X. We assume that the base power is normally distributed and therewith remove all points in $\mathbf{P}_{baseLx}$ that are further away than σ from the mean value and calculate $\mathbf{P}_{baseLx}$ as the mean from the cleaned signal.

Figure 13 shows the apparent power consumption including the lighting and the estimated base power with a remaining Root-Mean-Square Error (RMSE) of 17 V A.

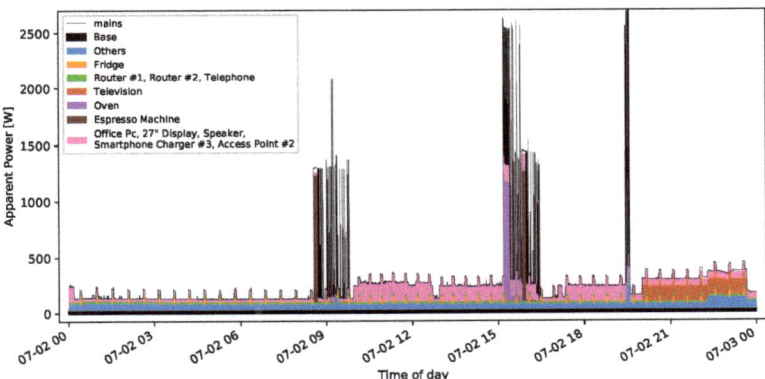

Figure 13. The power consumption of the apartment over one full day (2nd July 2020). The power is down-sampled to one sample every 3 s. The black line indicates the power consumption recorded by the SmartMeter. The contribution of the six top-most consumers is shown as stacked colored blocks. The consumption of all remaining appliances and the reconstructed consumption of the apartment's lighting are aggregated and shown as the blue block *others*. The black base block represents the apartment's base power estimated as 26.66 V A on average for this day.

3.3.3. Availability

The data availability over the complete recording duration was 99.96 %. The missing data amounts to 1405 min and is mainly due to a reliability reset which we perform each day at 12:00 a.m. accounting for approximately 20 s of missing data per meter and day. Figure 14 shows the availability of each metering device over the complete recording period. The reset each day at midnight can be clearly seen in the plot. Occasionally, due to WiFi connection outages and an erroneous implementation of the TCP/IP stack on the ESP micro-controller, some data packets can be lost. However, a packet only accounts for less than 20 ms of data. Once detected, we replace missing samples with zeros to maintain the correct timestamps for all remaining samples. It is still possible to identify these time periods as voltage, and current zero plateaus cannot occur in other situations. *PowerMeter14* and *PowerMeter22* show this behavior more frequently compared to all other meters. The reason might be an unstable WiFi condition as the RSSI values reported by both meters were the lowest of all.

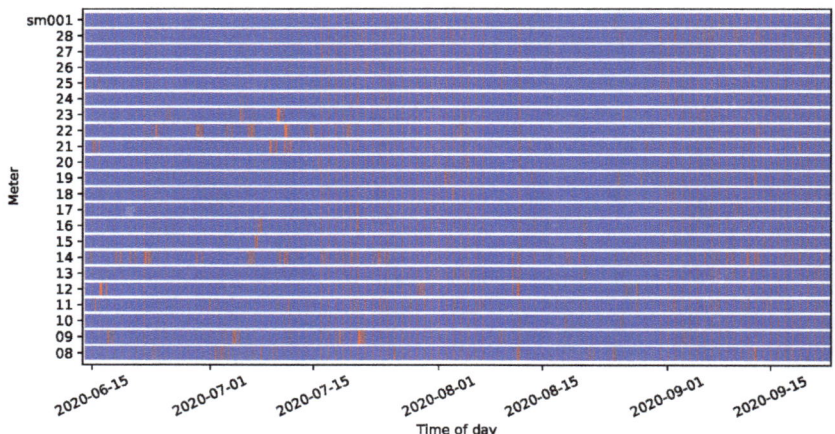

Figure 14. Time periods when measurement data are available. Each individual meter is shown on the y-axis (ID 08-28) while the time period of the complete recording is shown on the x-axis. Red areas indicate gaps in the dataset. Please note that gaps are plotted with at least one pixel width for better visibility and therewith appear significantly larger.

3.3.4. Clock Synchronization

In Section 2, we defined one challenge as "simultaneous recordings with high temporal accuracy". Our framework is comprised of numerous meters and sensors distributed across the home. To maintain precise timestamps for the measured data and the event labels precise clock synchronization techniques are required.

We also observed non-negligible clock drifts in initial experiments originating from clock inaccuracies of the used ADC and microcontroller which vary depending on the meter and also over time for the same meter due to temperature and aging effects. To remove these drifts, we equipped each metering device (aggregated meter and distributed meters) with an RTC to synchronize the internal ADC clock. A NTP server in the recording network is utilized to sync each microcontroller's system time every 120 s. We calculate the system time as the NTP response plus half the time it takes for the NTP request to be answered by the server. Therewith, we can obtain the time from a NTP request which took e.g., 10 ms with an accuracy of ±5 ms. Hence, the accuracy of each device's system time depends on the network latency, which is typically below 10 ms. As we can measure the NTP response time, which is a good indication for the network latency, we only utilize NTP requests with response times better than ±10 ms to sync the system time. If we detect a time drift after an NTP synchronization, it is slowly reduced by removing or adding samples. The used technique adds only a minor jitter to the data ($1/samplingrate$ seconds or 125 µs for 8 kHz data).

Figure 7 highlights the achieved clock synchronization. The figure shows the voltage and current traces of PowerMeter15 and the SmartMeter. PowerMeter15 shows a rapid rise in the current consumption due to a heating element in the connected espresso machine. The corresponding increase can be also observed in the data of L_3 of the SmartMeter. Both signals are shifted by around 10 ms highlighting the achieved clock synchronization. Such a time shift allows for syncing the voltage and current waveforms with sub-cycle precision

3.4. Data and Code Availability

The FIRED dataset is available under the creative common license from our servers. The code which has been used to generate all plots in this work is provided as open source. The code and further information on how to download the data can be found at: https://github.com/voelkerb/FIRED_dataset_helper.

4. Discussion

In this work, we proposed a set of challenges which need to be addressed to record datasets which can be used to evaluate a wide variety of electricity related algorithms (especially event-based NILM). These challenges are summarized in Table 1. We further proposed a framework to record and label datasets which meet the defined challenges. It is comprised of the required hardware and software components to record the data, an algorithm to automatically find and label events in the recorded data and a tool to visually inspect the data and adjust the labels.

Using the framework, we recorded and labeled the FIRED dataset which features 101 days of electricity measurements (C3) of a residential apartment in Germany. This is significantly longer than most existing high frequency datasets such as REDD or BLUED. Aggregated level data are available as 8 kHz voltage and current waveforms while individual appliance data are available at 2 kHz for 21 appliances (C1, C2). While the aggregated sampling rate is matched or even exceeded by other datasets, we are currently unaware of any other residential dataset which features high frequency individual appliance recordings. The data is further time synced with an accuracy of around 10 ms (C5) and shows a coverage of 99.96 % over the complete recording time period (C3). Other datasets such as REDD or UK-DALE show a significant amount of missing samples due to bad wireless communication. Our framework also provides a 1 Hz and 50 Hz summary with derived active, reactive, and apparent power measurements. All data is stored in Matroska multimedia containers (C6) with included metadata information such as timestamps and measurands. Additional *CSV* files are included in the dataset which provide information about the apartments lighting states, room temperature, and device operation states (C4). Event positions and state labels have been added for two weeks of the data in a semi-automatic way using the presented Annoticity labeling tool (C4). No other dataset known to us includes such information. The dataset itself and the tools to process it are provided as open source (C6).

5. Conclusions

In summary, this work offers three main contributions to the community of smart meter data analytics:

1. We defined a set of challenges that an electricity dataset needs to address so that it can be used to evaluate a large set of electricity and smart meter related algorithms such as event-based and event-less Non-Intrusive Load Monitoring.
2. We proposed an expandable framework comprised of the hardware and software components required to record datasets that meet these challenges.
3. We introduced and evaluated a novel dataset to the community, which, compared to other residential electricity datasets such as BLUED, REDD or UK-DALE, features simultaneous high frequency recordings of the aggregated mains' signal and of individual household appliances as well as two weeks of fully labeled appliance events.

The high sampling rates, the achieved clock synchronization, and the marked events allow for using datasets that have been recorded using our framework (like FIRED) to evaluate event-based and event-less NILM algorithms. Additional data like detailed textual labels and additional sensor readings allow for developing disaggregation algorithms that utilize multi-modal information. Offering simultaneous, high frequency aggregated and individual appliance recordings will allow researchers to develop hybrid load monitoring systems which use individual appliance recordings in a semi-supervised fashion to aid the laborious training process of supervised NILM systems.

Besides the mentioned advantages, the framework is currently optimized for use in the residential domain as, e.g., plug level meters and WiFi are being used. While it should be possible to move the overall concepts to the commercial or industrial domain, such specialized environments may require additional adaption. Besides temperature and humidity readings, other environmental information such as occupancy or light sensors may be of interest and are theoretically supported by the framework but have not been installed while generating the FIRED dataset. Further increasing the sampling rate of the meters or the sheer number of meters is theoretically possible; however, in our findings it reduced the reliability of our framework mainly due to bandwidth problems. One suggestion to overcome such a limitation in the future is to compress the data before it is sent over the bandwidth limited WiFi channel.

We provide electricity datasets and the software and hardware to record these so that researchers can set their focus on improving load monitoring and eco-feedback techniques. These have shown tremendous potential in saving our earth's energy resources.

Author Contributions: Conceptualization, B.V.; Writing—review and editing, M.P.; Writing—review and editing, P.M.S.; Supervision, B.B. All authors have read and agreed to the published version of the manuscript.

Funding: This research received no external funding.

Institutional Review Board Statement: Not applicable.

Informed Consent Statement: Not applicable.

Data Availability Statement: Data available in a publicly accessible repository that does not issue DOIs.

Conflicts of Interest: The authors declare no conflict of interest.

Abbreviations

The following abbreviations are used in this manuscript:

ILM	Intrusive Load Monitoring
NILM	Non-Intrusive Load Monitoring
RTC	Real Time Clock
FIRED	Fully-labeled hIgh-fRequency Electricity Disaggregation
CSV	Comma-Separated Values
SRT	SubRip Text
RMSE	Root-Mean-Square Error
TP	True Positives
FP	False Positives
FN	False Negatives
NTP	Network Time Protocol

References

1. Nations, U. About the Sustainable Development Goals—United Nations Sustainable Development. Available online: https://www.un.org/sustainabledevelopment/sustainable-development-goals (accessed on 29 August 2020).
2. Ehrhardt-Martinez, K.; Donnelly, K.A.; Laitner, S. *Advanced Metering Initiatives and Residential Feedback Programs: A Meta-Review for Household Electricity-Saving Opportunities*; American Council for an Energy-Efficient Economy: Washington, DC, USA, 2010.
3. Bundesministerium für Wirtschaft und Energie (BMWi). Gesetz zur Digitalisierung der Energiewende. 2016. Available online: https://www.bmwi.de/Redaktion/DE/Downloads/Gesetz/gesetz-zur-digitalisierung-der-energiewende.png (accessed on 29 August 2020).
4. Hart, G.W. Nonintrusive appliance load monitoring. *Proc. IEEE* **1992**, *80*, 1870–1891. [CrossRef]
5. Norford, L.K.; Leeb, S.B. Non-intrusive electrical load monitoring in commercial buildings based on steady-state and transient load-detection algorithms. *Energy Build.* **1996**, *24*, 51–64. [CrossRef]

Bundesministerium für Wirtschaft und Energie (BMWi). Information on the Funding Program Entitled 'Smart Energy Showcases—Digital Agenda for the Energy Transition' (SINTEG). 2016. Available online: https://www.bmwi.de/Redaktion/EN/Downloads/bmwi-papier-sinteg-kernbotschaften.png?__blob=publicationFile&v=3 (accessed on 29 August 2020).

Armel, K.C.; Gupta, A.; Shrimali, G.; Albert, A. Is disaggregation the holy grail of energy efficiency? The case of electricity. *Energy Policy* **2013**, *52*, 213–234. [CrossRef]

Hsu, C.Y.; Zeitoun, A.; Lee, G.H.; Katabi, D.; Jaakkola, T. Self-supervised learning of appliance usage. In Proceedings of the 8th International Conference on Learning Representations (ICLR), Addis Ababa, Ethiopia, 26–30 April 2020.

Kolter, J.Z.; Johnson, M.J. REDD: A public data set for energy disaggregation research. In Proceedings of the Workshop on Data Mining Applications in Sustainability (SIGKDD), San Diego, CA, USA, 21–24 August 2011; Volume 25, pp. 59–62.

Batra, N.; Kelly, J.; Parson, O.; Dutta, H.; Knottenbelt, W.; Rogers, A.; Singh, A.; Srivastava, M. NILMTK: An open source toolkit for non-intrusive load monitoring. In Proceedings of the 5th International Conference on Future Energy Systems, Cambridge, UK, 11–13 June 2014; pp. 265–276.

Kelly, J.; Knottenbelt, W. The UK-DALE dataset, domestic appliance-level electricity demand and whole-house demand from five UK homes. *Sci. Data* **2015**, *2*, 150007. [CrossRef]

Kriechbaumer, T.; Jacobsen, H.A. BLOND, a building-level office environment dataset of typical electrical appliances. *Sci. Data* **2018**, *5*, 180048. [CrossRef] [PubMed]

Anderson, K.; Ocneanu, A.; Benitez, D.; Carlson, D.; Rowe, A.; Berges, M. BLUED: A fully labeled public dataset for event-based non-intrusive load monitoring research. In Proceedings of the 2nd KDD Workshop on Data Mining Applications in Sustainability (SustKDD), Beijing, China, 8 December 2012; Volume 7.

Belley, C.; Gaboury, S.; Bouchard, B.; Bouzouane, A. An efficient and inexpensive method for activity recognition within a smart home based on load signatures of appliances. *Pervasive Mob. Comput.* **2014**, *12*, 58–78. [CrossRef]

Alcalá, J.M.; Ureña, J.; Hernández, Á.; Gualda, D. Assessing human activity in elderly people using non-intrusive load monitoring. *Sensors* **2017**, *17*, 351. [CrossRef] [PubMed]

Völker, B.; Pfeifer, M.; Scholl, P.M.; Becker, B. Annoticity: A Smart Annotation Tool and Data Browser for Electricity Datasets. In Proceedings of the 5th International Workshop on Non-Intrusive Load Monitoring, Yokohama, Japan, 18 November 2020; pp. 1–5.

Völker, B.; Pfeifer, M.; Scholl, P.M.; Becker, B. FIRED: A Fully-labeled hIgh-fRequency Electricity Disaggregation Dataset. In Proceedings of the 7th ACM International Conference on Systems for Energy-Efficient Buildings, Cities, and Transportation, Yokohama, Japan, 19–20 November 2020; pp. 294–297.

Beckel, C.; Kleiminger, W.; Cicchetti, R.; Staake, T.; Santini, S. The ECO data set and the performance of non-intrusive load monitoring algorithms. In Proceedings of the 1st ACM Conference on Embedded Systems for Energy-Efficient Buildings, Memphis, TN, USA, 4–6 November 2014; pp. 80–89.

Makonin, S.; Popowich, F.; Bartram, L.; Gill, B.; Bajić, I.V. AMPds: A public dataset for load disaggregation and eco-feedback research. In Proceedings of the 2013 IEEE Electrical Power & Energy Conference, Halifax, NS, Canada, 21–23 August 2013; pp. 1–6.

Makonin, S.; Wang, Z.J.; Tumpach, C. RAE: The rainforest automation energy dataset for smart grid meter data analysis. *Data* **2018**, *3*, 8. [CrossRef]

Pereira, L. Developing and evaluating a probabilistic event detector for non-intrusive load monitoring. In Proceedings of the 2017 Sustainable Internet and ICT for Sustainability (SustainIT), Funchal, Portugal, 6–7 December 2017; pp. 1–10.

Völker, B.; Scholls, P.M.; Schubert, T.; Becker, B. Towards the Fusion of Intrusive and Non-intrusive Load Monitoring: A Hybrid Approach. In Proceedings of the Ninth International Conference on Future Energy Systems, Karlsruhe, Germany, 12–15 June 2018; pp. 436–438.

Völker, B.; Scholl, P.M.; Becker, B. Semi-Automatic Generation and Labeling of Training Data for Non-Intrusive Load Monitoring. In Proceedings of the Tenth International Conference on Future Energy Systems, Phoenix, AZ, USA, 25–28 June 2019.

Völker, B.; Pfeifer, M.; Scholl, P.M.; Becker, B. A Versatile High Frequency Electricity Monitoring Framework for Our Future Connected Home. In Proceedings of the International Conference on Sustainable Energy for Smart Cities, Braga, Portugal, 4–6 December 2019; pp. 221–231.

Analog Devices. *ADE9000—High Performance, Multiphase Energy, and Power Quality Monitoring IC*; ; Analog Devices: Norwood, MA, USA, 2017; Rev. A.

STMicroelectronics. *STPM32, STPM33, STPM3—ASSP for Metering Applications with up to Four Independent 24-bit 2nd Order Sigma-Delta ADCs, 4 MHz OSF and 2 Embedded PGLNA*; STMicroelectronics: Geneva, Switzerland, 2016; Rev. 5.

OASIS. The MQTT 5.0 standard—A Machine-to-Machine (M2M) "Internet of Things" Connectivity Protocol. 2020. Available online: https://www.mqtt.org/ (accessed on 29 August 2020).

Pereira, L.; Nunes, N.J. Semi-automatic labeling for public non-intrusive load monitoring datasets. In Proceedings of the Sustainable Internet and ICT for Sustainability (SustainIT), Madrid, Spain, 14–15 April 2015; pp. 1–4.

Matroska, N.P.O. The Matroska File Format. 2020. Available online: https://www.matroska.org/ (accessed on 29 August 2020).

Scholl, P.M.; Völker, B.; Becker, B.; Van Laerhoven, K. A multi-media exchange format for time-series dataset curation. In *Human Activity Sensing*; Springer: Berlin, Germany, 2019; pp. 111–119.

WavPack. Hybrid Lossless Audio Compression. 2020. Available online: http://www.wavpack.com (accessed on 29 August 2020).

32. Group, T.H. THE HDF5 LIBRARY & FILE FORMAT. 2020. Available online: https://www.hdfgroup.org/solutions/hdf (accessed on 11 December 2020).
33. co2online. Der Stromspiegel für Deutschland 2019. 2020. Available online: https://www.stromspiegel.de/stromverbrauch-verstehen/stromverbrauch-3-personen-haushalt/ (accessed on 29 August 2020).
34. Ag, C.E. Voltcraft Energy Logger 4000. 2020. Available online: http://www.voltcraft.ch/index.html (accessed on 29 August 202

Article

A Dataset for Non-Intrusive Load Monitoring: Design and Implementation [†]

Douglas Paulo Bertrand Renaux [1,*], Fabiana Pottker [1], Hellen Cristina Ancelmo [1], André Eugenio Lazzaretti [1], Carlos Raiumundo Erig Lima [1], Robson Ribeiro Linhares [1], Elder Oroski [1], Lucas da Silva Nolasco [1], Lucas Tokarski Lima [1], Bruna Machado Mulinari [1], José Reinaldo Lopes da Silva [1], Júlio Shigeaki Omori [2] and Rodrigo Braun dos Santos [2]

[1] LIT-Laboratory of Innovation and Technology in Embedded Systems and Energy, Universidade Tecnológica Federal do Paraná-UTFPR, Sete de Setembro, 3165, Curitiba 80230-901, Brazil; fpottker@utfpr.edu.br (F.P.); hellen@alunos.utfpr.edu.br (H.C.A.); lazzaretti@utfpr.edu.br (A.E.L.); erig@utfpr.edu.br (C.R.E.L.); linhares@utfpr.edu.br (R.R.L.); oroski@utfpr.edu.br (E.O.); lucasnolasco.5@gmail.com (L.d.S.N.); tokarski.lima@gmail.com (L.T.L.); brunamachadomulinari@gmail.com (B.M.M.); joses@alunos.utfpr.edu.br (J.R.L.d.S.)

[2] COPEL-Companhia Paranaense de Energia, José Izidoro Biazetto, 158, Curitiba 82305-100, Brazil; julio.omori@copel.com (J.S.O.); rodrigo.braun@copel.com (R.B.d.S.)

* Correspondence: douglasrenaux@utfpr.edu.br

[†] This paper is an extended and improved version of our paper published at the VIII Brazilian Symposium on Computing Systems Engineering (SBESC), Salvador, Brazil, 6–9 November 2018; pp. 243–249; 20th International Conference on Intelligent System Application to Power Systems (ISAP), New Delhi, India, 10–14 December 2019; pp. 1–7; 2019 IX Brazilian Symposium on Computing Systems Engineering (SBESC), Natal, Brazil, 19–22 November 2019; pp.1–8.

Received: 21 August 2020; Accepted: 1 October 2020; Published: 15 October 2020

Abstract: A NILM dataset is a valuable tool in the development of Non-Intrusive Load Monitoring techniques, as it provides a means of evaluation of novel techniques and algorithms, as well as for benchmarking. The figure of merit of a NILM dataset includes characteristics such as the sampling frequency of the voltage, current, or power, the availability of indications (ground-truth) of load events during recording, the variety and representativeness of the loads, and the variety of situations these loads are subject to. Considering such aspects, the proposed LIT-Dataset was designed, populated, evaluated, and made publicly available to support NILM development. Among the distinct features of the LIT-Dataset is the labeling of the load events at sample level resolution and with an accuracy and precision better than 5 ms. The availability of such precise timing information, which also includes the identification of the load and the sort of power event, is an essential requirement both for the evaluation of NILM algorithms and techniques, as well as for the training of NILM systems, particularly those based on Machine Learning.

Keywords: Non-Intrusive Load Monitoring (NILM); NILM datasets; power signature; electric load simulation

1. Introduction

Non-Intrusive Load Monitoring (NILM) techniques are under development, globally, as part of the effort to improve Electrical Energy Efficiency. To support this development, specific datasets have been elaborated, particularly during the last decade. A NILM dataset consists of a collection of samples taken over time; these may include voltage, current, active power, and reactive power. As NILM techniques concern disaggregation of loads, typically, the samples in a NILM dataset comprise aggregated current and power.

According to the International Energy Agency [1], the worldwide electricity demand is currently 29,000 TWh and will increase to 42,000 TWh in 2040 at about 2.1% per year. Under current Stated Policies, less than 50% of this energy will come from renewable sources: mainly solar, wind, and hydro. Improvements in this scenario may be attainable by reducing waste and improper use, as well as reducing the electrical energy needs, mainly by improving the efficiency of electrical devices. Hence, the importance of Electrical Energy Efficiency, which aims at the reduction in power and energy demands of electrical systems without affecting their functionality. Most importantly, reducing energy needs has a significant effect on the environment by reducing the world's carbon emissions.

NILM techniques are based on a centralized measurement of electrical energy consumption and, through a disaggregation process, determination of the individual consumption of each electrical load. Typically, NILM uses a database of known power signatures of devices to analyze the aggregated power consumption and identify the contribution of each load. Therefore, NILM is a low-cost, easily deployed, flexible, and, therefore, viable solution that provides consumers with detailed information about their energy consumption [2]. NILM provides essential information for use in Smart Grids, in Energy Management Systems, and for Energy Efficiency initiatives.

The rationale for a NILM dataset is fivefold:

1. A dataset provides a stable set of input data that can be used to compare the performance of different solutions. As such, research under development by different groups can be compared over the same conditions.
2. Collecting data for a dataset requires a significant amount of effort and time. Making a dataset publicly available is a means of supporting researchers globally and accelerating results.
3. The development of new NILM techniques requires a thorough understanding of the problem domain. A comprehensive NILM dataset provides support for such an understanding.
4. As new NILM techniques and algorithms evolve, performance must be compared incrementally. A dataset provides a framework for consistent comparisons as well as for debugging.
5. NILM datasets can be used for training, i.e., to feed event identification and load classification methods to build an initial signature database that is key to many NILM techniques.

To support our ongoing research on a NILM solution [3] a dataset with particular requirements was needed. Since the available NILM datasets did not match these requirements, we decided to pursue the development of a new dataset [4], named after our laboratory, by using an engineering development process starting with requirements elicitation. During this development, a testing jig was constructed to allow recording in a framework where up to eight loads could be individually controlled (turned on or off) and register their waveforms (samples of voltage and current) in a controlled load shaping scenario, named Synthetic load shaping subset. Power detection devices were also built and connected to each load, in a residential or research lab environment, to provide precise event records in a scenario of recording in a real (thus, not-controlled) environment (this subset was named Natural load shaping). To these two subsets, a third one was added, consisting of Simulated loads. In this case, scenarios that are hard to obtain in the real world, such as short circuits, can be included.

The taxonomy of NILM datasets may be organized by (1) sample frequency with low-frequency being up to 1 Hz and high-frequency when above that [5]; (2) the event-aware datasets being those that register the occurrence of each load event, while the event-free datasets do not; and (3) the presence (or not) of ground-truth information, either by indicating which loads caused each event or by registering the individual consumption of each load over time. The LIT-Dataset samples voltage and aggregated current at 15 kHz (256 samples per 60 Hz-mains cycle); records single and multiple concurrent loads and registers each load event to provide ground truth.

The organization of the following sections is as follows: Section 2 describes the publicly available datasets; Section 3 lists the requirements for the proposed dataset; Section 4 describes the three subsets that compose the LIT-Dataset: synthetic load shaping, simulated loads, and natural load shaping; Section 5 presents and analyses the results obtained, and Section 6 presents the conclusions to the work presented here.

Previous Research Contributions

The LIT-Dataset has been developed under an ongoing research project, funded by COPEL and ANEEL. Previous publications and patent requests, resulting from this research project, are listed in Table 1.

Table 1. Related publications from the same research project.

Ref.	Year	Pub.	Description
–	2017	INPI	Patent application for NILM system
[6]	2018	PEAC	Initial proposal of HCApP and multi-agent architecture
[7]	2018	EECS	Initial Prony-based method proposal.
[8]	2018	EECS	Improved Multi-agent architecture.
[9]	2018	SBESC	Synthetic subset proposal.
[10]	2019	ISAP	Simulated subset proposal.
[11]	2019	ISAP	Prony-based proposal and comparisons.
[12]	2019	ISGT-LA	Steady-state and transient V-I features proposal.
[13]	2019	SBESC	Natural dataset proposal.
[3]	2020	Energies (MDPI)	Validated multi-agent architecture.

Three of the publications listed in Table 1 concern the LIT-Dataset, presenting preliminary results. In [9], the dataset proposal, jig's design, and initial results for the Synthetic Subset were presented, emphasizing the control mechanism for load switching and the acquisition circuit with its respective instrumentation. In [10], subsequently, the initial results with Simulated Subset were discussed, demonstrating the validation of load models and the automation procedure for generating waveforms. Finally, in [13], the architecture for a Natural Subset was presented, with a focus on the low-cost proposal of a time synchronization mechanism among nodes.

The other publications [3,6–8,11,12] detail the power signature analysis methods proposed in the same research project, using the LIT-Dataset and other recent datasets presented in the literature. Particularly in [3], a multi-agent architecture was presented and validated for event detection, feature extraction, and load classification, using different publicly available datasets. Some of the results were only possible due to the original features of the proposed LIT-Dataset. For instance, agents trained in a single load scenario and tested in another scenario with multiple concurrent loads were only possible because the LIT-Dataset includes such waveforms with single loads and different load combinations. A sample-level comparison for event detection was also only feasible due to the accurate labeling of the LIT-Dataset. This precise annotation of occurrence of each event is also primordial to allow the extraction of transient features from waveforms during the training stage, and, consequently, make use of the different feature extraction agents proposed in that work.

2. Related Work

The subject of the NILM dataset can be placed in the broader area of energy-related datasets and the associated means of data sensing and recording. Concerning data acquisition technologies, according to [14], there are five technologies classes employed to gather data and associated modeling methodologies: (1) energy consumption quantification, based on electricity meters; (2) indoor environmental measurements, based on ambient sensors, e.g., temperature, humidity, CO_2 concentration, among others; (3) occupant behavior statistics that are estimated using cameras, Passive InfraRed (PIR) sensing, and similar sensors; (4) status sensors, including doors and windows status readers; (5) others, combining different elements, as Radio Frequency IDentification (RFID) or Ultra Wide Band (UWB) sensors.

Concerning NILM datasets and NILM systems, electrical energy data is usually collected directly by low-cost voltage and current sensors. In [15], voltage AC sensors, Hall-effect based current sensors, and analog-to-digital converters were employed for load monitoring purposes. With respect to the communications infrastructure, [15] employed Ethernet, while [16] used a 433 MHz wireless sensor network gathering AC voltages and currents from individual devices.

A taxonomy for datasets of power consumption in buildings is presented in [17]. On a first level, datasets are classified as Appliance Level versus Aggregated Level. An Appliance Level dataset contains individualized information of energy consumption of every appliance, while an Aggregated Level dataset contains aggregated power consumption data of a whole residence or building. On a second level, seven application purposes are listed: energy savings, appliance recognition, occupancy detection, preference detection, energy disaggregation, demand prediction, and anomaly detection. A survey with 32 datasets is presented, comparing their characteristics and application purposes.

In the following sections, NILM datasets described in the literature are presented in two classes: (1) low-frequency datasets (sampling frequency up to 1 Hz); (2) high-frequency datasets.

2.1. Low-Frequency Datasets

The following low-frequency datasets were analyzed and compared: Smart [18], HES [19], Tracebase [20], DataPort [21], AMPds [22], iAWE [23], GREEND [24], REFIT [25], and RAE [26].

The relevance of these datasets is due to the characteristics of the installed measuring devices. However, many of the possible strategies for feature extraction, that can be used for NILM classification, are restricted due to low sampling frequency.

A comparison between some characteristics of the different types of low-frequency NILM datasets can be seen in Table 2. Where f_s represents the sampling frequency; DCD stands for Data Collection Duration; NoC corresponds to Number of Appliance Classes; NoA represents the Number of Appliances; and Res., Lab., Com. and Ind. are short forms for: Residential, Laboratory, Commercial and Industrial installations, respectively. Since the sampling occurs at very low rates (once a minute to once a second) the recordings can take place for very long times (weeks to years).

Table 2. Comparison between low-frequency Non-Intrusive Load Monitoring (NILM) datasets.

Dataset	Date	Environment	f_s	DCD	NoC	NoA
Smart	2012	Res.	1 Hz	3 Months	25	25
HES	2012	Res.	8.33 mHz	1 year/1 Month	~20	251
Tracebase	2014	Res.	1 Hz	1 day	43	158
Dataport	2013	Res. + Com. + Ind.	16.67 mHz–1 Hz	4 years	~70	>1200
iAWE	2013	Res.	1 Hz	73 days	33	-
GREEND	2014	Res.	1 Hz	3–6 months	-	-
AMPds	2015	Res.	16.67 mHz	2 years	19	-
REFIT	2017	Res.	125 mHz	2 years	9	20
RAE	2018	Res.	1 Hz	72 days	24	-

2.2. High-Frequency Datasets

The following high-frequency datasets were analyzed and compared (Table 3): REDD (Reference Energy Disaggregation dataset), BLUED (Building-Level fUlly-labeled dataset for Electricity Disaggregation), PLAID (Plug Load Appliance Identification Dataset), HFED (High-Frequency Energy Data), UK-DALE (United Kingdom recording Domestic Appliance-Level Electricity), COOLL (Controlled On/Off Loads Library), SusDataED (Sustainable Data for Energy Disaggregation), WHITED (Worldwide Household and Industry Transient Energy Dataset), BLOND (Building-Level Office enviroNment Dataset), and SynD (Synthetic energy Dataset).

REDD [27] is a residential dataset intended for research on disaggregation methods. REDD contains measurements from 6 different houses obtained over several months. The house input AC mains voltage and aggregated current are monitored at a sample rate of 15 kHz. Furthermore, the voltages and currents at individual circuits are monitored at a sample rate of 0.5 Hz, and plug-level monitors at a sample rate of 1 Hz. Similar to several of the datasets analyzed here, REDD provides ground truth data by presenting energy samples of individual appliances (monitored at plug-level) and of subsets (monitored at circuit level) of the total load.

Similarly, BLUED [28] is a dataset obtained from a single-family residence. This dataset registers the AC mains voltage and aggregated current. The sampling rate is 12 kHz, and the measurements were performed for 1 week. Every state transition of the 43 appliances is labeled and time-stamped, providing ground truth for event detection algorithms.

PLAID [29] is a public and crowd-sourced dataset consisting of one-second voltage and current waveforms for different residential appliances. The goal of this dataset is to provide a public library for high-frequency (30 kHz) measurements that can be integrated into existing or novel appliance identification algorithms. PLAID currently contains measurements for more than 200 different appliances, grouped into 11 appliance classes, and totaling over a thousand records.

UK-DALE [16] is a publicly available dataset comprising records from 5 different houses. It contains AC mains voltage and aggregated current, as well as voltage and current of individual loads, hence, providing ground-truth for testing disaggregation and training algorithms. The sampling rate is 16 kHz for the house input, while the individual sensors are sampled every 6 s. There are more than 4 years of data in this dataset and it is continuously updated.

HFED [30] is a high-frequency Electromagnetic Interference (EMI) dataset comprising high-frequency measurements of EMI, emanated from electronic appliances, propagated through the power infrastructure, and measured at a single point. HFED includes 24 appliances connected over four different test setups (in lab settings and one test setup in home settings). EMI measurements are taken over a frequency range of 10 kHz to 5 MHz.

COOLL [31] is a publicly available home appliance dataset containing 42 appliances grouped into 12 classes. The AC mains voltage and current are monitored for each appliance at a sample rate of 100 kHz for 6 s, which includes turn-ON and turn-OFF transients. For each appliance, there are 20 measurements on different power-on angles of the mains cycle. Each appliance is measured individually; hence, there is no aggregated current data registered in the dataset.

SusDataED [32] is an extended version of the dataset SusData [33]. This dataset is composed of measurements taken from a single-family residence in Portugal. Samples of 17 distinct appliances were taken at a sampling rate of 12.8 kHz for ten days.

WHITED [34] is a dataset of appliance measurements from several locations (households and small industries) around the world. The voltage and current waveforms are recorded with the first 5 s of the appliance start-ups for 110 different appliances, amounting to 47 different appliance types. This dataset aims to provide a broad spectrum of different appliance types in different regions around the world.

BLOND [15] is a dataset with waveforms collected at a typical office building in Germany. It is a fully-labeled ground truth dataset, with 53 appliances distributed in 16 classes of devices, sampled at 50 kHz during 213 days.

SynD [35] is a synthetic dataset composed of residential loads. This dataset is the result of a 180 days custom simulation of a residential environment that relies on power traces of real household appliances. SynD is composed of measurements taken from 21 appliances in Austria, with a sampling rate of 5 Hz, during 180 days.

Table 3 shows a comparison between these high-frequency NILM datasets. It includes information on the environment (if data was collected in a Residential, Commercial, or Industrial environment); the Duration of the period of Data Collection (DCD); if the dataset includes scenarios of Multiple Simultaneous Loads (MSL); the sampling frequency (f_s); if Ground Truth is recorded, either as the recordings of current/power of individual loads or as recordings of events (at a given Load Event Resolution—LER); the Number of Appliance Classes (NoC); and the Number of Appliances (NoA).

Table 3. Comparison between high-frequency NILM Datasets.

Dataset	Date	Nature	DCD	MSL	f_s	Ground Truth Resolution (LER)	NoC	NoA
REDD	2011	Res.	119 days (10 houses)	yes	15 kHz	3 s	8	24
BLUED	2012	Res.	8 days (1 house)	yes	12 kHz	640 ms	9	43
PLAID	2014	Res.	1094 waveforms (of 1 s each)	no	30 kHz	>1 cycle	12	235
HFED	2015	Res. + lab.	-	yes	10 kHz–5 MHz	-	-	24
UK-DALE	2015	Res.	655 days	yes	16 kHz	6 s	16	54
COOLL	2016	Res.	840 waveforms (of 6 s each)	no	100 kHz	20 ms	12	42
SustDataED	2016	Res.	10 days	yes	12.8 kHz	2 s	-	17
WHIETED	2016	Res. + Ind.	5123 waveforms (of 5 s each)	no	44.1 kHz	-	47	110
BLOND	2018	Res.	50–213 days	yes	50–250 kHz	-	16	53
SynD	2020	Res.	180 days	yes	5 Hz	0.2 s	-	21

2.3. Evaluation of Datasets

The analysis of the datasets, both high-frequency and low-frequency, presented above indicates that: (1) the majority of NILM datasets contains data collected in a residential environment; (2) the majority of high-frequency datasets register 200 or more samples per mains cycle, a notable exception being SynD whose sampling frequency is 5 Hz; (3) the majority of the datasets register multiple simultaneous loads. Concerning the unique characteristics of each dataset it can be observed that: (1) the highest sampling frequency is used by COOLL (100 kHz); (2) while most low-frequency datasets do not provide ground-truth information, the high-frequency datasets provide ground truth by recording at a much lower rate (typically bellow 1 Hz) samples for individual loads.

2.4. Tools for NILM Datasets

The NILM Toolkit (NILMTK) [36] is an open-source toolkit designed to allow the comparison between NILM algorithms. It provides a Python API that operates on input and output binary files, therefore facilitating compatibility with data from NILM datasets. The input files used by NILMTK must be converted to the NILMTK-DF (data format), which is a data structure inspired on the dataset REDD comprising disaggregated power data (i.e., separate sample sets for each of the loads in a dataset) as well as metadata annotations about the sample set.

3. The Design of a Novel Dataset

Since none of the evaluated datasets had all the required characteristics for our research project, a new dataset development took place, with the first activity being requirements elicitation.

The LIT-Dataset is composed of three subsets: Synthetic, Simulated, and Natural. The Synthetic subset is obtained by a programmable power sequencing to a given set of loads in a controllable laboratory setup, so that repeatable scenarios can be obtained. In the Simulated subset, data is collected by simulating a circuit operation, allowing to test different scenarios and to control parameters that otherwise would not be possible or would be unsafe. The Natural subset is composed of voltage and current samples collected in a real-world uncontrolled environment; furthermore, apart from recording the aggregated current and the AC mains voltage, power sensors monitoring each load identify and record when each load event occurs.

Concerning the taxonomy presented in [17], the LIT-Dataset is an Aggregated Level dataset whose main application is Energy Disaggregation but is also applicable to energy saving, appliance recognition, and anomaly detection.

One of the requirements of the LIT-Dataset is that it includes multiple loads, as a NILM system must identify the loads that compose an aggregated current signal. Another requirement is that it must include precise indications of every load event (load on and load off), with a resolution better than one mains cycle, and have a high sample rate.

The Stakeholder requirements of the LIT-Dataset are based on the needs of the authors' NILM project, as well as on the requirements common to other NILM datasets. The LIT-Dataset Stakeholder requirements are listed below, as well as the rationale for each requirement:

DSReq 1. Data collection from loads connected to a single-phase 127 V, 60 Hz mains (the Brazilian power grid standard).
R: Due to power grid availability in our lab. Considering that 127 V, 60 Hz, is a standard used in many countries around the world, such a requirement does not restrict the usage of the LIT-Dataset elsewhere.

DSReq 2. Comprised of residential, commercial, and low-voltage industrial loads.
R: A NILM dataset should include a variety of loads related to these environments so that NILM systems can be evaluated and compared to distinct scenarios.

DSReq 3. Include loads of five types: LT1 to LT5, defined below.
R: A NILM dataset should include a variety of loads types so that NILM systems can be evaluated and compared over the range of loads available in the real-world.

DSReq 4. Waveform recordings of voltage and aggregated current of multiple simultaneous loads.
R: The purpose of a NILM system is to disaggregate the individual loads from an aggregated signal (current/power/...); hence, a NILM dataset should provide data of aggregated acquisitions representing actual scenarios where NILM is used.

DSReq 5. Accurate indication of load events (accuracy better than 5 ms).
R: A high-frequency NILM dataset can be used by NILM algorithms that evaluate the waveform of the current in each mains cycle to determine accurately the occurrence of load events. Ground-truth indications of such events with an accuracy better than one mains semicycle provide information to validate such algorithms. 5 ms is a typical switching time for relays used to energize the loads of a dataset.
Remark: concerning this requirement, accuracy is the measure of the error between the instant were the actual load event occurred, and when the event is reported (labeled).

DSReq 6. The minimum sampling rate is 15,360 Hz, corresponding to 256 samples along one mains cycle.
R: In high-frequency datasets, there is a trade-off between sampling frequency and storage requirements. Based on the analysis of datasets with sampling frequencies up to 100 kHz, the spectral densities of frequencies above 5 kHz in the aggregated signal, and the waveforms reconstructed from samples at 256 samples per cycle, this sampling rate was determined as an adequate trade-off selection.

DSReq 7. Recordings over a mix of loads so that low-power load-events (<5 W) occur while high power (>800 W) are energized.
R: Switching a low-power load when high-power loads are energized poses a challenging scenario for NILM systems; hence, the LIT-Dataset should include such scenarios for evaluation of these systems.

For the Synthetic subset:

DSReqSy 1. Synthetic load shaping of up to eight concurrent loads.
R: As a NILM system must disaggregate loads, a dataset should have aggregated data collected from loads energized concurrently. As there is a trade-off between

cost/complexity of the data collecting infra-structure and the number of concurrent loads, eight loads were selected as an adequate trade-off.

DSReqSy 2. The duration of each recording must be longer than 10 seconds and must include at least one power-ON and one power-OFF event.
R: By examining the data from other datasets, 10 s was determined as a sufficient duration so that the stable periods occur between transient periods due to power-ON and power-OFF.

For the Simulated subset:

DSReqSim 1. Recording at multiple power levels for each type of simulated load.
R: To explore the flexibility due to simulation allowing multiple loads to be employed by just changing the component values.

DSReqSim 2. Different scenarios of the AC Mains must include wiring stray inductance, as well as harmonics and white noise added to the mains voltage.
R: To simulate multiple actual environments considering wiring stray inductance, harmonics, and noise.

For the LIT Natural subset:

DSReqN 1. Minimum monitoring time for naturally shaped loads (for each monitoring file): 1 day.
R: Considering the daily seasonality typically present in the load shaping of the Natural subset, a day-long acquisition records such seasonality.

The taxonomy presented by Hart [37], from the perspective of power switching, was extended, resulting in these types of loads:

LT 1. On/Off. Such as a resistive load.
LT 2. State-Machine based. Such as electronic equipment (e.g., printer).
LT 3. Asymmetric. A load whose positive and negative semi-cycles are distinct, such as a drill in which the lower velocity employs a half-wave rectifier.
LT 4. Continuously variable. Such as a motor with speed control.
LT 5. Random. Loads in which the power consumption varies randomly.

As per requirement DSReq 3, all these types of loads are required in the LIT-Dataset.

In [38], the authors present 17 suggestions to dataset providers to improve dataset interoperability and comparability. Since these suggestions were published after the LIT-Dataset requirements were specified, we present in Table 4, the coverage of the LIT-Dataset requirements with respect to the presented suggestions.

Table 4. Coverage of Klemenjak's [38] suggestions.

Sugg.	Cov.	Comment	Sugg.	Cov.	Comment
1	Yes	Contains raw samples of V and I	10	Yes	Individualized labeling
2	Yes	Microscopic data is collected	11	No	Not in our requirements
3	Yes	Can calculate power from samples	12	Yes	Metadata and scripts made available
4	Yes	Sampled at 15,360 Hz	13	No	Not in our requirements
5	Yes	Data was validated	14	Yes	Documented data formats
6	Yes	Transducers accuracy checked	15	Yes	University's cloud
7	Yes	Some long term recording is planned where applicable	16	Yes	Publicly available
8	Yes	Individualized labeling	17	N.A.	
9	Partial	Photos on web site			

4. Proposed Dataset

In this section, the three subsets that compose the LIT-Dataset are presented.

4.1. Synthetic Subset

The Synthetic subset is named in relation to its load shaping being defined by a controller that switches the loads ON or OFF in a programmed pattern. To collect data for the synthetic subset, a single-phase 1 kW jig was designed and built according to the requirements of Section 3.

4.1.1. Data Collecting Jig Hardware Design

The Jig block diagram is presented in Figure 1. The protection block consists of an emergency stop button, a circuit breaker, and a fast fuse. The current sensors, for the operator's protection sake, are connected in the neutral line, while the loads are switched in the phase line. Two current sensors, shunt, and hall, are used to compare the hall performance to the shunt, in loads with high derivatives of current as well as asymmetrical loads. To sense the AC mains voltage, a resistor divider is used. An oscilloscope provides a performance benchmark for the current and voltage sensors.

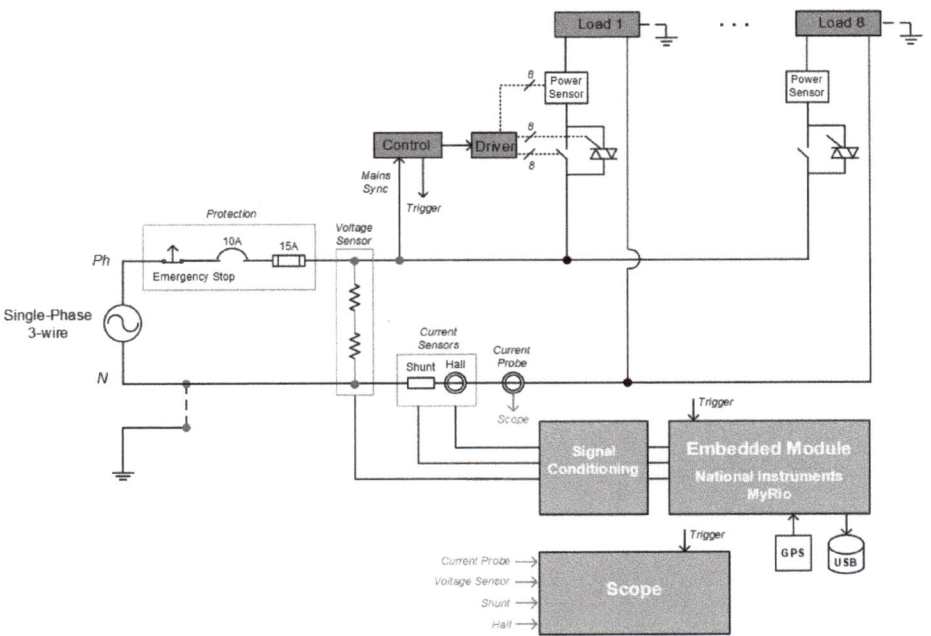

Figure 1. Block diagram of data collecting jig. adapted from [9].

The control module senses the AC mains zero crossing; therefore, a precise timing is achieved in every power ON or OFF event. The load power control is provided by a relay in parallel to a TRIAC, each with its independent driver module. The TRIAC provides precise power switching at a giving point of the AC mains cycle while the relay operates as a conventional switch. As the relay presents a delay between the relay driver signal and the actual opening/closing of the relay contacts, a power sensor provides a precise indication of when a load is actually powered. It is possible to trigger a load only with the relay or the TRIAC, as well as with both simultaneously. Triggering only the TRIAC makes it possible to obtain a dimmer effect on the load(s).

The signal conditioning module has low pass filters and differential amplifiers so that the sensors' signals are adequate to the embedded module analog to digital converter. The embedded module uses a National Instruments MyRio [39] board with the following functionalities: A/D conversion, load event registering obtained by the trigger signal sent by the control module, and dataset storage in a non-volatile storage device (such as a USB flash disk).

Figure 2 shows the synthetic LIT-Dataset Jig. An aluminum structure and an acrylic panel were used to support the jig's components. The wiring is inside PVC ducts. On the left are the power connection, auxiliary power supplies, and the sockets for the jig's equipment power supplies. The protection board is on the left side. The eight sockets for the monitored loads are on the top right, and below are the relays and TRIACs, and the control and driver boards. In the center of the board are the voltage and current sensors. On the bottom right are the conditioning boards and the MyRio Embedded Module.

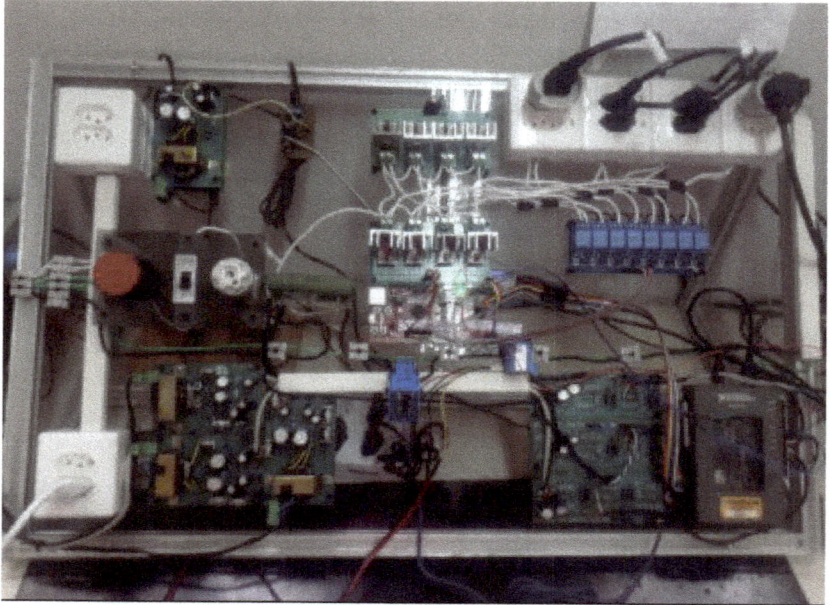

Figure 2. Data collecting jig.

4.1.2. Data Collecting Jig Software Architecture

The embedded module block diagram is presented in Figure 3. The signal conditioning board delivers the conditioned signals from the voltage and current sensors to the MyRio module. The MyRio FPGA implements an acquisition loop that operates at 15,360 Hz, as per DSReq 6 s. On every cycle of this loop, a set of three 12-bit samples is obtained, corresponding to the A/D conversions. A GPS receiver sends a Pulse Per Second (PPS) signal, which is grouped in a data tuple with the sample signals, indicating precise 1 s periods, typically less than 100 ns jitter. An 8-bit ID and an event notification signal are also grouped in this data tuple to indicate the samples when the control module commands a load event.

The real-time application runs into the CORTEX A9 processor, composed of three NI LabView timed loops, which act as independent periodic threads. The tuple is sent by the FPGA to the real-time application via DMA. An external GPS receiver sends the NMEA strings to the GPS Parsing Timed Loop. The absolute time fields are decoded from a specific NMEA message (GPRMC) and converted to a 32-bit time-stamp value. The Sample Processing Timed Loop uses this 32-bit time-stamp, together with the event information contained in the tuple received from the FPGA, to store the time-stamped events into the event annotation file. This loop also shifts each 12-bit sample of the tuple one bit to the left and adds the PPS signal as its least significant bit, thus, allowing a precise identification of the samples during which the 1 s transitions occurred. The resulting 13-bit sample data is sent to the Storage Timed Loop via an internal FIFO, which stores the data for each of the sample inputs as a separate field of a NI Technical Data Management Streaming (TDMS) file into the USB flash disk attached to the MyRIO. The TDMS file may be processed by a PC application, to add its collected data to the LIT-Dataset.

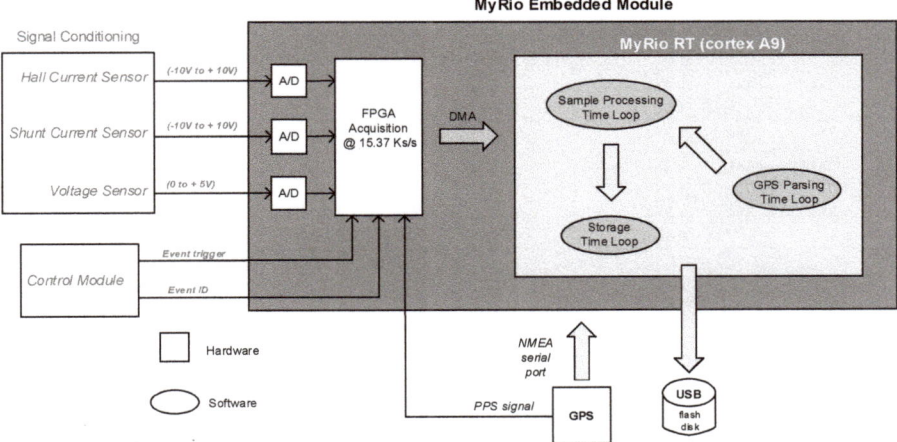

Figure 3. Block diagram of data acquisition using the MyRio device. Adapted from [9].

4.1.3. Collected Data

For the synthetic subset, 26 different load configurations, divided into 16 load classes (Table 5), were used, as well as their combinations (2, 3, and 8 loads). "Load configurations" means that one load may have more than one power level and/or that more than one equipment of the same class was used (e.g., two appliances of class LED lamp).

Linear and non-linear loads, ranging in power from 4 W up to 1.5 kW. The loads are powered on at different angles of the mains cycle, as per Table 6. Different turn-on trigger angles affect the loads inrush current, resulting in distinct waveform acquired at each angle. Each acquisition is commanded by the control board and consists of 16 voltage/current waveforms at the specified angles. The number of individual and multiple-loads acquisitions are presented in Table 7, in a total of 104 acquisitions, corresponding to 1664 waveforms. The number of acquisitions for multiple loads were limited by the jig's maximum power (1 kW).

Table 5. Characteristics of the synthetic subset of the LIT-dataset.

Class	Class Description	Power (W)	Num. of Appliances	Num. of Load Configurations	Num. of Waveforms in LIT-SYN-1
1	Microwave Oven (standby and on)	4.5/950	1	2	32
2	Hairdryer (two fan speed levels)	365/500 600/885	2	4	64
3	Hairdryer (two power levels)	660/1120	1	2	32
4	LED Lamp	6	2	2	32
5	Incandescent Lamp	100	1	1	16
6	CRT Monitor	10	1	1	16
7	LED Monitor	26	1	1	16
8	Fume Extractor	23	1	1	16
9	Phone Charger	38 50	2	2	32
10	Laptop Charger	70 90	2	2	32
11	Drill (two speed levels)	165/350	1	2	32
12	Resistor	80	1	1	16
13	Fan	80	1	1	16
14	Oil Heater (two power levels)	520/750	1	2	32
15	Soldering Station	40	1	1	16
16	Air Heater	1500	1	1	16
	Total		20	26	416

Table 6. Turn-on trigger angles.

ID	Trigger Angle (°)	ID	Trigger Angle (°)
0	0	8	180
1	22.5	9	202.5
2	45	10	225
3	67.5	11	247.5
4	90	12	270
5	112.5	13	292.5
6	135	14	315
7	157.5	15	337.5

Table 7. Loads combination.

Loads Combined (Sets of 16 Waveforms)	Acquisitions
Single	26
2	42
3	30
8	6

The sequence of events (ON and OFF) for the single and multiple loads are presented in Figure 4.

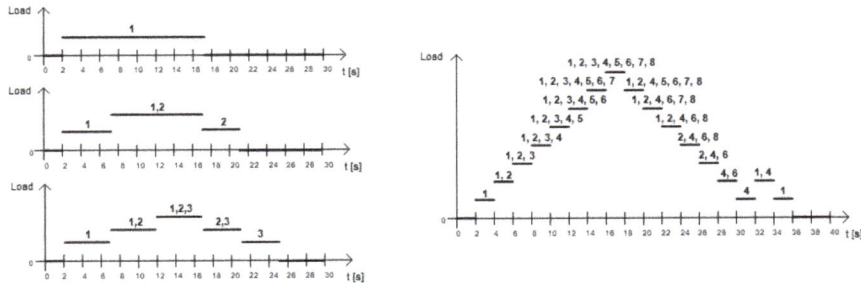

Figure 4. Data collecting jig: single, two, three, and eight loads ON and OFF events.

4.1.4. Accuracy of the Jig

The jig's hardware initially went through a calibration process, and then its accuracy was evaluated based on a comparison with laboratory-grade measurement equipment.

The calibration process consisted of collecting data from resistive loads that were measured with an HP bench multimeter with a 5-digit-resolution and a precision better than 0.1 %. Since the voltage and current waveforms are bipolar (positive and negative values) but the 12-bit measurements of the A/D converters from the MyRio are unipolar (0 to 4095), an offset value corresponding to inputs at zero must be determined; as well as the gain factor to convert a binary value produced by the ADC to a voltage or current value (in Volts or Amperes). This calibration process is performed before every acquisition on the Jig, and the calibration values (Ki, Kv, ZeroOffsetI, ZeroOffsetV) are reported in the file config_processed available in every acquisition folder of the LIT-Dataset.

The determination of the jig's accuracy was performed by connecting an oscilloscope (Agilent Infiniium 54830D) and a current probe (Tektronix A6302) during the acquisitions. A total of 28 acquisitions with different loads were performed while data were simultaneously acquired by the Jig and by the scope. Data from both sources were stored as spreadsheets and imported into MATLAB for comparison. Over the 28 acquired voltage and current waveforms, the maximum error was 3.2 % with a mean value of 2.1 %. This value of accuracy was considered as acceptable for a NILM dataset. Most datasets do not provide an accuracy evaluation for comparison.

4.2. Simulated Subset

The simulated subset consists of data collected from twenty-eight different simulated loads grouped into seven kinds of electrical models, each one containing up to four power variations. The loads, waveform generation, and simulated subset settings are detailed as follows.

4.2.1. Loads

In this subset of LIT-Dataset, the electrical circuits are: (a) resistor; (b) resistor and inductor; (c) diode rectifier with a resistor; (d) diode full-wave bridge rectifier with resistor and capacitor; (e) thyristor rectifier with resistor; (f) thyristor rectifier with resistor and inductor; and (g) universal motor. The load templates were chosen according to the load profile of electrical appliances commonly found in consumer units [40], such as drill (universal motor), mobile phone charger (different types of rectifiers), fan (universal motor), hairdryer (universal motor), LED lamp (different types of rectifiers), incandescent lamp (resistor), router (different types of rectifiers), and vacuum cleaner (simplified by resistor and inductor).

The diagram of the simulated subset is shown in Figure 5, in which each block represents a different load. The switching control of each load is automated, and the trigger time can be previously adjusted.

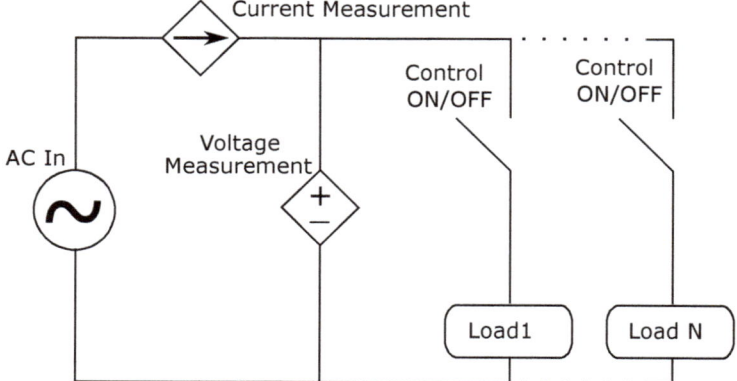

Figure 5. General diagram of the simulated subset.

To implement each set of loads with different electrical power from circuits (a)–(f), the Power Eletronics Library from Matlab/Simulink was used, as shown in Figure 6: (a) resistor, (b) resistor and inductor, (c) diode rectifier with resistor, (d) diode bridge with resistor and capacitor, (e) thyristor rectifier with resistor, (f) thyristor rectifier with resistor and inductor.

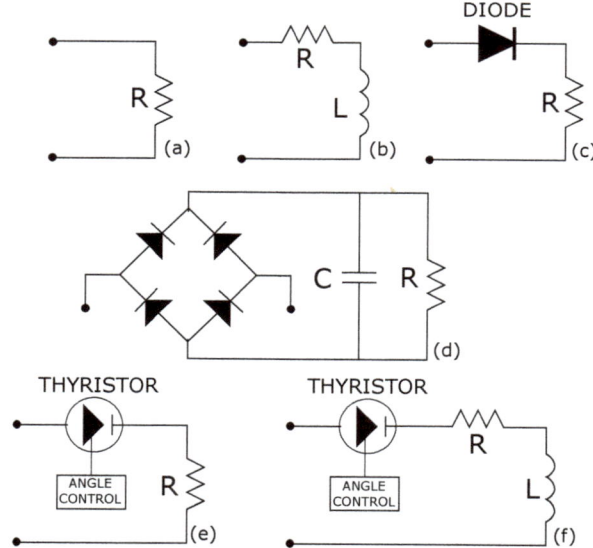

Figure 6. Diagrams of simulated loads, being (**a**) resistor, (**b**) resistor and inductor, (**c**) diode rectifier with resistor, (**d**) diode bridge with resistor and capacitor, (**e**) thyristor rectifier with resistor, (**f**) thyristor rectifier with resistor and inductor.

For the implementation of the universal motor (g), a mathematical model based on [41] was used. Figure 7 shows the diagram that represents this model, in which the following parameters are included: rated power, rated terminal voltage, rated speed, armature winding inductance (L_{aq}), series field winding inductance (L_{se}), rated frequency of supply voltage, armature winding resistance (R_a), series field winding resistance (R_{se}), rotor inertia (J), speed at which magnetization curve data was taken (ω_{mo}). To connect the math model with other circuits, the generated signal was connected to a current source generator and sent to other blocks, i.e., electrical—mathematical interface in MATLAB-Simulink.

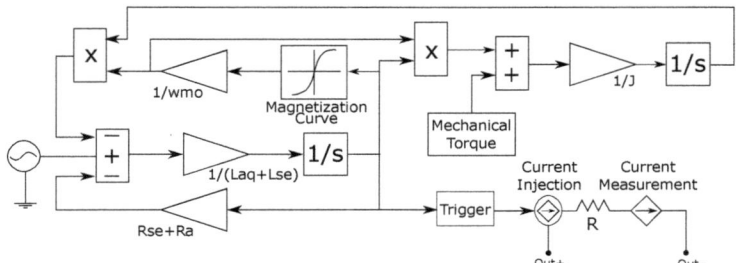

Figure 7. MATLAB-Simulink block diagram of Universal Motor Model.

4.2.2. Waveform Generation

To automate the waveform generation, an automatic parameter variation method was implemented. With that method, it is possible to vary: up to seven different load combinations for each waveform; up to four different values of the electrical components for each circuit; the total time of the simulation; the load combination; and the trigger time of the circuits, with three different options of switching (turn-ON and turn-OFF) angles: 0, 45, and 90 degrees. The four parameters variations, resulting in four different rated power levels for each set of loads, are detailed in Table 8, where: (a) resistor; (b) resistor and inductor; (c) diode rectifier with a resistor; (d) diode full-wave bridge rectifier with resistor and capacitor; (e) thyristor rectifier with resistor; (f) thyristor rectifier with resistor and inductor; (g) universal motor.

Table 8. Load parameters.

(a)	$R_1 = 2.5\ \Omega$ $R_2 = 5\ \Omega$ $R_3 = 8\ \Omega$ $R_4 = 10\ \Omega$	(e)	$R_1 = 10\ \Omega$ $R_2 = 12\ \Omega$ $R_3 = 15\ \Omega$ $R_4 = 20\ \Omega$
(b)	$R_1 = 10\ \Omega$ $R_2 = 20\ \Omega$ $R_3 = 30\ \Omega$ $R_4 = 30\ \Omega$ $L_1 = 10\ \text{mH}$ $L_2 = 10\ \text{mH}$ $L_3 = 10\ \text{mH}$ $L_4 = 100\ \text{mH}$	(f)	$R_1 = 10\ \Omega$ $R_2 = 5\ \Omega$ $R_3 = 30\ \Omega$ $R_4 = 105\ \Omega$ $L_1 = 1\ \text{mH}$ $L_2 = 1\ \text{mH}$ $L_3 = 1\ \text{mH}$ $L_4 = 1\ \text{mH}$
(c)	$R_1 = 5\ \Omega$ $R_2 = 10\ \Omega$ $R_3 = 20\ \Omega$ $R_4 = 50\ \Omega$	(g)	$w_1 = 1500\ \text{rev/min}$ $w_2 = 1200\ \text{rev/min}$ $w_3 = 1000\ \text{rev/min}$ $w_4 = 800\ \text{rev/min}$
(d)	$R_1 = 2\ \text{k}\Omega$ $R_2 = 1\ \text{k}\Omega$ $R_3 = 500\ \Omega$ $R_4 = 300\ \Omega$	$C_1 = 100\ \mu$ $C_2 = 100\ \mu$ $C_3 = 200\ \mu$ $C_4 = 330\ \mu$	

4.2.3. Configuration of Simulation Scenarios

To create different simulation scenarios, six configuration settings can be used, as follows: ideal (DB-1); with stray inductance, representing the equivalent of the electrical network (DB-2); with stray inductance and harmonics (DB-3); and with stray inductance, harmonics, and additive white gaussian noise (AWGN), with 60 dB, 30 dB, and 10 dB of SNR (DB-4, DB-5, and DB-6), as shown in Table 9. Each of the six configurations are applied to all the loads, resulting in 4824 waveforms, being 804 for each configuration. Therefore, it is possible to evaluate the impact of harmonics and noise (with different intensities) and to compare to an ideal scenario, to the performance of detection, feature extraction, and classification methods. This type of analysis can support the proposal of more robust and applicable methods in different NILM scenarios.

Table 9. Simulated subset settings.

Settings	Parameters
Setting (DB-1)	ideal
Setting (DB-2)	stray inductance
Setting (DB-3)	stray inductance and harmonics
Setting (DB-4)	stray inductance harmonics and AWGN with SRN 60 dB
Setting (DB-5)	stray inductance harmonics and AWGN with SRN 30 dB
Setting (DB-6)	stray inductance harmonics and AWGN with SRN 10 dB

The first scenario (DB-1), was an ideal setting, without stray inductance and harmonic content in the voltage waveform. The second one (DB-2), include stray inductance. The characteristics of the electrical network in our laboratory were used to select the values of the inductor and resistor, resulting in L = 1 µH and R = 2 mΩ. The third scenario (DB-3), includes stray inductance and a voltage source with harmonics, based on the voltage acquisition in our laboratory. The last three include a voltage source with harmonics, stray inductance, and different levels of AWGN.

4.3. Natural Subset

The Natural subset of the LIT-Dataset consists of recording where a natural load shaping occurs, in the sense that waveforms are registered in a real-world environment (residential, research lab, commercial, industrial) over longer periods of time. To precisely detect and record the load events, sensors that detect power-ON, power-OFF, and power-level-changes are attached to each load, therefore, while the aggregated current and voltage are recorded, so are the individual load events.

4.3.1. Natural Subset—Data Collection Architecture and Implementation

Accurate time synchronization is an important requirement in this scenario, in which time-stamped data should be provided by distributed nodes and then correlated with a limited jitter among them. Concerning specifically the development of the Natural subset of the LIT-Dataset, an infrastructure composed of a centralized acquisition device and a large number (50+ units) of networked wireless sensors is required. These nodes are attached to each load to detect load events, such as ON-OFF transient, change of state and power variations, and send the event data to the centralized acquisition element so that they can be later consolidated and correlated with the acquired voltage and current data.

This infrastructure, from this point on referred to as Natural Subset Acquisition System (NSAS), depends on time synchronization with accuracy and precision of at least 1 ms, to facilitate the correlation between the events obtained by the distributed event detection modules and the voltage and current samples obtained by the centralized acquisition element. Additionally, considering the large number of modules to be installed and their distributed characteristic, they are required to be built with low-cost components. In this sense, even though there are several techniques and protocols that address the precise time synchronization issue, most of them rely on specialized hardware and/or software solutions, thus incurring a relatively high cost to deploy the synchronization network [42,43].

An overview of the architecture used to collect a dataset of traces with natural load shaping is presented in Figure 8. It is important to notice that the voltage and current traces for the aggregate of the loads are collected at a single point, namely at the sensors next to the fuse box. The distributed nodes only detect power events (ON, OFF, and power changes) and record the occurrence of such events locally. It is this recording that requires a millisecond timing accuracy, achieved through the synchronization mechanism implemented by the NSAS.

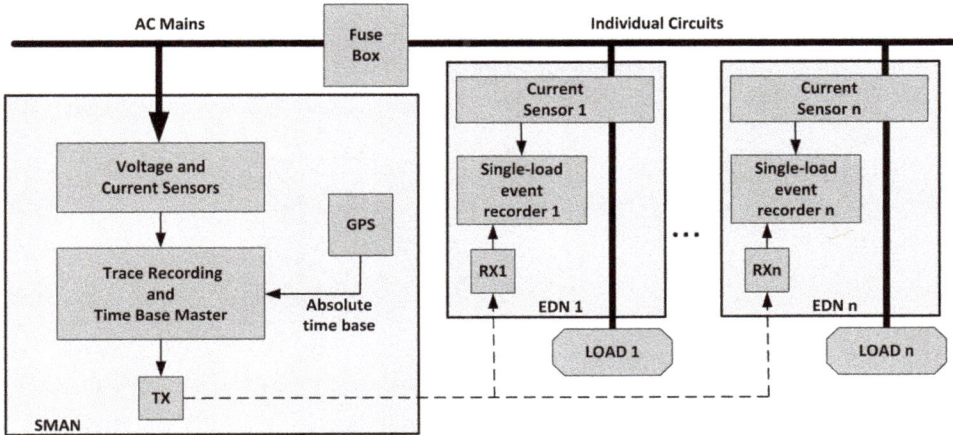

Figure 8. Overview of the architecture for collecting a natural load shaping subset.

The principle of operation of this low-cost synchronization network is to have a time base master, with a GPS based real-time clock, to periodically broadcast a two-byte synchronization packet to all nodes in the synchronization network.

To avoid delays imposed by complex packet-based protocols, an approach that implements the synchronization task right before the PHY is used. This is performed using a low-cost, byte-based RF 433 MHz transmitter-receiver pair [44], similar to the one used in [45] for an application with similar requirements. The typical reception delay for this solution is about 300 µs, which meets NSAS timing requirements of 1 ms.

Furthermore, the main contribution of this proposed architecture is its low cost (about one dollar for the receiver), in a way that its impact on the cost of the whole NSAS is minimized. The block diagram of the Natural Subset Acquisition System is presented in Figure 9.

The Synchronization Master and Acquisition Node (SMAN), on the top of the block diagram, is implemented by using a National Instruments MyRIO module [39] attached to a GPS module and the 433 MHz RF transmitter [44]. The MyRIO module is connected to the other NSAS modules via a WLAN and is programmed, via LabView, to perform the SMAN main tasks. The RF transmitter receives a digital timing synchronization signal as input and broadcasts it in the 433 MHz band at a rate of up to 2400 bps. The EDNs (Event Detection Nodes) consist of ESP32 Heltec WiFi modules, as well as 433 MHz RF receivers. The ESP32 Heltec kit is a low-cost development board, which is programmable using the Arduino IDE and corresponding libraries to perform the EDN tasks. It connects to the other NSAS modules via a WiFi-based WLAN. The RF receivers are responsible for receiving the signal that is broadcast by the RF transmitter of the SMAN. Each EDN is physically connected to a Power circuit connection element (interrupter, outlet, etc.), so it can perform the sampling of current to detect variations that indicate a load switch event (ON, OFF, or other state change such as changing from standby to active mode).

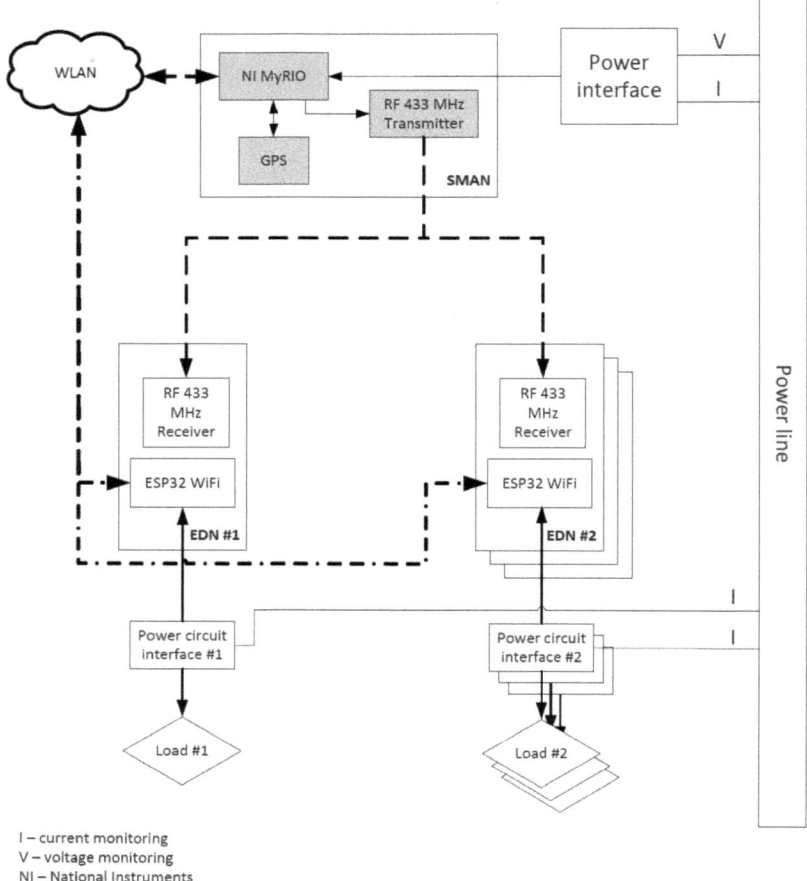

Figure 9. Block diagram of the Natural Subset Acquisition System (NSAS).

The SMAN is responsible for acquiring the voltage and current samples at a frequency of 15,384 Hz, which is slightly above the minimum 15,360 Hz frequency specified for the LIT-Dataset due to the MyRio-timer configuration options. The SMAN is also responsible for collecting and storing the event data sent by the EDNs via the WLAN. The GPS module provides the SMAN with an absolute time reference on every second employing the PPS (pulse per second) signal, whose typical jitter is of hundreds of nanoseconds. This time reference is used to ensure that the millisecond data used by the SMAN to synchronize the EDNs is synchronized to an absolute reference, regardless of potential clock drifts presented by the SMAN itself (typically 10 ppm).

Upon detection of an event on a load connected to a monitored power circuit connection element, the corresponding EDN sends the event data to the SMAN via the WLAN and waits for the event acknowledgment. If the acknowledgment times out, the event is sent again. The EDNs also communicate with the SMAN by means of "abs time req" messages, which are sent during EDN initialization. The SMAN responds with an "abs time resp" message containing the absolute time and date, with a resolution of one second, obtained from the GPS receiver. This transaction is responsible for performing a relatively coarse synchronization (i.e., with an accuracy of one second) between the SMAN and the EDNs. The synchronization between EDNs and SMAN is improved to millisecond-accuracy upon reception of an RF message, broadcast by the SMAN, which consists of a 16-bit synchronization code. The SMAN sends the code at a rate of 1000 bps (i.e., one bit per millisecond) on every second boundary (1 Hz). Hence, upon completion of the reception and validation of the code, every EDN shall (re)adjust the millisecond's field of its current time to 16, corresponding to the 16 ms that have passed from the latest second boundary to the end of reception of the last bit of the synchronization code.

As the typical clock drift for the EDN hardware is 10 ppm, a drift of 0.5 ms would occur every 50 s; therefore the resynchronization rate of 1 Hz is, theoretically, widely sufficient to ensure that the EDNs remain synchronized with the SMAN even if 98% (49 of 50) of the RF synch messages are lost. Additionally, the typical jitter of the RF link (300 µs) is small enough not to introduce indeterminism on the millisecond value to be adjusted into the EDNs.

However, it is observed that some EDNs present much higher drift rates than the typical case; in some cases, more than 1000 ppm have been observed under operating conditions, which would compromise the millisecond precision required by the system. Therefore, it is necessary to implement an extra strategy to prevent desynchronization between the several EDNs and the SMAN that compose the NSAS.

The drift correction strategy consists of the algorithm shown in Figure 10a. Initially, the timer tick is set to 1000 µs (1 ms), which is the default period for time-stamp updates. Upon reception of a synch word (i.e., on every second), the EDN compares the millisecond on which the synch word has been effectively received with the millisecond on which it should have been completely received (16, because of the 16-bit synch word sent at 1000 bps starting from 0 ms at the SMAN) (line 6). The more positive the difference between the former and the latter, the more this EDN's specific tick is being advanced in relation to the nominal tick frequency (1 kHz) because of its clock drift; the same happens when the difference is negative, meaning that the clock drift is causing the EDN tick to be delayed. Next, the EDN timer period is proportionally adjusted (lines 9 and 10), so the next ticks can compensate the clock drift by an increase (or decrease) of the programmed tick frequency.

```
1  On initialization:
2     adjust timer tick to 1000 us;
3     tick_us = 1000;
4  Drift correction strategy algorithm
5     wait for synch reception;
6     drift = (current_msec - 16) /
7             (current_abs_time_in_sec - last_synch_time_in_sec);
8     last_synch_time_in_sec = current_abs_time_in_sec;
9     tick_us = tick_us + drift;
10    adjust timer to tick to tick_us;
```

(a)

```
1  Spurious synch management algorithm
2     calculate drift using drift correction algorithm;
3     if (!first_sync && abs(drift) > 5)
4     {
5        spurious_synch_counter++;
6        if (spurious_synch_counter > 3)
7           perform synchronization;
8     }
9     else
10    {
11       spurious_synch_counter = 0;
12       perform synchronization;
13    }
```

(b)

Figure 10. Algorithms used in Event Detection Nodes (EDNs). (**a**) Drift correction algorithm. (**b**) Spurious synch management algorithm.

Another algorithm, shown in Figure 10b, is implemented to take into account possible spurious synchronization words that can be received due to noise at the RF link. This is a real concern, as the 433 MHz radios used for the NSAS are very susceptible to such noise, and the implemented synchronization algorithm, which is supposed to be simple and deterministic, does not make use of any software checking mechanisms to improve data reception reliability.

The spurious sync management algorithm analyzes the calculated drift obtained from the algorithm of Figure 10a. If this is the first synchronization, the calculated drift is probably correct, as there is no previous synchronization between the SMAN and this EDN. If this is not the first synchronization, and the absolute calculated drift value is greater than a specified limit of 5, corresponding to 5000 ppm. Since 5000 ppm is significantly larger than the typical 10 ppm drift, or even the 1000 ppm drift occasionally detected, a spurious sync word has likely been received on a random time, leading to a drift miscalculation; in this case, the EDN ignores the spurious sync unless it has already been received more than three times in sequence (as tested in line 6). If that happens, the first received sync was probably spurious, and thus, the new sync is assumed to be the correct one.

4.3.2. Natural Subset—Collected Data

For the natural subset, 14 different load configurations, divided into 11 load classes (Table 10), were used, as well as their combinations. The 3-load combination has 30 s of duration and 6 events. The 7-load combinations have 2 h of duration and 20 events or more. The load configurations mean either that one load has more than one state or that more than one device of the same class was used.

Table 10. Characteristics of the natural subset of the LIT-dataset.

Class	Class Description	Power (W)	Num. of Appliances	Num. of Load Configurations	Num. of Waveforms in LIT-NAT
1	Aquarium Digital Thermostat	380	1	1	4
2	Aquarium Light Fish Lamp 1	100	1	1	4
3	Aquarium Light Fish Lamp 2	170	1	1	4
4	Hot-air hand tool	1400	1	1	4
5	LED Lamp	25	1	1	2
6	Incandescent Lamp	100	1	1	2
7	Oil Heater	600/900	1	2	4
8	Fan	140	1	1	2
9	Laptop Charger	140	1	1	2
10	Drill	160/680	2	2	4
11	Hairdryer (two power levels)	150/300	1	2	4
	Total		12	14	36

4.4. LIT-Dataset Integration to NILMTK

As NILMTK uses an internal data format (NILMTK-DF), a data format conversion function must be implemented such as those already available for REDD, Smart, and UK-Dale [36]. Such a function was implemented for the LIT-Dataset; hence, its waveforms can be processed in NILMTK. Figure 11a,b presents one of the LIT-Dataset waveforms, an incandescent light bulb that is also presented in Section 5.

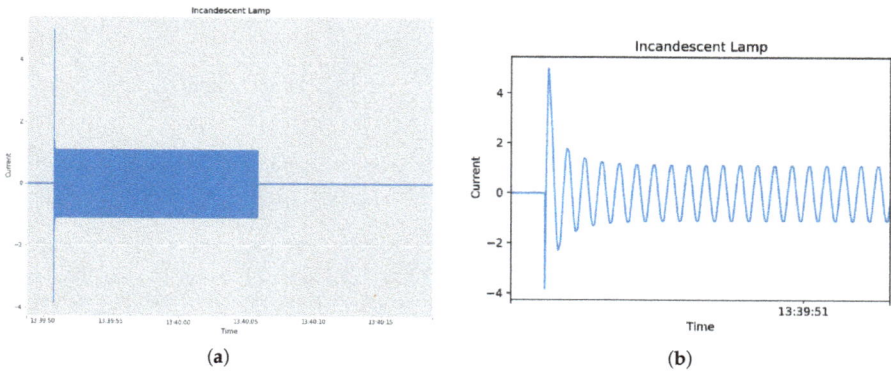

Figure 11. A LIT-Dataset waveform of an incandescent lamp presented in NILMTK. (**a**) Complete waveform. (**b**) Image zoomed into power-ON event.

5. Results and Analysis

The results of data collection for each subset and the corresponding analysis are detailed as follows.

5.1. Synthetic Subset

The original aspects of the Synthetic subset include multiple concurrent loads of distinct classes, with precise turn-ON and turn-OFF control and annotations of these events with an accuracy better than 5 ms. These annotations (labels) can later be used to validate event detection, transient feature extraction, and load classification methods.

The synthetic subset is composed of 1664 acquisition for single, double, threefold, and eight-fold concurrent loads. For every load or load combinations, acquisitions are made for 16 distinct turn-on trigger angles.

In Figure 12a, the acquisition of an incandescent lamp with a turn-on trigger angle of 90 degrees is shown, while Figure 12b,c present a detailed (zoomed-in) view of the turn-ON and turn-OFF events, respectively. The high inrush current is due to the variation of the filament resistance of the lamp, as its temperature rises. The inrush current is also dependent on the turn-on trigger angle. This unique transient response may be beneficial to the detection as well as the classification methods. In these figures, the up-arrow indicates a turn-ON event while the down-arrow indicates a turn-OFF event.

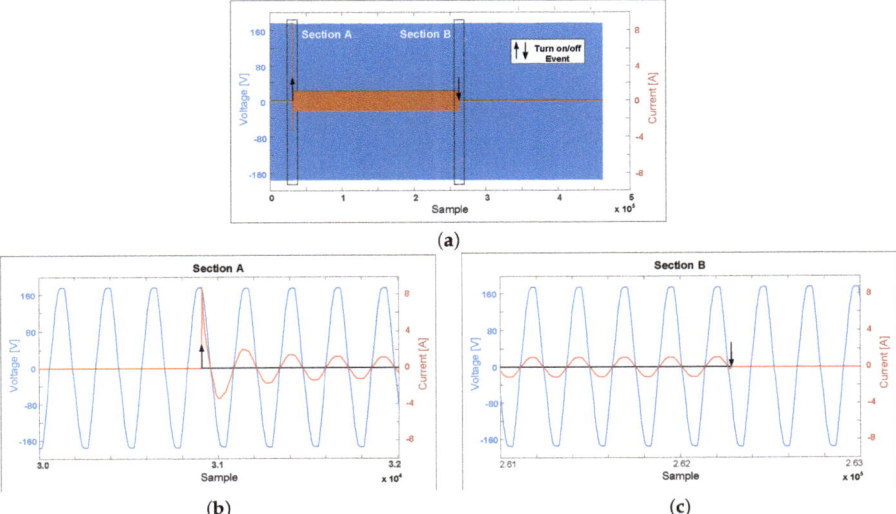

Figure 12. Jig acquisition of a single load: incandescent lamp—trigger angle of 90 degrees. (**a**) Complete acquisition—AC mains voltage and current. (**b**) Turn-ON event. (**c**) Turn-OFF event.

A single load acquisition of a laptop power supply is presented in Figure 13a, for a turn-on trigger angle of 45 degrees. A detail of the turn-ON and turn-OFF events are presented in Figure 13b,c. Typically a power supply first stage consists of a diode rectifier followed by a capacitor. The inrush current depends on the capacitance and the turn-on trigger angle and is very high compared to the steady-state peak current. This transient response is very rich in detecting an event and classify the load. The steady-state low power may be challenging to detect and classification steady-state based methods.

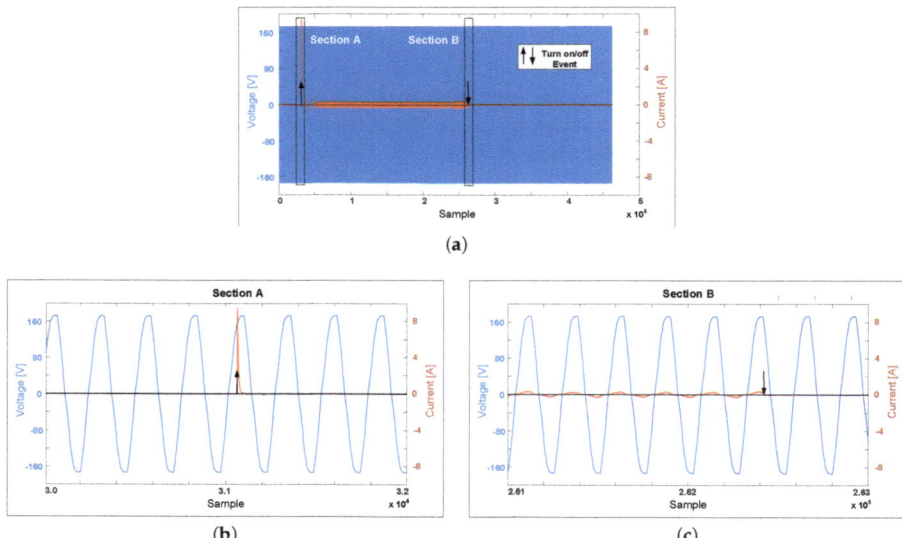

Figure 13. Jig acquisition of a single load: laptop power supply—trigger angle of 45 degrees. (**a**) Complete acquisition—AC mains voltage and current. (**b**) Turn-ON event. (**c**) Turn-OFF event.

An example of a double load acquisition is presented in Figure 14. An oil heater (520 W) is turned-on (trigger angle of 135 degrees), and then a LED lamp (6 W) is turned on, also at trigger angle of 135 degrees. Later, the heater is turned off, and finally, the lamp is turned off. This is an interesting combination of linear and non-linear loads of significantly different power levels. Details of the turn-ON and turn-OFF events are presented in Figure 15a–d. As the oil heater has a higher power, the turn-ON event of the LED lamp may be challenging to detect, as Figure 15b shows, likewise, the turn-OFF event of the LED lamp, as shows Figure 15d.

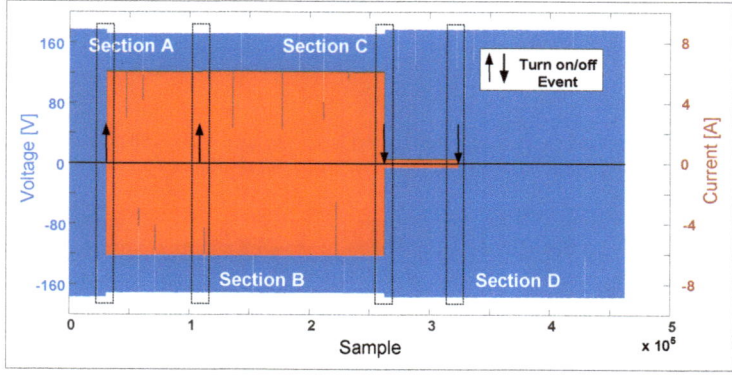

Figure 14. Jig acquisition of two loads: oil heater and LED lamp—trigger angle of 135 degrees.

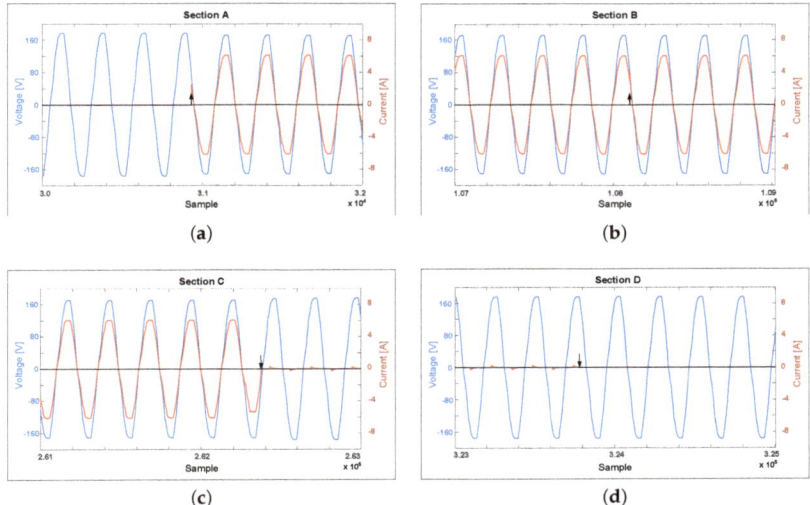

Figure 15. Jig acquisition of two loads: oil heater and LED lamp—trigger angle of 135 degrees. (**a**) Turn-ON event section A: heater turned on. (**b**) Turn-ON event section B: LED lamp turned on. (**c**) Turn-OFF event section C: heater turned off. (**d**) Turn-OFF event section D: LED lamp turned off.

A three loads combination composed of a hairdryer, a LED lamp, and a drill is presented in Figure 16, with a turn-on trigger angle of 225 degrees. The hairdryer, at the low power level setting, has a half-wave diode rectifier, hence, an asymmetrical load. The drill high inrush current may also be observed.

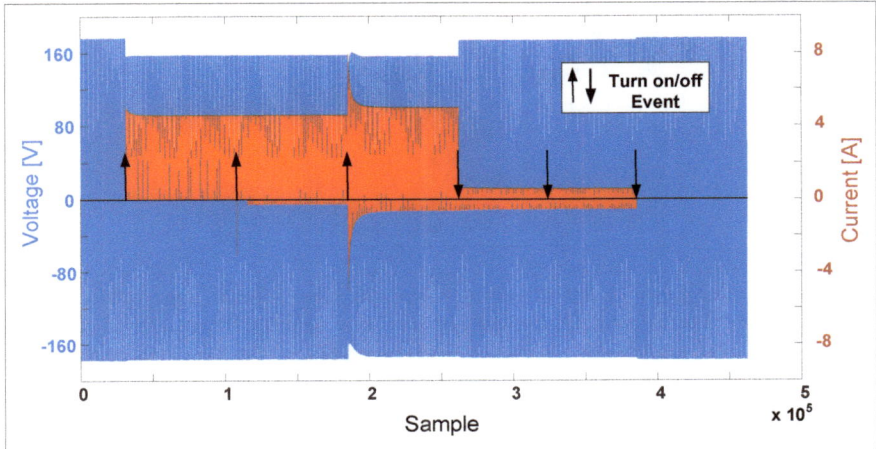

Figure 16. Jig acquisition of three loads: hairdryer at low power level, LED lamp, and drill—trigger angle of 225 degrees. AC mains voltage and current.

Finally, Figure 17 shows an example of eight loads combination: a LED lamp, a laptop power supply, a microwave, a cell phone charger, a soldering station, an incandescent lamp, an oil heater, and a smoke extractor with turn-on events triggered at 270 degrees. The combination of eight loads with different power levels, linear and non-linear characteristics, is important to evaluate detection and classification methods.

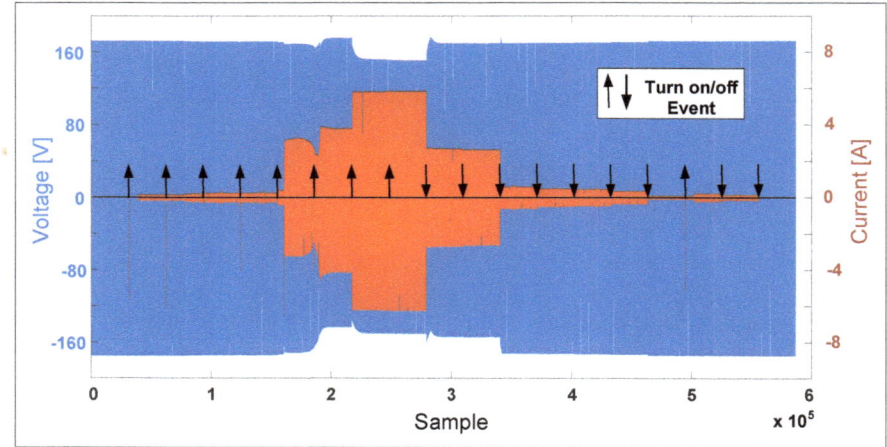

Figure 17. Jig acquisition of eight loads: LED lamp, laptop power supply, microwave, cell phone charger, soldering station, incandescent lamp, oil heater, smoke extractor—trigger angle of 270 degrees. AC mains voltage and current.

5.2. Simulated Subset

The circuits with (a) resistor; (b) resistor and inductor: (c) diode rectifier with resistor; (d) diode full-wave bridge rectifier with resistor and capacitor; (e) thyristor rectifier with resistor; and (f) thyristor rectifier with resistor and inductor were evaluated with a test bench. The load current and mains voltage were acquired using voltage and current probes and an oscilloscope, with the simulated loads configured as presented in Table 11.

Table 11. Parameters of real components in the test bench.

Circuit	Values
(a)	R = 50 Ω
(b)	R = 100 Ω L = 1 H
(c)	R = 100
(d)	R = 300 Ω C = 600 µF
(e)	R = 100
(f)	R = 100 Ω L = 1 H

In addition to the voltage and current measurements using the test bench, the amplitude and phase of each harmonic of the waveform of the voltage of the power network was measured. These values, presented in Table 12, were included in the voltage source block in the simulator and used in all simulations that included harmonic contents (configurations DB-3 to DB-6 in Table 9). The amplitude is presented with respect to the fundamental component: 60 Hz and peak voltage of 179 V).

The parameters presented in Tables 11 and 12 were used in the simulation framework developed in Matlab/Simulink. Then, the measured and simulated waveforms were compared, as exemplified in Figure 18.

One way to validate the simulation is by comparing the waveform's electrical parameters, such as transient and steady-state current and voltage peaks, power factor (PF), and mean squared error (MSE) of the samples of the measured and simulated waveforms, as suggested in [46]. Therefore, Tables 13 and 14 present such comparisons for the simulations proposed in this work. The results presented in these tables validate the presented simulation approach for circuits (a) to (f).

Table 12. Measurement of voltage and phase of harmonics in the power network.

Harm.	Amplitude (%)	Phase (Radians)
3°	2.0	0.3
5°	3.0	0.4
7°	1.0	3.1
9°	1.0	−2.5
11°	0.1	−1.0
13°	0.3	−1.9
15°	0.4	1.3
17°	0.1	−0.2
19°	0.1	2.2
21°	0.1	−1.4
23°	0.1	1.5
25°	0.1	1.0
27°	0.1	3.0
29°	0.1	2.6
31°	0.1	−1.6
33°	0.1	0.7
35°	0.1	0.3
37°	0.1	1.5
39°	0.1	1.6

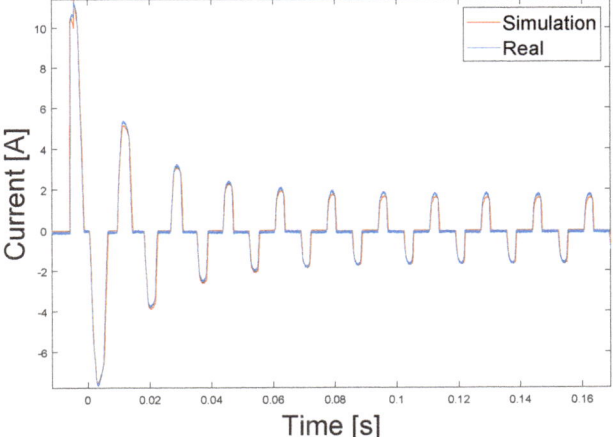

Figure 18. Measured and simulated current of the full-wave bridge rectifier load. Adapted from [10].

Table 13. Validation parameters (absolute current and voltage peak differences).

Circuit	Absolute Voltage Peak Difference (V)	Absolute Current Peak Difference (A)
Load (a)	0.1	0.001
Load (b)—Transient	1.5	0.01
Load (b)—Steady-State	1.5	0.004
Load (c)	1.9	0.01
Load (d)—Transient	0.8	0.1
Load (d)—Steady-State	0.8	0.05
Load (e)	4.9	0.001
Load (f)	2.5	0.01

Table 14. Validation parameters (power factor (PF) and mean squared error (MSE)).

	PF Real	Sim	MSE Voltage	Current
(a)	1	1	2.46×10^{-4}	1.80×10^{-4}
(b)	0.34	0.32	4.26×10^{-4}	0.0042
(c)	1	0.999	3.35×10^{-4}	1.01×10^{-4}
(d)	0.848	0.887	1.49×10^{-4}	0.0074
(e)	0.54	0.4756	3.64×10^{-4}	0.0041
(f)	0.18	0.16	0.0025	0.0041

Finally, the validation of the last circuit (g), the universal motor, was performed using an electric drill of 750 W_{peak} as a reference, with two-speed selection. The procedure was conducted in two stages. Firstly, the current and voltage signals of the electric drill were acquired in different scenarios, i.e., switching angle and load conditions. Then, the acquired signals were used to obtain the field and rotor resistances and inductances of the model presented in Figure 6. The final obtained values used in this model were:

- Rated power = 325 W;
- Rated terminal voltage = 120 Vrms;
- Rated speed = 2800 rev/min;
- Armature winding inductance, L_{aq} = 10 mH;
- Series field winding inductance, L_{se} = 26 mH;
- Rated frequency of supply voltage = 60 Hz;
- Armature winding resistance, R_a = 0.6 Ω;
- Series field winding resistance, R_{se} = 0.1 Ω;
- Rotor inertia, J = 0.0015 kg·m^2;
- Speed at which mag. curve data was taken $\omega_{mo} = 2\pi w/60$, with $w = 1500$ rev/min.

Secondly, with these parameters, the comparison of the real and simulated current of a universal motor is presented in Figure 19. As can be observed, the waveforms present similar values in transient and steady-state. The MSE between these waveforms is 0.03 A^2.

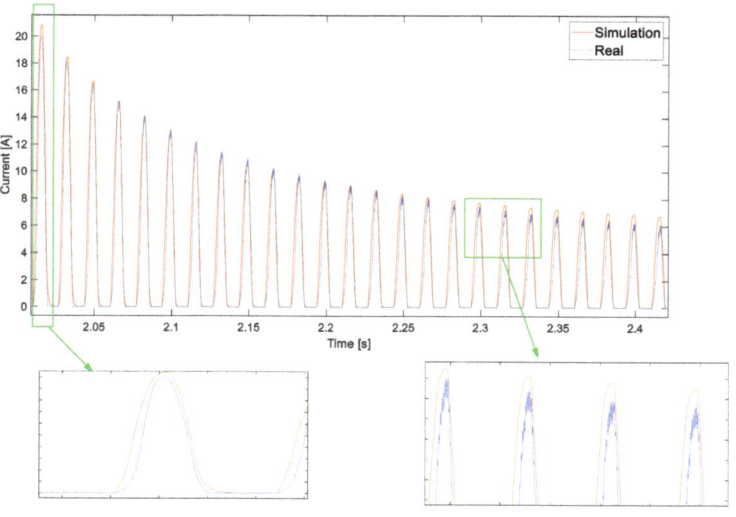

Figure 19. Comparison of the measured and simulated waveforms of a drill. Adapted from [10].

Concerning the generation of the waveforms that compose the Simulated subset, Figure 20 presents an example of the current waveform generated in the proposed subset (DB-5), in a double load scenario where the first load is a resistor and inductor circuit (section A, Figure 21a and section C, Figure 21c) and the second load is the universal motor (section B, Figure 21b and section D, Figure 21d).

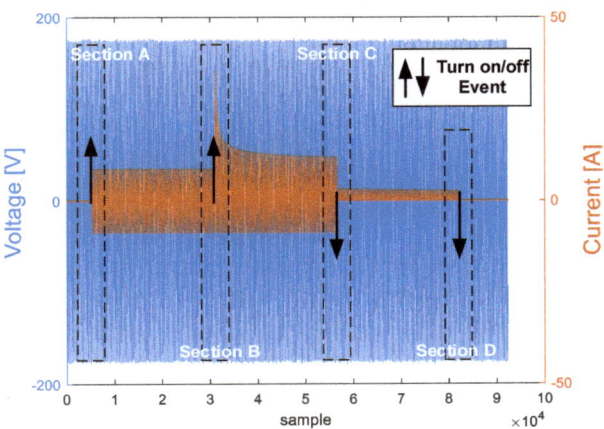

Figure 20. Simulated subset complete acquisition: RL and universal motor. AC mains voltage and current.

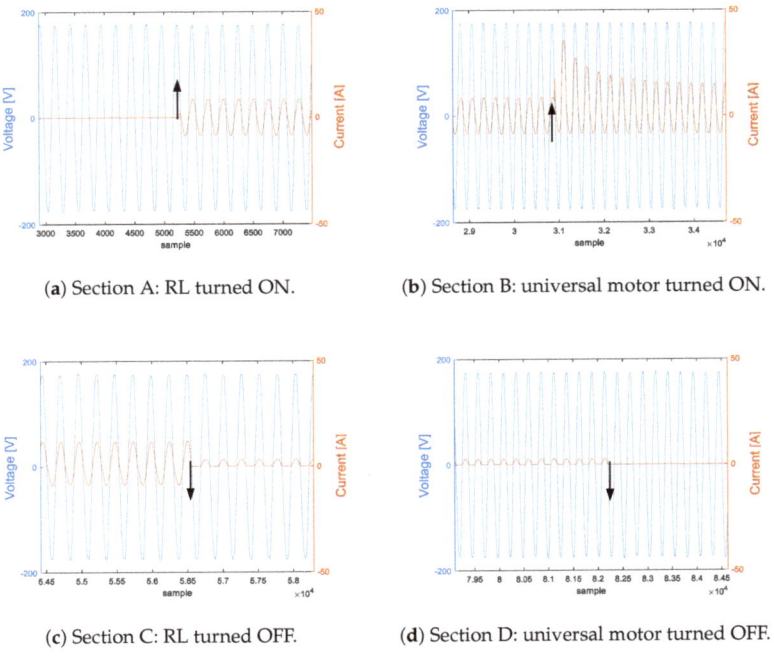

(a) Section A: RL turned ON.

(b) Section B: universal motor turned ON.

(c) Section C: RL turned OFF.

(d) Section D: universal motor turned OFF.

Figure 21. Sections (a–d): RL and universal motor.

The sampling frequency for the Simulated subset is also 15,360 Hz and switching instants (ON or OFF) are precisely controlled at the sample level. For each switching-event, the load is also properly labeled, allowing the correct use of supervised classifiers and transient feature extraction methods.

The simulator's functionality allows for the generation of single-load waveforms as well as the combination of two, three, four, five, six, and seven loads. Such combinations can be accomplished using a MATLAB script that automates the waveform generation, using pre-defined trigger instants and types of loads that are selected in each simulation (The MATLAB-Simulink template to generate this dataset is made publicly available at https://github.com/hellenancelmo/Simulated-LIT-dataset). Hence, this subset can be extended to other types of residential, commercial, and low-voltage industrial loads.

5.3. Natural Subset

To illustrate the validation of the data collection system of the Natural subset (NSAS—Figure 9) a 3-load combination recording is presented. In this case, the EDNs are identified by a code transmitted in the data package to the Synchronization Master and Acquisition Node (SMAN). Table 15 presents the three loads used, the number of the respective EDN, and the corresponding identification code for power-ON and power-OFF events.

Table 15. Natural subset Devices (Acquisition Example).

EDN	Load	ON Code Id	OFF Code Id
8	Incandescent Lamp	35	32
2	LED Lamp	11	08
5	Drill	23	20

The following sequence of load switching occurred: (i) turn-ON incandescent lamp; (ii) turn-OFF incandescent lamp; (iii) turn-ON LED lamp; (iv) turn-OFF LED lamp; (v) turn-ON drill; (vi) turn-OFF drill, generating the voltage and current curves represented in Figure 22a.

Table 16 presents all six events. The actual instant of the power-ON/power-OFF events can be obtained by analyzing the waveforms, as represented in Figure 22b, where the turn-ON Event 1 at sample 194,266 is shown. These values are presented in column "Event observed in waveform (in samples)" and represent the ground truth. The EDN data packet contains the time stamp and event code in the format $YYYY:MM:DD:HH:MM:SS, sample\_number\_after\_SS, Code\_Id$. From the data packet, the reported time of the event (in samples) is obtained using the time-stamp of the first sample in the waveform. The corresponding error, measured in the number of samples, corresponds to the distance (in samples) between ground-truth and the detected event. The sampling frequency is 15,360 Hz; hence, each sample corresponds to 65.1 µs.

Table 16. Events representations file.

EDN Data Packet	Event Description	Event Reported by EDN (in Samples)	Event Observed in Waveform (in Samples)	Error (in Samples)
2020:04:16:15:12:06, 1631, 35	turn-ON incandescent lamp	193,899	194,266	367
2020:04:16:15:12:14, 2892, 32	turn-OFF incandescent lamp	318,232	318,294	62
2020:04:16:15:12:19, 9815, 11	turn-ON LED lamp	402,075	402,425	350
2020:04:16:15:12:27, 2538, 08	turn-OFF LED lamp	517,870	518,092	222
2020:04:16:15:12:33, 1261 ,23	turn-ON drill	608,897	609,303	406
2020:04:16:15:12:40, 2892, 20	turn-OFF drill	718,216	718,564	348
Time-stamp of first sample	1,587,049,915			

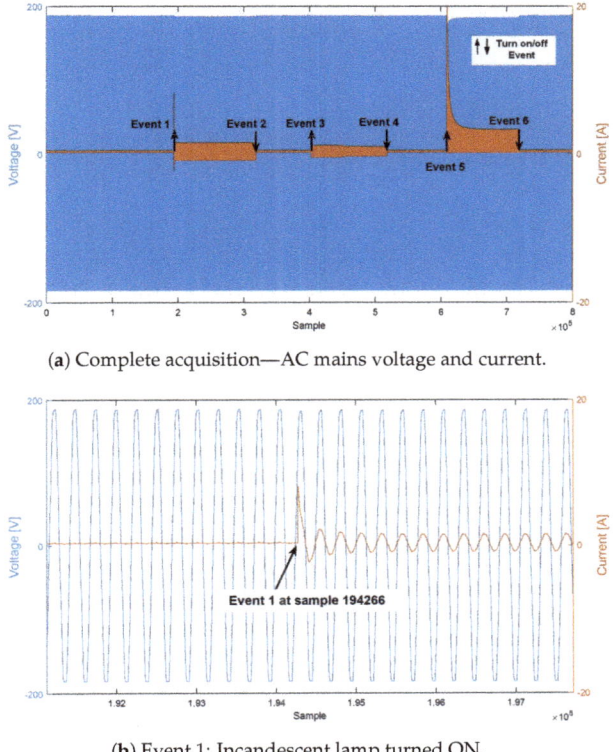

(a) Complete acquisition—AC mains voltage and current.

(b) Event 1: Incandescent lamp turned ON.

Figure 22. Natural subset: incandescent lamp, LED lamp, and drill.

5.4. Analysis of the Results

In this section, the analysis of the three subsets is presented. Table 17 illustrates the high-frequency datasets (initially presented in Section 2.2, Table 3), now including the three LIT-Dataset subsets.

From the results presented previously in this section and by the comparative analysis summarized in Table 17, the distinct features of the LIT-Dataset are:

- It consists of three subsets, each one with multiple concurrent loads of distinct types, including those found in residential, commercial, and low-voltage industrial environments.
- The Synthetic subset contains waveforms that were collected on a jig, with precise control of turn-ON and turn-OFF of up to eight loads (thus, synthetic load shaping).
- The Simulated subset contains waveforms that were collected by simulation; hence, the simulated circuits can be easily modified to match several distinct real-world scenarios.
- The Natural subset contains waveforms that were collected in a real-world environment; hence, representing what a NILM system would actually monitor and analyze.
- Ground Truth, an essential requirement for the evaluation of NILM algorithms and techniques, is achieved by labeling, at sample level, the load events, i.e., when each load has a change in power, due to power-ON, power-OFF of power-level-change. For each load event, the corresponding load and event type is recorded in the label.
- The resolution of the load event labeling is better than 5 ms; hence, identifying the mains semi-cycle where the load event occurs.

Table 17. Comparison of high-frequency NILM datasets.

Dataset	Date	Nature	DCD	MSL	f_s	Ground Truth Resolution (LER)	NoC	NoA
REDD	2011	Res.	119 days (10 houses)	yes	15 kHz	3 s	8	24
BLUED	2012	Res.	7 days (1 house)	yes	12 kHz	640 ms	9	43
PLAID	2014	Res.	1094 waveforms (of 1 s each)	no	30 kHz	>1 cycle	12	235
HFED	2015	Res. + lab.	-	yes	10 kHz–5 MHz	-	-	24
UK-DALE	2015	Res.	655 days	yes	16 kHz	6 s	16	54
COOLL	2016	Res.	840 waveforms (of 6 s each)	no	100 kHz	20 ms	12	42
SustDataED	2016	Res.	10 days	yes	12.8 kHz	2 s	-	17
WHIETED	2016	Res.+ Ind.	5123 waveforms (of 5 s each)	no	44.1 kHz	-	47	110
BLOND.	2018	Res.	50–213 days	yes	50–250 kHz	-	16	53
SynD	2020	Res.	180 days (21 households)	yes	5 Hz	0.2 s	-	22
LIT SYNTHETIC	2020	Res. + Com. + Ind.	1664 waveforms (30 s to 40 s each)	yes	15.36 kHz	<5 ms	16 *	19 *
LIT SIMULATED	2020	Res. + Com. + Ind.	4824 waveforms (2.5 s to 16 s each)	yes	15.36 kHz (dec. from 1 MHz)	<65 µs	7 *	28 *
LIT NATURAL	2020	Res. + Com. + Ind.	2 h *	yes	15.36 kHz	<5 ms	11 *	12 *

(*)—The LIT-Dataset is still in the process of data collection, particularly for the Natural subset that is currently in the early phases of data collection.

5.5. Considerations on the Design Process

The LIT-Dataset, composed of three subsets, is the result of a design process that started with Requirements Engineering, including requirements harvesting of the Stakeholders Requirements, identification of the source and derived requirements, data format design, development process of the Enabling Systems (Jig, Simulator, and network of NSAS), validation of the enabling systems, data collection, data validation, and publication of the dataset files and user support documentation.

Following such a well-defined design process was beneficial to keep the project on track and according to established project planning. Even so, as with most engineering projects, some difficulties presented themselves during the process. The most relevant and time-consuming were:

- Due to cost restrictions of each NSAS module, an ESP32 processor was selected that we had no previous experience on. It turned out that the particular model selected had a design fault that causes the interference of the WiFi and ADCs. Some extra effort on the project was required until this problem was identified and solved (by disabling the WiFi and reconfiguration of the ADC before every A/D conversion).
- The low-cost transmitter-receiver 433 MHz RF synchronization network resulted in a relatively high packet loss rate. Again, an extra effort was required to identify the problem and design an algorithm to cope with such high packet losses.

- Certainly, the most unexpected difficulty was to finish the project, on-time, during the COVID-19 Pandemic. Significant changes in the work environment, basically moving all activities to home office, required an unexpected amount of extra work.

As the initial planning included very little slack time to cope with such difficulties, the solution to keep the original schedule of the project was to increase the weekly work effort of the participants. The collection of data for the Natural subset is somewhat delayed. The aim is to continue data collection for all subsets.

6. Conclusions

The LIT-Dataset was presented chronologically. Its rationale in supporting our own NILM development as well as making it publicly available. Its conception; its requirements elicitation and specification, based on an evaluation of available NILM datasets and the additional needs. Its design, structuring the LIT-Dataset into three subsets, each exploring a different load-set context. The design and implementation of the supporting systems for each of the subsets: jig, simulator, and NSAS. Its evaluation and validation, based on the comparison of simulated loads to real-world loads as well as its usage in NILM techniques. Finally, its publication (the LIT-Dataset is publicly available, upon free registration, at http://dainf.ct.utfpr.edu.br/~douglas/LIT_Dataset), with detailed documentation and usage scripts.

The three subsets consider the scenarios of (1) a set of up to eight loads that are controlled (on and off switching) individually during the recording of aggregated current and load events; (2) a set of simulated loads that are recorded under conditions that would be difficult in real-world situations, either because they are uncommon or due to hazardous scenarios such as short-circuits; and (3) a set of loads monitored during their daily use. The first subset is the named Synthetic load shaping, as the "on" and "off" events are controlled, the second is named Simulated, and the third is named Natural load shaping as there is no influence on the loads during the recording period.

Among the distinct features of the LIT-Dataset, as described in Section 5.4, is the labeling of the load events at sample level resolution and with an accuracy better than 5 ms; the availability of such precise timing information that also includes the identification of the load and of the sort of power event is an essential requirement both for the evaluation of NILM algorithms and techniques, as well as, for training of NILM systems, particularly those based on Machine Learning.

Our contribution is to make publicly available a new dataset whose combination of features makes it unique. These features are: (1) the availability of load-event labels, with an accuracy better than 5 ms, providing ground-truth information of the load events, (2) the availability of three subsets (as described above), (3) recording scenarios with up to eight concurrent loads, (4) combination of residential, commercial and low-voltage industrial loads, and (5) load shaping scenarios with low-power loads being switched when high-power loads are energized.

To summarize the benefits of these contributions, concerning the availability of load-event labels, the LIT-Dataset achieved the best accuracy among the datasets that were analyzed (Table 17). This is an important characteristic for those using a dataset to validate event detection and load classification algorithms. Having loads recorded individually and concurrently also provides the required information for training as well as for evaluating the performance of NILM algorithms. Furthermore, scenarios where low-power loads switching when higher-power loads are powered-on, provides challenging test cases for these NILM algorithms.

The LIT-Dataset was presented here, from its conception to implementation, analysis of results, and publication. However, data collection is in progress as new loads, and new scenarios are frequently recorded and added to the dataset.

Author Contributions: Conceptualization, D.P.B.R., F.P., H.C.A., and C.R.E.L.; data curation, D.P.B.R., F.P., H.C.A., R.R.L., L.d.S.N., L.T.L., B.M.M., and J.R.L.d.S.; formal analysis, D.P.B.R., A.E.L., R.R.L., and E.O.; funding acquisition, D.P.B.R., A.E.L., J.S.O., and R.B.d.S.; investigation, D.P.B.R., F.P., H.C.A., A.E.L., E.O., and L.d.S.N.; methodology, D.P.B.R., F.P., H.C.A., A.E.L., C.R.E.L., and R.R.L.; project administration, D.P.B.R., J.S.O., and R.B.d.S.; resources, D.P.B.R., C.R.E.L., and R.R.L.; software, D.P.B.R., H.C.A., C.R.E.L., R.R.L., L.T.L., and B.M.M.; supervision, D.P.B.R., J.S.O., and R.B.d.S.; validation, D.P.B.R., F.P., H.C.A., A.E.L., C.R.E.L., E.O., L.d.S.N., and L.T.L.; visualization, D.P.B.R., L.d.S.N., and J.R.L.d.S.; writing—original draft preparation, D.P.B.R., F.P., H.C.A., A.E.L., C.R.E.L., R.R.L., and E.O.; writing—review and editing, D.P.B.R., F.P., and E.O. All authors have read and agreed to the published version of the manuscript.

Funding: This study was fully financed by Agência Nacional de Energia Elétrica (ANEEL) and Companhia Paranaense de Energia Elétrica (COPEL) under the research and development program (project PD2866-0464/2017).

Acknowledgments: The authors would like to thank COPEL and ANEEL for the support and promotion in the research project PD2866-0464/2017.

Conflicts of Interest: The authors declare no conflict of interest. The funders had no role in the design of the study; in the collection, analyses, or interpretation of data; the authors affiliated to the funder company had the role of manuscript revision and evaluating the request for publication.

Abbreviations

The following abbreviations are used in this manuscript:

AC	Alternating current	MSL	Multiple Simultaneous Loads
AMPds	Almanac of Minutely Power Dataset	NILM	Non-Intrusive Load Monitoring
ANEEL	Agência Nacional de Energia Elétrica	NILMTK	NILM Toolkit
API	Application Programming Interface	NMEA	National Marine Electronics Association
AWGN	Additive White Gaussian Noise	NoC	Number of Appliance Classes
BLUED	Building-Level fUlly-labeled dataset for Electricity Disaggregation	NoA	Number of Appliances
BLOND	Building-Level Office enviroNment Dataset	NSAS	Natural Subset Acquisition System
Com	Commercial	PC	Personal Computer
COOLL	Controlled On/Off Loads Library	PF	Power Factor
COPEL	Companhia Paranaense de Energia	PIR	Passive InfraRed
DB	Database	PLAID	Plug Load Appliance Identification Dataset
DCD	Data Collection Duration	PPS	Pulse Per Second
DF	Data Format	PVC	Polyvinyl Chloride
DMA	Direct Memory Access	RAE	Rainforest Automation Energy
DSReq	Dataset Stakeholder Requirements	REDD	Reference Energy Disaggregation Dataset
EMI	Electromagnetic Interference	Res	Residential
EDN	Event Detection Node	RF	Radio Frequency
FIFO	First In, First Out	RFID	Radio Frequency IDentification
FPGA	Field-Programmable Gate Array	SNR	Signal-to-Noise Ratio
GPS	Global Positioning System	SMAN	Synchronization Master and Acquisition Node
HES	Household Electricity Survey	SusDataED	Sustainable Data for Energy Disaggregation
HFED	High-Frequency Energy Data	SynD	Synthetic energy Dataset
iAWE	Indian Dataset for Ambient Water and Energy	TDMS	Technical Data Management Streaming
IDE	Integrated Development Environment	TRIAC	Triode for Alternating Current
Ind	Industrial	USB	Universal Serial Bus
Lab	Laboratory	UK-DALE	United Kingdom recording Domestic Appliance-Level Electricity
LED	Light-Emitting Diode	UWB	Ultra Wide Band
LER	Load Event Resolution	WHITED	Worldwide Household and Industry Transient Energy Dataset
LIT	Laboratory of Innovation and Technology in Embedded Systems and Energy	WLAN	Wireless Local Area Network
MSE	Mean Squared Error		

References

1. International Energy Agency. *World Energy Outlook*; OECD Publishing: Paris, France, 2019; p. 810.
2. Zoha, A.; Gluhak, A.; Imran, M.A.; Rajasegarar, S. Non-intrusive Load Monitoring approaches for disaggregated energy sensing: A survey. *Sensors* **2012**, *12*, 16838–16866. [CrossRef] [PubMed]

3. Lazzaretti, A.; Renaux, D.; Lima, C.; Mulinari, B.; Ancelmo, H.; Oroski, E.; Pottker, F.; Linhares, R.; Nolasco, L.; Lima, L.; et al. A Multi-Agent NILM Architecture for Event Detection and Load Classification. *Energies* **2020**, *13*, 4396. [CrossRef]
4. LIT. LIT-Dataset: A Dataset of Voltage and Current Waveforms on a Variety of Single and Multiple Loads. 2019. Available online: http://dainf.ct.utfpr.edu.br/~douglas/LIT_Dataset (accessed on 15 August 2020).
5. Ruano, A.; Hernandez, A.; Ureña, J.; Ruano, M.; Garcia, J. NILM techniques for intelligent home energy management and ambient assisted living: A review. *Energies* **2019**, *12*, 2203. [CrossRef]
6. Renaux, D.P.B.; Lima, C.R.E.; Pottker, F.; Oroski, E.; Lazzaretti, A.E.; Linhares, R.R.; Almeida, A.R.; Coelho, A.O.; Hercules, M.C. Non-Intrusive Load Monitoring: An Architecture and its evaluation for Power Electronics loads. In Proceedings of the 2018 IEEE International Power Electronics and Application Conference and Exposition (PEAC), Shenzhen, China, 4–7 November 2018; pp. 1–6.
7. Ancelmo, H.C.; Grando, F.L.; Costa, C.H.D.; Mulinari, B.M.; Oroski, E.; Lazzaretti, A.E.; Pottker, F.; Renaux, D.P.B. Automatic Power Signature Analysis using Prony's Method and Machine Learning-Based Classifiers. In Proceedings of the 2nd European Conference on Electrical Engineering and Computer Science (EECS), Bern, Switzerland, 20–22 December 2018; pp. 65–70.
8. Pottker, F.; Lazzaretti, A.E.; Renaux, D.P.B.; Linhares, R.R.; Lima, C.R.E.; Ancelmo, H.C.; Mulinari, B.M. Non-Intrusive Load Monitoring: A Multi-Agent Architecture and Results. In Proceedings of the 2nd European Conference on Electrical Engineering and Computer Science (EECS), Bern, Switzerland, 20–22 December 2018; pp. 328–334.
9. Renaux, D.; Linhares, R.; Pottker, F.; Lazzaretti, A.E.; Lima, C.; Coelho Neto, A.; Campaner, M. Designinga Novel Dataset for Non-intrusive Load Monitoring. In Proceedings of the VIII Brazilian Symposium on Computing Systems Engineering (SBESC), Curitiba, Brazil, 7–10 November 2018; pp. 243–249.
10. Ancelmo, H.C.; Mulinari, B.M.; Pottker, F.; Lazzaretti, A.E.; Bazzo, T.d.P.M.; Oroski, E.; Renaux, D.P.B.; Lima, C.R.E.; Linhares, R.R.; Gamba, A.R.d.A. A New Simulated Database for Classification Comparison in Power Signature Analysis. In Proceedings of the 20th International Conference on Intelligent System Application to Power Systems (ISAP), New Delhi, India, 10–14 December 2019; pp. 1–7.
11. Ancelmo, H.C.; Grando, F.L.; Mulinari, B.M.; da Costa, C.H.; Lazzaretti, A.E.; Oroski, E.; Renaux, D.P.B.; Pottker, F.; Lima, C.R.E.; Linhares, R.R. A Transient and Steady-State Power Signature Feature Extraction Using Different Prony's Methods. In Proceedings of the 20th International Conference on Intelligent System Application to Power Systems (ISAP), New Delhi, India, 10–14 December 2019; pp. 1–6.
12. Mulinari, B.M.; de Campos, D.P.; da Costa, C.H.; Ancelmo, H.C.; Lazzaretti, A.E.; Oroski, E.; Lima, C.R.E.; Renaux, D.P.B.; Pottker, F.; Linhares, R.R. A New Set of Steady-State and Transient Features for Power Signature Analysis Based on V-I Trajectory. In Proceedings of the 2019 IEEE PES Innovative Smart Grid Technologies Conference—Latin America (ISGT Latin America), Gramado, Brazil, 15–18 September 2019; pp. 1–6.
13. Linhares, R.R.; Lima, C.R.E.; Renaux, D.P.B.; Pottker, F.; Oroski, E.; Lazzaretti, A.E.; Mulinari, B.M.; Ancelmo, H.C.; Gamba, A.; Bernardi, L.A.; et al. One-millisecond low-cost synchronization of wireless sensor network. In Proceedings of the IX Brazilian Symposium on Computing Systems Engineering (SBESC), Natal, Brazil, 19–22 November 2019; pp. 1–8.
14. Jia, M.; Srinivasan, R. Occupant behavior modeling for smart buildings: A critical review of data acquisition technologies and modeling methodologies. In Proceedings of the 2015 IEEE Winter Simulation Conference (WSC), Huntington Beach, CA, USA, 6–9 December 2015.
15. Kriechbaumer, T.; Jacobsen, H.A. BLOND, a building-level office environment dataset of typical electrical appliances. *Sci. Data* **2018**, *5*, 180048. [CrossRef] [PubMed]
16. Kelly, J.; Knottenbelt, W. The UK-DALE dataset, domestic appliance-level electricity demand and whole-house demand from five UK homes. *Sci. Data* **2015**, *2*, 1–14. [CrossRef] [PubMed]
17. Himeur, Y.; Alsalemi, A.; Bensaali, F.; Amira, A. Building power consumption datasets: Survey, taxonomy and future directions. *Energy Build.* **2020**, *227*, 110404. [CrossRef]
18. Barker, S.; Mishra, A.; Irwin, D.; Cecchet, E.; Shenoy, P.; Albrecht, J. Smart: An Open Data Set and Tools for Enabling Research in Sustainable Homes. In Proceedings of the 2012 Data Mining Applications in Sustainability (SustKDD), Beijing, China, 12–16 August 2012.
19. Zimmermann, J.P.; Evans, M.; Griggs, J.; King, N.; Harding, L.; Roberts, P.; Evans, C. *Household Electricity Survey: A Study of Domestic Electrical Product Usage*; Technical Report; Intertek: Oxford, UK, 2012.

20. Reinhardt, A.; Baumann, P.; Burgstahler, D.; Hollick, M.; Chonov, H.; Werner, M.; Steinmetz, R. On the accuracy of appliance identification based on distributed load metering data. In Proceedings of the 2012 Sustainable Internet and ICT for Sustainability (SustainIT), Pisa, Italy, 4–5 October 2012; pp. 1–9.
21. Parson, O.; Fisher, G.; Hersey, A.; Batra, N.; Kelly, J.; Singh, A.; Knottenbelt, W.; Rogers, A. Dataport and NILMTK: A building data set designed for non-intrusive load monitoring. In Proceedings of the 2015 IEEE Global Conference on Signal and Information Processing (GlobalSIP), Orlando, FL, USA, 14–16 December 2015; pp. 210–214.
22. Makonin, S.; Ellert, B.; Bajic, I.; Popowich, F. Electricity, water, and natural gas consumption of a residential house in Canada from 2012 to 2014. *Sci. Data* **2016**, *3*, 160037. [CrossRef] [PubMed]
23. Batra, N.; Gulati, M.; Singh, A.; Srivastava, M.B. It's Different: Insights into Home Energy Consumption in India. In Proceedings of the 5th ACM Workshop on Embedded Systems for Energy-Efficient Buildings, Rome, Italy, 14–15 November 2013; Association for Computing Machinery: New York, NY, USA, 2013; pp. 1–8.
24. Monacchi, A.; Egarter, D.; Elmenreich, W.; D'Alessandro, S.; Tonello, A.M. GREEND: An energy consumption dataset of households in Italy and Austria. In Proceedings of the 2014 IEEE International Conference on Smart Grid Communications (SmartGridComm), Venice, Italy, 3–6 November 2014; pp. 511–516.
25. Murray, D.; Stankovic, L.; Stankovic, V. An electrical load measurements dataset of United Kingdom households from a two-year longitudinal study. *Sci. Data* **2017**, *4*, 1–12. [CrossRef] [PubMed]
26. Makonin, S.; Wang, Z.; Tumpach, C. RAE: The Rainforest Automation Energy Dataset for Smart Grid Meter Data Analysis. *Sci. Data* **2018**, *3*, 8. [CrossRef]
27. Kolter, J.; Johnson, M. REDD: A Public Data Set for Energy Disaggregation. *Res. Artif. Intell.* **2011**, *25*, 59–62.
28. Anderson, K.; Ocneanu, A.; Benitez, D.; Carlson, D.; Rowe, A.; Berges, M. BLUED: A fully labeled public dataset for event-based non-intrusive load monitoring research. In Proceedings of the 2nd KDD Workshop on Data Mining Applications in Sustainability (SustKDD), Beijing, China, 12–16 August 2012; pp. 1–5.
29. Gao, J.; Giri, S.; Kara, E.; Bergés, M. PLAID: A public dataset of high-resolution electrical appliance measurements for load identification research. In Proceedings of the 1st ACM Conference on Embedded Systems for Energy-Efficient Buildings, New York, NY, USA, 4–6 November 2014; pp. 198–199. Available online: https://dl.acm.org/doi/10.1145/2674061.2675032 (accessed on 15 August 2020).
30. Gulati, M.; Ram, S.S.; Singh, A. An in Depth Study into Using EMI Signatures for Appliance Identification. In Proceedings of the 1st ACM Conference on Embedded Systems for Energy-Efficient Buildings, New York, NY, USA, 4–6 November 2014; Association for Computing Machinery: New York, NY, USA, 2014; pp. 70–79. Available online: https://dl.acm.org/doi/10.1145/2674061.2674070 (accessed on 15 August 2020).
31. Picon, T.; Nait-Meziane, M.; Ravier, P.; Lamarque, G.; Novello, C.; Le Bunetel, J.C.; Raingeaud, Y. COOLL: Controlled On/Off Loads Library, a Public Dataset of High-Sampled Electrical Signals for Appliance Identification. *arXiv* **2016**, arXiv:1611.05803.
32. Ribeiro, M.; Pereira, L.; Quintal, F.; Nunes, N. SustDataED: A Public Dataset for Electric Energy Disaggregation Research. In Proceedings of the 2016 ICT for Sustainability, Amsterdam, The Netherlands, 29 August–1 September 2016; Atlantis Press: Amsterdam, The Netherlands, 2016; pp. 244–245.
33. Pereira, L.; Quintal, F.; Gonçalves, R.; Nunes, N.J. SustData: A Public Dataset for ICT4S Electric Energy Research. In Proceedings of the 2014 Conference ICT for Sustainability, Stockholm, Sweden, 24–27 August 2014; Atlantis Press: Amsterdam, The Netherlands, 2014; pp. 359–368.
34. Kahl, M.; Haq, A.; Kriechbaumer, T.; Jacobsen, H.A. WHITED—A Worldwide Household and Industry Transient Energy Data Set. In Proceedings of the 3rd International Workshop on Non-Intrusive Load Monitoring, Vancouver, BC, Canada, 14–15 May 2016.
35. Klemenjak, C.; Kovatsch, C.; Herold, M.; Elmenreich, W. A synthetic energy dataset for non-intrusive load monitoring in households. *Sci. Data* **2020**, *7*, 1–17. [CrossRef] [PubMed]
36. Batra, N.; Kelly, J.; Parson, O.; Dutta, H.; Knottenbelt, W.; Rogers, A.; Singh, A.; Srivastava, M. NILMTK: An open source toolkit for non-intrusive load monitoring. In Proceedings of the 5th ACM International Conference on Future Energy Systems, Cambridge, UK, 11–13 June 2014.
37. Hart, G.W. Nonintrusive appliance load monitoring. *Proc. IEEE* **1992**, *80*, 1870–1891. [CrossRef]

38. Klemenjak, C.; Reinhardt, A.; Pereira, L.; Makonin, S.; Bergés, M.; Elmenreich, W. Electricity Consumption Data Sets: Pitfalls and Opportunities. In Proceedings of the 6th ACM International Conference on Systems for Energy-Efficient Buildings, Cities, and Transportation, New York, NY, USA, 13–14 November 2019; Association for Computing Machinery: New York, NY, USA, 2019; pp. 159–162.
39. National Instruments. *NI myRIO-1900 User Guide and Specification*; National Instruments: Austin, TX, USA, 2018.
40. Collin, A.J.; Tsagarakis, G.; Kiprakis, A.E.; McLaughlin, S. Development of Low-Voltage Load Models for the Residential Load Sector. *IEEE Trans. Power Syst.* **2014**, *29*, 2180–2188. [CrossRef]
41. Ong, C.M. *Dynamic Simulation of Electric Machinery: Using MATLAB/SIMULINK*; Prentice Hall PTR: Upper Saddle River, NJ, USA, 1998.
42. Mahmood, A.; Gaderer, G.; Trsek, H.; Schwalowsky, S.; Kerö, N. Towards high accuracy in IEEE 802.11 based clock synchronization using PTP. In Proceedings of the 2011 IEEE International Symposium on Precision Clock Synchronization for Measurement, Control and Communication, Munich, Germany, 12–13 September 2011; pp. 13–18.
43. Carbone, P.; Cazzorla, A.; Ferrari, P.; Flammini, A.; Moschitta, A.; Rinaldi, S.; Sauter, T.; Sisinni, E. Low Complexity UWB Radios for Precise Wireless Sensor Network Synchronization. *IEEE Trans. Instrum. Meas.* **2013**, *62*, 2538–2548. [CrossRef]
44. Mantech Electronics. *433 Mhz RF Transmitter with Receiver Kit for Arduino ARM MCU Wireless*; Mantech Electronics: Johannesburg, South Africa, 2018.
45. Ferreira, D.; Ribeiro, L. Analysis of RF 433 MHz communication in home monitoring prototype. *J. Eng. Technol. Ind. Appl.* **2018**, *4*, 24–30. [CrossRef]
46. Bacca, I.; Mendonça, M.; Tavares, C.; Gondim, I.; Oliveira, J. ATP-MODELS Language to Represent Domestic Refrigerators Performance with Power Quality Disturbances. *Renew. Energy Power Qual. J.* **2009**, *1*, 1–11. [CrossRef]

Publisher's Note: MDPI stays neutral with regard to jurisdictional claims in published maps and institutional affiliations.

© 2020 by the authors. Licensee MDPI, Basel, Switzerland. This article is an open access article distributed under the terms and conditions of the Creative Commons Attribution (CC BY) license (http://creativecommons.org/licenses/by/4.0/).

Article

Synthetic Data Generator for Electric Vehicle Charging Sessions: Modeling and Evaluation Using Real-World Data

Manu Lahariya [1,*], Dries F. Benoit [2] and Chris Develder [1]

1. IDLab, Ghent University—Imec, Technologiepark Zwijnaarde 126, 9052 Ghent, Belgium; chris.develder@ugent.be
2. Center for Statistics, Ghent University, Tweekerkenstraat 2, 9000 Ghent, Belgium; dries.benoit@ugent.be
* Correspondence: manu.lahariya@ugent.be

Received: 15 July 2020; Accepted: 12 August 2020; Published: 14 August 2020

Abstract: Electric vehicle (EV) charging stations have become prominent in electricity grids in the past few years. Their increased penetration introduces both challenges and opportunities; they contribute to increased load, but also offer flexibility potential, e.g., in deferring the load in time. To analyze such scenarios, realistic EV data are required, which are hard to come by. Therefore, in this article we define a synthetic data generator (SDG) for EV charging sessions based on a large real-world dataset. Arrival times of EVs are modeled assuming that the inter-arrival times of EVs follow an exponential distribution. Connection time for EVs is dependent on the arrival time of EV, and can be described using a conditional probability distribution. This distribution is estimated using Gaussian mixture models, and departure times can calculated by sampling connection times for EV arrivals from this distribution. Our SDG is based on a novel method for the temporal modeling of EV sessions, and jointly models the arrival and departure times of EVs for a large number of charging stations. Our SDG was trained using real-world EV sessions, and used to generate synthetic samples of session data, which were statistically indistinguishable from the real-world data. We provide both (i) source code to train SDG models from new data, and (ii) trained models that reflect real-world datasets.

Keywords: smart grid; electric vehicle; synthetic data; exponential distribution; Poisson distribution; Gaussian mixture models; mathematical modeling; machine learning; simulation

1. Introduction

The growth of electric vehicles (EVs) in the past decade has induced significant modifications in city-wide electric grids. More than one million plug-in EVs were registered in Europe in 2018, and multiple charging stations have been installed to facilitate this growth. This rise provides opportunities to collect EV session data and use it to exploit flexibility, balance load and create responsive grids. Companies can use the data generated from charging stations to understand consumer behavior, provide incentives and make pricing decisions.

Session data collected from city-wide EV charging stations can be used for both academic and industrial purposes: the increased inflow of data has huge impacts on the energy informatics field [1]. Previous studies of different EV datasets include (i) statistical analyses of data collected in the Netherlands by ElaadNL [2,3], (ii) analysis of energy consumption of EVs on data collected by the US department of energy [4] and (iii) multiple studies on the socioeconomic effects of switching to EVs in day to day use [5,6]. However, studies require reliable session data for understanding behaviors and exploring flexibility. The scarcity of reliable data has been discussed previously [7], and its necessity has been pointed out for further research purposes. Where data are available, they may still be protected under confidentiality by private data collectors, and not freely available for academic or public use. The lack of availability and difficulty in accessibility of EV charging session data poses a significant hurdle to further research in the field.

1.1. Related Work

EV session data contains the session duration and charging requirements of each EV. Previous studies studying the flexibility provided in the power grid [8], and in individual sessions [9], offer a statistical modeling methodology with which to understand EV sessions. Arrivals of EVs can be considered as events on a time scale, where session duration and charging load are dependent on each EV arrival event.

A probabilistic time series model using a generative adversarial network (GAN) has been used previously to generate synthetic samples in [10]; they modeled energy consumption for users. However, consumption can be represented as a continuous time series, which is not the case when we consider *EV arrivals* as discrete events in time. Another method used to model data was implemented and validated in [11]; they used a Markov chain model to generate load profiles only in individual charging stations, based on a Swedish dataset. This does not satisfy the need to model EV arrivals jointly for a set of charging stations. Statistical characterization of the session plug in times was also explored: Flammini et al. [12] used beta mixture models to represent the multi-modal distributions. They analyzed the distribution of arrival times during the day, but did not provide a synthetic sample generation process that includes a temporal component. Statistical representation of EV arrivals throughout the day using GMMs can also be used to randomly sample arrivals, e.g., in [3], for which they took data for 221 EVs to create day long profiles. Other methods include using a stochastic simulation methodology to generate a schedule of EVs for a population [13]. Aforementioned works only implemented temporal modeling on continuous time series collected from smart grids, which is not the case with arrival times of EVs. Arrival times of EVs are discrete events in time, and hence difficult to model.

The *departure time* of EV is dependent on the arrival time, so the connection times become conditional on arrivals. Departure time modeling has been explored exhaustively in [14], for both uni-modal and multi-modal data distributions. The underlying assumption is that in the 24 h duration, the probability of the event occurring is a time-varying function. A mixture of multiple distributions can be used to estimate this function. For EV connection times, these conditional probability distributions have been modeled using Abe–Ley mixtures [15], and a cylindrical WeiSSVM distribution [16]. Both Abe–Ley mixtures and the WeiSSVM distribution offer good alternatives for initializing the number of mixtures and their properties. Beta mixture models have also been used; an estimation method was suggested in [12] to estimate the departure profiles. However, generation and evaluation of samples from these mixtures were not included. The dependency of connection times on arrival times introduces a complexity that has not been addressed so far.

For predicting *charging demand*, a k-nearest neighbors algorithm was evaluated in [17], to predict the charging requirements of EVs at individual charging stations. However, it did not include the effect of EV session durations. Other methods including auto-regressive models [18] have also been explored for smart grids datasets, which can be used to synthetically generate smart meter data. A combination of arrival times, departure times and charging requirements of EVs have not been studied, and modeling them together provides an opportunity to generate synthetic samples of EV session data.

1.2. Contribution

In this paper, we present a state of the art model for generating samples of EV session data that will generate synthetic samples of (i) arrival times, (ii) connection times and (iii) charging load, for each EV. We describe this model as synthetic data generator (SDG), as defined in our previous work [19]. This includes temporal statistical modeling of arrivals and modeling of conditional distributions for departures and the energy required for charging the EV. This differs from [3], in the sense that we generate data on each session level, whereas they have only studied charging matrices. Herein, we also define and release trained parametric SDG models that can be used to generate session data, which were not provided in [3]. In comparison to [11], wherein load profiles were modeled using a spatial Markov chain model for five charging stations, our study includes temporal modeling of EV sessions arrivals for the joint set of multiple charging stations, derived from a large-scale real-world dataset comprising about 2000 charging stations. Along with this, we also include methods to jointly model the arrival and departure times of EVs for a large number of charging stations. Compared to [12], where the arrivals of EVs were characterized for weekends and weekdays, we propose a modeling method that can be used for any set of days that have similar properties, and adopt different statistical models. Our approach also gives us further insights into consumer behavior, by providing us the rates of EV arrivals for different hours, days and months. These generated arrivals will be used to generate the departures and required energy for each session. Our main contributions from this paper include:

- A novel approach to generating synthetic data for EV sessions over a group of charging stations defined as the SDG (Section 2).
- Training of the SDG using a real-world dataset. An analysis of statistical properties of real-world data is also included (Sections 3 and 4).
- Generation of synthetic samples, and evaluation of similarity with the real-world data. We compare results from different models that can be used in SDG (Sections 5–7).
- Trained models and code are provided in GitHub (https://github.com/mlahariya/EV-SDG). Python was used for the models developed in this article (see Appendix A);

2. Modeling Methodology

We define the synthetic data generator (SDG) in this section. We define a parametric model (SDG) that can be used to generate synthetic samples of EV session data, and its inputs. We assume that each session can be described using three parameters: (i) arrival time (t_a), (ii) connection time (t_c) and (iii) required energy (E). The departure time can be calculated using $t_d = t_a + t_c$. E represents the charging load that an EV has requested (based on measured charging power throughout the full session). Session parameters for date d can be generated using Equations (1)–(3).

$$t_a = AM(d) \tag{1}$$

$$t_c = MM_c(t_a, d) \tag{2}$$

$$E = MM_e(t_a, d) \tag{3}$$

In what follows, we define (i) the arrival model (AM), (ii) the mixture model for connection times (MM_c) and (iii) the mixture model for required energy (MM_e). Trained SDG models can be used to generate a sample of data. Data generation is a two step process.

Step 1. Arrivals: We generate the arrival of EVs (t_a) for all dates in the input horizon. This horizon is the period of time for which the data needs to be generated, and can be defined using the first date (starting date) and the last date (ending date) of this period.

Step 2. Connected time and energy required: Once we have the arrivals of EVs, we generate the connected time (t_c) and energy required (E) for that particular EV arrival.

AM, MM_c, MM_e is trained for a set of dates (**S**). Dates present in **S** will have similar daily properties (e.g., arrival profiles), and we can define **S** by assuming a grouping criteria for days, e.g., we can assume that each month will have similar arrival profiles, i.e., the grouping criteria for dates is months m. For each month m, all dates of that month will be the elements of set **S**. Details about defining **S** in practice, in particular for a real-world dataset, are included in Section 4.

2.1. Arrival Models

Arrivals of EVs in a group of charging stations (poles) can be considered as events over time. For a large number of poles, we can assume that the inter-arrival times (IATs, Δt) of EVs follow an exponential distribution (which we validate in Section 4.2). Based on this assumption, one method to model arrival times of EVs is to model the time in between arrivals (Δt). A second method is to model the total number of EV arrivals in a time interval. Both these methods are defined below.

2.1.1. Inter-Arrival Time Models

To model inter-arrival times (Δt) we use the exponential distribution, which is characterized by a rate parameter λ (rate of EV arrivals). Inter-arrival time (IAT) models are defined as follows:

$$t_i = t_{i-1} + \Delta t \tag{4}$$

$$PDF(\Delta t) = \lambda_{i-1} e^{-\lambda_{i-1} \Delta t} \tag{5}$$

$$\lambda = f_{\mathbf{S}}(t) \tag{6}$$

where the ith EV arrives at time t_i, PDF represents the probability distribution function and t is time of day. The rate parameter λ is dependent on time, and $f_{\mathbf{S}}$ defines the profile of λ with respect to t for the type of days present in $\mathbf{S}$. We can use different methods to fit $f_{\mathbf{S}}$: The **mean model** is based on average values of λ for given timeslot t_s. This results in a discontinuous mapping between λ and t, with a sudden change in λ at the boundaries of each timeslot t_s. To have continuous λ throughout the day, we use regression-based methods: either a **polynomial model** using polynomial regression, or a **localized regression model**. Training these models is explained in detail in Section 4.1.1. In Algorithm 1, we outline the pseudocode to generate arrivals over a given horizon. We use the date (d) to retrieve the appropriate $f_{\mathbf{S}}$, and predict λ. The IAT between the current and new arrival is generated as a random sample from the exponential distribution with rate λ. Arrivals are generated throughout the horizon for each date.

Algorithm 1: Inter-arrival time (IAT) model.

Input : H (Horizon, initial to final date)
Output : T (List of EV arrival times in H)
for $d \in H$ **do**
  $f_{\mathbf{S}}$ = get arrival rate model for d;
  $t = 0$;
  while $t < 24$ **do**
    $\lambda = f_{\mathbf{S}}(t)$;
    Δt = sample from exponential distribution with rate λ;
    $t = t + \Delta t$;
    append t to list T;

2.1.2. Arrival Count Models

Instead of generating the next arrival of EV, here we focus on generating the number of arrivals in a given t_s (timeslot, e.g., slots of 60 min). The number of arrivals N in t_s can be generated as a random sample from a discrete probability distribution Equation (7). This distribution can be characterized using parameters $\mathbf{P}$, and Equation (6) can be modified to Equation (8), wherein we model these parameters. We distribute N arrivals uniformly over the duration of timeslot t_s. Arrival count (AC) models can be defined as follows:

$$PDF(N) = f(\mathbf{P}) \tag{7}$$

$$\mathbf{P} = f_\mathbf{S}(t_s) \qquad (8)$$

We model the parameters **P** of the discrete distribution for each t_s using the function $f_\mathbf{S}$. Our underlying assumption that the IATs of EVs follow an exponential distribution amounts to assuming a Poisson distribution for the number of arrivals N in such a timeslot. Yet, for the Poisson distribution, the variance is equal to the mean of the distribution, while the number of arrivals may have a larger variance. In such case we need to include other discrete probability distributions that describe counts data [20]: we propose using the negative binomial model. In summary, we have two options to model the arrival counts (AC):

(1) **Poisson model:** Assuming that N follows a Poisson distribution (characterized by rate parameter λ; i.e., **P** is λ).
(2) **Negative binomial model:** Assuming that the N follows a negative binomial distribution (**P** is (μ, α)).

Pseudocode for generation of arrivals of EVs using the Poisson model is given in Algorithm 2 (adaptation to the negative binomial model for sampling N is straightforward).

Algorithm 2: Arrival count (AC) model.

Input : H (Horizon, initial to final date)
Output : T (List of EV arrival times in H)
for $d \in H$ **do**
  $f_\mathbf{S}$ = get arrival rate model for d;
  for $t_s = 1, 2, \ldots 24$ **do**
    $\lambda = f_\mathbf{S}(t_s)$;
    N = sample from Poisson distribution with rate λ;
    A = evenly space N points in t_s;
    append all $t \in A$ to list T;

2.2. Mixture Models (MM_c, MM_e)

The connection time of each plugged-in EV depends on what time the EV arrived, i.e., its arrival time. We can model the probability distribution, $PDF_{t_a}(t_c)$ using gaussian mixture models (GMM), where t_c can be generated as a random sample from the probability distribution, Equation (9), once we know the value of t_a. We can group dates of a month (or daytype) into the same type of day, for which we use the same model. These dates then form a set **S** (set of dates). Similarly to the connected times, GMMs can be fitted for required energy (charging load).

$$MM_c : PDF_{t_a, \mathbf{S}}(t_c) \qquad (9)$$

$$MM_e : PDF_{t_a, \mathbf{S}}(E) \qquad (10)$$

The steps for data generation using SDG are summarized in Figure 1b. We used a trained SDG model and horizon as inputs. As seen in Figure 1a, we provided the methodology to train the models from a raw dataset. In Section 3 we describe the data cleaning and prepossessing, and session clustering steps. Then come the details of training and evaluation in Section 4.

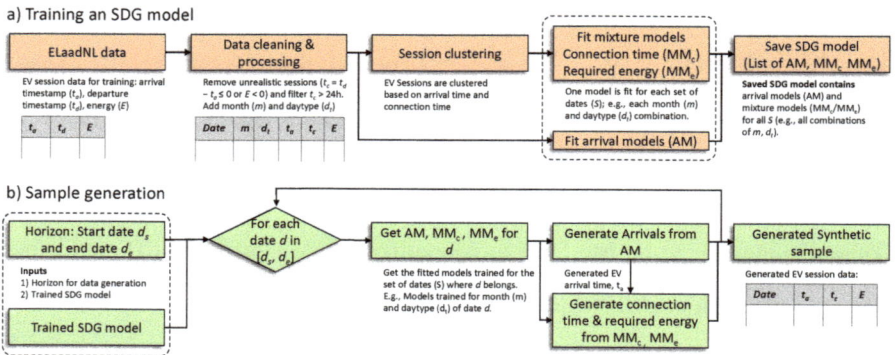

Figure 1. Modeling methodology for (**a**) training SDG models, and (**b**) generating synthetic samples.

In this section we define and outline the inputs of SDG, by defining AM for EV arrivals, and MM_c and MM_e for connection times and required energy. Inputs are simply the dates d (and arrival times t_a in case of MM_c and MM_e). We also summarize the parameters of SDG) by characterizing models using the parameters of the underlying probability distributions.

3. Dataset

The data used here were collected from ELaadNL (https://www.elaad.nl/), which is the knowledge and innovation center mutually associated with providers of charging infrastructure for the grid, to prepare for a future with electric mobility and sustainable charging. Operating since 2009, it has established a network of approximately 2000 public charging stations across The Netherlands. The EV session data collected by ELaadNL are not publicly available, and we obtained them based on an agreement. Furthermore, ELaadNL was not involved in the study, and acted only as a data provider. People interested in the dataset are encouraged to contact us. In this section, we provide the details of the data cleaning and processing, and session clustering steps (Figure 1a).

3.1. SDG Training Data

The EV sessions' time series data were prepared for training the SDG (the training process is detailed in Section 4). These data contain: the date d, month m, type of day d_t, arrival time t_{arr}, arrival timeslot t_s, connection time t_c and required energy E, as shown in Table 1.

Timeslots have values ranging from 1 to 24, where 1 indicates the timespan 00:00–00:59, 2 indicates 01:00–01:59, etc. Further, t_a and t_c are real numbers ($\in [0, 24)$); e.g., 1.5 means 01:30 A.M. More than 98% of the sessions have t_c under 24 h, so we safely assumed that the maximum connection time was 24 h (and removed data points with $t_c > 24$). In the real world, we will have sessions where the EV departs before it is fully charged. However, the collected data do not include the charging load that was unmet before the EV departed. Lacking such information, we resorted to assuming the measured charging load represents fully charging the EV. We represent this charging load, or energy required by E in kWh.

Further, the training data were properly cleaned, which included removing impractical or incorrect sessions parameters (where $E < 0$ or $t_a = t_d$).

Table 1. Processed session data. Each row corresponds to an EV session.

d Date	m Month	d_t Day Type	t_a Arrival Time (h)	t_s Arrival Time Slot	t_c Connection Time (h)	E Required Energy (kWh)
01/01/2015	1	0	0.15	1	4.3	3
...	...	...	...	...	...	...

3.2. Charging Stations Analysis

The full ELaadNL dataset contains 1.8 million sessions from January 2012 till June 2018. The infrastructure consists of charging stations of 10 different types, divided by manufacturer type, charging speed and other factors. In 2016 the EVnetNL (the infrastructure provider associated with ELaadNL) stations were transformed to integrate smart charging capability. Hardware and software of the charging stations (poles) were updated based on the station type. In 2017, more than 50% of EVnetNL stations were taken over by other charging station operators. Due to those two factors, we observed a sudden drop in the number of daily active charging stations in 2016 and 2017 (Figure 2). The years prior to 2014 have a very steep growth curve in terms of active poles, while from 2016 onwards, the active poles become unpredictable because of market factors. As we wanted our model to reflect charging behavior, and not be influenced by infrastructure changes, we selected the training data from the reasonably stable year 2015. The data used for training our SDG were from January to December 2015 of the ElaadNL dataset. This data contains 365,000 sessions. In 2015, the number of used poles amounted to 1677, out of which 1645 poles were active before and after 2015. We used the data from these 1645 poles for our analysis. Thus, we considered a constant number of poles to construct our SDG model, and avoided the effects of a changing number of EV charging stations.

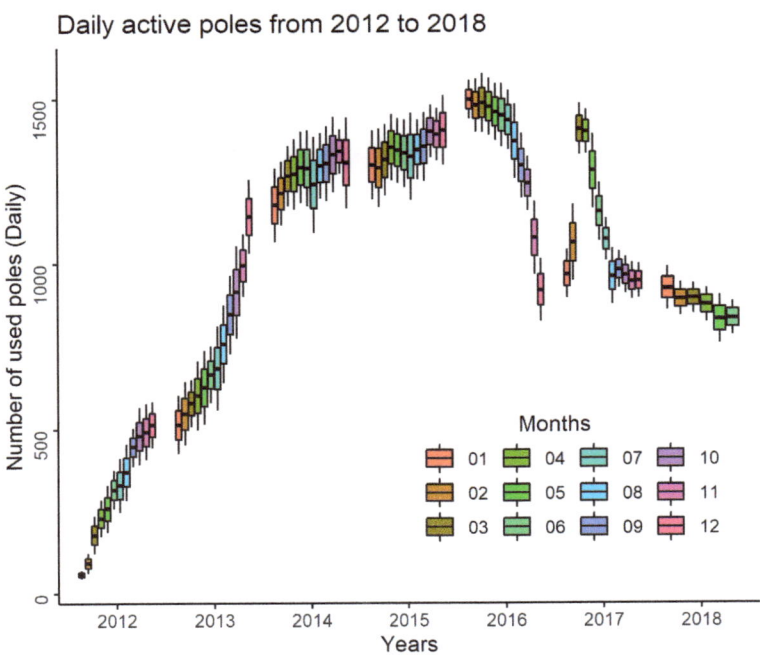

Figure 2. Number of used poles per day, from 2012 to 2018. Each boxplot represents data for 1 month. The y-axis represents the number of daily active poles.

3.3. Clustering

We used the expectation maximization (EM) algorithm for training the GMM used in MM_c (defined in Section 2.2). EM algorithm can be initialized with a realistic number of mixtures (along with mean and variance for each mixture) for it to converge to a practical solution. To achieve this practical solution, we initialized each GMM with session clusters. EV sessions were clustered based on arrival and connection times, and here we outline the different types of sessions that are observed in the real-world data.

Sessions clusters: In our previous work [2], on the same data, we discussed three types of sessions. Namely, (i) **Park to charge**: arrivals throughout the day; (ii) **Charge near home**: arrivals during evenings, and staying till late at night; (iii) **Charge near work**: arrivals during early morning, and staying till evenings. The largest cluster was the park to charge cluster (60% of sessions), followed by the charge near home (29% of sessions) and the charge near work clusters (11% of sessions). The DBSCAN algorithm was used to determine these clusters, which is a density based clustering algorithm. We could see a similar distribution of sessions in the 2015 dataset, after clustering the sessions. The resulting session clusters are shown in Figure 3. These clusters are only based on 2015 data, contrary to the previous work, which combined the full data for 2012–2016. Please refer to [2] for further details.

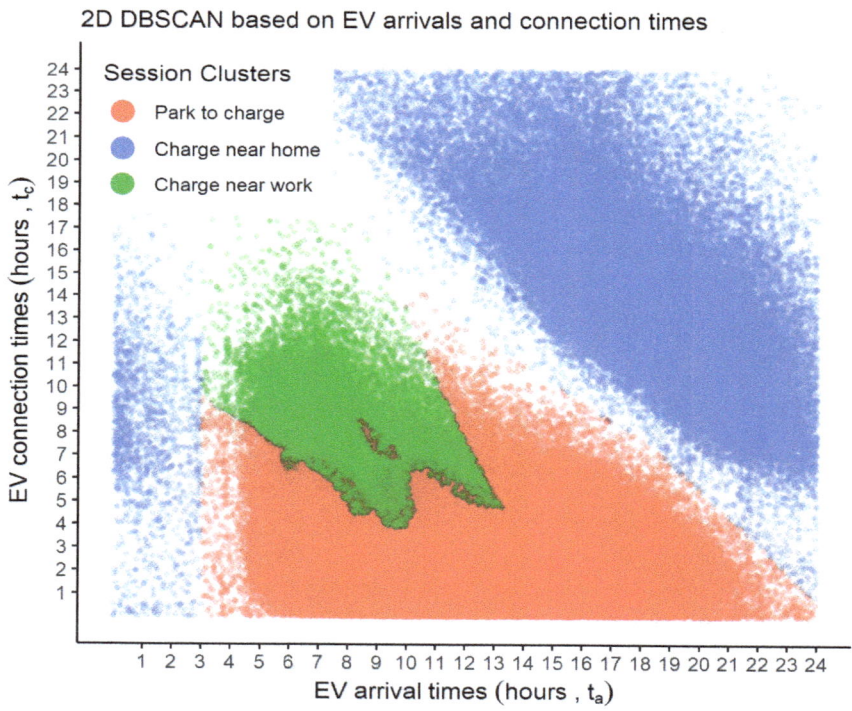

Figure 3. Session clusters for 2015. We used DBScan to cluster EV sessions on a monthly basis, and combine the data for all months.

4. Training Additionally, Evaluation

4.1. Training

We used the training data of 2015, as described in Section 3.1, for training and evaluating our models. Aggregating EV sessions on a monthly basis reveals a higher number of sessions during the winter months, compared to the summer months. In the case of daily EV sessions, it has been noticed that all weekdays have similar profiles, which are different from weekends [12]. Thus, we could assume that days belonging to the same month (m) and daytype (d_t, weekday vs. weekend) have similar profiles, and defined a set of dates (**S**) as pairs (m, d_t) (e.g., for m = January and d_t = weekday, **S** will have all dates that are weekdays from January). Training data for the model for a (m, d_t) combination comprises the session data for dates present in the respective **S**. We trained 24 individual models, one for each of the (m, d_t) combination.

4.1.1. Arrival Model

Inter-arrival time models: For IAT models, we modeled the daily profiles of λ, which can be fitted using (i) a mean model, (ii) a polynomial regression model or (iii) a localized regression model (outlined in Section 2.1.1). We can rewrite Equation (6) in terms of (m, d_t) as in Equation (11). We calculated the EV arrival rates (λ) for each day and t_s (24 timeslots of 60 min each), by fitting the inter-arrival time to an exponential distribution.

$$\lambda_{m,d_t} = f_{m,d_t}(t) \tag{11}$$

For the mean model, the fitted value for each t_s is the average λ (across all days). Accordingly, each t_s has a single value of λ. This results in a discontinuous mapping between λ and t, for which the function in Equation (11) becomes discontinuous at the boundaries of t_s and we see a sudden change in λ.

For regression methods, we transform λ by taking the logarithm and applying min-max normalization for each day. This transformation is necessary to correctly fit the peak hours, during which inter-arrival times are very low (high λ). We take the logarithm in order to more accurately model the values of λ during the night hours (00:00–06:00), which have few arrivals (low λ). Normalization is used to scale the arrival rates of all days in **S** to the same levels. We represent this transformed λ using λ_t, and we use s to represent the re-scaling parameter for predicted values. Equation (13) can be used to get the fitted λ from the regression models $f_{m,d_t}(t)$.

$$(\lambda_t)_{m,d_t} = f_{m,d_t}(t), \quad 0 < \lambda_t \leq 1 \tag{12}$$

$$\lambda_{m,d_t} = e^{s f_{m,d_t}(t)} \tag{13}$$

For the polynomial regression model we modeled the relationship in Equation (12) using a grid search for the best polynomial degree ($\in \{1, \ldots, 50\}$). Mean squared error (MSE) was used as the error metric during grid searching. It provides a strong penalty for large errors, which was necessary to fit the sharp morning peaks during weekdays. This resulted in a continuous and differentiable function of λ in terms of t.

For the localized regression model, polynomials of degree 1 and 2 with $\alpha \in \{0.125, 0.25, 0.5\}$ were tested. We noticed that the best results sdfd generated for degree 1 and α = 0.125. This resulted in a piecewise, continuous and differentiable profile of λ throughout the day.

Scale treatment and randomization: For regression based methods, for which we model λ_t, we may encounter a situation wherein $f(t) < 0$. In this case, λ becomes very low, which can cause the sampled inter-arrival time (Δt) of EVs to be very large. Since the next EV arrival is calculated relative to the past arrival Equation (4), such high Δt may cause the next arrival to be very late, thereby skipping a large period of time. This becomes problematic when this period covers times with high values of λ (hence a high number of EV arrivals, which however will not be generated). For practical purposes, we impose a lower limit of 1 on λ (meaning we have at least 1 arrival in each t_s).

When we transformed λ, we appiedy a min-max normalization on $\ln(\lambda)$ for each day (where the minimum value of $\ln(\lambda)$ is 0, because $\lambda \geq 1$). Each day in the training data has its own maximum value of $\ln(\lambda)$. These values can be saved as an array of re-scaling parameters (represented by s in Equation (13)). When generating session arrivals, we randomly selected a value from this array to re-scale the predicted values. This helped in introducing variance in to the otherwise smooth profiles of the predicted λ.

Arrival count models: We mapped each t_s to the parameters **P** that characterize the discrete distribution of the number of arrivals in that timeslot. Similarly to the IAT models, we have a model for each (m, d_t) combination. Our training data for each model are the numbers of EV arrivals at t_s for each of the days of the respective combination (m, d_t).

$$f_{m,d_t}: \quad t_s \quad \rightarrow \quad \mathbf{P}_{m,d_t} \quad \quad (14)$$
$$\{1,\ldots,24\} \rightarrow \{\mathbf{P}_{m,d_t,1},\ldots,\mathbf{P}_{m,d_t,24}\}$$

For the Poisson model, the average number of EV arrivals λ was calculated per t_s. In the Poisson distribution, the mean is equal to its variance, a restriction that is not present in the negative binomial distribution.

In case of the negative binomial model (with parameters $\mathbf{P} = \{\mu, \alpha\}$), μ is the average number of EV arrivals per timeslot ($=\lambda$), and α is the dispersion parameter, which can be used to define the variance of the distribution (var $= \mu + \alpha \mu^2$). A negative binomial distribution model thus allows one to introduce more variability in the generated number of EV arrivals, compared to a Poisson model. Both α and μ were fitted for each individual (m, d_t) combination. It is possible that the estimated α for an (m, d_t) combination is extremely low, in which case the underlying distribution is more likely to be Poisson. It is also possible that during night hours (low EV arrivals), the estimation process of α might result in impractical values (less than 0). To adjust for this, we can use a Poisson distribution wherein the estimated values of α are negative, or set a lower limit on α (e.g., $\alpha \geq 0.1$).

In IAT models, the time of the next EV arrival is the sum of the time of previous EV arrival and randomly sampled Δt. As previously stated, this dependency becomes troublesome if Δt is very large (due to the low λ, the next EV arrival may be very late, skipping a large time interval) or very low (high λ, large number of EV arrivals in a small amount of time). Due to that, fitting λ as a function of t requires caution in IAT models. However, we do not face this problem when using the AC modeling approach, wherein the number of arrivals are generated separately for each t_s. Indeed, a low/high λ in the previous t_s will not affect the number of arrivals in next t_s. For practical uses, we can also assume that night hours with low numbers of EV arrivals are similar, and combine $t_s = 1$–6 into a single timeslot. The fitted value of λ is then associated with the time from 00:00 to 06:00.

4.1.2. Mixture Models (MM_c, MM_e)

Similarly to the arrival models, we verified that days belonging to a (month, daytype) combination have similar distributions in terms of departure times and charging loads, and thus fit models for each (m, d_t) combination. For each t_s we fit a Gaussian mixture model to the real-world data, and modified Equation (9) as follows.

$$\begin{aligned} P_{t_a=t_s,m,d_t}(t_c) &= GMM_{m,d_t,t_s} \\ &= \left(\sum_{k=1}^{K} \phi_k \mathcal{N}(\mu_k, \sigma_k^2)\right)_{m,d_t,t_s} \end{aligned} \quad (15)$$

This resulted in a GMM fitted for each (m, d_t, t_s) combination, with trained values for (i) mixing probabilities (ϕ_k), (ii) mixture means (μ_k) and (iii) mixture variances (σ_k^2), for each mixture. We used expectation minimization to fit the GMM.

Expectation maximization for fitting GMM requires initialization of mixtures (μ_k, σ_k^2). The number of mixtures also needs to be chosen for each model (representing a m, d_t, t_s combination). We initialized each GMM based on session clusters (see Section 3.3). We grouped the sessions observed in m, d_t, t_s, into their respective session clusters. We used the number of clusters obtained to initialize the K for the GMM, and calculated the μ_k, σ_k^2 from the EV sessions in the respective groups.

4.2. Evaluation

Exponential distribution: We performed a Kolmogorov–Smirnov (KS) goodness-of-fit test to validate the assumption that inter-arrival times of EV sessions follow the exponential distribution.

Arrival models: Once the AM was trained using the 2015 EV session data, a synthetic sample for 2015 could be generated. This sample was to generate EV arrivals from January 1, 2015 to December 31, 2015. We generate 10 samples for each modeling method (three IAT models and two AC models). EV arrivals were aggregated on an hourly and daily basis. Since the aggregated values represent count data, we used a non-parametric Wilcoxon test to assess similarity between the generated samples and the actual data. We performed the test on a monthly basis for the daily aggregated data and on an hourly basis for hourly aggregated data. We provide plots for visual comparison.

Mixture models: Connection times were sampled from the fitted GMMs, for the actual EV arrivals. Density plots were created to evaluate whether the peaks of the conditional probability distributions were modeled correctly. A similar evaluation was preformed for required energy.

SDG: Final generated data (and actual data) were 3-dimensional, with each session defined by (t_a, t_c, E). The actual data comprised 350,000 sessions, and the numbers of sessions in the generated samples were of the same order. Since two-sample similarity tests for high dimensional data become unreliable as the data size increases, we used a kernel density estimation (KDE) test [21] and a multidimensional version of the KS test [22,23]. We did those tests for (t_a, t_c) and (t_a, E) combinations.

In this section we defined different methods for fitting the parameters of SDG. Depending on the modeling method, the parameters of SDG will also change (λ in case of the exponential distribution, and (μ, α) in case of the negative binomial distribution).

5. Results

5.1. Assumptions

KS test p-values are greater than 0.05 for each hour of the day, as plotted in Figure 4. This validates that the inter-arrival times of EV sessions are exponentially distributed (Section 2.1), and thus supports our chosen models AM of the arrival times.

5.2. Distribution of Arrival Rates λ

To understand how the SDG parameters change with inputs, we have plotted the profiles of λ for weekend and weekdays for 2015 in Figure 5. We see a similar pattern for all months. Arrival models were fitted to approximate this behavior of λ. On weekdays, we see two peaks in the profile of λ that represent high frequencies of EV arrivals.

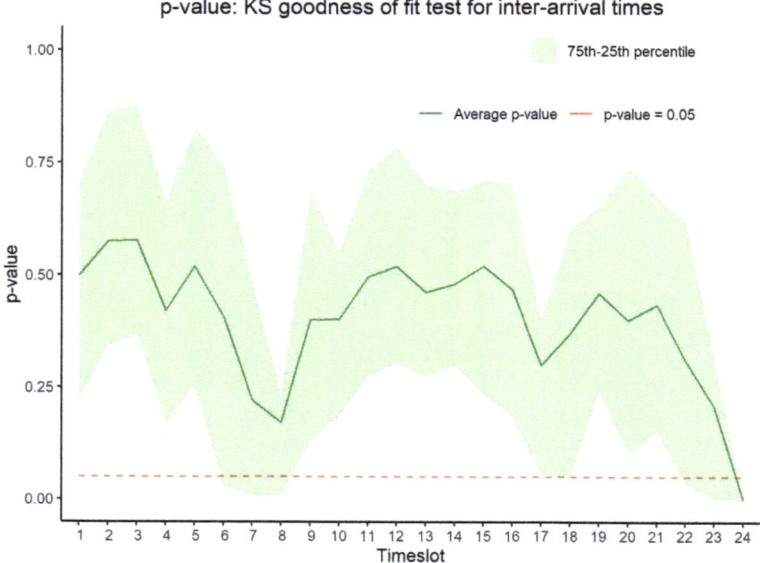

Figure 4. KS test p-values: For each (m, d_t) combination, 24 KS tests were performed for each timeslot (t_s). High p-values indicate that null hypotheses (IATs are exponentially distributed) could not be rejected.

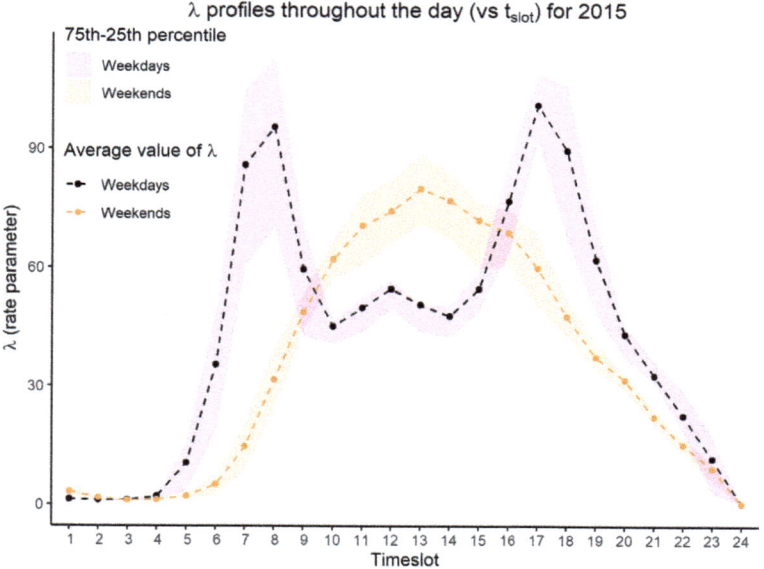

Figure 5. Daily λ profiles for 2015: For each (m, d_t) combination, we calculated the average arrival rate λ. The dotted line represents the average over those 12 months; the shaded areas indicate the percentile range.

5.3. Arrival Models (AM)

We generated 10 samples of arrivals of EVs for 2015 for both inter-arrival time (IAT) and arrival count (AC) models. The total number of arrivals per day was calculated and plotted in Figure 6. Similarly, Figure 7 shows the aggregated hourly EV arrivals. Both these plots are for weekdays, and similar results were observed in case of weekends. We can clearly see that the generated data are very similar to the actual data. We further quantitatively compared the values of the actual EV arrivals with the generated EV arrivals using a Wilcoxon test. The null hypothesis was that the means of these are equal. High p-values (> 0.05) indicate that the daily generated arrivals are statistically similar to the actual data. This is represented by ns in the figure, implying that the difference between the two samples is not significant. The results presented are for comparisons between one month of actual data to 10 samples of the same month of generated data. We got similar results when we compare the real-world data to a single sample.

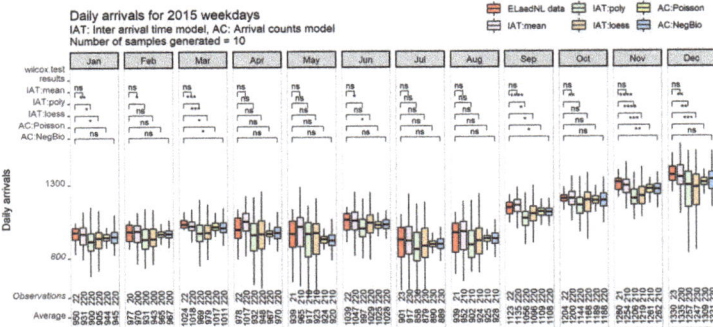

Figure 6. Daily aggregated EV arrivals (2015, weekdays). Significance was calculated based on Wilcoxon tests (ns: not significant, p-value > 0.05; * p-value ≤ 0.05; ** p-value ≤ 0.01; *** p-value ≤ 0.001; **** p-value ≤ 0.0001).

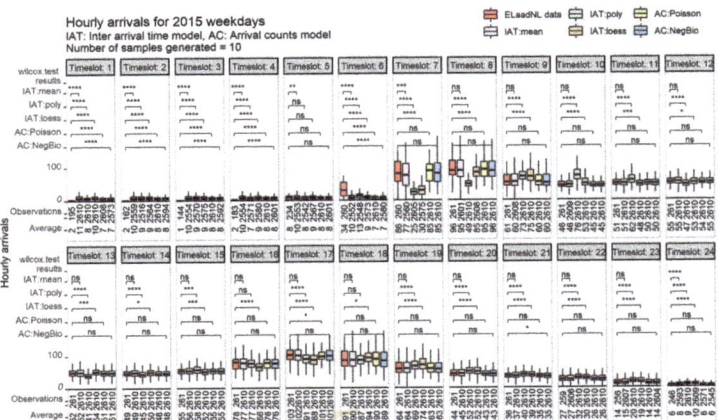

Figure 7. Hourly aggregated EV arrivals (2015, weekdays). Significance was calculated based on Wilcoxon tests (ns: not significant, p-value > 0.05; * p-value ≤ 0.05; ** p-value ≤ 0.01; *** p-value ≤ 0.001; **** p-value ≤ 0.0001).

5.4. Mixture Models (MM_c, MM_e)

Conditional distributions for connection times (hours) and energy required (kWh) are plotted in Figures 8 and 9 respectively. The plots on the left were created from the real-world data, and those on the right were created from the data generated from mixture models (MM_c, MM_e). These figures are for weekdays, and similar plots were generated for weekends. Connection times (and energy required) were generated using GMM and real-world EV arrivals. Vertical divisions in the generated data for each times slot can be seen, because we use one GMM per $\{m, dt, t_s\}$ combination.

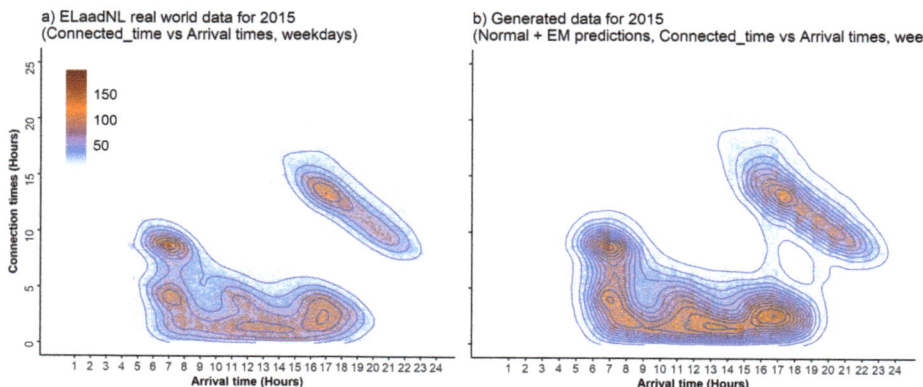

Figure 8. Density plots for connection times (2015, weekdays). Generated data represent sampled connection times for real-world EV arrivals. Each point represents a bin (10 min by 10 min), and is colored based on the number of EV sessions in the bin (bins with less than 5 arrivals were not plotted to keep the graph readable).

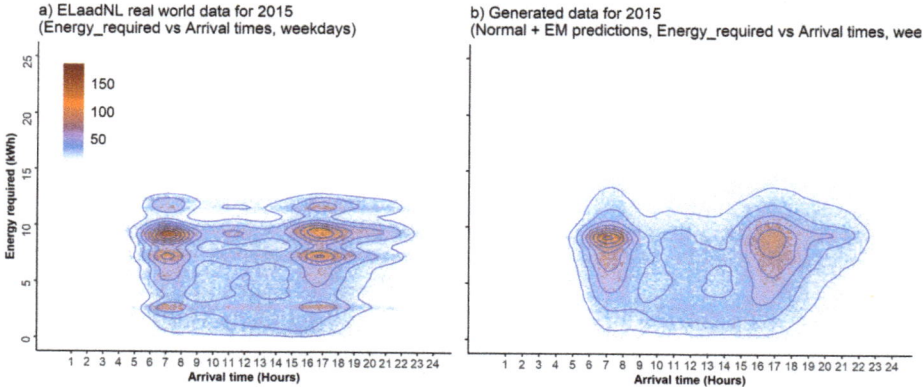

Figure 9. Density plots for required energy (2015, weekdays). Generated data represent sampled energy requirements for real-world EV arrivals. Each point represents a bin (10 min by 0.16 kWh), and is colored based the number of EV sessions in the bin (bins with less than 5 arrivals were not plotted to keep the graph readable).

5.5. Synthetic Data Generator (SDG)

We generated full session samples including generated arrival times, connection times and required energy for all the models. These generated samples were compared with the real-world session data for 2015. To compare the models' synthetic samples with real-world data, 2-sample KDE tests were performed. In Figure 10a, we show the KDE test for (t_a, t_c), and Figure 10b shows the KDE test results on (t_a, E). As we can see, the mean model for the IAT, and both models for AC have average p-values > 0.05, indicating that the generated data are similar to the real-world EV session data. We observed similar results in the multidimensional KS test. These results conclude that the generated samples are statistically similar to real-world data.

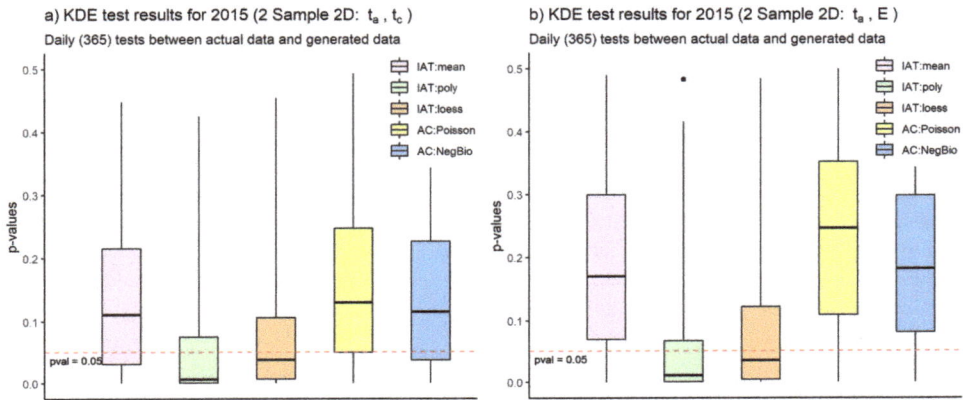

Figure 10. KDE test p-values: Daily 2 sample 2 dimensional KDE tests to compare real-world and generated data. Total of 365 tests performed for each model. p-value > 0.05 means datasets are similar. (**a**) Results for (arrival times t_a, connection times t_c). (**b**) Results for (arrival times t_a, energy required E).

6. Discussion

In this paper, we proposed a synthetic data generator (SDG) to create samples of realistic EV session data. Each session is defined by arrival time, departure time and required energy. We described two modeling methodologies to generate arrivals, assuming that inter-arrival times follow exponential distribution. Different methods for modeling the daily profiles of the parameter λ were tested. For connection times and required energy, mixture models were trained to estimate the probability distributions. Our real-world dataset was used to train the SDG, and multiple samples of session data were generated.

Inter-arrival times followed an exponential distribution, which was validated by KS test results. Wilcoxon tests were used to compare daily and hourly EV arrivals from the generated samples and real-world data (Figure 4).

Arrival count (AC) models performed better compared to inter-arrival time (IAT) models. The negative binomial model from the AC models outperformed all the other models for generating EV arrivals. Samples generated by the IAT model exhibited high variance, which was introduced during the scale treatment and randomization step (Section 4.1.1). In the IAT models, we note that regression methods failed to capture the morning peak in the hourly arrivals. This occurred because the regression curves failed to capture the sharp increase in the number of arrivals. We indeed see that the polynomial model (IAT:poly) and localized regression model (IAT:loess) generated very low numbers of samples during morning peaks ($t_s = 7, 8$). In contrast, the AC models were able to capture both morning and evening peaks, as can be seen in Figure 7. The AC models also captured the variance in number of arrivals throughout all t_s. During night hours, we noticed a difference between the generated and real-world data. However, for practical purposes, this difference can be neglected as the average number of arrivals is very low. We can see that the negative binomial model performs best for both daily and hourly generation. This makes it ideal for both short and long-term data generation.

In our mixture models, Gaussian mixture models (GMM) were able to properly capture peaks of the conditional probability distributions. We clustered EV sessions based on arrival times and connection times, and each peak in the conditional probability distribution corresponds to one session cluster. Two peaks in required energy distribution represent the morning and evening demand. For the connection times, we can see that after generating the data, all three session cluster peaks were captured (Figure 8). In case of required energy, both morning and evening peaks were captured (Figure 9). Since we were able to capture the probability distributions of t_c and E, GMM-based mixture models could be used for fitting the conditional distributions.

In case of weekdays, during some spring and summer months we observed a very high variance in the number of daily arrivals in actual data (in Figure 6). The reason is that there are multiple holidays during May, July and August that have very low numbers of arrivals. In retrospect, we found that many of these holidays (on weekdays) have arrival profiles similar to weekends. Due to that, arrival models trained for weekdays are unable to capture this variance. Introducing holidays as another daytype (d_t) or modeling holidays as weekends should help to overcome this limitation.

We modeled the data under the assumption that the number of active charging stations would remain constant during the time in which the EV session data are collected. Furthermore, the generated sample is representative of EV sessions that might occur on this constant number of active charging stations. Hence, the proposed methodology does not model the effect of changing the number of charging stations, where future research is possible.

7. Conclusions

EV session data collected from charging stations on a electricity gird can be used for flexibility analysis, making pricing decisions, etc., and are essential for advancement in the field of smart grids.

We defined a synthetic data generator (SDG) to generate samples of EV session data collected on charging stations. We modeled arrival times of EVs using inter-arrival time (IAT) and arrival counts (AC) methods. For generating the connection times and required energy, we used mixture models based on GMM. The generated sample of session data is statistically indistinguishable from the real-world data, as seen from the KDE test results. We can conclude that our proposed SDG is suited for generating a synthetic sample of EV session data.

This generated data sample will have the properties of a real-world EV sessions, and can be used for purposes such as flexibility analysis. We will release the trained SDG models that can be used to generate new samples of EV session data. Complete code for training and evaluating the SDG models is open source, and can be used to fit the models on a new EV session data (see Appendix A). These models can be shared without violating the privacy concerns of the real data collection companies.

For future work, further exploration is required in studying reduced variance in daily arrivals in AC models. IAT models for arrival times misses the first peak of weekdays, wherein improvements are possible. A deeper dive into the mixture models for estimating the conditional distribution of required energy can also provide an improvement to results.

Author Contributions: M.L. developed the methodology and performed data curation, validation, visualization, software and writing of the original draft; C.D. handled resources and funding acquisition; D.F.B. and C.D. jointly supervised the work presented, and performed review and editing of the paper. All authors have read and agreed to the published version of the manuscript.

Funding: This research received funding from the Flemish Government under the "Onderzoeksprogramma Artificiële Intelligentie (AI) Vlaanderen" programme.

Acknowledgments: We thank Nazir Refa from ElaadNL for the real-world EV session data.

Conflicts of Interest: The authors declare no conflict of interest.

Appendix A. Code

Code for training SDG models is open source, and can be accessed on GitHub: https://github.com/mlahariya/EV-SDG. SDG models trained with a real-world dataset are also included with the code. These can be used to generate a random sample of EV session data using script

SDG_sample_generate.py. Trained models that can be used as default models to generate samples with are located at modeling/default_models, and include:

- **SDG Model (IAT,mean)**: IAT model based on mean model.
- **SDG Model (IAT,poly)**: IAT model based on polynomial regression model.
- **SDG Model (IAT,loess)**: IAT model based on localized regression model.
- **SDG Model (AC,poisson_fit)**: AC model based on Poisson distribution.
- **SDG Model (AC,neg_bio_reg)**: AC model based on negative binomial distribution.

Users can also employ our code to fit AM, MM_c, and MM_e to their own datasets. For training a SDG model from scratch, this process will be followed: (i) Clean real-world EV session data (*preprocess*). (ii) Generate session and pole clusters (*preprocess*). (iii) Prepare data for SDG training (*preprocess*). (iv) Train AM, MM_c and MM_e models (*modeling*). (v) Save the model along with a log file in the 'res/' folder. A command line callable script SDG_fit.py can be used to fit the models.

Please visit the repository for further details.

References

1. Watson, R.T.; Boudreau, M.C.; Chen, A.J. Information systems and environmentally sustainable development: Energy informatics and new directions for the IS community. *MIS Q.* **2010**, *34*, 23–38. [CrossRef]
2. Develder, C.; Sadeghianpourhamami, N.; Strobbe, M.; Refa, N. Quantifying flexibility in EV charging as DR potential: Analysis of two real-world data sets. In Proceedings of the 2016 IEEE International Conference on Smart Grid Communications (SmartGridComm), Sydney, Australia, 6–9 November 2016; pp. 600–605.
3. Quirós-Tortós, J.; Espinosa, A.N.; Ochoa, L.F.; Butler, T. Statistical representation of EV charging: Real data analysis and applications. In Proceedings of the 2018 Power Systems Computation Conference (PSCC), Dublin, Ireland, 11–15 June 2018; pp. 1–7.
4. Islam, E.; Moawad, A.; Kim, N.; Rousseau, A. *An Extensive Study on Sizing, Energy Consumption, and Cost of Advanced Vehicle Technologies*; Argonne National Lab. (ANL): Argonne, IL, USA, 2018.
5. Hanke, C.; Hüelsmann, M.; Fornahl, D. Socio-Economic Aspects of Electric Vehicles: A Literature Review. In *Evolutionary Paths Towards The Mobility of the Future*; Hüelsmann, M., Fornahl, D., Eds.; Springer: Berlin/Heidelberg, Germany, 2014; pp. 13–36. [CrossRef]
6. Li, X.; Chen, P.; Wang, X. Impacts of renewables and socioeconomic factors on electric vehicle demands: Panel data studies across 14 countries. *Energy Policy* **2017**, *109*, 473–478. [CrossRef]
7. Pevec, D.; Babic, J.; Podobnik, V. Electric vehicles: A data science perspective review. *Electronics* **2019**, *8*, 1190. [CrossRef]
8. Sadeghianpourhamami, N.; Deleu, J.; Develder, C. Definition and evaluation of model-free coordination of electrical vehicle charging with reinforcement learning. *IEEE Trans. Smart Grid* **2019**, *11*, 203–214. [CrossRef]
9. Mies, J.; Helmus, J.; van den Hoed, R. Estimating the charging profile of individual charge sessions of electric vehicles in the Netherlands. *World Electr. Veh. J.* **2018**, *9*, 17. [CrossRef]
10. Zhang, C.; Kuppannagari, S.R.; Kannan, R.; Prasanna, V.K. Generative adversarial network for synthetic time series data generation in smart grids. In Proceedings of the 2018 IEEE International Conference on Communications, Control, and Computing Technologies for Smart Grids (SmartGridComm), Aalborg, Denmark, 29–31 December 2018.
11. Shepero, M.; Munkhammar, J. Data from Electric Vehicle Charging Stations: Analysis and Model Development. In Proceedings of the 1st E-Mobility Power System Integration Symposium, Berlin, Germany, 23 October 2017.
12. Flammini, M.G.; Prettico, G.; Julea, A.; Fulli, G.; Mazza, A.; Chicco, G. Statistical characterisation of the real transaction data gathered from electric vehicle charging stations. *Electr. Power Syst. Res.* **2019**, *166*, 136–150. [CrossRef]
13. Brady, J.; O'Mahony, M. Modelling charging profiles of electric vehicles based on real-world electric vehicle charging data. *Sustain. Cities Soc.* **2016**, *26*. [CrossRef]
14. Gadda, S.; Kockelman, K.M.; Damien, P. Continuous departure time models: A bayesian approach. *Transp. Res. Rec.* **2009**, *2132*, 13–24. [CrossRef]

15. Sadeghianpourhamami, N.; Benoit, D.; Deschrijver, D.; Develder, C. Bayesian cylindrical data modeling using Abe-Ley mixtures. *Appl. Math. Model.* **2018**, *68*, 629–642. [CrossRef]
16. Sadeghianpourhamami, N.; Benoit, D.; Deschrijver, D.; Develder, C. Modeling real-world flexibility of residential power consumption: Exploring the cylindrical WeiSSVM distribution. In Proceedings of the Ninth International Conference on Future Energy Systems, Karlsruhe, Germany, 12–15 June 2018; pp. 408–410.
17. Majidpour, M.; Qiu, C.; Chu, P.; Gadh, R.; Pota, H.R. Fast prediction for sparse time series: Demand forecast of EV charging stations for cell phone applications. *IEEE Trans. Ind. Informatics* **2015**, *11*, 242–250. [CrossRef]
18. Iftikhar, N.; Liu, X.; Danalachi, S. A scalable smart meter data generator using spark. In Proceedings of the OTM Confederated International Conferences On the Move to Meaningful Internet Systems, Rhodes, Greece, 23–28 October 2017.
19. Lahariya, M.; Benoit, D.; Develder, C. Defining a synthetic data generator for realistic electric vehicle charging sessions. In *Proceedings of the Eleventh ACM International Conference on Future Energy Systems*; Association for Computing Machinery: New York, NY, USA, 2020; p. 406–407. [CrossRef]
20. Cameron, A.; Trivedi, P. Count Panel Data. In *The Oxford Handbook of Panel Data*; Baltagi, B.H., Ed.; Oxford University Press: Oxford, UK, 2015. [CrossRef]
21. Duong, T.; Goud, B.; Schauer, K. Closed-form density-based framework for automatic detection of cellular morphology changes. *Proc. Natl. Acad. Sci. USA* **2012**, *109*, 8382–8387. [CrossRef] [PubMed]
22. Fasano, G.; Franceschini, A. A multidimensional version of the Kolmogorov–Smirnov test. *Mon. Not. R. Astron. Soc.* **1987**, *225*, 155–170. [CrossRef]
23. Peacock, J.A. Two-dimensional goodness-of-fit testing in astronomy. *Mon. Not. R. Astron. Soc.* **1983**, *202*, 615–627. [CrossRef]

© 2020 by the authors. Licensee MDPI, Basel, Switzerland. This article is an open access article distributed under the terms and conditions of the Creative Commons Attribution (CC BY) license (http://creativecommons.org/licenses/by/4.0/).

Article

Privacy-Functionality Trade-Off: A Privacy-Preserving Multi-Channel Smart Metering System

Xiao-Yu Zhang [1,*], Stefanie Kuenzel [1,*], José-Rodrigo Córdoba-Pachón [2] and Chris Watkins [3]

1. Department of Electronic Engineering, Royal Holloway, University of London, London TW20 0EX, UK
2. School of Business and Management, Royal Holloway, University of London, London TW20 0EX, UK; J.R.Cordoba-Pachon@rhul.ac.uk
3. Department of Computer Science, Royal Holloway, University of London, London TW20 0EX, UK; C.J.Watkins@rhul.ac.uk
* Correspondence: xiaoyu.zhang.2018@live.rhul.ac.uk (X.-Y.Z.); stefanie.kuenzel@rhul.ac.uk (S.K.)

Received: 8 April 2020; Accepted: 18 June 2020; Published: 21 June 2020

Abstract: While smart meters can provide households with more autonomy regarding their energy consumption, they can also be a significant intrusion into the household's privacy. There is abundant research implementing protection methods for different aspects (e.g., noise-adding and data aggregation, data down-sampling); while the private data are protected as sensitive information is hidden, some of the compulsory functions such as Time-of-use (TOU) billing or value-added services are sacrificed. Moreover, some methods, such as rechargeable batteries and homomorphic encryption, require an expensive energy storage system or central processor with high computation ability, which is unrealistic for mass roll-out. In this paper, we propose a privacy-preserving smart metering system which is a combination of existing data aggregation and data down-sampling mechanisms. The system takes an angle based on the ethical concerns about privacy and it implements a hybrid privacy-utility trade-off strategy, without sacrificing functionality. In the proposed system, the smart meter plays the role of assistant processor rather than information sender/receiver, and it enables three communication channels to transmit different temporal resolution data to protect privacy and allow freedom of choice: high frequency feed-level/substation-level data are adopted for grid operation and management purposes, low frequency household-level data are used for billing, and a privacy-preserving valued-add service channel to provide third party (TP) services. In the end of the paper, the privacy performance is evaluated to examine whether the proposed system satisfies the privacy and functionality requirements.

Keywords: smart grids; smart energy system; smart meter; GDPR; data privacy; ethics

---

1. Introduction

The smart grid is a worldwide modernization of electrical power systems in the 21st century. Two-way communication networks enable smart grids to collect real-time data from both the electricity supply (i.e., power stations) and demand (i.e., households) sides, and further boost the power system's reliability, availability, and efficiency.

As an essential enabler and prerequisite of the smart grid, smart meters are being installed country- and world-wide at single houses to collect real-time data on energy consumption. Smart meters offer an opportunity to consumers to play an active role in household energy consumption management. Based on these advantages, the UK government is working to ensure 80% of households install smart meters by 2020, paving the way for future smart grid construction [1].

However, with the EU's mandate to install smart meters, worries about privacy intrusions caused by smart meters are rising. Researchers point out that private household information can be revealed by smart meters [2–4]. Through continuously monitoring the real-time smart meter data, third parties

(TP) could have an inside view of household activities and behaviours (e.g., how many residents live in the house, when people leave the home, what the residents are doing at particular times, such as sleeping, bathing, watching TV, washing clothes, etc.). Although data collection may be justified on ethical grounds of utilitarianism (i.e., ensuring the greater, collective good of energy efficiencies in smart grids), the intrusion into privacy could also have negative ethical social consequences, including the conditional shaping of freedom and behaviour of individuals and households [5,6].

It is urgently expected that a more reliable smart metering system should be proposed to improve privacy and security. To do this, there could be three operational methods to protect households' privacy: (a) user demand shaping, (b) data manipulation, and (c) encryption techniques. User demand shaping approaches modify electricity data using methods such as energy storage systems or rechargeable batteries in households [7]. This requires the installation of extra devices, which is expensive. Data manipulation approaches modify energy data before sending it to TP (i.e., utility companies) by employing strategies like data obfuscation, data aggregation [8], data down-sampling, encryption protocols [3], or data anonymization. However, these methods sacrifice functionalities to protect privacy. Encryption techniques include homomorphic encryption (HE) and multi-party computation (MPC); these techniques encrypt the input data and can still implement essential operations with encrypted data, but techniques such as HE also cause computing overhead, increasing the budget.

At a legal level, the General Data Protection Regulation (GDPR) has been in force since 25 May 2018 [9]. Covering all European countries, the purpose of GDPR is to protect all EU citizens from privacy and data violation, providing more power to individuals to control their personal information. With these operational and legal operational possibilities, it is also important to consider 'soft' ethical strategies that use them to contribute to protect household privacy, potentially enabling households to be more in control of their digital data [10]. One such strategies is that of considering different stakeholders involved or affected by digital data gathering [11].

In this paper, we extend the approach from [8,12,13] to the combined use of existing data aggregation and data down-sampling techniques to design a privacy-preserving smart metering system. The system follows an operational and ethically (consequentialist) driven trade-off strategy and model which could contribute to increase functionalities of current smart metering devices in smart grids whilst ensuring that digital privacy intrusion is minimised and protected if not appropriately governed. In addition, the system provides three different communication channels for data collection to enable diverse data granularity transmission to TP, with each channel also providing required functionalities (time-of-use billing, grid operation and management, and TP services). We present our system and discuss the results of testing it with implications for the future design or management of smart meters by TP and households.

The paper is organized as follows: A presentation of smart grids and smart metering systems with ethical concerns about privacy intrusion is offered in Section 2. A review of current operational strategies to deal with privacy intrusion is presented in Section 3. In Section 4 our main contribution is proposed: a trade -off strategy is discussed with a proposed new smart metering system model to support it. A simulation work to quantify the privacy boundary is given in Section 5. The conclusion, implications, and future work are drawn in the last sections of the paper.

2. Background

2.1. Smart Grids

Smart grids are physical networks that use cutting-edge technologies and equipment, enabling the interconnection of different components through two-way networks that could achieve real-time optimizations to deliver electricity more reliably and efficiently. Smart grids contain not only electricity interfaces, but also communication interfaces. With these, other stakeholders (utility companies) or domains (electricity markets) can be included for analysis and management. Future smart grids can enable better operation and control, better network planning and maintenance, advanced smart metering infrastructure (AMI), and overall energy efficiency for countries [14].

2.2. Advanced Smart (Metering) Infrastructures

Within smart grids, AMI systems are integrations of smart meters, communication networks, and data management systems [14,15]. With the adaptation of narrowband(NB) powerline communication (PLC) technologies (e.g., Powerline Intelligent Metering Evolution (PRIME) and G3-PLC standards adopted in Europe), AMI enable real-time bidirectional communication between the TP (suppliers) and electricity consumers [16–18]. Smart meters are the most vital components within AMI. As smart energy sensors are installed in consumers' residences (households), smart meters can gather and transmit data including power consumption and electricity/gas bills on a real-time basis.

As illustrated in Figure 1, stakeholders of the smart metering system can include the consumers, energy suppliers, network operators, and data and communications companies (DCC) and TP. With smart meters, consumers can obtain near real-time and more accurate power usage data and bills, which helps them manage their energy usage. Energy suppliers (ES) are the utility companies that buy electricity from the wholesale market, then sell it to consumers. For instance, large British ES include British Gas, E-On, and SSE. Network operators (NO) (i.e., transmission system operators or TSOs, distribution system operators, etc.) construct, maintain, and operate the energy network, ensuring normal operation. These latter stakeholders can also benefit from smart meters: firstly, with the near real-time communication networks, utility companies can save money initially used for manual billing; secondly, network operators can implement demand-based management responses, in particular under peak load times; and finally, operators and companies can detect fraud or electricity theft and thus improve their efficiencies.

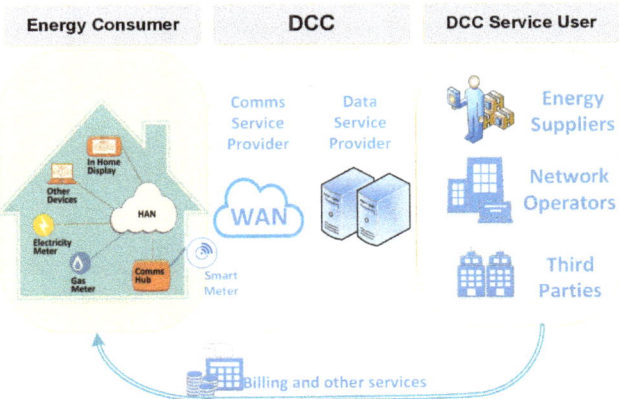

Figure 1. The structure of the current smart metering system.

The DCC collect energy consumers' data through the wide area network (WAN). Processed data are then sent to energy suppliers and network operators. DCC are responsible for ensuring compliance with the GDPR directive [19]. However, there is a lack of clarity about other stakeholders' responsibilities, making smart meters potentially fragile to a few privacy risks and concerns. These need to be identified and further explored if improvements in smart metering systems are to be made by or affecting several or all the stakeholders presented in the above figure. Before doing so, basic functions of smart meters are presented as following subsections.

2.3. Functions of the Smart Metering System

The European Commission identifies the 13 main functions of a smart meter and classifies them into five categories [20]. The most significant functions are listed in [21], which are billing correctness, grid operation and management, and additional consumer services. In addition, an emerging function is that of time-of-use (TOU) tariff.

2.3.1. Billing

The most vital function of the smart meter is providing accurate billing for consumers. Any data protection method which influences the accuracy of billing is useless. The current sample interval of the smart meter is 15 min up to a few seconds; however, consumers do not need such high-frequency billings, weekly or monthly basis billing is enough [21].

2.3.2. Grid Operation and Management

The smart meter contributes to the smart grid by improving the efficiency and stability of the whole power system. The real-time two-way communication networks provided by the smart metering system can measure, analyse, and control the energy consumption data, and further support the smart grid to implement demand side response services and power system estimation. For grid operators, the measurement of every individual household smart meter is not compulsory. Instead, they have more interest in high aggregated level data, such as measurement at feeder level or substation level [22].

2.3.3. Value-Added Services

The consumers can order additional services provided by the utility or TP. The additional consumer services could be awareness, (e.g., sending a warning for exceeding power), or scheduling and control (scheduling for controllable appliances, peak shaving) [23]. Demand-side response and non-intrusive load monitoring (NILM) have received the most attention. Demand-side response [21,24,25] optimizes the strength of the grid and enhances the power quality by utilizing power plants, distributed generators, and loads and energy storages. In demand-side response, consumers can also participate in the response process by accepting the bids provided by grid operators. By turning off appliances such as air conditioners and heaters, the load would be shed during peak time. NILM is a technique to disaggregate consumers' power consumption curve into individual appliance usage. The consumer can have an insight into how electricity is consumed and can better manage their home appliances to save energy and reduce carbon dioxide emissions [26]. Normally, value-added services require consumers to submit their energy data to a server; the server would use a pre-trained model to evaluate the data and send the results back to consumers. The difficulty exists in how to share personal data with TP while guaranteeing privacy at the same time.

In [21], two privacy-preserving value-added schemes are proposed. The naive scheme down-samples the original data into multiple interval resolution data, referring to the requirement of different services. Then the different resolution data are sent to different TPs with a key [13]. The second solution is to enable services at consumers' own devices (personal computer, mobile phone) via a home area network (HAN), however, TPs have the risk of revealing their models/algorithms [21].

2.3.4. Time-of-Use Tariff

The TOU tariff determines the electricity price during different periods. Consumers benefit from the TOU tariff by shifting their electricity usage habits to enjoy a cheaper bill, while the energy suppliers can also reduce the power plant capacity as a result [14], so TOU is becoming the mainstream method for billing in the UK. With the installation of the smart meter, the TOU moves closer to the real-time pricing tariff, allowing it to better represent the true conditions [27]. Although TOU tariff needs high-frequency data, the price volatility is determined by the aggregated power usage of the whole area rather than individuals. So, consumers do not have to send high-frequency data to realize the TOU tariff.

2.4. Privacy Intrusion Issues

Currently, smart metering systems could easily suffer from internal [28] and external attacks [29] and be subject to privacy intrusion [2]. All privacy intrusion issues related to the smart meters fall into two categories:

Category (i) Data sensitivity. Personal energy data that cannot be measured by a conventional electricity meter. While the traditional electricity meter measures the power consumption with a low resolution (e.g., one month) and can only provide the energy consumption information in kWh, the smart meter measures the power consumption with a high frequency (ranging from every second to every half hour, and normally every 15 min [21]), and more parameters are recorded, such as real-time active/reactive power, voltage, current, TOU tariff, etc. The high granularity data provide adversaries enough information to intrude on personal information.

Category (ii) Algorithm sensitivity. Advanced algorithms/mechanisms to intrude on privacy-sensitive features that could not be extracted from raw data using traditional data processing mechanisms. With the implementation of smart meters in smart grids to meet the above functions, and the increasing development of new services and applications by TP based on big data and artificial intelligence (AI) (e.g., Machine Learning (ML), Deep Neural Network (DNN), cloud computing), more and more sophisticated data could become available [30]. New services to better understand and monitor household behaviour include NILM [26], short-term load forecasting (STLF), distributed data mining, and others [14,30]. These advanced techniques are a double-edged sword to the consumers. The benefits espoused to consumers described above (e.g., managing their own energy consumption) and the adoption of a utilitarian ethic (i.e., ensuring the greater, collective good of energy efficiencies in smart grids) need to be weighed against potential privacy intrusion risks. Privacy intrusion would mean that not only individual but also collective freedom is compromised, given that household behaviour would be shaped and constrained by the perceived presence of digital surveillance [10,11,31].

Moreover, referring to the US National Institute of Standards and Technology (NIST) guideline NIST IR 7628v2 [32], the above two categories can be divided into four aspects as follows:

2.4.1. Behaviour Patterns Identification

Behaviour patterns identification belongs to category i; it aims to identify the appliances used. The smart meter and AMI communication network enables the utility and TP to access individual energy data continuously [15]. The high granularity data can reveal information about specific appliances at certain times and locations inside the home. Based on this information, operators can further infer the activities inside the house [32]. Potential usage of the appliance information may include that the retailers would adjust the warranty policy or use the information for advertising and marketing purposes.

2.4.2. Real-Time Surveillance

Real-time surveillance means that by regularly accessing energy data via smart meters, power system operators/TPs can have an overall picture of the activities inside a house, and even the entire life cycles of all residents (waking/sleeping pattern, number of residents, when people leave their home). This privacy concern belongs to category ii; the surveillance relies on advanced techniques such as data mining, machine learning/deep learning algorithms [26,30,32]. This information could be abused by hackers and stolen for criminal purpose [33].

2.4.3. Fraud

Fraud represents the potential risks of personal energy data being modified without authority, either to increase/decrease energy consumption or attribute the energy consumption to another house [32]. This risk belongs to category ii; the AMI enables more opportunities for adversaries to implement fraud than conventional meters since the vulnerabilities of the real-time communication network would be abused.

2.4.4. Non-Grid Commercial Uses of Data

This privacy risk falls into category ii. The smart meter data may be used by TP to make a profit from the data; activities include advertising and insurance that are not welcomed by consumers [32]. Companies would sell their products to residents according to the personal preference information revealed by the energy data. Even sensitive information, such as employment information [30], can be inferred from energy data with machine learning algorithms. This information can be used by adversaries to estimate the income of the target family.

3. Related Work for Privacy Intrusion Protection

The state-of-the-art methods dealing with the above smart meter privacy issues can be divided into two categories: user demand shaping and data manipulation. Both these techniques try to reduce the privacy loss by decreasing the probability to infer individual appliance signatures from the overall power data [34].

3.1. Demand Shaping

User demand shaping uses external energy storage devices (such as a large rechargeable battery (RB) [7,35–42], renewable energy system (RES) [43–46]), or load shifting [47–49] to distort the real power consumption curves. The RB and RES method can be treated as a noise-adding approach at the physical layer, as the original power demand is distorted, the utility cannot infer sensitive information from the smart meter data. An RB system contains a smart meter, a battery, and an energy management unit (EMU), the EMU controls battery to implement optimal energy management policy (EMP), with the injection of power from the RB B_t, the mismatching between the power supplied by the grid Y_t and consumers' power demand X_t provide privacy guarantee to consumers. The works conclude that the larger the battery capacity size, the better privacy can be guaranteed. However, the RB is a finite capacity energy storage device with capacity ranges from 2 kWh to 20 kWh [50], therefore there exists a lower and upper bound ($\hat{P}_c$ and $\hat{P}_d$) to limit the performance of the mechanism. The optimal EMP, such as best-effort (BE) algorithm [7], water-filling algorithm [51], Q-Learning algorithm [1], non-intrusive load-leveling (NILL) algorithm [52] are introduced to optimize the charging/discharging process, these algorithms control the battery either hide, smooth, or obfuscate the load signature [7]. NILL algorithms are designed to blind the NILM [52], instead of only one target load, the NILL has two states, a steady-state and recovery state if the battery capacity cannot enable the load to maintain steady-state, the load is switched to the recovery state. A privacy-versus-cost trade-off strategy considering the TOU tariff is proposed by Giaconi et.al in 2017 [53]. Instead of a constant load target, a piecewise load target referring to the current TOU price is generated, the cost of the electricity is minimized, and the consumers can also sell extra energy to the grid to reduce the cost further.

RES utilizes rooftop PV, small wind turbine, and even Electric Vehicle (EV) [54] to replace the conventional battery. To overcome the difficulty to roll-out expensive RES and RB facilities, Reference [55] proposed a multiuser shared RSE strategy that enables serval users to share one RES and one EMU. The EMU control the RES to allocate the energy from the RES to each user. In this case, the target of the system is to minimize the overall privacy loss of all users rather than an individual user. EV is another scheme to reduce the reliance of the RB [54] since the charging period is almost overlapping with the peak load, it can mask other appliance signatures. However, the EV can only be used when the consumers are at home, the consumers are still under real-time surveillance since the adversary would obtain information when the residents leave their home.

To summarize, in RB/RES methods, researchers view the identification information of the load curve as the variation of the load measurements of two neighboring measure points $Y_t - Y_{t-1}$. The ideal situation for the grid curve is a constant value C_t which will not reveal any sensitive features of the demand, the modified load curve Y_t is then compared to C_t, the more similarities between these two curves, better privacy can be guaranteed. To quantify the privacy loss, Mean Squared-Error (MSE) [53], Mutual Information (MI), Fisher Information (FI) [35], KL divergence [7], Empirical MI [37] are adopted in related works. However, user demand shaping also has drawbacks: Firstly, extra energy storage systems and renewable energy sources are required to implement the demand shaping strategy; these devices are prohibitively expensive and can be difficult to roll-out, the batteries need to be renewed frequently. Secondly, the energy storage system blinds demand response, which is one of the most important functions of the smart grid.

As the drawbacks of RB/RES methods are obvious, another demand shaping method named load shifting is proposed to replace the RB/RES techniques. This method hides sensitive information by shifting the controllable loads [47–49]. The loads can be divided into uncontrollable loads (e.g., lighting, microvan, kettle) and controllable loads (e.g., heating, ventilation, and air conditioning (HVAC) systems, EV, dishwasher, washing machine). The operation time and model of the controllable loads can be scheduled by consumers to prevent occupancy detection. In [49], combined heat and privacy (CHPr) are proposed, thermal energy storage such as electric water heater is adopted to mask occupancy. Compared with the RB approach, CHPr neither requires expensive devices nor increase electricity cost. There are several limitations of the load shifting technique, firstly, some of the controllable loads have limited operation modes and cannot be interrupted; secondly, there are restrictions for the thermal loads to store energy.

3.2. Data Manipulation

Different from the demand shaping approach, data manipulation aims to modify the smart meter data before sending it to the utility. Data aggregation, data obfuscation, data down-sampling, and anonymization all belong to this category.

Data obfuscation, which is also called data distortion, tries to add noise to the original smart meter data to cover the real power consumption [56–60]. Like demand shaping technique, data obfuscation also reduces the privacy loss by distorting the smart meter data, but on the network layer. Noises such as Gaussian noise [56,60], Laplace noise [56], gamma noise [57] are added into the original smart meter data to distort the load curve. These noise-adding mechanisms follow normal distributions with mean μ equals to zero, hence the noise would cancel out if enough readings are added up together, P. Barbosa, et al. [59] conclude that these probability distributions would not influence the relationship between the utility and privacy, so all distributions can achieve similar performance in protecting privacy. Moreover, to guarantee the billing correctness, serval schemes are proposed: Reference [56] proposes a power consumption distribution reconstruction methods by adding another Gaussian distribution into the data, but the method does not quantify how much noise should be added to recover the original curve; Reference [59] sends a filtered profile to the utility rather than masked profile, then result shows that the error of the overall power consumption is reduced in this way. However, they also find that the error during different periods (peak period, off-peak period) is significantly different, which provides new challenge. In summary, although the data distortion scheme shows efficient performance in reducing privacy loss, there are serval problems which should be discussed in future studies: (1) The TOU tariff is unavailable. Although the noise would be zero-mean, but the multiplier for TOU pricing is not. Hence the sum of TOU bills would be influenced. (2) Although from the signal processing and information-theoretic viewpoint that a zero-mean noise would not influence the result, we should notice that the power system is operating on a real-time basis. The power system operator manages the grid with the real-time data sent from the smart meter, even a minor error between the ground truth with the distorted data could result in serious faults, even the collapse of the whole system.

Data aggregation reduces the privacy loss by constructing aggregators to collecting the data from a few smart meters together, so the utility is unable to detect the electricity events in a single house [8,22,60–62]. The data aggregation technique is divided into aggregation with trusted third parties (TTP) [60] and aggregation without TTP [8,22]. J.-M Bohli, et al. [60] propose data aggregation with TTP, the data aggregator (DA) operated by the TTP is responsible for gathering the data from neighbouring smart meters and then sending the aggregated data to ES. At the end of every month, the DA also generates energy consumption of individual consumer for billing purpose. However, there are several concerns about involving TTP. Above all, a TTP could try to infer the personal information, so the TTP itself may bring extra privacy risks to the system. Secondly, with the increasing numbers of smart meters being installed, it is unrealistic for the TTP to build enough DA to satisfy the demand, and the maintenance and development budget would be unaffordable to EP and NO.

References [8,22,61–63] introduces data aggregation mechanisms without TTP. Instead, encryption techniques such as HE, MPC [8,22,62,63] are employed. Both HE and MPC encrypt personal smart meter data before sending it to the utility/TP. However, differently from conventional encryption techniques, HE and MPC enable TPs to manipulate the data without knowing the detail of it. F. Li, et al. [8] and R. Lu, et al. [22] independently proposed an aggregation method with HE separately. By encrypting smart meter data, the DA can implement aggregation without knowing the data details. In this way, there are no concerns that the TTP may infer sensitive information without permission. However, the drawbacks of data aggregation technology are twofold. Firstly, after aggregating, it is impossible for the utility to obtain the power usage information of an individual consumer. Secondly, complex encryption would cause high computational overhead. MPC requires low computing ability but involves several servers to deal with the data [63]. In MPC, each server holds a part of the input data and they cannot infer the whole information. MPC has been successfully adopted in smart metering services such as TOU billing. However, complex value-added services, such as load forecasting and online energy disaggregation, require an advanced cloud server to implement these algorithms. So, the availability of MPC in these services should be discussed. The privacy boundary of aggregation size is also investigated in T.N. Buescher, et al. work [61]. They investigated the aggregation size referring to a privacy metric named 'privacy game'. Referring to the data-driven evaluation, a conclusion is made that even a DA with over 100 houses can still reveal private information. But the privacy measure they adopt is abstract and just simply measures the difference between the individual load curve and the aggregated curve, a more detailed privacy measure should be proposed to reflect whether advanced algorithms (such as NILM) can infer personal information from the aggregated data.

References [56–58] combines data aggregation with noise-adding technique together, to enable differential privacy to the aggregated data. Differential privacy is employed as privacy guarantee, the concept of differential privacy is through adding noise to a largescale dataset, any two neighboring datasets (only one data in these two datasets is different) should be indistinguishable [64]. In other aggregation mechanisms, N smart meters are aggregated at first, then a distributed Laplacian Perturbation Algorithm (DLPA) is applied to the aggregated data. By adjusting the parameters ε and δ, then we can say (ε, δ)-differential privacy is achieved (ε is the parameter to show the strength of privacy guarantee, and the δ is the failure probability, the closer ε and δ to 0, the better privacy can guarantee). The security and privacy performance are analysed in [56], two denoising filter attacks, the linear mean (LM) filter, and the non-local mean (NLM) filter are employed to evaluate the original. The results convince that attackers cannot infer the original load curve from the distorted one.

Data Anonymization mechanism [12,65,66] reduces privacy loss by replacing the real smart meter identification with pseudonyms. C. Efthymiou and G. Kalogridis proposed a data anonymization method with a TTP escrow in 2010 [12]. They suggested that two IDs are attached to each smart meter, LFID for sending attributable low frequency and HFID for sending anonymous high-frequency data, while the HFIDs are kept by a TTP, making it unknown to the utility. The low-frequency data are used for billing purposes while the high-frequency information is for network management. However, the workload of the TTP is high, and the development costs increase since all anonymous IDs are processed here. Moreover, with the introduction of the TTP escrow, the privacy risks are not eliminated but just shift from the utility to TTP.

The down-sampling method is a naive approach that aims to reduce sensitive information by reducing the interval resolution of the metered data [13,33,66]. However, like other methods, functions such as demand response and TOU billing would be sacrificed. Moreover, value-add services that require high-resolution data are unavailable as well. To quantify the privacy loss with different interval data, G. Eibl and D. Engel adopt NILM as adversary to the extract of personal information. They apply an edge detection NILM to smart meter data and examine the performance of 15 appliances via F-score values and the proportion of appliances. They conclude that 15-min interval data already protect most appliances. We would like to have an in-depth research based on the research by implanting more powerful NILM algorithm (such as deep learning based NILM) since deep learning has shown distinctive ability to extract features than conventional approaches.

To sum up, solutions either require the installation of expensive devices (rechargeable battery or RES) or employ complex and high computing algorithms (data distortion and data aggregation). Moreover, some schemes introduce TTP into the smart metering system, which just moves the privacy risk from one party (ES) to another one (TTP). Most importantly, unlike other communication networks, the physical connections of the electricity grid already aggregate load consumptions at feeder level or substation level without privacy concerns, the construction of the data aggregator is superfluous. And no existing solution emphasizes the availability of value-added services, which is the vital functionality the smart meter brings to consumers. Comparing the two solutions listed above, the proposed scheme is simpler and more efficient: The proposed scheme is based on existing physical facilities (the smart meter, private platform, distribution substation) and does not require any extra RES or high computation encryption. In the proposed scheme, the smart meter only communicates with the private platform (PC or smartphone) inside the house via Home Area Network (HAN). A multi-channel smart metering system enables the private platform to communicate with other stakeholders (e.g., ES, TP) with different data granularity, which takes the advantages of both data aggregation mechanism to enable grid operation and management and data down sampling mechanism to provide accurate TOU bills. Furthermore, a privacy preserving NILM algorithm is designed to enable value-added services.

4. A Proposed Privacy-Functionality Trade-off Strategy and Model

Given the scale of smart meter roll-out processes in countries and worldwide, the above risks and operational strategies could be dismissed or subordinated to utilitarian market logic, with the responsibility for their implementation and subsequent privacy protection of consumers (i.e., households) delegated to third parties, many of whom might not have privacy protection as a priority in their agendas. Moreover, and as stated before, there is a lack of clarity about such responsibilities. Furthermore, whilst smart grids could be conceived as necessary technologies to regulate the conduct of individuals in our societies [67], what could be more concerning is that privacy intrusion could also generate negative social consequences [31]. Consumers can be left powerless or socially isolated to devise their own strategies to counteract intrusion to their privacy, becoming mere means rather than ends [5].

It might be possible, however, for stakeholders to exert their creativity even in the face of privacy intrusion and existing regulations (i.e., GDPR directive) [5,10,68]. This would help households comply with the functionalities that digital technologies establish for them [68] whilst socially protecting or enhancing their sense of authentic household 'hood' [6].

To meet this, we thus propose a trade-off strategy that attends to both the operational and ethical concerns for smart meters and smart grids raised in this paper. The strategy adopts these principles:

I. Ensuring the compulsory functions of smart meters as previously described and complemented.
II. Data minimization.
III. Protection from inner or outer attacks (to be explained in Section 5).

The operation of the strategy is shown in the proposed model, Figure 2, as follows. The model components are consumers, DCC, energy supplier (ES), a network operator (NO), third parties (TP), and the distribution-level substations (Sub) which supply electricity to households.

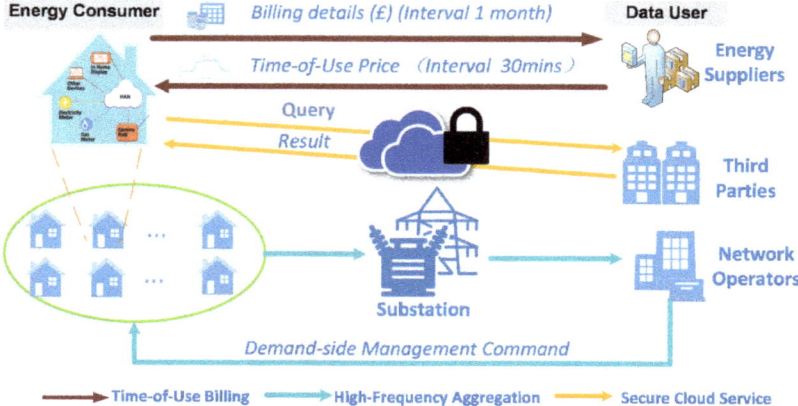

Figure 2. Proposed smart metering system.

4.1. Compulsory Functions

The interval resolution and categories of data of compulsory functions are listed in Table 1 below. For billing purposes, the frequent transmission of the power consumption data would put consumers under the monitoring of the utility. For grid operation and management, although the utility requires high interval resolution data (seconds to 100 Hz), it is unnecessary to access every individual's power consumption; aggregated data of a defined area is more desirable. Most additional services provided by TP only require a specific part of the power consumption data (a certain period, a specific appliance power consumption, etc.), so, all TPs are to obey the data minimization principle (explained below), and only collect the minimum data required to complete the service.

Table 1. Summary of data granularity of different functionalities.

Functionalities	Sample Frequency Required	Data Required
Billing	Low (weekly or monthly)	Usage of every single household smart meter
Grid Operation and Management	Very High (sometimes more than 50 Hz)	Active/Reactive power, voltage of selected area
Value-added Services	Depending on specific services	Depending on specific services
TOU tariff	High (15 min–1 h)	The TOU price from the electricity market

In the proposed strategy and model, it is not considered appropriate to authorize TTP (described earlier) to be responsible for data aggregation, since as a potential inner adversary TTP can still acquire valuable information during the aggregating process. Rather, substations could be a better choice for data aggregation. In countries such as the US and China, there is already an installation of substation-level supervisory control and data acquisition (SCADA) systems [69]. This provides evidence that substation-level smart metering or intelligent substation data would be a trend of the future smart grid system.

Moreover, in contrast to conventional smart metering systems that can only transmit a single temporal resolution trace, this novel scheme contains three communication channels to support multi-temporal resolutions data. These three channels are a high-frequency aggregated data channel, to transmit 100 Hz aggregated data measured at the distribution level substation; a TOU billing channel, to send dynamic TOU price information to smart meters and send bills to the ES monthly; and an additional service channel, to transmit selected data to support additional services. The smart meter in the scheme plays the role of the assistant processor rather than the information sender and receiver; it has basic computation ability to calculate billing inside the house rather than sending individual power consumption near real-time.

4.2. Data Minimisation and Protection

As one of the most vital principles of data protection, data minimization is mentioned in five separate sections in GDPR (Article 5 (Chapter II), Article 25 (Chapter IV), Article 47 (Chapter V), Article 89 (Chapter IX)) [9]. It highlights that limitations should be set on the measurement of personal data implemented by organizations; only the minimized information necessary to complete specific required purposes can be collected. More specifically, the data minimization principle for the smart grid is recommended in the US National Institute of Standards and Technology (NIST) Guidelines [32]. To deal with privacy risks caused by smart meters, strict limitations need to be set; only the data which are essential for smart grid operation should be collected (e.g., billing, demand-side management, grid planning). Data minimization is strongly related to protection from inner and outer attacks, as will be presented in the following section of the paper.

4.3. Mathematical Model

4.3.1. The Smart Meter

The smart meter in the proposed system does not communicate with the energy supplier directly, the measurements of the smart meter will be uploaded to a private platform (PC or smartphone) via HAN. The private platform has basic storage and computation ability to save power consumption and calculate the bills. Assume the area involves a smart meter group $SM = \{sm_1, \ldots sm_i, \ldots sm_N\} (i \in [1, N])$. The smart meter can measure power consumption with interval T (normally 15 min), marked as $P_{T,i}$. The smart meter data are encrypted to prevent consumers from modifying the power consumption data. There is no backdoor when the smart meter is manufactured, so manufacturers or energy suppliers cannot illegally access the smart meter data, and all data transmission between consumers and the utility is monitored by the DCC. In the proposed system, the smart meter reports the monthly energy consumption E_{month} and monthly bills B_{month}.

4.3.2. Protection from Inner and Outer Attacks—Adversary Element

Using a consequentialist perspective to ensure that stakeholders are held to identification and account [11], in our model all stakeholders could adopt an "honest-but-curious" ethic [21]. They follow functional protocols properly and provide expected services to consumers ("honest"), but at the same time, they keep inferring sensitive information from the consumers ("curious"). In the proposed system, household adversaries could access aggregated power consumption P_{AGG} (kW) and monthly energy consumption of smart meter i E_{month} (kW·h). Their purpose would be to obtain data. They could have a high computational ability to disaggregate the obtained data into individual appliance power consumption data by applying methods like the NILM algorithm, leading them to potentially use data for unethical or illegal purposes.

4.4. High-Frequency Aggregated Data Channel

The high-frequency aggregated data channel transmits the aggregated power consumption data to the DCC. We install a substation-level smart meter inside the distribution-level substation. The substation contains all consumers' power consumption in the local area without requiring every individual smart meter to send data to it, so it plays the role of an "aggregator," but without collecting the power consumption data from every single house. The measurement frequency of substations in our research is selected as $f_{hf} = 100$ Hz, which is twice the British power system frequency. The high interval resolution data is used for grid operation and management since near real-time data is vital for demand-side management to deal with unexpected incidents such as a blackout.

The reason that the distribution-level substation can play the role of "aggregator" is twofold. Firstly, substations already exist. No extra facilities like data aggregators need to be constructed, so the development investment can be saved. Secondly, no TTP or homomorphic encryption is involved in this scheme, so the concerns of inner attacks from TTP and computation overhead raised by complex encryption are eliminated. Table 2 shows three typical feeder models summarized by GridLAB-D's feeder taxonomy [70], these three models represents feeders at light rural area, heavy suburban, and moderate urban respectively. The house units under each feeder can be estimated by adding up household-level data to match the feeder model [71]. From the table, the light rural area consists around 408 houses. In Section 5, an evaluation is implemented whether feeder/substation level measurement at light rural area satisfies the privacy requirement.

Table 2. House units under different feeder models [70,71].

Feeder Model	Active Power/kW	Description	Units under the Feeder
R4-25.00-1	948	Light rural	408
R1-12.47-4	5334	Heavy suburban	2299
R2-25.00-1	17,021	Moderate urban	7336

4.5. Time-of-Use Billing Channel

The TOU channel enables the dynamic TOU tariff, see Figure 3. In the conventional smart metering system, the smart meter should report the energy consumption at each charging point to obtain TOU bills. The more charging points the utility sets, the more detailed information about an individual is obtained by the utility, and the more it is possible that privacy is breached.

In our TOU billing channel, the direction of information transmission is the opposite. The algorithm of calculating the TOU billing is shown in Algorithm 1.

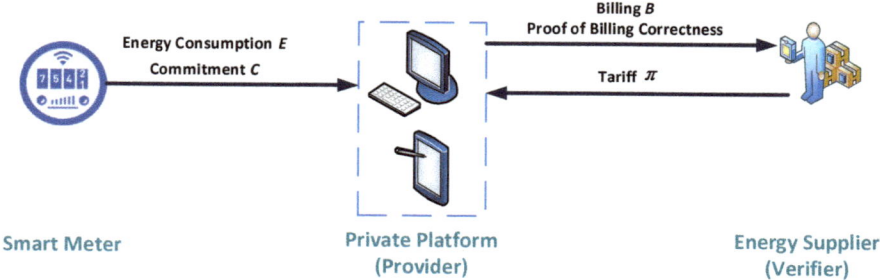

Figure 3. Time-of-use billing channel and billing correctness verification.

Algorithm 1 Dynamic TOU billing program.

Input: Half-Hourly Energy Consumption $E_{d,t}$, Half-Hourly TOU tariff $\pi_{d,t}$;
For d = 1; d ≤ 30 (d is the day of month) do
   for t = 1; t ≤ 48 (t is the time of the day) do
     Record and storage $E_{d,t}$ and π_t during t.
   End
End
While d ≥ 30 do:
   Calculate $E_{month} = \sum_{d=1}^{30} \sum_{t=1}^{48} E_{d,t}$.
   Calculate $B_{month} = \sum_{d=1}^{30} \sum_{t=1}^{48} \pi_{d,t} E_{d,t}$.
End
Return E_{month}, B_{month}.

Output: Monthly Energy Consumption E_{month}, Monthly Bills B_{month}.

- Step 1: Data storage. The ES sends the TOU price π to the smart meter every 30 min. The smart meter stores the energy consumption of the past half-hour with the current TOU price in pairs
- Step 2: Bills calculation. The total TOU bills in £ are calculated at the end of each month, then sends the bill data to the ES, the ES then assigns a bill to the consumers.
- Step 3. Billing correctness verification. A zero-knowledge proof [72] is utilized to verify the billing correctness. A detailed description is shown in Appendix A.

4.6. Additional Service Channel

The additional service channel is designed for TP to provide additional services to the consumers. The "third party" refers to non-licensed energy service companies. They bring profits and innovation to the smart grid industry. The consumers have the freedom of choice to select wanted services. Currently, services include a warning for exceeding power thresholds, monitoring for seniors/children, monitoring the operating condition of selected appliances. From the consultation documents of Department of Energy & Climate Change (DECC) [19], there are strict limitations on TPs' access to data, an agreement is required among TP, DCC, and consumers, and the TP can only access the consumers' smart meter data when they have consumer consent.

Referring to the privacy-functionality trade-off strategy mentioned in Section 4, the value-added service channel follows a data minimization principle, preventing personal data "leaving" consumer's house. Rather than sending personal data to the server of TPs, TPs send algorithms and models to consumers' private platforms, including their personal computers and mobile phone, then the consumer can use the model to obtain the result on these platforms. However, there are two concerns related to this method referring to [21]:

- By sending algorithms/model parameters to consumers, the model's parameters and training dataset would be stolen by users, while these models and datasets are confidential.
- It is difficult to implement a privacy-preserving algorithm to complex models such as machine learning/deep learning-based services.

To settle the above two concerns, particularly relevant to our work [73], a value-added service channel utilizing a privacy-preserving deep learning algorithm is proposed. The algorithm adds noise into gradient descents of deep neural network parameters to reduce the sensitivity of single training data, and further preserve the privacy of both neural network model and training dataset [74]. The privacy-preserving deep learning combines two advanced techniques, differential privacy, and deep learning together.

As shown in Figure 4, the process of the value-added service channel consists of the following steps, and all steps can be divided into two categories depending on the network (WAN or HAN):
Operations via WAN:

- Step 1: the model owner trains the service model with a private database using a noisy algorithm.
- Step 2: the trained model is sent to consumers' private platform (personal computer, mobile phone).

Operations via HAN:

- Step 3: the smart meter uploads the measurements to the private platforms.
- Step 4: the consumer starts a query to the platform; then the platform returns a result to the consumer.

The data flow in Figure 4 shows that the consumer's energy data are shared inside HAN and are never sent to the utility, but the services are enabled. The enabled services include NILM, STLF, and demand response. The detail of the privacy-preserving deep learning NILM algorithm is shown in our paper [73].

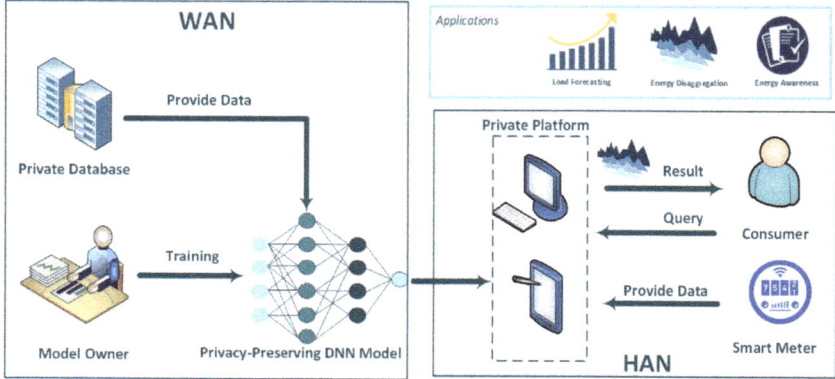

Figure 4. Privacy-preserving valued-added services.

5. Evaluation

In this section, an evaluation is implemented to discuss whether the proposed scheme satisfies the requirements in Section 2 considering both functionality and privacy.

5.1. Privacy Measure

The privacy measures are adopted to the smart meter data shared with stakeholders (high-frequency aggregated data and down-sampled individual data). Referring to privacy intrusion categories highlighted in Section 2.4, all privacy intrusion issues belong to two categories: data sensitivity and algorithm sensitivity. In many previous works, researchers view the sensitive information from the smart meter as the variation of the power consumption curve, a constantly changing curve would reveal more private information than a flattening curve. While the advanced NILM algorithms are used to infer individual appliance signatures, state-of-the-art algorithms such as DNN, ML, Hidden Markov Model (HMM) are proposed.

Hence, in this paper, both above two privacy intrusion issues are studied, Mutual Information and Mean Squared Error (MSE) is utilized to quantify the sensitivity of the data, and a NILM adversary, named NILMTK, is adopted to examine whether the proposed scheme can blind the algorithms.

Moreover, since a noise-adding deep learning approach is applied to the value-added services, privacy performance is evaluated to determine whether the system can provide differential privacy guarantees.

5.1.1. Mean Square Error (MSE) as a Privacy Measure

Mean squared error (MSE) is a naïve metric to evaluate the error between two groups of data. In this paper, MSE is adopted to quantify the difference between original consumption data and the modified data:

$$\text{MSE} = \frac{\sum_{i=1}^{N}(\hat{y}-y)^2}{N} \quad (1)$$

5.1.2. Mutual Information (MI) as a Privacy Measure

Mutual information (MI) is employed as a privacy measure to quantify privacy loss in [44,46,51,53,75]. MI $I(X^n; Y^n)$ measures the dependence between two random variable sequences X^n and Y^n [76]. In other word, MI can explain the reduction of the original load sequence X^n given knowledge of the modified sequence Y^n:

$$\begin{aligned} I(X^n; Y^n) &= H(X^n) - H(X^n|Y^n) \\ &= H(X^n) + H(Y^n) - H(X^n, Y^n) \\ &\approx -\tfrac{1}{n}logp(Y^n) - \tfrac{1}{n}logp(X^n) + \tfrac{1}{n}logp(X^n, Y^n) \end{aligned} \quad (2)$$

where $H(X^n)$ and $H(Y^n)$ are the marginal entropies, which measures the uncertainty about the random variable; $H(X^n|Y^n)$ is the conditional entropies, and (X^n, Y^n) is the joint entropy of $H(X^n)$ and $H(Y^n)$. In this paper, a variant MI named Normalized Mutual Information (NMI) is adopted to show the normalized results between 0 and 1 (0 represents no mutual information, 1 represents perfect correlation).

5.1.3. NILM Performance as a Privacy Measure

NILM is used as a privacy measure in previous works [33,35,56,61], the NILM plays the role of a powerful adversary to evaluate the privacy loss of the smart metering system. The adversary can adopt a state of the art NILM algorithms to obtains individual appliance signatures from the measured demand, hence the NILM is a desirable privacy measure to quantify the privacy loss. In this paper, the NILMTK toolbox [77] in Python is used to implement the NILM algorithm, we utilize the deep neural network model proposed in [78]. Five appliances, Air conditioner (Air), EV, refrigerator, stove, and dryer are investigated in this paper. Confusion matrix and F-score are used to evaluate the performance of the adversary, see Table 3.

$$F - measure = \frac{1}{1 + (\text{FN} + \text{FP})/(2\text{TP})} \quad (3)$$

Table 3. Confusion matrix.

	Actual Positive	Actual Negative
Predicted Positive	True Positive (TP)	False Positive (FP)
Predicted Negative	False Negative (FN)	True Negative (TN)

5.1.4. Differential Privacy as Privacy Guarantee

As a state of the art notion of privacy, differential privacy is proposed by Dwork in 2006 [64], the adversary cannot distinguish two neighboring datasets with only one pair of data that are different. Normally, differential privacy is achieved by adding noise into the data (e.g., Laplacian noise [64], Gaussian noise [79], exponential noise). A (ε, δ) differential privacy is obtained, while ε denotes the amount of noise added to the data, and δ represents the threshold to break the privacy.

Definition 1. *(ε-Differential Privacy) A randomized function $\mathcal{R}$ satisfies (ε, δ) privacy $\mathbb{P}_\mathbb{R}$ for any two neighboring datasets β and β'*

$$\mathbb{P}_\mathbb{R}[\mathcal{R}(\beta) \in \xi] \leq e^\varepsilon \mathbb{P}_\mathbb{R}[\mathcal{R}(\beta') \in \xi] + \delta \quad (4)$$

where ξ denotes all possible outcomes in range R, and δ is the possibility that the differential privacy is broken, in this paper, we select 10^{-5} as δ.

Definition 2. *(Global Sensitivity) For a random function f, the global sensitivity, S_f, is the maximum difference between the outputs of two neighboring datasets β and β'. S_f also determines the overall noise to be added into the DP mechanism:*

$$\Delta f = \max_{d(\beta,\beta')=1} \|f(\beta) - f(\beta')\| \tag{5}$$

5.2. Dataset Description and Data Preprocessing

We adopted the Dataport [80] as the dataset. As the world's largest residential electricity consumption dataset, the dataset contains electricity data from 722 houses in the US. The interval resolution of the data is 1 min. We delete the data from 11 pm to 6 am since fewer electricity activities occur during this period.

5.3. The High-Frequency Aggregated Channel Satisfies Privacy Requirement

In this case study, the privacy performance of the high-frequency aggregated data is evaluated. As shown in Figure 5, with the increasing aggregation level, the curve of power consumption becomes smoother, and the details of individual appliance signature become difficult to extract. The dataset used for simulation is the Dataport [80] during 2018, the dataset contains both total power consumption as well as the details of each appliance. Different aggregation sizes are investigated (1 house, 2 houses, 5 houses, 10 houses, and 50 houses, respectively). The following will evaluate the privacy loss from both data sensitivity and algorithm sensitivity aspects.

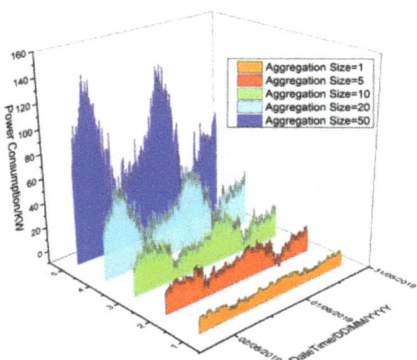

Figure 5. Single house power consumption versus different aggregation sizes of power consumption.

5.3.1. Influence of Aggregation Size on Data Sensitivity

The data sensitivity of the aggregated smart meter is evaluated in this subsection. In this scenario, we wanted to find out whether the adversary can still infer the individual's power usage data P_{real} from the high-frequency aggregated data P_{AGG}. Figure 6 shows the value of MI and MSE with different aggregation sizes. A reduction of the MI value is observed, from 1 at a single house to 0 at 10 houses, and the MI value would remain 0. The MSE value increases from 0 to 10^4 kW^2 when the aggregation size changes from a single house to 100 houses. The result shows that when aggregation size AGG is larger than 10 houses, P_{real} and P_{AGG} are totally independent, and no knowledge about the P_{real} would be revealed from P_{AGG}.

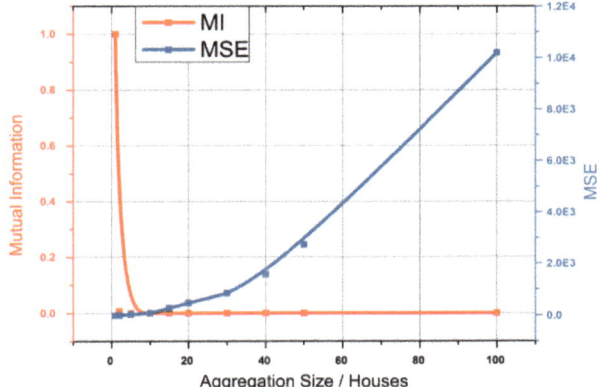

Figure 6. Mutual information and MSE of different aggregation sizes.

5.3.2. Influence of Aggregation Size on Algorithm Sensitivity

The algorithm sensitivity of the aggregated data P_{AGG} is evaluated via NILMTK tool, the target of the algorithm is inferring the appliance signature inside house i given aggregated load P_{AGG}. From Figure 7, when implementing NILMTK to a single house, the adversary can infer the appliance signatures with F-score between 80–100%, presenting that the NILMTK has perfect performance. When the aggregation size AGG reaches 2, the performance of NILMTK on most appliances such as EV, fridge, stove, and dryer has been influenced greatly, especially the F-score of EV reduces from 100% to 0. By continuously increasing AGG to 50 houses, the F-score of all appliances decreases to zero. From the result, it is concluded that at least 50 houses need to be aggregated to blind the NILM adversary.

To summarize, when AGG is larger than 50, both privacy intrusion issues can be prevented.

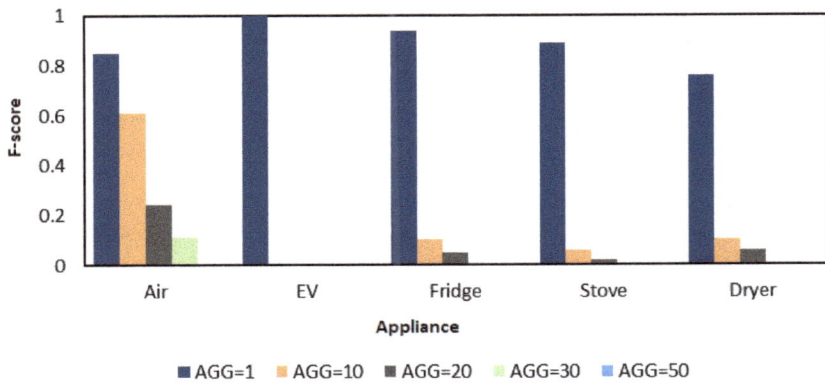

Figure 7. F-score of the NILMTK performance on appliances from different aggregation sizes.

5.4. The TOU Tariff Channel Satisfies Privacy Requirement

The temporal resolution level is another vital parameter that influences the privacy loss. In this case study, we take the data with 1 min interval as the P_{real}, and then downsample P_{real} to the lower interval T by taking the average values of all sampling points of P_{real} duration interval T (in this study, T ranges from 5 min to 1 month).

5.4.1. Privacy Measure of Data Sensitivity

This scenario tries to find out whether the adversary can still infer the individual's power usage data P_{real} from the downsampled data, P_T. Figure 8 shows the value of MI and MSE with the increase of interval resolution T. A dramatic reduction in MI is observed when T increases from 1 min to 180 min (3 h), the reduction of MI then becomes gentle when T continuously increases. The F-score drops to 0 when T reaches 1440 min (24 h) when only one data is recorded each day under this interval resolution. In contrast to MI, the value of MSE increases from 0 to 12.8, showing that the increase of $\hat{T}$ would reduce the knowledge of the original load curve.

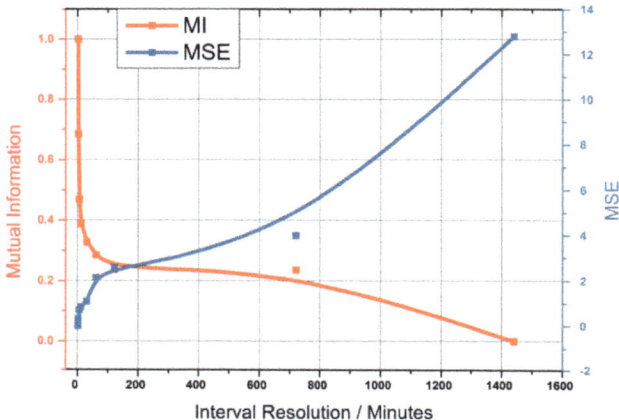

Figure 8. F-score of the NILMTK performance on appliances from different interval resolution.

5.4.2. Privacy Measure of Algorithm Sensitivity

The algorithm sensitivity on smart meter data different interval resolution is evaluated in this subsection, as shown in Figure 9. The F-score shows how the NILMTK adversary infers appliance information from the overall power consumption. While the NILMTK has a good performance with 1-min interval resolution data (achieving an F-score of 80–100%), the F-score drops gradually when the interval resolution increases. Taking air conditioner as an example, the NILMTK adversary achieves an F-score with 83%, representing that most of the operation duration of the air conditioner is detected. When interval resolution T increases to 1 h, the F-score drops to 42%. Furthermore, the F-score decreases to 0 when T equals to 24 h, meaning that the NILMTK is blinded totally. Most importantly, it is observed that even with 6-h interval resolution, the NILMTK achieves an estimation with 36%, 21%, and 20% F-score in EV, fridge, and dryer respectively, showing that a large interval (such as 6 h) still cannot guarantee the privacy.

Based on the above two discussions, to completely reduce the privacy loss, a 24 h smart meter data is required.

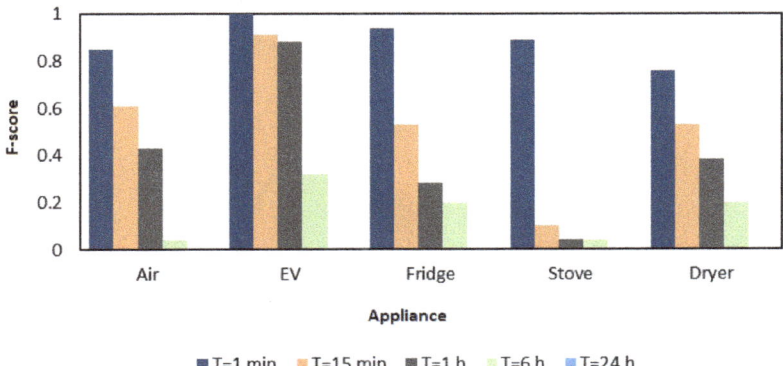

Figure 9. F-score of the NILMTK performance on appliances from different interval resolution.

5.4.3. ES Can Verify Billing Correctness

A detailed proof is given in Appendix A. The private platform generates a bill to ES monthly according to the stored TOU tariff and energy consumption, then the private platform sends a series of commitments to ES. Given TOU record and bill, ES can open commitments and verify if the commitments match the received bill.

5.5. Value-Added Service Channel Satsifies Privacy Requirements and Provides Differential Privacy to ES

Referring to the Demonstration in Appendix B, the value-added service channel provides a $(2\frac{L}{N}\varepsilon\sqrt{T}, \delta)$-differential privacy guarantee to ES, hence the model parameters and training dataset for the service is protected. As for consumer, since the service is implemented inside HAN and completed by private platform, the private information never be shared with other parties.

5.6. Comparison of the Proposed System with Related Schemes

In this Subsection, a comprehensive comparison is made between the proposed smart metering system and other related operational strategies (e.g., rechargeable battery, data aggregation, data down-sampling) from the aspects of both functionalities and privacy protection. Referring to Sections 2.3 and 2.4, the four compulsory functionalities: billings, TOU tariff, grid management and operation, and value-added services. While the four privacy intrusion risks cover data sensitivity and algorithm sensitivity, and can be further divided into fraud, real-time surveillance, behaviour patterns identification, non-grid commercial uses of data four categories. These strategies cover the private information by modifying the load curves or encrypting the consumer's energy data. And privacy evaluations employed in these would either assess the performance of data sensitivity (MI, FI, KL-Divergence, etc.) or the performance of algorithm sensitivity (NILM as adversary). As shown in Table 4, it is observed that most strategies settle both privacy intrusion problems, but some strategies sacrifice conclusory functionalities: data distortion adds noise to the original data making the modified data different from the real energy consumption, as a consequence the TOU billing is unavailable; the data aggregation method adds dozens of smart meters' data together and then sends it to the utility, which also prevents the utility from obtaining individual bills' information; and the data down-sampling technique reduces the sampling interval of the smart meter, which would influence grid management and value-added services. Moreover, HE and rechargeable battery approaches require either extra expensive energy storage systems or extremely high computation ability, which is unrealistic to roll-out. The proposed system enables different granularity data to be transmitted between the smart meter and the utility/TP, depending on the required functionalities. What the adversaries can obtain is high-frequency but aggregated data (substation/feeder level) and household-level but down-sampled data, both these two information streams would not reveal useful personal information (see Section 6 for demonstration). In addition, instead of adopting a TTP, the proposed method installs smart meter besides feeders or substations directly, so the worries about the privacy risks brought by TTP are solved.

Especially, a Distribution Network Operator (DNO) would benefit from the proposed smart metering system from both economic and technical aspects. As for economic benefits, the proposed smart metering system provides a more cost-effective network for DNO. By monitoring the real-time substation/feeder level demand, DNO has an insight view about the operation condition of the distributed network. The improvement of the visibility help DNO implements better and prioritized management to feeder voltage, and the energy loss is reduced as a result. Moreover, substation/feeder level smart meters help DNO understand peak demand patterns of the local area, the DNO takes advantages of these patterns when designing and planning networks. In this case, DNO can save the unnecessary cost of networks and enable the network to operate just above the maximum peak load. As for the technical aspect, the proposed smart metering system provide high-resolution electricity data to DNO, DNO can utilize the collected data for following technical tasks: (1) Load forecasting and feeder-level energy disaggregation. With feeder/substation level historical and real-time smart meter data, DNO can forecast the variation of load demand accurately, and the load components under the substation can be evaluated via ML or DNN algorithms, paving way for demand-side management; (2) Batter manage distributed generation. The continuously increasing of the distributed generation (such as solar panel and wind turbine) bring high reserve power flow to the low-voltage (LV) network, which causes stability issues such as voltage spikes. The proposed smart metering system help DNO identify the reserve power flow, and DNO can employ operation to maintain the stability of the power system.

As for the cost of the proposed system, the system is mostly constructed based on existing smart metering infrastructure, except for the installation of feeder/substation-level smart meters and utilization of the private platform to store the historic energy data. The rechargeable battery/energy storage system method requires each house to install a mini energy storage system or EV to flatten the power consumption curve [7] and that they should change the battery frequently, while the cost of each battery can reach thousands of pounds [81]. When we are comparing the cost at substation scale (each substation contains hundreds of houses), the cost of rechargeable batteries is much higher than the proposed system. As for encryption techniques, traditional encryption techniques such as symmetric encryption can only guarantee the security of data transmission from the consumer side to energy suppliers/third parties' side. However, energy suppliers and third parties are potential adversaries as well, the privacy of consumer's data cannot be ensured. As for encryption methods such as HE, they enable TP to process/manipulate data without knowing the detail of the data. However, the disadvantages of HE is also obvious, HE requires extremely high computation ability to encrypt/decrypt the data. Considering memory usage, 1 Mb of data results in more than 10 Gb of encrypted data [82]. As far as computation, multiplication takes over 5 s per multiplication. The above is just the cost of one smart meter when we move to the whole smart metering system it contains millions of smart meters, the cost would be an astronomical figure.

Table 4. Comparison of the performance between proposed method and related operational strategies.

Operational Strategies		Functionalities of the Smart Metering System				Privacy Intrusion Issues Protection		Comments
		Billing	TOU	Grid Management	Value-Added Service	Data Sensitivity	Algorithm Sensitivity	
Proposed method		✓	✓	✓	✓	✓	✓	
Demand Shaping	Rechargeable battery [7,35–42]	✓	✓	✓	✗	✓	✓	Battery is expensive
	Load shifting [47–49]	✓	✓	✓	✗	✓	Unknown	
	Energy Storage System [43–46]	✓	✓	✓	✗	✓	✓	Device is expensive
Data Manipulation	Data obfuscation [56–60]	✓	✗	✗	✗	✓	✓	Data is useless to grid
	Data aggregation with TTP [60]	✗	✗	✓	✗	✓	✓	TTP brings new privacy issues
	Data aggregation without TTP [8,22,61–63]	✗	✗	✓	✗	✓	✓	Computation overhead
	Data aggregation with noise-adding [56–58]	✗	✗	✓	✗	✓	✓	
	Data down-sampling [13,33,66]	✓	✗	✗	Unknown	✓	✓	
	Data anonymization [12,65,66]	✓	✗	✓	Unknown	✓	✓	

6. Conclusions and Future Work

6.1. Conclusions

In this paper, we have presented a smart metering scheme (strategy and model) to prevent privacy risks (operational and ethical) raised by the smart meter. The proposed scheme has three communication channels to enables power system management and operation, TOU billing, and value-added services three functionalities. The different channel transmits different interval resolution data. As for privacy aspects, we divide all privacy issues related to the smart meter into two categories, data sensitivity, and algorithm sensitivity.

There are two main contributions of this paper to existing operational methods to deal with privacy intrusion. Firstly, in the high-frequency aggregation channel, we adopt the distribution-level substation as "aggregator", the substation supplies power to over 400 houses in a light rural area and over 7000 houses in a moderate urban area. In this way, we eliminate the risk of an inner attack from the TTP. Secondly, we use the private platform as a data processor, only reporting billing details monthly without frequently sending individual energy consumption data to the utility. Thirdly, privacy preserving NILM algorithm is employed to the value-added services to protect both consumers and ES's privacy. Finally, an evaluation is implemented to the system which demonstrates the proposed system satisfies all privacy requirements. From the evaluation, the conclusion is made a dataset with aggregation size over 50, and interval resolution larger than 24 h can overcome both data sensitivity and algorithm sensitivity.

6.2. Implications for Policy

Current smart metering systems always share the real-time household-level smart meter data with the utility. Smart metering system policymakers (e.g., the Department for Business, Energy & Industrial Strategy (BEIS) in the UK) should be aware of the trade-off between functionalities and privacy when operating the system and should have a clear idea about the data granularity required by different stakeholders. [83] suggests that the policymaker should classify the smart meter data into different openness categories, ranging from open data (the data can be totally open to the public) to closed data (private data that is confidential). In this case, the operators can maximize the value of data and minimize the privacy and security issues. Different stakeholders (e.g., NO, ES, TP) should access different granularity of smart meter data, while the granularity of the data includes the interval resolution, the aggregation size, etc.

Policymakers could also find it difficult to sacrifice functionalities to protect individual consumers' (i.e., households') privacy. The importance of the smart metering system is to provide accurate real-time reading and further reduce energy costs. Policymakers could carefully implement methods such as noise-adding or load curve distortion. Although these methods would reduce the sensitivity of personal information and thus risks of privacy intrusion, the usability and the value of the data would decrease as well, potentially undermining the achievement of benefits for stakeholders. The proposed strategy and model suggest that it might be possible to balance demands and benefits without compromising household privacy; rather other opportunities could emerge if policy considers freedom from digital surveillance and analysis as a creative situation. In this regard, and through the inclusion of adversaries and aggregators as potentially valuable 'stakeholders' of smart meters, it might be possible to help households comply with societal functionalities whilst retaining their sense of freedom and using it creatively for other purposes than energy efficiencies.

6.3. Future Work

In this paper, only the overall smart metering system is proposed. However, the efficient of the proposed smart metering system should be evaluated in practice, a pilot network with small groups of residence to be built to validation the availability of proposed system. Secondly, the functionality for grid management and operation in the proposed system should be verified via simulation. Thirdly, since we adopt multi-frequency communication channels in the proposed system, the noise would be generated in data transmission; we would propose a further study to investigate the influence on the quality of data. Finally, we could also devise participative methods to continue exploring the ethical consequences of smart meters for different (digital and non-digital) stakeholders.

Author Contributions: X.-Y.Z. and S.K. did the methodology, simulation, and validation. X.-Y.Z. did the analysis and wrote the paper. Writing-Original Draft Preparation, X.-Y.Z. and S.K.; Writing-Review Editing, X.-Y.Z., S.K., J.-R.C.-P., and C.W.; Supervision, S.K., J.-R.C.-P., and C.W. All authors have read and agreed to the published version of the manuscript.

Funding: This research is funded by The Leverhulme Trust. This research is a part of the project "Ethics and Laws of Algorithms" funded by Royal Holloway, University of London.

Acknowledgments: The author acknowledges the support of the Department of Electronic Engineering, Department of Computer Science, School of Business and Management, Royal Holloway, University of London. Thanks to the anonymous peer-reviewers and editors whose comments helped to improve and clarify this manuscript.

Conflicts of Interest: The authors declare no conflict of interest.

Abbreviations and Notations

The following abbreviations are used in this manuscript:

TOU	Time-of-use
HE	Homomorphic Encryption
TP	Third party
MPC	Multi-party computation
GDPR	General Data Protection Regulation
PRIME	Powerline Intelligent Metering Evolution
PLC	Powerline communication
NILM	Non-intrusive Load Monitoring
ES	Energy suppliers
NO	Network operators
DCC	Data and communications companies
DNN	Deep Neural Network
MI	Mutual Information
SCADA	Supervisory control and data acquisition
HAN	Home Area Network
WAN	Wide Area Network
TTP	Trusted third parties
P_{real}	Active power of individual house
P_{AGG}	High frequency aggregated active power
P_T	Down sampled power consumption data
E_{month}	Monthly energy consumption of smart meter
B_{month}	Monthly bills
$E_{d,t}$	Half-Hourly Energy Consumption
$\pi_{d,t}$	Half-Hourly TOU tariff

Appendix A. Billing Verification with Zero-Knowledge Proof

After the bill is generated by the private platform, the ES should verify the correctness of the bill. Zero-knowledge proof (ZKP) is a proof approach that can verify the correctness of the provided information without revealing the details of the information [72,84]. ZKP protocol involves two parties, provider (P) and verifier (V), P interacts with V to convince that a secret is true, but V has no knowledge about the secret P wants to prove. Pedersen Commitment is employed in this verification process [85]. A commitment with secret x is generated as $c = Commit(x, r)$, and x is difficult to infer from c, the commitment can be opened with c, x, r, marked as $Open(c, x, y)$. The $Open(c, x, y)$ would return True if c is the commitment of secret x, otherwise False is returned. They have homomorphic features of commitments are important to verify the billing correctness:

$$Commit(x, y) \times Commit(m, n) = Commit(x + m, y + n) \tag{A1}$$

$$Commit(x, y)^n = Commit(x \times n, y \times n) \tag{A2}$$

In the proposed TOU billing channel, the private platform is P, and the ES is V, Public Key Infrastructure (PKI) assigns a series of commitments COM with the power consumption sequence $E = E_{1,1}, E_{1,2}, \ldots E_{d,t} \ldots E_{30,48}$, and random value sequences $R = r_{1,1}, r_{1,2} \ldots r_{d,t} \ldots E_{30,48}$ to the private platform:

$$Commit_{d,t} = Commit(E_{d,t}, r_{d,t}) \tag{A3}$$

$$COM = (Commit_{1,1}, Commit_{1,2} \ldots Commit_{d,t} \ldots Commit_{30,48}) \tag{A4}$$

The private platform will receive the commitment c and energy consumption sequence E from the smart meter, and the TOU tariff sequence $TARIFF = \pi_{1,1}, \pi_{1,2} \ldots \pi_{d,t} \ldots \pi_{30,48}$ from the ES. So, the private platform can calculate monthly price:

$$B_{month} = \sum_{d=1}^{30} \sum_{t=1}^{48} \pi_{d,t} E_{d,t} \tag{A5}$$

referring to (A1) and (A2), the ES multiplies COM with $TARIFF$ to obtain the commitment of monthly bill:

$$Commit_{Bill} = \prod_{d=1, t=1}^{30 \times 48} Commit(E_{d,t}, r_{d,t})^{\pi_{d,t}} \tag{A6}$$

By opening the commitment of monthly bill, ES can verify the correctness of the bill:

$$Open_{Bill} = (Commit_{Bill}, B_{month}, r_{Bill}) \tag{A7}$$

If $Commit_{Bill}$ is really the commitment of B_{month}, ES accepts the generated bill, otherwise, ES reject the bill.

Appendix B. Privacy-Preserving Deep Learning-Based NILM Algorithm

Deep learning is an important branch of machine learning, it is based on an artificial neural network (ANN) with multi-layers (normally one input layer, one output layer, and several hidden layers between the input and output), each layer contains a number of neurons. A neuron is a mathematic function that computes the sum of weighted input and obtain nonlinear output via an activation function (e.g., ReLU, Sigmoid, Tanh). Compared with a shallow ANN, a deep neural network (DNN) contains several hidden layers, which enables much more complex computation tasks. The expression for a typical N-layer DNN with input x is shown in (A8):

$$a_N(x; \theta_{1,\ldots,N}) = f_N(f_{N-1}(\ldots f_1(x, \theta_1), \theta_{N-1}), \theta_N) \tag{A8}$$

where f_i is the activation function of ith layer, and θ_i is the weight of ith layer.

A loss function $\mathcal{L}$ is adopted to calculate the mismatch between ground truth y and a_N. The purpose of the DNN is to find the optimal parameters of the model θ^* that minimize the $\mathcal{L}$ throughout the whole training process:

$$\mathcal{L}(\theta) = \frac{1}{N} \sum_{(x,y) \in (X,Y)} \mathcal{L}(y, a_N(x; \theta_{1,\ldots,N})) \tag{A9}$$

$$\theta^* \leftarrow argmin_\theta \mathcal{L}(\theta) \tag{A10}$$

Differential privacy-stochastic gradient descent (DP-SGD) provides ε-differential privacy to the DNN model by adding noise to the SGD [74]. The hyperparameters of the model are shown in Table A1. Different from conventional SGD, at each time to calculate the gradient, the gradient is clipped and then a random Gaussian noise is added to the gradient. By adding the noise, a (ε, δ)-differential privacy is enabled each step. Since the number of training steps are large (range from hundreds to thousands), referring to the Composition Theorem stated as follow, the overall privacy guarantee is extremely large.

Theorem A1. *(Composition Theorem) If f is (ε_1, δ_1)-differential privacy and g is (ε_2, δ_2)-differential privacy, then*

$$f(D), g(D) \text{ is } (\varepsilon_1 + \varepsilon_2, \delta_1 + \delta_2) - Differential\ Privacy \qquad (A11)$$

To minimize the value of $\varepsilon_{total} = T\varepsilon, \delta_{total} = T\delta$ (T is the total training steps), a Moments Accountant Theorem is proposed by M. Abadi, et al. [74]:

Theorem A2. *The privacy preserving NILM algorithm provides a $(2\frac{L}{N}\varepsilon \sqrt{T}, \delta)$-differential privacy guarantee to ES.*

Proof of Theorem 1. A detailed proof is given in [74]. □

Algorithm A1. Privacy-preserving Deep Neural Network Algorithm

Input: Model input $X = \{x_1 x_2 \ldots x_N\}$.
Initialisation of weights θ_0;
For training time step t ≤ total training time T:
  Computing Loss function $\mathcal{L}(\theta) = \frac{1}{N}\sum_{(x,y)\in (X,Y)} \mathcal{L}(y, a_N(x;\theta_{1,\ldots,N}))$
  Calculating Gradient $g_B = \frac{1}{B}\sum_{x\in B}\nabla_\theta \mathcal{L}(\theta, x)$;
  Clipping Gradient $\overline{g}(x_i) = \frac{g(x_i)}{\max(1,\ \|g(x_i)\|_2/C)}$;
  Adding Random Noise to the Gradient $\widetilde{g}(x_i) = \frac{1}{L}(\sum_{x\in L}\overline{g}(x_i) + \mathcal{N}(0, \Delta f^2 \sigma^2))$;
  Update Parameters after each training step t $\theta_{t+1} = \theta_t - \alpha\widetilde{g}_t(x_i)$.
End
Output: Output Result Model weights θ and privacy cost (ε, δ).

As shown in Figure A1, the target of the NILM services is to evaluate the individual appliance consumption (output) from overall power consumption measured by the smart meter (input). Three hidden layers are linked between input layer and output layer to extract features. DP-SGD algorithm is applied in the neural network to calculate the gradient of each training step.

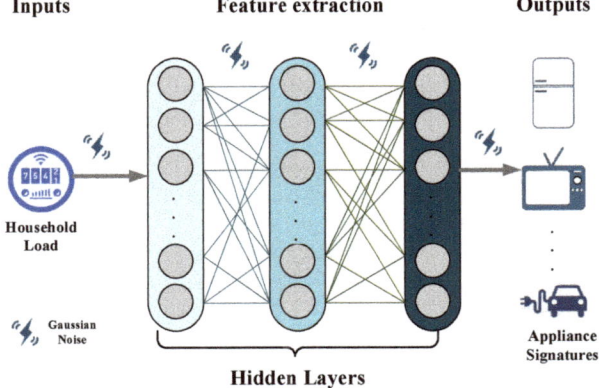

Figure A1. Privacy-preserving deep learning NILM model.

Figure A2 shows the relation between the performance of NILM algorithm and privacy level ε [73], the smaller the value ε, the better privacy it provides. It is found that without using DP-SGD, the accuracy of NILM is near 90%. By increasing the noise scale σ (a decrease of ε), and the accuracy of the algorithm decreases. So, the TPs should carful choose a noise scale that follows privacy-utility trade-off to settle both services and privacy.

Table A1. Hyperparameters of the Model [74,86].

Hyperparameters	Value	Description
Learning rate α	0.05–0.3	The steps to adjust θ according to errors.
Hidden layers number	3	The total number of hidden layers.
Lot size L	$\sqrt{N}$	The group size for computing gradient.
Batch size B	$\ll L$	The number of training examples utilized in one iteration.
Activation function f	ReLU	$f_{ReLU} = \max[0, z]$.
Gradient clipping norm C	1–10	Control the range of each gradient value, ensure that no gradient is much different with others.
Noise Scale σ	$\frac{\sqrt{2\log 1/\delta}}{\varepsilon}$	The amount of noise added.

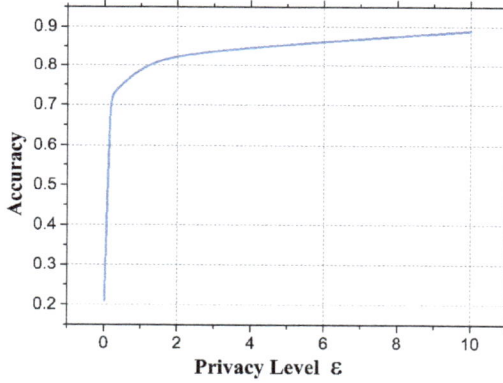

Figure A2. Privacy–accuracy curve of privacy-preserving NILM (Data from [73]).

References

1. *Smart Meters: A Guide*; Department for Business (EIS): London, UK, 2018.
2. King, N.J.; Jessen, P.W. Smart metering systems and data sharing: Why getting a smart meter should also mean getting strong information privacy controls to manage data sharing. *Int. J. Law Inf. Technol.* **2014**, *22*, 215–253. [CrossRef]
3. Molina–Markham, A.; Shenoy, P.; Fu, K.; Cecchet, E.; Irwin, D. Private memoirs of a smart meter. In Proceedings of the 2nd ACM Workshop on Embedded Sensing Systems for Energy-Efficiency in Building, New York, NY, USA, 2 October 2010; pp. 61–66.
4. Quinn, E.L. Smart Metering and Privacy: Existing Laws and Competing Policies. *SSRN Electron. J.* **2009**. [CrossRef]
5. Autili, M.; Di Ruscio, D.; Inverardi, P.; Pelliccione, P.; Tivoli, M. A software exoskeleton to protect and support citizen's ethics and privacy in the digital world. *IEEE Access* **2019**, *7*, 62011–62021. [CrossRef]
6. Taylor, C. *The Ethics of Authenticity*; Harvard University Press: Cambridge, MA, USA, 1992.

7. Kalogridis, G.; Efthymiou, C.; Denic, S.Z.; Lewis, T.A.; Cepeda, R. Privacy for smart meters: Towards undetectable appliance load signatures. In Proceedings of the 2010 First IEEE International Conference on Smart Grid Communications, Gaithersburg, MD, USA, 11–13 October 2010; pp. 232–237.
8. Li, F.; Luo, B.; Liu, P. Secure Information aggregation for smart grids using homomorphic encryption. In Proceedings of the 2010 First IEEE International Conference on Smart Grid Communications, Gaithersburg, MD, USA, 11–13 October 2010; pp. 327–332.
9. Voigt, P.; Bussche, A.V.D. *The EU General Data Protection Regulation (GDPR)*; Springer Science and Business Media LLC: Berlin, Germany, 2017.
10. Floridi, L. Soft ethics and the governance of the digital. *Philos. Technol.* **2018**, *31*, 1–8. [CrossRef]
11. Introna, L.D.; Pouloudi, A. Privacy in the information age: Stakeholders, interest and values. *J. Bus. Ethic* **1999**, *22*, 27–38. [CrossRef] [PubMed]
12. Efthymiou, C.; Kalogridis, G. Smart Grid Privacy via Anonymization of Smart Metering Data. In Proceedings of the 2010 First IEEE International Conference on Smart Grid Communications, Gaithersburg, MD, USA, 11–13 October 2010; pp. 238–243.
13. Knirsch, F.; Eibl, G.; Engel, D. Multi-resolution privacy-enhancing technologies for smart metering. *EURASIP J. Inf. Secur.* **2017**, *2017*, 807. [CrossRef]
14. Hernández-Callejo, L.; Callejo, H.-A. Comprehensive review of operation and control, maintenance and lifespan management, grid planning and design, and metering in smart Grids. *Energies* **2019**, *12*, 1630. [CrossRef]
15. Kochański, M.; Korczak, K.; Skoczkowski, T. Technology innovation system analysis of electricity smart metering in the European Union. *Energies* **2020**, *13*, 916. [CrossRef]
16. Fahim, M.; Sillitti, A. Analyzing load profiles of energy consumption to infer household characteristics using smart meters. *Energies* **2019**, *12*, 773. [CrossRef]
17. Llano, A.; Angulo, I.; De La Vega, D.; Marron, L. Virtual PLC lab enabled physical layer improvement proposals for PRIME and G3-PLC standards. *Appl. Sci.* **2020**, *10*, 1777. [CrossRef]
18. Uribe-Pérez, N.; Angulo, I.; Hernández-Callejo, L.; Arzuaga, T.; De La Vega, D.; Arrinda, A. Study of unwanted emissions in the CENELEC-A band generated by distributed energy resources and their influence over narrow band power line communications. *Energies* **2016**, *9*, 1007. [CrossRef]
19. *The Smart Metering System Leaflet*; Department of Energy & Climate Change: London, UK, 2014.
20. *A Joint Contribution of DG ENER and DG INFSO towards the Digital Agenda, Action 73: Set of Common Functional Requirements of the SMART METER*; European Commission: Brussels, Belgium, 2011.
21. Asghar, M.R.; Dán, G.; Miorandi, D.; Chlamtac, I. Smart Meter Data Privacy: A Survey. *IEEE Commun. Surv. Tutor.* **2017**, *19*, 2820–2835. [CrossRef]
22. Lu, R.; Liang, X.; Li, X.; Lin, X.; Shen, X. EPPA: An efficient and privacy-preserving aggregation scheme for secure smart grid communications. *IEEE Trans. Parallel Distrib. Syst.* **2012**, *23*, 1621–1631. [CrossRef]
23. Pitì, A.; Verticale, G.; Rottondi, C.; Capone, A.; Schiavo, L.L. The role of smart meters in enabling real-time energy services for households: The Italian case. *Energies* **2017**, *10*, 199. [CrossRef]
24. Torriti, J. *Peak Energy Demand and Demand Side Response*; Informa UK Limited: London, UK, 2015.
25. Qadrdan, M.; Cheng, M.; Wu, J.; Jenkins, N. Benefits of Demand-Side response in combined gas and electricity networks. *Appl. Energy* **2017**, *192*, 360–369. [CrossRef]
26. Ruano, A.; Hernández, Á.; Ureña, J.; Ruano, M.; Domínguez, J.J.G. NILM techniques for intelligent home energy management and ambient assisted living: A review. *Energies* **2019**, *12*, 2203. [CrossRef]
27. Hogan, W.W. Fairness and dynamic pricing: Comments. *Electr. J.* **2010**, *23*, 28–35. [CrossRef]
28. Parks, R.C. *Advanced Metering Infrastructure Security Considerations*; Sandia Report; Sandia National Laboratories: Albuquerque, New Mexico, 2007.
29. McCullough, J. AMI Security Considerations. Elster. 2010. Available online: https://silo.tips/download/ami-security-considerations (accessed on 30 April 2020).
30. Montanez, C.A.C.; Hurst, W. A machine learning approach for detecting unemployment using the smart metering infrastructure. *IEEE Access* **2020**, *8*, 22525–22536. [CrossRef]
31. Rachels, J. Why privacy is important. In *Privacy*; Informa UK Limited: London, UK, 2017; pp. 11–21.
32. Chan, A.C.; Zhou, J. On smart grid cybersecurity standardization: Issues of designing with NISTIR 7628. *IEEE Commun. Mag.* **2013**, *51*, 58–65. [CrossRef]

33. Eibl, G.; Engel, D. Influence of Data Granularity on Smart Meter Privacy. *IEEE Trans. Smart Grid* **2014**, *6*, 930–939. [CrossRef]
34. Sultan, S. Privacy-preserving metering in smart grid for billing, operational metering, and incentive-based schemes: A survey. *Comput. Secur.* **2019**, *84*, 148–165. [CrossRef]
35. Farokhi, F.; Sandberg, H. Fisher information as a measure of privacy: Preserving privacy of households with smart meters using batteries. *IEEE Trans. Smart Grid* **2018**, *9*, 4726–4734. [CrossRef]
36. Li, S.; Khisti, A.; Mahajan, A. Privacy-optimal strategies for smart metering systems with a rechargeable battery. In Proceedings of the 2016 American Control Conference (ACC), Boston, MA, USA, 6–8 July 2016; pp. 2080–2085.
37. Varodayan, D.; Khisti, A. Smart meter privacy using a rechargeable battery: Minimizing the rate of information leakage. In Proceedings of the 2011 IEEE International Conference on Acoustics, Speech and Signal Processing (ICASSP), Prague, Czech, 22–27 May 2011; pp. 1932–1935.
38. Zhang, Z.; Qin, Z.; Zhu, L.; Weng, J.; Ren, K. Cost-friendly Differential Privacy for Smart Meters: Exploiting the Dual Roles of the Noise. *IEEE Trans. Smart Grid* **2016**, *8*, 619–626. [CrossRef]
39. Yang, L.; Chen, X.; Zhang, J.; Poor, H.V. Cost-Effective and Privacy-Preserving energy management for smart meters. *IEEE Trans. Smart Grid* **2014**, *6*, 486–495. [CrossRef]
40. Rajagopalan, S.R.; Sankar, L.; Mohajer, S.; Poor, H.V. Smart meter privacy: A utility-privacy framework. In Proceedings of the 2011 IEEE International Conference on Smart Grid Communications (SmartGridComm), Prague, Czech, 22–27 May 2011; pp. 190–195.
41. Zhu, L.; Zhang, Z.; Qin, Z.; Weng, J.; Ren, K. Privacy Protection Using a Rechargeable Battery for Energy Consumption in Smart Grids. *IEEE Netw.* **2017**, *31*, 59–63. [CrossRef]
42. Backes, M.; Meiser, S. Differentially private smart metering with battery recharging. In *Data Privacy Management and Autonomous Spontaneous Security*; Springer: Berlin/Heidelberg, Germany, 2013; pp. 194–212.
43. Tan, O.; Gunduz, D.; Poor, H.V. Increasing smart meter privacy through energy harvesting and storage devices. *IEEE J. Sel. Areas Commun.* **2013**, *31*, 1331–1341. [CrossRef]
44. Giaconi, G.; Gunduz, D.; Poor, H.V. Smart meter privacy with renewable energy and an energy storage device. *IEEE Trans. Inf. Forensics Secur.* **2018**, *13*, 129–142. [CrossRef]
45. Sun, Y.; Lampe, L.; Wong, V.W. Smart meter privacy: Exploiting the potential of household energy storage units. *IEEE Internet Things J.* **2017**, *5*, 69–78. [CrossRef]
46. Gunduz, D.; Vilardebò, J.G. Smart meter privacy in the presence of an alternative energy source. In Proceedings of the 2013 IEEE International Conference on Communications (ICC), Beijing, China, 9–13 June 2013; pp. 2027–2031.
47. Egarter, D.; Prokop, C.; Elmenreich, W. Load hiding of household's power demand. In Proceedings of the 2014 IEEE International Conference on Smart Grid Communications (SmartGridComm), Venice, Italy, 3–6 November 2014; pp. 854–859.
48. Chen, D.; Irwin, D.; Shenoy, P.; Albrecht, J.; Albrecht, J. Combined heat and privacy: Preventing occupancy detection from smart meters. In Proceedings of the 2014 IEEE International Conference on Pervasive Computing and Communications (PerCom), Budapest, Hungary, 24–28 March 2014; pp. 208–215.
49. Chen, D.; Kalra, S.; Irwin, D.; Shenoy, P.; Albrecht, J. Preventing Occupancy Detection from Smart Meters. *IEEE Trans. Smart Grid* **2015**, *6*, 2426–2434. [CrossRef]
50. Giaconi, G.; Gunduz, D.; Poor, H.V. Privacy-Aware smart metering: Progress and challenges. *IEEE Signal Process. Mag.* **2018**, *35*, 59–78. [CrossRef]
51. Tan, O.; Vilardebò, J.G.; Gunduz, D. Privacy-Cost Trade-offs in Demand-Side management with storage. *IEEE Trans. Inf. Forensics Secur.* **2017**, *12*, 1458–1469. [CrossRef]
52. McLaughlin, S.; McDaniel, P.; Aiello, W. Protecting consumer privacy from electric load monitoring. In Proceedings of the 18th ACM Conference on Computer and Communications Security, Chicago, IL, USA, 17–21 October 2011; pp. 87–98.
53. Giaconi, G.; Gunduz, D.; Poor, H.V. Optimal demand-side management for joint privacy-cost optimization with energy storage. In Proceedings of the 2017 IEEE International Conference on Smart Grid Communications (SmartGridComm), Dresden, Germany, 23–27 October 2017; pp. 265–270.
54. Sun, Y.; Lampe, L.; Wong, V.W.S. Combining electric vehicle and rechargeable battery for household load hiding. In Proceedings of the 2015 IEEE International Conference on Smart Grid Communications (SmartGridComm), Miami, FL, USA, 2–5 November 2015; pp. 611–616.

55. Gomez-Vilardebo, J.; Gunduz, D. Smart meter privacy for multiple users in the presence of an alternative energy source. *IEEE Trans. Inf. Forensics Secur.* **2014**, *10*, 132–141. [CrossRef]
56. He, X.; Zhang, X.; Kuo, C.-C.J. A Distortion-Based Approach to Privacy-Preserving metering in smart grids. *IEEE Access* **2013**, *1*, 67–78. [CrossRef]
57. Eibl, G.; Engel, D. Differential privacy for real smart metering data. *Comput. Sci. Res. Dev.* **2016**, *32*, 173–182. [CrossRef]
58. ÁCS, G.; Castelluccia, C. I have a dream!(differentially private smart metering). In *International Workshop on Information Hiding*; Springer: Berlin/Heidelberg, Germany, 2011; pp. 118–132.
59. Barbosa, P.; Brito, A.; Almeida, H.; Clauß, S. Lightweight privacy for smart metering data by adding noise. In Proceedings of the 29th Annual ACM Symposium on Applied Computing, Gyeongju, Korea, 24–28 March 2014; pp. 531–538. [CrossRef]
60. Bohli, J.-M.; Sorge, C.; Ugus, O. A privacy model for smart metering. In Proceedings of the 2010 IEEE International Conference on Communications Workshops, Gaithersburg, MD, USA, 11–13 October 2010; pp. 1–5.
61. Büscher, N.; Boukoros, S.; Bauregger, S.; Katzenbeisser, S. Two is not enough: Privacy assessment of aggregation schemes in smart metering. In Proceedings of the Privacy Enhancing Technologies, Minneapolis, MN, USA, 18–21 July 2017; Volume 2017, pp. 198–214.
62. Kursawe, K.; Danezis, G.; Kohlweiss, M. Privacy-friendly aggregation for the smart-grid. In Proceedings of the International Symposium on Privacy Enhancing Technologies Symposium, Waterloo, ON, Canada, 27–29 July 2011; pp. 175–191.
63. Thoma, C.; Cui, T.; Franchetti, F. Secure multiparty computation based privacy preserving smart metering system. In Proceedings of the 2012 North American Power Symposium (NAPS), Champaign, IL, USA, 9–11 September 2012; pp. 1–6. [CrossRef]
64. Dwork, C. Differential privacy: A survey of results. In Proceedings of the International Conference on Theory and Applications of Models of Computation, Xi'an, China, 25–29 April 2008; pp. 1–19.
65. Martínez, S.; Sebé, F.; Sorge, C. Measuring privacy in smart metering anonymized data. *arXiv Preprint* **2020**, arXiv:04863 2020.
66. Cardenas, A.; Amin, S.; Schwartz, G.A. Privacy-Aware Sampling for Residential Demand Response Programs. Available online: http://www.eecs.berkeley.edu/schwartz/HiCons2012ASG.pdf (accessed on 30 April 2020).
67. Foucault, M. *The Foucault Effect: Studies in Governmentality*; University of Chicago Press: Chicago, IL, USA, 1991.
68. Córdoba-Pachón, J.-R. *Managing Creativity: A Systems Thinking Journey*; Routledge: Abingdon, IK, USA, 2018.
69. Antonijevic, M.; Sučić, S.; Keserica, H. Augmented reality applications for substation management by utilizing standards-compliant scada communication. *Energies* **2018**, *11*, 599. [CrossRef]
70. Schneider, K.P.; Chen, Y.; Chassin, D.P.; Pratt, R.G.; Engel, D.W.; Thompson, S.E. *Modern Grid Initiative Distribution Taxonomy Final Report*; Pacific Northwest National Lab (PNNL): Richland, WA, USA, 2008.
71. Ledva, G.S.; Balzano, L.; Mathieu, J.L. Real-time energy disaggregation of a distribution feeder's demand using online learning. *IEEE Trans. Power Syst.* **2018**, *33*, 4730–4740. [CrossRef]
72. Jawurek, M.; Johns, M.; Kerschbaum, F. Plug-In privacy for smart metering billing. In *Intelligent Tutoring Systems*; Springer Science and Business Media LLC: Berlin, Germany, 2011; Volume 6794, pp. 192–210.
73. Zhang, X.-Y.; Kuenzel, S. Differential Privacy for Deep Learning-based Online Energy Disaggregation System. In Proceedings of the 2020 IEEE PES Innovative Smart Grid Technologies Europe (ISGT-Europe), The Hague, Netherlands, 25 October 2020.
74. Abadi, M.; Chu, A.; Goodfellow, I.; Mcmahan, H.B.; Mironov, I.; Talwar, K.; Zhang, L. Deep Learning with Differential Privacy. In Proceedings of the 2016 ACM SIGSAC Conference on Computer and Communications Security, Vienna, Austria, 24–28 October 2016; pp. 308–318.
75. Zhao, J.; Jung, T.; Wang, Y.; Li, X. Achieving differential privacy of data disclosure in the smart grid. In Proceedings of the IEEE INFOCOM 2014-IEEE Conference on Computer Communications, Toronto, ON, Canada, 27 April–2 May 2014; pp. 504–512.
76. Cover, T.M.; Thomas, J.A. *Elements of Information Theory*; John Wiley & Sons: Hoboken, NJ, USA, 2012.
77. Batra, N.; Kelly, J.; Parson, O.; Dutta, H.; Knottenbelt, W.; Rogers, A.; Singh, A.; Srivastava, M. NILMTK: An open source toolkit for non-intrusive load monitoring. In Proceedings of the 5th International Conference on Future Energy Systems, New York, NY, USA, 11 June 2014; pp. 265–276.

78. Kelly, J.; Knottenbelt, W. Neural nilm: Deep neural networks applied to energy disaggregation. In Proceedings of the 2nd ACM International Conference on Embedded Systems for Energy-Efficient Built Environments, Seoul, Korea, 4–5 November 2014; pp. 55–64.
79. Mironov, I. Rényi Differential Privacy. In Proceedings of the 2017 IEEE 30th Computer Security Foundations Symposium (CSF), Santa Barbara, CA, USA, 21–25 August 2017; pp. 263–275.
80. Parson, O.; Fisher, G.; Hersey, A.; Batra, N.; Kelly, J.; Singh, A.; Knottenbelt, W.; Rogers, A. Dataport and NILMTK: A building data set designed for non-intrusive load monitoring. In Proceedings of the 2015 IEEE Global Conference on Signal and Information Processing (GlobalSIP), Orlando, FL, USA, 14–16 December 2015; pp. 210–214.
81. Bmz Gmbh li-io Ess 3.0 Lithium-Ionen-Energy Storage System 3.0 for Sma. Available online: https://www.off-grid-europe.com/bmz-gmbh-li-io-ess-3-0-lithium-ionen-energy-storage-system-3-0-for-sma?gclid=CjwKCAjwv41BRAhEiwAtMDLsuqBrUnfvzcJRUb7IOyCkaH1XWJZAQY7XuNHR5qNVUYk5S9grA7aHxoC1qYQAvD_BwE (accessed on 30 April 2020).
82. Chillotti, I.; Gama, N.; Georgieva, M.; Izabachène, M. TFHE: Fast fully homomorphic encryption over the torus. *J. Cryptol.* **2019**, *33*, 34–91. [CrossRef]
83. Sandys, L. Energy Data Taskforce Report: A Strategy for a Modern Digitalised Energy System. 2019. Available online: https://es.catapult.org.uk/wp-content/uploads/2019/06/Catapult-Energy-Data-Taskforce-Report-A4-v4AW-Digital.pdf (accessed on 30 April 2020).
84. Blum, M.; Calif, U.O.; Feldman, P.; Micali, S. Mit Non-interactive zero-knowledge and its applications. In *Providing Sound Foundations for Cryptography: On the Work of Shafi Goldwasser and Silvio Micali*; Association for Computing Machinery (ACM): New York, NY, USA, 2019; pp. 329–349.
85. Pedersen, T.; Petersen, B. Explaining gradually increasing resource commitment to a foreign market. *Int. Bus. Rev.* **1998**, *7*, 483–501. [CrossRef]
86. Schmidhuber, J. Deep learning in neural networks: An overview. *Neural. Netw.* **2015**, *61*, 85–117. [CrossRef]

© 2020 by the authors. Licensee MDPI, Basel, Switzerland. This article is an open access article distributed under the terms and conditions of the Creative Commons Attribution (CC BY) license (http://creativecommons.org/licenses/by/4.0/).

Review

Watt's up at Home? Smart Meter Data Analytics from a Consumer-Centric Perspective

Benjamin Völker [1], Andreas Reinhardt [2], Anthony Faustine [3] and Lucas Pereira [4,*]

[1] Chair of Computer Architecture, University of Freiburg, 79110 Freiburg, Germany; voelkerb@informatik.uni-freiburg.de
[2] Department of Informatics, TU Clausthal, 38678 Clausthal-Zellerfeld, Germany; andreas.reinhardt@tu-clausthal.de
[3] Center for Artificial Intelligence (CeADAR), University College of Dublin, D04 V1W8 Dublin 4, Ireland; anthony.faustine@ucd.ie
[4] ITI, LARSyS, Técnico Lisboa, 1049-001 Lisboa, Portugal
* Correspondence: lucas.pereira@tecnico.ulisboa.pt

Abstract: The key advantage of smart meters over traditional metering devices is their ability to transfer consumption information to remote data processing systems. Besides enabling the automated collection of a customer's electricity consumption for billing purposes, the data collected by these devices makes the realization of many novel use cases possible. However, the large majority of such services are tailored to improve the power grid's operation as a whole. For example, forecasts of household energy consumption or photovoltaic production allow for improved power plant generation scheduling. Similarly, the detection of anomalous consumption patterns can indicate electricity theft and serve as a trigger for corresponding investigations. Even though customers can directly influence their electrical energy consumption, the range of use cases to the users' benefit remains much smaller than those that benefit the grid in general. In this work, we thus review the range of services tailored to the needs of end-customers. By briefly discussing their technological foundations and their potential impact on future developments, we highlight the great potentials of utilizing smart meter data from a user-centric perspective. Several open research challenges in this domain, arising from the shortcomings of state-of-the-art data communication and processing methods, are furthermore given. We expect their investigation to lead to significant advancements in data processing services and ultimately raise the customer experience of operating smart meters.

Keywords: smart metering; smart power grids; power consumption data; energy data processing; user-centric applications of energy data

1. Introduction

After the invention of electricity meters over a century ago, billions of such devices have been installed worldwide [1]. They are found in private households, commercial buildings, industrial sites, and all other domains that require electrical energy consumption to be tracked. Initially realized as rotating-disc meters (also known as Ferraris meters) with mechanical displays, transferring the actual energy consumption data used to be a labor-intensive manual process. However, with the rise of digital metering devices, so-called smart meters, the collection of electrical power consumption at much more fine-grained spatial and temporal resolutions has become possible. The digital communication interface to report the collected data represents a major advancement over rotating-disc meters in particular. As a result, meter data have started to become available at previously unimaginable temporal resolutions on the order of seconds to minutes, on building- or even apartment-level. Moreover, the resultant digital data can be easily transmitted to online data centers for storage and further data processing. This opens up unprecedented opportunities to analyze, compare, and combine such data and provide novel energy-based services to customers and grid operators alike [2].

Numerous research activities have accompanied the global roll-out of smart meters [seeking to exploit the information content of the collected data to its fullest extent. I leveraging and combining signal processing techniques from a wide range of domai (digital signal processing, stochastic analysis, artificial intelligence, and many other various indicators and identifying features can be detected and extracted from smart me data. They allow for the realization of plentiful use cases that benefit consumers, util companies, or other stakeholders. It is noteworthy, however, that existing works ha focused mainly on smart meter data analytics methods to the benefit of the power gr as a whole [4,5]. In contrast to this application domain, equally great potential lies in t provision of user-centric services based on electrical consumption data. We dedicated survey such use cases in this work, not least because we anticipate many more servic that provide direct benefits to electricity consumers to be developed in the near future.

This review of the state of the art in smart meter data analytics applications is target to be a concise introduction that seeks to provide an overview of the range of user-cent applications for smart meter data as well as highlighting promising future research avenu in this domain. It is organized as follows. Section 2 sets the definition of smart me data used in this paper and highlights frequently used data (pre)processing steps. V survey consumer-centric applications based on smart meter data in Section 3 including t provision of electricity related user feedback, the recognition of patterns or anomalies, t recognition of flexible loads, vital improvements in single home demand forecasting, ar finally the comparison and correlation of consumers based on their load profiles. Section discusses the most widely encountered obstacles in developing customer-centric sma meter data services such as missing standardization, mediocre-performing algorithms, ar privacy concerns. To surmount these obstacles, we formulate the corresponding resear challenges and finally conclude this review paper in Section 5.

2. Smart Meter Data Collection and Preprocessing

Metering the consumption of primary energy is commonplace and an everyd experience for most people. When refueling the storage tanks of oil- or gas-powered centr heating systems or vehicles powered by combustion engines, measuring fuel quantiti is ubiquitous to be billed only for the amount added. Consumption metering has al manifested itself for commodities beyond fuels, such as running water, district heating, pressurized air. However, none of these fields has seen the same enormous increase in da analytics research as the field of electrical power consumption monitoring. In fact, besides single work on interpreting natural gas consumption [6], only the evaluation of water flov has seen scientific consideration in related works [7–11], primarily seeking to infer us activities based on the corresponding water demands. The reason for the surge of electric consumption analytics is simple: With the rise of smart meters, electrical consumptic data have become available in unprecedented temporal and spatial resolutions. This n only makes longitudinal analyses much easier to conduct, but the high penetration of t building stock with smart meters has also created the foundation to run data analyti at scale. The large variety of electrical consumers [12,13], coupled with their freque use in everyday activities and the ensuing potentials to save energy, makes them viab candidates for analysis. Before surveying possible use cases and their practical implicatio in the following section, however, let us first revisit the definition of smart meter data ar delineate them from other ways of consumption measurements in electrical power grid

As shown in Figure 1, smart meters are located at the entry-point of a building's ele trical grid connection. All power flows between the (smart) power grid and the applianc in the (smart) home can be captured at this point, thus smart meters can effectively lea to benefits on both sides. Their primary use case lies in billing consumers for the exa amount of electrical energy taken from the power grid or balancing between consumptic and generation of *prosumers*, i.e., grid-connected entities with local generation faciliti respectively. Smart meters are thus distinct from both customer-side monitoring systen such as circuit-level or even plug-level power monitors, which rarely exhibit the san accuracy as smart meters but rather serve as data sources for smart home installations or Building Management Systems (BMS). Likewise, monitoring devices exist in transmission and distribution grids, yet their data can generally not be unambiguously attributed to a single customer. Coupled with their generally smaller number when compared to the scale at which smart meters have been rolled out, we also exclude such grid-level monitors (such as phasor measurement units) from our analysis in this paper. We specifically wish to highlight, however, that the presented data processing mechanisms and correspondingly enabled use cases can likely also find application on such devices or smart meters for quantities beyond electrical energy.

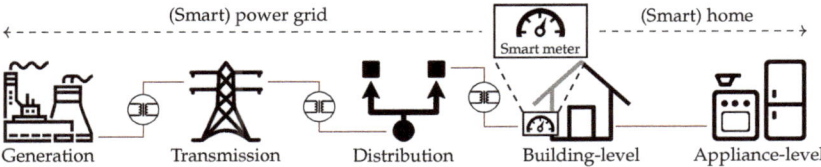

Figure 1. Location of smart meters within the electrical power grid (icons by Icon Fonts; CC BY 3.0).

Smart electricity meters represent the state-of-the-art solution to collect, process, and forward load information to all stakeholders involved. Through direct connections to the Internet, or indirect connection using smart meter gateways [14], access to metered data is ubiquitously possible. Incentive schemes and policymakers in many countries furthermore contribute to the increasing market penetration of smart meters. This enables numerous user-centric use cases beyond billing, which we survey and categorize in Section 3. Before documenting how the full potential of smart meter data can be unleashed, we would like to note that the enablement of these use cases frequently relies on data preprocessing steps to isolate characteristic features from the stream of raw measurements provided by smart meters.

2.1. Data (Pre-)Processing

Data collected by smart meters are not always directly usable for the provision of user-centric services. At least some preprocessing steps are generally needed to create a uniform and error-free foundation for data analytics. On the one hand, many services rely on processed input data, such as a building's energy consumption during a specific period, rather than raw readings of electrical voltage levels and current flows. On the other hand, errors introduced during the sampling process, the analog-to-digital conversion step, and the transmission over communication channels raise the possibility of errors and signal falsifications that need to be eliminated. Proper preprocessing thus serves to transform the collected data into a unified and interpretable format, based on which user-centric services can be provided reliably. To establish the foundation for the data preprocessing steps required to realize the use cases surveyed in Section 3, we list typical data preprocessing steps preceding the actual data analysis as follows.

First, obviously erroneous values are generally eliminated. These primarily occur due to faulty storage devices, unreliable communication channels, or buffer overflows on the transmitting or receiving devices. Readings that do not represent valid number representations and infeasible values (e.g., current flows exceeding the nominal circuit breaker limits by a large factor) are thus removed. Unless a long sequence of wrong data is being reported, the imputation of values and the interpolation of gaps in the sampled data (e.g., by using the *impyute* library [15]) is an effective means to prepare the data for further processing.

The fundamental mode of operation of smart meters is to measure raw voltage (V) and current (I) waveforms at sampling rates that allow for the computation of Root Mean Square

(RMS) values, $V_{RMS} = \sqrt{\frac{1}{T} \int_{t_0}^{t_0+T} V(t)^2 dt}$ and $I_{RMS} = \sqrt{\frac{1}{T} \int_{t_0}^{t_0+T} I(t)^2 dt}$, with T denoting the duration of one or more mains periods and $V(t)$ and $I(t)$ being the voltage and current waveform signals, respectively. However, raw data are rarely communicated beyond the local system boundary due to their sheer size and their highly redundant information content [16]. Instead, smart meters typically process the raw samples locally and return one or multiple of the following parameters: RMS voltage (V_{RMS}), RMS current (I_{RMS}), phase angle between voltage and current ($cos \Phi$), active power (P), reactive power (Q), apparent power (S), and/or the consumed electrical energy (E). In multi-phase electric installations, parameters are either returned individually for all phases or merely available in an aggregated fashion. If a particular parameter is required but not directly provided by the smart meter, it may still be possible to calculate it from the provided parameters; this is, again, a part of the preprocessing step.

To demonstrate the variability of data reported by practical smart meter deployments, Table 1 provides a brief overview of the attributes, sampling rate, and communication interface of smart meters and custom-built meters, which have been used to record publicly released electrical consumption datasets. The diversity of the provided data highlights why general data preprocessing is required to create a uniform data representation to realize consumer-centric use cases independently of the specific underlying smart meter hardware.

Table 1. Metering devices used and parameters provided in a selection of electricity datasets.

Dataset	Smart Meter Model	Captured Parameters	Sampling Rate	Interface
Dataport [17]	EG3000 + EG201X [a]	$V_{RMS}, I_{RMS}, P, Q, S, cos \Phi$	1 Hz	Modbus
iAWE [18]	EM6400 [b]	$V_{RMS}, I_{RMS}, P, cos \Phi$	1 Hz	Modbus
AMPds [19]	Powerscout18 [c]	$V_{RMS}, I_{RMS}, P, Q, S, E, cos \Phi$	1/60 Hz	Modbus
RAE [20]	Powerscout24 [c]	$V_{RMS}, I_{RMS}, P, Q, S, E, cos \Phi$	1 Hz	Modbus
ECO [21]	E750 [d]	V_{RMS}, I_{RMS}, P	1 Hz	SyM^2
REDD [22]	custom design	V, I	16.5 kHz	USB
SustDataED [23]	custom design	V, I	12.8 kHz	USB
BLOND [24]	custom design	V, I	250 kHz	TCP

[a] eGauge. [b] Schneider Electric. [c] DENT Instruments. [d] Landis + Gyr.

Table 1 highlights one more aspect of heterogeneity in smart meter data, which is also confirmed in [25]: the temporal resolution at which the parameters are being reported. Reducing the rate at which values are being made available, i.e., *downsampling* smart meter data, is usually trivial and computationally lightweight, as long as the original data have undergone low-pass filtering to avoid aliasing artifacts. Commonly used methods to downsample data include subsampling, averaging, and interpolation [26,27]. Conversely, increasing the temporal resolution of data is not as trivial, but it may be required for smart meter data reported at very low sampling rates. Interpolation techniques such as *super-resolution* [28] have been shown to achieve good performance during preliminary tests on the Dataport [17] dataset. As the sampling rate is frequently limited by the smart meter's communication channel and processing power, finding the optimal sampling rate for various electricity load analysis algorithms has been investigated in numerous works (e.g., [29–32]). Similarly, lossy compression mechanisms [33,34]), and pattern recognition methods [16] have been investigated as candidates to maintain high temporal resolution while reducing the extent of exchanged data.

2.2. Extracting Higher-Level Information

While inspecting conditioned smart meter data may be of interest for tech-savvy users or grid operators, it has been shown to provide little benefit to the average consumer, according to Serrenho et al. [35]. Consumer relevant information such as provided in Section 3 must first be inferred from the consumption data by extracting higher level information. This includes signal features, transient events, or individual appliance consumption data. Calculating these features from the consumption data is a widely used preprocessing step that goes beyond the data cleansing and adaptation steps described in Section 2.1. Instead, it is used to eliminate redundant information and only retain the most informative features about the consumption data. Besides this, it also generally leads to implicit data compression, e.g., to utilize the available communication channels optimally or to reduce the input size for machine learning algorithms. Domain experts have introduced and compared numerous features in related works [36–38]. For example, Kahl et al. [36] evaluated 36 features such as the *voltage and current trajectory* or the *harmonic energy distribution* for their suitability to serve as distinctive higher-level features for the enablement of user-centric services. Because of their virtually ubiquitous usage, we survey a selection of methods to extract higher-level features from smart meter data as follows.

Many user-centric use cases for smart meter data rely on the analysis of user-induced events, e.g., when electrical appliances are being switched on or off, or their mode of operation is changed. In Table 2, we summarize the number of such power events found in a selection of publicly available electricity datasets. The average of the tabulated values is approximately 275 events per day, i.e., approximately one event every 6 min. As such, the Switch Continuity Principle (SCP), first introduced by Hart [39] and confirmed to hold by Makonin [40], states that the total number of events is small compared to the number of samples in the overall signal. In other words, events can be assumed to be anomalies in the signal, which makes it possible to utilize a range of known methods for their detection [41].

Table 2. Summary of the number of events detected in publicly released electricity datasets.

Dataset	# Events	Timespan	Source of Event Count
UK-DALE [42]	5440	7 days	Pereira and Nunes [43]
REDD [22]	1944	8 days	Völker et al. [44]
REDD [22]	1258	7 days	Pereira and Nunes [45]
BLUED [46]	2335	8 days	Anderson et al. [46]
FIRED [47]	4379	14 days	Völker et al. [47]
BLOND-50 [24]	3310	30 days	Kahl et al. [48]
AMPds [19]	651	7 days	Pereira and Nunes [45]
SustDataED [23]	2196	11 days	Pereira et al. [49]

In practice, event detection algorithms span the range from computationally lightweight solutions (e.g., using thresholds between successive power samples [39,50,51]) to the application of probabilistic models and voting methods [52–54]. More recently, the application of even more complex filters to electrical signals was proposed in order to suppress minor fluctuations while emphasizing actual events. Trung et al. [55] used a CUmulative SUM (CUSUM) filter to clean the power signal, while Wild et al. [56] applied a Kernel Fisher Discriminant Analysis (KFDA) on harmonics of the current signal. De Baets et al. [57] used spectral components of the current signal which have been smoothed using an inverse *Hann* window in the *Cepstral* domain, and the method of Cox et al. [58] solely uses the voltage signal and extracts the spectral envelope of the first and third harmonics.

Data collection from smart meters implies that data are only available on the scale of buildings or apartments (cf. Figure 1). Consequently, the energy consumption of individual electrical consumers is not directly identifiable within the reported (aggregate) data. The concept of Non-Intrusive Load Monitoring (NILM) thus refers to the process of disaggregating a composite electrical load into the contributions of all individual consumers. NILM methods frequently utilize machine learning techniques or neural networks to this end [59–69]. This makes their execution on current-generation smart meters largely impossible. However, it is generally possible to send collected data to external entities that offer the required storage and processing capabilities to perform NILM and thus provide appliance-level consumption values. As will become apparent in Section 3, several use cases can benefit from the availability of appliance-level data. The use of NILM, whi[ch] comes at the advantage of requiring no additional metering devices to be deployed, is th[us] a widely usable data preparation method to enable additional user-centric use cases wh[en] smart meter data is available.

3. Consumer-Centric Use Cases of Smart Meter Data

While it is crucial for the operators of electrical power grids to understand the lo[ad] and generation characteristics [5] in order to ensure grid stability and avoid power outag[es], electrical parameters can also be used to provide services to the benefit of the custome[r]. Figure 2 depicts the primary services that can be realized when smart meter data and t[he] corresponding higher-level information are available. We provide more details about t[he] enabled use cases as follows.

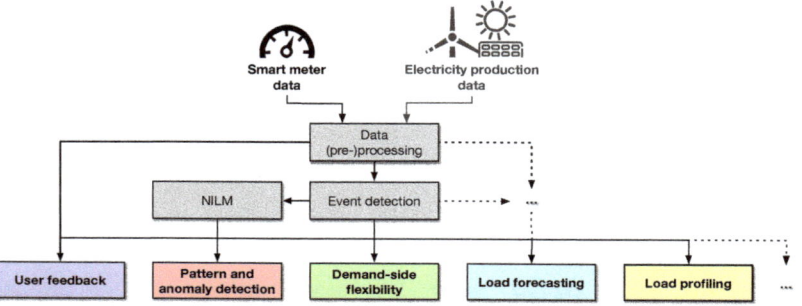

Figure 2. Overview of consumer-centric services enabled by smart meter data and their proper da[ta] (pre-)processing. Dashed lines indicate the possible existence of other potential (pre-)processing ste[ps] or use cases beyond those covered in this work.

3.1. Providing User Feedback

One of the vital value propositions of smart meter deployments is providing ne[w] real-time and historical information on electricity consumption to the customers. Havi[ng] access to such information is expected to result in the adoption of more sustainable co[n]sumption behavior, and thus to ultimately lead to energy savings [70–72]. Feedback [on] electricity consumption has been provided in numerous ways, including In-Home Displa[ys] (IHDs) [73,74], ambient displays [75,76], web and mobile applications [77–79], and pub[lic] displays [80,81]. While the majority of the works focused on providing information only [to] the home residents, other studies also looked at the potential of social pressure by enabli[ng] direct comparisons between individual consumers or consumer groups [82,83].

A meta-review of 118 studies that involved providing feedback on electricity co[n]sumption is presented in [35]. In general, the surveyed studies report that feedback ca[n] reduce a household's energy consumption from 5 % to 10 %, particularly in cases where t[he] deployed systems are able to provide consumption information of individual applianc[es]. The potential of feedback to energy savings was also confirmed in [84], where 12 studies [on] the efficacy of disaggregated feedback were examined. Again, an average energy reducti[on] of 4.5 % was reported across the surveyed studies. Even though there are no reports [on] long-term results on how to sustain the accomplished energy savings, many works ha[ve] identified that, without proper engagement strategies, once habituation sets in (after as lit[tle] as four weeks), there is a considerable loss of interest from the end-users in the feedba[ck] devices (e.g., [85–88]). However, it is evident from the literature that, through visualizi[ng] smart meter data in a timely and intuitive way, consumers become increasingly literate [in] understanding their domestic energy consumption, and in particular on how unintention[al] behavior can lead to unnecessary consumption [89,90].

With increasing distributed Renewable Energy Sources (RES), such as rooftop Pho[to]voltaic (PV) installations, it also becomes increasingly important to aid users in alignin[g] their consumption habits to their local generation [91,92]. As a result of this trend, energy feedback has received renewed interest to enable prosumers, i.e., consumers with local production facilities, to interact with the power grid optimally. Even at larger scales (e.g., smart microgrids [93]), the emergence of Peer-to-Peer (P2P) energy markets requires prosumers to have an understanding of the saving potentials and the consequences of their actions, both of which can be conveyed through feedback systems [94–96]. One such use case is practically studied in [97], confirming that user feedback was consistently utilized throughout the entire duration of the study (4.5 months) in order to make or defer consumption decisions.

3.2. Recognizing Patterns and Anomalies

Finding patterns that do not conform to the expected behavior indicated through abnormal electrical energy consumption is another consumer-centric use case for smart meter data. Even though detecting anomalies in smart meter data is challenging, signal processing and machine learning techniques can efficiently be utilized for this purpose. For example, detecting anomalies in smart meter data can be used to enable Ambient Assisted Living (AAL), where consumption patterns are indicative of the Activities of Daily Livings (ADLs) executed by the residents [98–101]. Detecting unusually short or long ADLs, or unexpected ADLs sequences, in general, are often suitable indicators of unusual user behavior. Knowledge of such situations can help to alert relatives early and thus contribute to safety and well-being [102]. Several different algorithmic approaches have been used to accomplish the recognition of patterns and anomalies. Clement et al. [98] presented a semi-Markov model that describes the daily use of appliances to detect human activity/behavior from smart meter data. In [99], smart meter data are analyzed to identify the behavioral patterns of the occupants, and Bousbiat et al. [100] proposed a framework for detecting abnormal ADLs from smart meter data.

Further use cases based on the application of machine learning for anomaly detection in smart meter data have emerged and manifested themselves in areas such as energy theft detection [103,104], detecting inaccurate smart meters [105], and detecting abnormal consumption behavior in general [106]. In [104], two anomaly detection schemes for detecting energy theft attacks and locating metering defects in smart meter data are presented. The work by Sial et al. [106] investigates heuristic approaches for identifying abnormal energy consumption from smart meter data, based on a combination of four distinct power-, energy-, and time-related features used in conjunction to detect anomalies. An even more sophisticated approach was presented by Liu et al. [105], who applied a deep neural network in detecting inaccurate meters to prevent the unnecessary replacement of smart meters, thus increasing their service life span. Lastly, the detection and quantification of anomalies in smart meter energy data play a crucial role in assessing the energy quality, which is essential for detecting faulty appliances, malfunctioning appliances, and non-technical losses [107–110].

3.3. Enabling Demand-Side Flexibility

Demand-side flexibility (DSF) refers to the portion of electricity demand that can be reduced, increased, or shifted within a specific time window. DSF plays a crucial role in the smart grid by facilitating the integration of RES and reducing peak load demand [111]. Traditionally provided by industrial consumers (e.g., refrigerated warehouses and steel mills [112]), flexibility can also be provided to operators by domestic and commercial consumers through controllable appliances and Electric Vehicles (EVs), e.g., by triggering them to change their consumption profiles [111]. While each consumer is only able to supply a limited amount of flexibility, once controllable consumers (and RES) of multiple dwellings are aggregated, their flexibility can add a significant volume of DSF to the grid. Ultimately, this leads to direct and indirect benefits to a larger group of consumers. On the one hand, it enables an additional revenue source by offering controllable loads to help make demand and supply meet. On the other hand, balanced power grids have a more favorable eco-footprint and an overall lower cost of generation, resulting in cheaper energy tariffs. Nevertheless, this flexibility is highly dependent on consumer behaviors, which correspondingly affects their willingness to provide flexible loads [113]. In this context, smart meter data are crucial to understand the potential of device-level flexibility on the consumer's premises [114–116].

In [114], the authors presented one of the first works that analyzed appliance-level consumption data in order to determine the device's flexibility and its relation to device operations and usage patterns. The work shows that a significant percentage (50 % on average) of the total energy demand for a house can be considered to provide flexibility. The results of a pilot study in Belgian households are reported in [115]. Five types of appliances available within residential premises were considered (washing machines, tumble dryers, dishwashers, domestic hot water buffers, and EVs) and assessed concerning their availability for DSF. The authors concluded that, except for EVs, the DSF potential is highly asymmetrical among appliances, possibly associated with user routines. The authors also estimated that EVs and water heaters have a flexibility potential that is much greater than that of wet appliances. In [116], the authors proposed and evaluated a data-driven approach to quantify the potential of flexible loads for participation in DSF programs. Their approach considered EVs, wet appliances (dryer, washing machine, and dishwasher), and Air Conditioning Unit (AC) loads and was evaluated on data from over 300 households from the Pecan Street project [117]. Analogous to previous works' results, the study confirms that variations in providing flexibility are considerable among households. Besides this, the results show that EVs and ACs provide higher levels of flexibility compared to wet appliances. As can be observed, in the context of DSF, EVs are of particular interest to the end-users since beyond sustainable transportation, they provide additional benefits like charging flexibility and a non-stationary energy storage solution [118,119].

While these and other works (e.g., [119–121]) assume that individual appliance consumption profiles are readily available, other researchers tried to assess the flexibility of domestic loads relying on NILM (cf. Section 2.2) to extract their individual consumption [122–124]. The main motivations for this approach are twofold: (1) avoid the costs of instrumenting the household with sensors in the individual appliances; and (2) protect the consumer privacy by not directly revealing data about individual appliance consumption (see Section 4.3). Ultimately, the obtained results show that it is possible to estimate and predict device-level flexibility from NILM outputs, even though a high disaggregation performance is necessary to reduce the uncertainty of the DSF estimation.

3.4. Forecasting Power Demand and Generation

The level of detail made available by smart meters opens several opportunities for load forecasting at the individual building level. Forecasting the electricity consumption using smart meter data plays a significant role in energy management for end-customers by enabling the possibility of linking current usage behaviors to future energy costs [125]. Similarly, anomaly detection (as discussed in Section 3.2) is often closely related to the comparison of actual and predicted consumption (or generation) behavior; as such, efficient and accurate forecasting techniques are required. Forecasting individual household demands is particularly challenging, however, due to many contributing factors. These include, but are not limited to, user behavior, appliance ownership, the considered time period(s), and/or external factors such as the prevailing weather conditions.

Against this background, researchers have proposed many forecasting approaches. For example, in [126], four of the most widely used machine learning methods, namely Multi-Layer Perceptron (MLP), Support Vector Machine (SVM), Classification and Regression Tree (CART), and Long Short-Term Memory (LSTM), are used to provide forecasts of both the daily consumption peak and the hourly energy consumption of domestic buildings using historical consumption data. It was found that MLPs and especially LSTM-based approaches can significantly improve the short term (24 h) demand forecasting as these models can capture the underlying non-linear relationships best. Several authors have tried to incorporate information from external factors into the forecasting algorithms. For instance, Amin et al. [127] proposed three different models Piecewise Linear Regression (PLR), Auto-Regressive Integrated Moving Average (ARIMA), and LSTM to forecast the electricity demand of a building leveraging smart meter data and weather information. A similar approach was followed by Gajowniczek and Ząbkowski [125]. However, instead of considering the effect of weather details, the authors focused on enhancing the forecasting algorithms by considering the impact of the residents' behavior patterns. The general consensus is that the combination of historical usage data and external features such as weather and household behavior can provide significant improvements to the forecasting results. Furthermore, these authors also confirm the suitability of LSTM models for short-term (24–48 h) forecasting. The work by Dinesh et al. [128] demonstrates a novel method to forecast the power consumption of a single house based on NILM and affinity aggregation spectral clustering. The presented work incorporates human behavior and environmental influence in terms of calendar and seasonal contexts to improve individual appliances' forecasting performance. The house-level forecast is thus obtained by the aggregation of the individual appliance-level forecasts.

Prosumers in general, but mainly when they own micro-production units (e.g., PV or wind generators) and Energy Storage Systems (ESS), can use forecasting to optimize and manage these resources. On the one hand, consumption forecasting techniques can help users to anticipate their future energy needs, so they can plan their local generation and optimize the operation of their ESS accordingly. On the other hand, users can also support the operation of the electricity grid by taking control actions to balance the electricity supply and demand while maximizing self-consumption and profiting from energy arbitrage (i.e., trading electricity by purchasing energy at times the price is low and selling it when it is expensive) [129,130]. For example, Hashmi et al. [129] proposed an algorithm to control the ESS in the presence of dynamic pricing, whereas Hashmi et al. [130] optimized the ESS to maximize the PV self-consumption in a scenario where there is no reward for feeding energy into the power grid. In either case, forecasting the future demand is necessary to decide when to charge or discharge the ESS. Particularly, if feeding surplus power into the power grid is not rewarded [130], an understanding of the residual load (i.e., the difference between consumption and production) is necessary, generally based on forecasts of the local production and demand, in order to avoid unintended grid injection or PV curtailment. Intuitively, these optimizations are sensitive to forecasting errors. For example, Kiedanski et al. [131] showed that when the optimizations are performed at higher sampling rates (every 15 min in this work), the negative implications of forecasting errors are limited. In contrast, the authors stated that lower sampling rates (e.g., a 12 h forecasting horizon) require almost perfect forecasts to unleash their full potential to optimize ESS operations.

With the increasing number of EVs sales and their high power consumption during charging, it is also necessary to forecast their charging needs, as this will allow for better scheduling and capacity planning [132,133]. Ai et al. [133] attempted to forecast household day-ahead charging needs using machine learning ensembles. Such forecasts gain particular importance if the EV owners are also prosumers, since in these cases their EVs also function as an ESS. The ability to increase self-consumption and reduce peak demand using EVs was studied by Fachrizal and Munkhammar [134], who showed that, in a single (Swedish) household, the self-consumption could be increased up to 8.7%. However, this result was obtained in the presence of perfect load demand and PV production forecasts, which again raises the question of sensitivity to forecasting errors. In sum, as more research works indicate that in general EV owners favor domestic over public charging infrastructures (e.g., [135–137]), it becomes evident that accurate load demand and production forecasts will gain increasing importance in the near future.

3.5. Load Profiling

Standard load profiles [138], i.e., averaged models of customer energy consumption over time, have traditionally found their application in power grid capacity planning. However, standard load profiles are only accurate when considering many connected customers in conjunction and generally do not adequately reflect individual consumer consumption characteristics. Smart meters can mitigate this situation and allow for capturing load profiles available in an unprecedented resolution. The enabled understanding of energy consumption profiles empowers users not only to better recognize how much energy they consume but also to compare their consumption profiles to the profiles of other dwellings [139]. This gives households greater control of their energy consumption and enables the adoption of more energy-efficient, and responsible behaviors [139–142]. Instead of considering the load profiles of buildings individually, it is often sufficient to know the *category* that better describes the dwelling. In other words, by categorizing the electrical power consumption, it is possible to approximate the load profile of a household sufficiently. It is therefore not unexpected that most of the existing load profiling techniques rely on clustering algorithms, such as *k*-means [143–145], fuzzy *k*-means [14], hierarchical clustering [143,146], Self-Organizing Maps (SOM) [143], neural network based clustering [147–149], Gaussian Mixture Models (GMM) [150,151], Density-Based Spatial Clustering (DBSCAN) [152], and agglomerative clustering [153]. Due to the high stochasticity and irregularity of household-level consumption, clustering techniques that analyze the variability and uncertainty of smart meter data have also been considered in the literature [150,151]. For example, the work by Lee et al. [143] proposes a two-stage (feature extraction and load pattern identification) *k*-means clustering for customers segmentation in residential demand response programs.

Load profiling results have been documented to find use in supporting and enhancing continuous energy audits in buildings that currently require multiple measurements [61,154]. Furthermore, the insights generated from load profiling can be used to enhance many other use cases of smart meter data. Eco-feedback techniques often utilize load profiles, e.g., to compare the consumption of individual days or different homes (see Section 3.). For instance, in [141], an algorithm for computing the carbon footprint derived from load curves is presented. Likewise, load profiles can be used along with load demand forecast (see Section 3.4) to generate optimal schedules of home appliance usage. This is presented in [155], in which a NILM-based energy management system was developed to schedule controllable loads taking into account customers' preferences and overall satisfaction.

In summary, load profiling at the end-user level enables and potentiates consumer centric services. Furthermore, as load profiling can play an essential role in assisting the smart grid, it will become even more relevant to the individual consumer when Distributed Energy Resources (DERs) and local energy communities become ubiquitous and require the active participation of citizens [156].

4. Open Research Challenges

The range of customer-centric use cases for smart meter data contributes to numerous areas of daily living. Besides allowing for monetary savings, grid-friendly appliance scheduling, and the detection of atypical and anomalous appliance operations, it has been shown to serve as the enabling technology for AAL as well as the integration of RES, ES, and EVs. Many of the underlying research challenges have been solved to a satisfactory degree to date, and corresponding commercial solutions are already on the market. During our survey of user-centric applications in Section 3, however, we identified obstacles to the enablement of the services, which potentially impact their widespread acceptance. We then summarize the most important observed challenges as follows.

4.1. Standardized Hardware and Data Formats

As stated in Section 2, there are no universally acknowledged definitions of: (1) the parameters to be reported by smart meters; (2) the temporal resolutions at which they are being made available; and (3) the interface using which service providers can access these data. As such, delivering the use cases to customers generally requires non-negligible adaptation efforts. A widespread approach to achieve compatibility nowadays is when the same company that rolls out and operates the smart meters also acts as the service provider. This "vendor lock-in", however, severely hampers the scale of services that can be provided, as well as their interoperability with other external services. Thus, creating an open ecosystem in which different stakeholders can synergistically combine their (often complementary) components to create an environment that leverages the full potential of smart meter data is currently not possible from a technological point of view. As one first step towards overcoming this obstacle, the International Electrotechnical Commission (IEC) has established a dedicated technical committee (*TC 85*) for the standardization of equipment, systems, and methods for the analysis of steady-state and transient electrical quantities. One of the committee's publications, *DIN IEC/TS 63297*, is a project report on "Sensing Devices for Non-Intrusive Load Monitoring" [157], seeking to unify the access to the required data from smart metering devices. Despite the ongoing standardization efforts, however, solutions that cater to the needs of metering and consumer-centric service providers alike remain to be found and widely adopted.

Besides the limited access to the electrical parameters measured by a smart meter, a second major impediment to the roll-out of services is the unavailability of a local execution environment for data processing code. This is particularly relevant to address privacy considerations, as detailed in Section 4.3. Most smart meters do not offer possibilities to run code apart from the device's (metering) firmware. While this is generally intentional, to prevent tampering with the reported data (e.g., to avoid electricity theft, cf. Section 3.2), it does not allow any of the user-centric services to be executed directly on the smart meter. As technology advances and embedded devices are gradually becoming more and more potent in processing power and the ability to run user code in dedicated sandboxes, the provision of an execution environment on smart metering devices appears as a promising and potentially groundbreaking approach. Retrofitting existing and coming smart meter generations will represent a challenge, given the expected operational life of smart meters (often more than a decade) and, more importantly, the expected evolution of software frameworks in the same period. One viable solution is to offload these services to dedicated data processing devices, such as local set-top boxes, edge-clouds, and Multi-access Edge Computing (MEC), or cloud computing in general. The widespread adoption of 5G networks (and more recent developments, such as 6G) will allow sending data of even finer temporal resolution to external processing devices. It can be expected that communication will no longer be the bottleneck, thus the execution of services can be distributed across the aforementioned range of possible processing devices in order to optimize consumer-specific expectations to reliability, privacy, and real-time needs. However, this requires standardized interfaces, (real-time) transport protocols, and data formats aligned with our above observations.

4.2. Innovative Consumer-Centric Data Processing Algorithms

Smart meter data analytics services are frequently based on the combination of data preprocessing with novel methods to find correlations, patterns, and outliers in the available input data. Although the concept of electrical load signature analysis has been investigated since 1985 [158], the underlying preprocessing methods (e.g., event detection and NILM) are still not perfect and yield mediocre disaggregation performances in certain settings [159]. Sometimes this limitation can be circumvented by enriching electrical data with other sensed parameters (e.g., ambient conditions) and combining the data collected from different dwellings. However, only when the full amount of information can be extracted from smart meter data, the complete spectrum of user-centric services can be realized. Increasing the data processing methods' reliability and accuracy is crucial for widespread user acceptance and remains a significant future research challenge.

However, the sole availability of a range of data processing services does not necessarily lead to their ubiquitous adoption. Rather, the selection of useful and necessary services is expected to differ significantly between users. Fitting all smart meters with the same processing methods will thus not only incur the excessive and unnecessary use of computational power but still not serve all customers' needs equally well. Ultimately, we expect customers to utilize services depending on their situations. This implies that the need to selectively decide which of the services are of relevance to them. Thus, helping users identify the required services and understand the privacy implications when sharing data with the service providers (cf. Section 4.3) is crucial. This is also well-aligned with Section 4.1, confirming that a more flexible configuration of services and corresponding data sources are needed.

Lastly, we would like to recall our statement from Section 2, in which we emphasize the enormous potential of analyzing *electrical* signals for the provision of user-centric services. In fact, smart meter data are primarily related to electrical quantities, for whose interpretation an in-depth understanding of electrical engineering and power engineering is required. Many of the use cases surveyed in Section 3 rely on the interpretation of time-series data (i.e., sequences of measurements), which, in turn, calls on the expertise of mathematicians and signal processing experts. Simultaneously, experts in artificial intelligence and machine learning (e.g., computer scientists) can contribute yet different data analytics methods, especially when the volume of data to process is enormous. Smart meter data analytics is thus not only a cross-domain challenge, but also transdisciplinary research communities are inevitable to apply the state-of-the-art methods on smart meter data and thus enable accurate service provision. Finally, we expect that the same methods that apply to smart meter data can also find their application to other metered commodities such as water or natural gas.

4.3. User Privacy Protection

The collection of smart meter data at high temporal resolutions bears the enormous potential to provide services to the electricity customers' benefit. Simultaneously, however, the appropriate protection of collected data against unauthorized third parties' access is strongly needed. The reason is straightforward: Any processing method applicable to captured data (cf. Section 3) can be equally well applied by an attacker, seeking to profile a building's inhabitants (see Section 3.5) or learn about their habits. Often, this includes learning about usual sequences of household activities from consumption profiles (as discussed in Section 3.2). Solutions to ensure the secure transmission of smart meter data and adequate user privacy preservation are thus indispensable.

One method to circumvent security and privacy implications from the transmission of smart meter data to centralized processing systems (e.g., cloud computing) is their purely local processing. Due to the high resource requirements of many (pre-)processing methods and services (see, e.g., Section 2.2), however, this approach cannot always be applied. Particularly, when the data processing methods depend on parameters unavailable locally, corresponding computations must be executed on remote systems. Collaborative data processing approaches, i.e., the local extraction of features and their forwarding (devoid of most sensitive information) to remote data processing centers, represent an important future research direction. As a side effect, this also increases the services' adherence to the "data minimization" and "purpose limitation" requirements of data protection laws (such as the European Union's General Data Protection Regulation (GDPR) [160]).

When users cannot exert full control over the data their smart meters report, covering up characteristics in smart meter traces to hide user actions/intentions may also be necessary to protect their privacy. Current approaches mostly realize this functionality employing operating controllable generators or consumers to obfuscate the operation of sensitive appliances (e.g., [161,162]) or by intentionally falsifying reported data [27]. The potentially negative impact on the achievable services based on smart meter data, however, needs to be weighed up individually by clients and their willingness to pay the "cost of privacy" [163].

5. Conclusions

The operation of smart electrical power grids has become unimaginable without the opportunity to capture the status of grid-connected consumers in real-time and at fine resolution. Processing smart meter data has traditionally been centered around use cases that benefit the operations of electricity providers and the stability of the power grid [5]. The range of services that are tailored to the needs of end-customers is still comparably small. In this review paper, we present and discuss the range of use cases that are enabled through the collection of smart meter data but primarily benefit the consumers of electrical energy. We believe that three major preconditions are crucial for the long-term establishment of user-centric service provision. First, smart meters and the corresponding data processing mechanisms must be capable of reporting accurate information. They must undergo continuous improvements in order to extract the information content to the fullest extent possible. Second, adequate measures must be provided to protect user privacy. Established methods to provide secure networking must be combined with meaningful local preprocessing steps to remove sensitive features before data leave the customers' premises. Third, not all services apply to all users in the same way. A dedicated ecosystem, such as an "app store" for energy-based services (similar to the proposition in [164]), thus represents a viable option to allow consumers to individually subscribe to their desired services and understand the ensuing privacy implications. The range of user-centric data analysis methods, as surveyed in this work, can then be executed either locally or with the help of remote execution environments. A corresponding ecosystem will ultimately make it possible for both developers and providers of smart meter data processing methods to easily offer novel services, and simultaneously lower the barrier for customers to consume these services and avail of their benefits.

Author Contributions: Conceptualization, A.R. and L.P.; methodology, A.R., L.P., B.V. and A.F.; formal analysis, L.P. and A.F.; investigation, B.V. and A.F.; resources, L.P.; data curation, B.V.; writing—original draft preparation, A.R. and L.P.; writing—review and editing, B.V.; visualization, B.V. and A.R.; supervision, A.R. and L.P.; project administration, L.P. and A.R.; and funding acquisition, A.R. All authors have read and agreed to the published version of the manuscript.

Funding: This work was supported by Deutsche Forschungsgemeinschaft grant No. RE 3857/2-1 and by the Portuguese Foundation for Science and Technology grants CEECIND/01179/2017 and UIDB/50009/2020.

Conflicts of Interest: The authors declare no conflict of interest.

Abbreviations

The following abbreviations are used in this manuscript:

AAL	Ambient Assisted Living
ADL	Activities of Daily Living
AC	Air Conditioning Unit
ARIMA	Auto-Regressive Integrated Moving Average
BMS	Building Management System
CART	Classification and Regression Tree
CUSUM	CUmulative SUM
DBSCAN	Density-Based Spatial Clustering
DER	Distributed Energy Resource
DSF	Demand-side flexibility
ESS	Energy Storage System
EV	Electric Vehicle
GDPR	General Data Protection Regulation
GMM	Gaussian Mixture Model

IEC	International Electrotechnical Commission
IHD	In-Home Display
KFDA	Kernel Fisher Discriminant Analysis
LSTM	Long Short-Term Memory
MEC	Multi-access Edge Computing
MLP	Multi-Layer Perceptron
NILM	Non-Intrusive Load Monitoring
P2P	Peer-to-Peer
PLR	Piecewise Linear Regression
PV	Photovoltaic
RES	Renewable Energy Source
RMS	Root Mean Square
SCP	Switch Continuity Principle
SOM	Self-Organizing Map
SVM	Support Vector Machine

References

1. Uribe-Pérez, N.; Hernández, L.; De la Vega, D.; Angulo, I. State of the Art and Trends Review of Smart Metering in Electric Grids. *Appl. Sci.* **2016**, *6*, 68. [CrossRef]
2. Wang, Y.; Chen, Q.; Hong, T.; Kang, C. Review of Smart Meter Data Analytics: Applications, Methodologies, and Challenges. *IEEE Trans. Smart Grid* **2019**, *10*, 3125–3148. [CrossRef]
3. Haney, A.B.; Jamasb, T.; Pollitt, M.G. Smart metering: technology, economics and international experience. In *The Future of Electricity Demand: Customers, Citizens and Loads*; Department of Applied Economics Occasional Papers, Cambridge University Press: Cambridge, UK, 2011; pp. 161–184.
4. Kuralkar, S.; Mulay, P.; Chaudhari, A. Smart Energy Meter: Applications, Bibliometric Reviews and Future Research Directions. *Sci. Technol. Libr.* **2020**, *39*, 165–188. [CrossRef]
5. Wang, Y.; Chen, Q.; Kang, C. *Smart Meter Data Analytics: Electricity Consumer Behavior Modeling, Aggregation, and Forecasting*; Springer: Berlin/Heidelberg, Germany, 2020; pp. 1–293. [CrossRef]
6. Alzaatreh, A.; Mahdjoubi, L.; Gething, B.; Sierra, F. Disaggregating high-resolution gas metering data using pattern recognition. *Energy Build.* **2018**, *176*, 17–32. [CrossRef]
7. Vu, T.T.; Sokan, A.; Nakajo, H.; Fujinami, K.; Suutala, J.; Siirtola, P.; Alasalmi, T.; Pitkanen, A.; Roning, J. Feature Selection and Activity Recognition to Detect Water Waste from Water Tap Usage. In Proceedings of the IEEE 17th International Conference on Embedded and Real-Time Computing Systems and Applications, Toyama, Japan, 29–31 August 2011; pp. 138–141.
8. Guyot, P.; Pinquier, J.; Valero, X.; Alías, F. Two-step detection of water sound events for the diagnostic and monitoring of dementia. In Proceedings of the IEEE International Conference on Multimedia and Expo (ICME), San Jose, CA, USA, 15–19 July 2013; pp. 1–6.
9. Fogarty, J.; Au, C.; Hudson, S.E. Sensing from the Basement: A Feasibility Study of Unobtrusive and Low-Cost Home Activity Recognition. In Proceedings of the 19th Annual ACM Symposium on User Interface Software and Technology, Montreux, Switzerland, 15–18 October 2006; pp. 91–100.
10. Fontdecaba, S.; Sánchez-Espigares, J.A.; Marco-Almagro, L.; Tort-Martorell, X.; Cabrespina, F.; Zubelzu, J. An approach to disaggregating total household water consumption into major end-uses. *Water Resour. Manag.* **2013**, *27*, 2155–2177. [CrossRef]
11. Froehlich, J.E.; Larson, E.; Campbell, T.; Haggerty, C.; Fogarty, J.; Patel, S.N. HydroSense: infrastructure-mediated single-point sensing of whole-home water activity. In Proceedings of the 11th International Conference on Ubiquitous Computing, Orlando, FL, USA, 30 September–3 October 2009; pp. 235–244.
12. Cabeza, L.F.; Ürge Vorsatz, D.; Palacios, A.; Ürge, D.; Serrano, S.; Barreneche, C. Trends in Penetration and Ownership of Household Appliances. *Renew. Sustain. Energy Rev.* **2018**, *82*, 4044–4059. [CrossRef]
13. Jones, R.V.; Lomas, K.J. Determinants of High Electrical Energy Demand in UK Homes: Appliance Ownership and Use. *Energy Build.* **2016**, *117*, 71–82. [CrossRef]
14. Förderer, K.; Lösch, M.; Növer, R.; Ronczka, M.; Schmeck, H. Smart Meter Gateways: Options for a BSI-Compliant Integration of Energy Management Systems. *Appl. Sci.* **2019**, *9*, 1634. [CrossRef]
15. Elton Law. Impyute—A Library of Missing Data Imputation Algorithms Written in Python 3. 2020. Available online: http://impyute.readthedocs.io (accessed on 28 December 2020).
16. Younis, R.; Reinhardt, A. A Study on Fundamental Waveform Shapes in Microscopic Electrical Load Signatures. *Energies* **2020**, *13*, 3039. [CrossRef]
17. Parson, O.; Fisher, G.; Hersey, A.; Batra, N.; Kelly, J.; Singh, A.; Knottenbelt, W.; Rogers, A. Dataport and NILMTK: A building data set designed for non-intrusive load monitoring. In Proceedings of the 2015 IEEE Global Conference on Signal and Information Processing (GlobalSIP), Orlando, FL, USA, 14–16 December 2015; pp. 210–214.
18. Batra, N.; Gulati, M.; Singh, A.; Srivastava, M.B. It's Different: Insights into home energy consumption in India. In Proceedings of the 5th ACM Workshop on Embedded Systems for Energy-Efficient Buildings, Rome, Italy, 13–14 November 2013; pp. 1–8.

Makonin, S.; Popowich, F.; Bartram, L.; Gill, B.; Bajić, I.V. AMPds: A public dataset for load disaggregation and eco-feedback research. In Proceeding of the Annual Electrical Power and Energy Conference (EPEC), Halifax, NS, Canada, 21–23 August 2013. [CrossRef]

Makonin, S.; Wang, Z.J.; Tumpach, C. RAE: The rainforest automation energy dataset for smart grid meter data analysis. *Data* **2018**, *3*, 8. [CrossRef]

Beckel, C.; Kleiminger, W.; Cicchetti, R.; Staake, T.; Santini, S. The ECO data set and the performance of non-intrusive load monitoring algorithms. In Proceedings of the 1st ACM Conference on Embedded Systems for Energy-Efficient Buildings, Memphis, TN, USA, 5–6 November 2014; pp. 80–89.

Kolter, J.Z.; Johnson, M.J. REDD: A Public Data Set for Energy Disaggregation Research. In Proceedings of the 1st KDD Workshop on Data Mining Applications in Sustainability (SustKDD), San Diego, CA, USA, 21 August 2011; pp. 1–6.

Ribeiro, M.; Pereira, L.; Quintal, F.; Nunes, N. SustDataED: A Public Dataset for Electric Energy Disaggregation Research. In *ICT for Sustainability 2016*; Advances in Computer Science Research; Atlantis Press: Amsterdam, The Netherlands, 2016; pp. 244–245. [CrossRef]

Kriechbaumer, T.; Jacobsen, H.A. BLOND, a building-level office environment dataset of typical electrical appliances. *Sci. Data* **2018**, *5*, 180048. [CrossRef] [PubMed]

Saputro, N.; Akkaya, K. Investigation of Smart Meter Data Reporting Strategies for Optimized Performance in Smart Grid AMI Networks. *IEEE Internet Things J.* **2017**, *4*, 894–904. [CrossRef]

Díaz García, J.; Brunet Crosa, P.; Navazo Álvaro, I.; Vázquez Alcocer, P.P. Downsampling methods for medical datasets. In Proceedings of the International Conferences Computer Graphics, Visualization, Computer Vision and Image Processing and Big Data Analytics, Data Mining and Computational Intelligence, Lisbon, Portugal, 21–23 July 2017; IADIS Press: Lisbon, Portugal, 2017; pp. 12–20.

Reinhardt, A.; Englert, F.; Christin, D. Averting the privacy risks of smart metering by local data preprocessing. *Pervasive Mob. Comput.* **2015**, *16*, 171–183. [CrossRef]

Kukunuri, R.; Batra, N.; Wang, H. An open problem: energy data super-resolution. In Proceedings of the 5th International Workshop on Non-Intrusive Load Monitoring, Yokohama, Japan, 18 November 2020; pp. 99–102.

Osathanunkul, K.; Osathanunkul, K. Different Sampling Rates on Neural NILM Energy Disaggregation. In Proceedings of the Joint International Conference on Digital Arts, Media and Technology with ECTI Northern Section Conference on Electrical, Electronics, Computer and Telecommunications Engineering (ECTI DAMT-NCON), Nan, Thailand, 30 January–2 February 2019; pp. 318–321.

Huang, B.; Knox, M.; Bradbury, K.; Collins, L.M.; Newell, R.G. Non-intrusive load monitoring system performance over a range of low frequency sampling rates. In Proceedings of the 6th IEEE International Conference on Renewable Energy Research and Applications (ICRERA), San Diego, CA, USA, 5–8 December 2017; pp. 505–509.

Huchtkoetter, J.; Reinhardt, A. On the Impact of Temporal Data Resolution on the Accuracy of Non-Intrusive Load Monitoring. In Proceedings of the 7th ACM International Conference on Systems for Energy-Efficient Buildings, Cities, and Transportation, Yokohama, Japan, 19–20 November 2020; pp. 270–273.

Huchtkoetter, J.; Reinhardt, A. A study on the impact of data sampling rates on load signature event detection. *Energy Inform.* **2019**, *2*, 24. [CrossRef]

Wang, Y.; Chen, Q.; Kang, C.; Xia, Q.; Luo, M. Sparse and Redundant Representation-Based Smart Meter Data Compression and Pattern Extraction. *IEEE Trans. Power Syst.* **2017**, *32*, 2142–2151. [CrossRef]

de Souza, J.C.S.; Assis, T.M.L.; Pal, B.C. Data Compression in Smart Distribution Systems via Singular Value Decomposition. *IEEE Trans. Smart Grid* **2017**, *8*, 275–284. [CrossRef]

Serrenho, T.; Zangheri, P.; Bertoldi, P. *Energy Feedback Systems: Evaluation of Meta-Studies on Energy Savings through Feedback*; EUR—Scientific and Technical Research Reports EUR 27992 EN; Publications Office of the European Union: Brussels, Belgium, 2015; ISBN 978-92-79-59778-7. [CrossRef]

Kahl, M.; Haq, A.U.; Kriechbaumer, T.; Jacobsen, H.A. A comprehensive feature study for appliance recognition on high frequency energy data. In Proceedings of the e-Energy 2017—8th International Conference on Future Energy Systems, Hong Kong, China, 16–19 May 2017; pp. 121–131. [CrossRef]

Liang, J.; Ng, S.K.; Kendall, G.; Cheng, J.W. Load signature study—Part I: Basic concept, structure, and methodology. *IEEE Trans. Power Deliv.* **2010**, *25*, 551–560. [CrossRef]

Sadeghianpourhamami, N.; Ruyssinck, J.; Deschrijver, D.; Dhaene, T.; Develder, C. Comprehensive feature selection for appliance classification in NILM. *Energy Build.* **2017**, *151*, 98–106. [CrossRef]

Hart, G.W. Nonintrusive appliance load monitoring. *Proc. IEEE* **1992**, *80*, 1870–1891. [CrossRef]

Makonin, S. Investigating the switch continuity principle assumed in Non-Intrusive Load Monitoring (NILM). In Proceedings of the 2016 IEEE Canadian Conference on Electrical and Computer Engineering (CCECE), Vancouver, BC, Canada, 15–18 May 2016; pp. 1–4.

Anderson, K.D.; Bergés, M.E.; Ocneanu, A.; Benitez, D.; Moura, J.M. Event detection for non intrusive load monitoring. In Proceedings of the 38th Annual Conference on IEEE Industrial Electronics Society (IECON), Montreal, QC, Canada, 25–28 October 2012; pp. 3312–3317.

42. Kelly, J.; Knottenbelt, W. The UK-DALE dataset, domestic appliance-level electricity demand and whole-house demand from five UK homes. *Sci. Data* **2015**, *2*, 150007. [CrossRef]
43. Pereira, L.; Nunes, N. An empirical exploration of performance metrics for event detection algorithms in Non-Intrusive Load Monitoring. *Sustain. Cities Soc.* **2020**, *62*, 102399. [CrossRef]
44. Völker, B.; Pfeifer, M.; Scholl, P.M.; Becker, B. Annoticity: A Smart Annotation Tool and Data Browser for Electricity Datasets. Proceedings of the 5th International Workshop on Non-Intrusive Load Monitoring, Yokohama, Japan, 18 November 2020; pp. 1–
45. Pereira, L.; Nunes, N.J. Semi-automatic labeling for public non-intrusive load monitoring datasets. In Proceedings of the Sustainable Internet and ICT for Sustainability (SustainIT), Madrid, Spain, 14–15 April 2015; pp. 1–4.
46. Anderson, K.; Ocneanu, A.F.; Benitez, D.; Carlson, D.; Rowe, A.; Bergés, M. BLUED: A Fully Labeled Public Dataset for Event-Based Non-Intrusive Load Monitoring Research. In Proceedings of the 2nd KDD Workshop on Data Mining Applications in Sustainability (SustKDD), Beijing, China, 12–16 August 2012; pp. 1 – 5.
47. Völker, B.; Pfeifer, M.; Scholl, P.M.; Becker, B. FIRED: A Fully-labeled hIgh-fRequency Electricity Disaggregation Dataset. Proceedings of the 7th ACM International Conference on Systems for Energy-Efficient Built Environments (BuildSys), Yokohama, Japan, 18–20 November 2020.
48. Kahl, M.; Kriechbaumer, T.; Jorde, D.; Ul Haq, A.; Jacobsen, H.A. Appliance Event Detection-A Multivariate, Supervised Classification Approach. In Proceedings of the 10th ACM International Conference on Future Energy Systems (e-Energy), Phoenix, AZ, USA, 25–28 June 2019; pp. 373–375.
49. Pereira, L.; Ribeiro, M.; Nunes, N. Engineering and deploying a hardware and software platform to collect and label non-intrusive load monitoring datasets. In *Sustainable Internet and ICT for Sustainability (SustainIT)*; IEEE/IFIP: Funchal, Portugal, 2017; pp. 1– [CrossRef]
50. Weiss, M.; Helfenstein, A.; Mattern, F.; Staake, T. Leveraging smart meter data to recognize home appliances. In Proceedings of the IEEE International Conference on Pervasive Computing and Communications, (PerCom 2012), Lugano, Switzerland, 19–23 March 2012; pp. 190–197. [CrossRef]
51. Meehan, P.; McArdle, C.; Daniels, S. An efficient, scalable time-frequency method for tracking energy usage of domestic appliances using a two-step classification algorithm. *Energies* **2014**, *7*, 7041–7066. [CrossRef]
52. Luo, D.; Norford, L.K.; Shaw, S.R.; Leeb, S.B. Monitoring HVAC equipment electrical loads from a centralized location–method and field test results/Discussion. *ASHRAE Trans.* **2002**, *108*, 841.
53. Pereira, L. Developing and evaluating a probabilistic event detector for non-intrusive load monitoring. In *Sustainable Internet and ICT for Sustainability (SustainIT)*; IEEE: Funchal, Portugal, 2017; pp. 1–10. [CrossRef]
54. Völker, B.; Scholl, P.M.; Becker, B. Semi-Automatic Generation and Labeling of Training Data for Non-Intrusive Load Monitoring. In Proceedings of the 10th ACM International Conference on Future Energy Systems (e-Energy), Phoenix, AZ, USA, 25–28 June 2019.
55. Trung, K.N.; Dekneuvel, E.; Nicolle, B.; Zammit, O.; Van, C.N.; Jacquemod, G. Event detection and disaggregation algorithms for nialm system. In Proceedings of 2nd International Non-Intrusive Load Monitoring (NILM) Workshop, Austin, TX, USA, 3 June 2014.
56. Wild, B.; Barsim, K.S.; Yang, B. A new unsupervised event detector for non-intrusive load monitoring. In Proceedings of the IEEE Global Conference on Signal and Information Processing (GlobalSIP), Orlando, FL, USA, 14–16 December 2015; pp. 73–77.
57. De Baets, L.; Ruyssinck, J.; Deschrijver, D.; Dhaene, T. Event detection in NILM using cepstrum smoothing. In Proceedings of the 3rd International Workshop on Non-Intrusive Load Monitoring, Vancouver, BC, Canada, 14–15 May 2016; pp. 1–4.
58. Cox, R.; Leeb, S.B.; Shaw, S.R.; Norford, L.K. Transient event detection for nonintrusive load monitoring and demand side management using voltage distortion. In Proceedings of the 21st Annual IEEE Applied Power Electronics Conference and Exposition, Dallas, TX, USA, 19–23 March 2006; p. 7.
59. Zoha, A.; Gluhak, A.; Imran, M.A.; Rajasegarar, S. Non-intrusive load monitoring approaches for disaggregated energy sensing: A survey. *Sensors* **2012**, *12*, 16838–16866. [CrossRef]
60. Bonfigli, R.; Squartini, S.; Fagiani, M.; Piazza, F. Unsupervised algorithms for non-intrusive load monitoring: An up-to-date overview. In Proceedings of the IEEE 15th International Conference on Environment and Electrical Engineering (EEEIC), Rome, Italy, 10–13 June 2015; pp. 1175–1180.
61. Faustine, A.; Mvungi, N.H.; Kaijage, S.; Michael, K. A survey on non-intrusive load monitoring methodies and techniques for energy disaggregation problem. *arXiv* **2017**, arXiv:1703.00785.
62. Alcalá, J.; Ure na, J.; Hernández, Á.; Gualda, D. Event-based energy disaggregation algorithm for activity monitoring from a single-point sensor. *IEEE Trans. Instrum. Meas.* **2017**, *66*, 2615–2626. [CrossRef]
63. Barsim, K.S.; Streubel, R.; Yang, B. Unsupervised adaptive event detection for building-level energy disaggregation. Proceedings of the Power and Energy Student Summt (PESS), Stuttgart, Germany, 22–24 January 2014.
64. Meziane, M.N.; Ravier, P.; Lamarque, G.; Le Bunetel, J.C.; Raingeaud, Y. High accuracy event detection for non-intrusive load monitoring. In Proceedings of the 2017 IEEE International Conference on Acoustics, Speech and Signal Processing (ICASSP), New Orleans, LA, USA, 5–9 March 2017; pp. 2452–2456.
65. Girmay, A.A.; Camarda, C. Simple event detection and disaggregation approach for residential energy estimation. In Proceedings of the 3rd International Workshop on Non-Intrusive Load Monitoring (NILM), Vancouver, BC, Canada, 14–15 May 2016.

Sethom, H.B.A.; Houidi, S.; Auger, F.; Ben, H.; Sethom, A.; Fourer, D.; Miègeville, L. Multivariate Event Detection Methods for Non-Intrusive Load Monitoring in Smart Homes and Residential Buildings. *Energy Build.* **2019**, *208*, 109624.

Gomes, E.; Pereira, L. PB-NILM: Pinball Guided Deep Non-Intrusive Load Monitoring. *IEEE Access* **2020**, *8*, 48386–48398. [CrossRef]

Faustine, A.; Pereira, L.; Bousbiat, H.; Kulkarni, S. UNet-NILM: A Deep Neural Network for Multi-tasks Appliances State Detection and Power Estimation in NILM. In Proceedings of the 5th International Workshop on Non-Intrusive Load Monitoring, Yokohama, Japan, 18 November 2020; pp. 84–88.

Faustine, A.; Pereira, L.; Klemenjak, C. Adaptive Weighted Recurrence Graphs for Appliance Recognition in Non-Intrusive Load Monitoring. *IEEE Trans. Smart Grid* **2020**. [CrossRef]

Fischer, C. Feedback on household electricity consumption: a tool for saving energy? *Energy Effic.* **2008**, *1*, 79–104. [CrossRef]

Froehlich, J.; Findlater, L.; Landay, J. The Design of Eco-feedback Technology. In Proceedings of the SIGCHI Conference on Human Factors in Computing Systems, Atlanta, GA, USA, 10–15 April 2010; ACM: New York, NY, USA, 2010; pp. 1999–2008. [CrossRef]

Armel, K.C.; Gupta, A.; Shrimali, G.; Albert, A. Is Disaggregation the Holy Grail of Energy Efficiency? The Case of Electricity. *Energy Policy* **2013**, *52*, 213–234. [CrossRef]

Choi, T.S.; Ko, K.R.; Park, S.C.; Jang, Y.S.; Yoon, Y.T.; Im, S.K. Analysis of energy savings using smart metering system and IHD (in-home display). In Proceedings of the 2009 Transmission Distribution Conference Exposition: Asia and Pacific, Seoul, Korea, 26–30 October 2009; pp. 1–4. [CrossRef]

Paay, J.; Kjeldskov, J.; Skov, M.B.; Lund, D.; Madsen, T.; Nielsen, M. Design of an appliance level eco-feedback display for domestic electricity consumption. In Proceedings of the 26th Australian Computer-Human Interaction Conference on Designing Futures: The Future of Design (OzCHI), Sydney, NSW, Australia, 2–5 December 2014; Association for Computing Machinery: New York, NY, USA, 2014; pp. 332–341. [CrossRef]

Broms, L.; Katzeff, C.; Bang, M.; Nyblom, A.; Hjelm, S.I.; Ehrnberger, K. Coffee maker patterns and the design of energy feedback artefacts. In Proceedings of the 8th ACM Conference on Designing Interactive Systems (DIS), Aarhus, Denmark, 16–20 August 2010; Association for Computing Machinery: New York, NY, USA, 2010; pp. 93–102. [CrossRef]

Rodgers, J.; Bartram, L. Exploring Ambient and Artistic Visualization for Residential Energy Use Feedback. *IEEE Trans. Vis. Comput. Graph.* **2011**, *17*, 2489–2497. [CrossRef] [PubMed]

Spagnolli, A.; Corradi, N.; Gamberini, L.; Hoggan, E.; Jacucci, G.; Katzeff, C.; Broms, L.; Jonsson, L. Eco-Feedback on the Go: Motivating Energy Awareness. *Computer* **2011**, *44*, 38–45. [CrossRef]

Costanza, E.; Ramchurn, S.D.; Jennings, N.R. Understanding Domestic Energy Consumption Through Interactive Visualisation: A Field Study. In Proceedings of the ACM Conference on Ubiquitous Computing, Pittsburgh, PA, USA, 5–8 September 2012; ACM: New York, NY, USA, 2012; pp. 216–225. [CrossRef]

Quintal, F.; Pereira, L.; Nunes, N.; Nisi, V.; Barreto, M. WATTSBurning: Design and Evaluation of an Innovative Eco-Feedback System. In *Human-Computer Interaction—INTERACT*; Lecture Notes in Computer Science; Springer: Berlin/Heidelberg, Germany, 2013; pp. 453–470. [CrossRef]

Moere, A.V.; Tomitsch, M.; Hoinkis, M.; Trefz, E.; Johansen, S.; Jones, A. Comparative Feedback in the Street: Exposing Residential Energy Consumption on House Façades. In *Human-Computer Interaction—INTERACT 2011*; Campos, P., Graham, N., Jorge, J., Nunes, N., Palanque, P., Winckler, M., Eds.; Lecture Notes in Computer Science; Springer: Berlin/Heidelberg, Germany, 2011; pp. 470–488. [CrossRef]

Quintal, F.; Barreto, M.; Nunes, N.; Nisi, V.; Pereira, L. WattsBurning on My Mailbox: A Tangible Art Inspired Eco-feedback Visualization for Sharing Energy Consumption. In *Human-Computer Interaction—INTERACT*; Lecture Notes in Computer Science; Springer: Berlin/Heidelberg, Germany, 2013; pp. 133–140. [CrossRef]

Foster, D.; Lawson, S.; Blythe, M.; Cairns, P. Wattsup? motivating reductions in domestic energy consumption using social networks. In Proceedings of the 6th Nordic Conference on Human-Computer Interaction: Extending Boundaries, Reykjavik, Iceland, 16–20 October 2010; Association for Computing Machinery: New York, NY, USA, 2010; pp. 178–187. [CrossRef]

Filonik, D.; Medland, R.; Foth, M.; Rittenbruch, M. A Customisable Dashboard Display for Environmental Performance Visualisations. In *Persuasive Technology*; Springer: Berlin/Heidelberg, Germany, 2013; pp. 51–62. [CrossRef]

Kelly, J.; Knottenbelt, W. Does disaggregated electricity feedback reduce domestic electricity consumption? A systematic review of the literature. *arXiv* **2016**, arXiv:1605.00962.

Pereira, L.; Quintal, F.; Barreto, M.; Nunes, N.J. Understanding the Limitations of Eco-feedback: A One-Year Long-Term Study. In *Human-Computer Interaction and Knowledge Discovery in Complex, Unstructured, Big Data*; Holzinger, A., Pasi, G., Eds.; Lecture Notes in Computer Science; Springer: Maribor, Slovenia, 2013; pp. 237–255.

Buchanan, K.; Russo, R.; Anderson, B. Feeding back about eco-feedback: How do consumers use and respond to energy monitors? *Energy Policy* **2014**, *73*, 138–146. [CrossRef]

Ma, G.; Lin, J.; Li, N. Longitudinal assessment of the behavior-changing effect of app-based eco-feedback in residential buildings. *Energy Build.* **2018**, *159*, 486–494. [CrossRef]

Pereira, L.; Nunes, N. Understanding the practical issues of deploying energy monitoring and eco-feedback technology in the wild: Lesson learned from three long-term deployments. *Energy Rep.* **2019**, *6*, 94–106. [CrossRef]

89. Schwartz, T.; Denef, S.; Stevens, G.; Ramirez, L.; Wulf, V. Cultivating Energy Literacy: Results from a Longitudinal Living L Study of a Home Energy Management System. In Proceedings of the SIGCHI Conference on Human Factors in Computi Systems, Paris, France, 27 April–2 May 2013; ACM: New York, NY, USA, 2013; pp. 1193–1202. [CrossRef]
90. Geelen, D.; Mugge, R.; Silvester, S.; Bulters, A. The use of apps to promote energy saving: A study of smart meter–relat feedback in the Netherlands. *Energy Effic.* **2019**, *12*, 1635–1660. [CrossRef]
91. Hansen, M.; Hauge, B. Prosumers and smart grid technologies in Denmark: Developing user competences in smart gr households. *Energy Effic.* **2017**, *10*, 1215–1234. [CrossRef]
92. Barreto, M.; Pereira, L.; Quintal, F. The Acceptance of energy monitoring technologies: The case of local prosumers. Proceedings of the 6th International Conference on ICT for Sustainability, Lappeenranta, Finland, 10–14 June 2019; CEUR-W Lappeenranta, Finland, 2019; Volume 2382.
93. Sobe, A.; Elmenreich, W. Smart Microgrids: Overview and Outlook. In Proceedings of the GI Informatik Smart Grid Worksh Braunschweig, Germany, 16–21 September 2012.
94. Meeuw, A.; Schopfer, S.; Ryder, B.; Wortmann, F. LokalPower: Enabling Local Energy Markets with User-Driven Engageme In *Extended Abstracts of the 2018 CHI Conference on Human Factors in Computing Systems*; Association for Computing Machine Montreal QC, Canada, 2018; pp. 1–6. [CrossRef]
95. Scuri, S.; Tasheva, G.; Barros, L.; Nunes, N.J. An HCI Perspective on Distributed Ledger Technologies for Peer-to-Peer Ener Trading. In *Human-Computer Interaction—INTERACT 2019*; Lamas, D., Loizides, F., Nacke, L., Petrie, H., Winckler, M., Zaphi P., Eds.; Lecture Notes in Computer Science; Springer: Cham, Switzerland, 2019; pp. 91–111. [CrossRef]
96. Scuri, S.; Nunes, N.J. PowerShare 2.0: A Gamified P2P Energy Trading Platform. In Proceedings of the International Conferer on Advanced Visual Interfaces, Lacco Ameno d'Ischia, Italy, 28 September–2 October 2020; Association for Computing Machine New York, NY, USA, 2020; pp. 1–3. [CrossRef]
97. Ableitner, L.; Tiefenbeck, V.; Meeuw, A.; Wörner, A.; Fleisch, E.; Wortmann, F. User behavior in a real-world peer-to-pe electricity market. *Appl. Energy* **2020**, *270*, 115061. [CrossRef]
98. Clement, J.; Ploennigs, J.; Kabitzsch, K. Smart Meter: Detect and Individualize ADLs. In *Ambient Assisted Living*; Spring Berlin/Heidelberg, Germany, 2012; pp. 107–122. [CrossRef]
99. Chalmers, C.; Fergus, P.; Curbelo Montanez, C.A.; Sikdar, S.; Ball, F.; Kendall, B. Detecting Activities of Daily Living and Routi Behaviours in Dementia Patients Living Alone Using Smart Meter Load Disaggregation. *IEEE Trans. Emerg. Top. Comput.* **20** [CrossRef]
100. Bousbiat, H.; Klemenjak, C.; Leitner, G.; Elmenreich, W. Augmenting an Assisted Living Lab with Non-Intrusive Load Monitori In Proceedings of the 2020 IEEE International Instrumentation and Measurement Technology Conference (I2MTC), Dubrovn Croatia, 25–28 May 2020; pp. 1–5. [CrossRef]
101. Reinhardt, A.; Klemenjak, C. Device-Free User Activity Detection using Non-Intrusive Load Monitoring: A Case Study. Proceedings of the 2nd ACM Workshop on Device Free Human Sensing (DFHS), Yokohama, Japan, 16 November 2020; pp. 1
102. Liao, J.; Stankovic, L.; Stankovic, V. Detecting Household Activity Patterns from Smart Meter Data. In Proceedings of t International Conference on Intelligent Environments, Shanghai, China, 2–4 July 2014; pp. 71–78. [CrossRef]
103. Jokar, P.; Arianpoo, N.; Leung, V.C.M. Electricity Theft Detection in AMI Using Customers' Consumption Patterns. *IEEE Tra Smart Grid* **2016**, *7*, 216–226. [CrossRef]
104. Yip, S.C.; Tan, W.N.; Tan, C.; Gan, M.T.; Wong, K. An anomaly detection framework for identifying energy theft and defecti meters in smart grids. *Int. J. Electr. Power Energy Syst.* **2018**, *101*, 189–203. [CrossRef]
105. Liu, M.; Liu, D.; Sun, G.; Zhao, Y.; Wang, D.; Liu, F.; Fang, X.; He, Q.; Xu, D. Deep Learning Detection of Inaccurate Sma Electricity Meters: A Case Study. *IEEE Ind. Electron. Mag.* **2020**, *14*, 79–90. [CrossRef]
106. Sial, A.; Singh, A.; Mahanti, A. Detecting anomalous energy consumption using contextual analysis of smart meter data. *Wi Netw.* **2019**, *8*. [CrossRef]
107. Buzau, M.M.; Tejedor-Aguilera, J.; Cruz-Romero, P.; Gómez-Expósito, A. Detection of Non-Technical Losses Using Smart Me Data and Supervised Learning. *IEEE Trans. Smart Grid* **2019**, *10*, 2661–2670. [CrossRef]
108. Persson, M.; Lindskog, A. Detection and localization of non-technical losses in distribution systems with future smart meters. Proceedings of the 2019 IEEE Milan PowerTech, Milan, Italy, 23–27 June 2019; pp. 1–6. [CrossRef]
109. Raggi, L.M.R.; Trindade, F.C.L.; Cunha, V.C.; Freitas, W. Non-Technical Loss Identification by Using Data Analytics and Custom Smart Meters. *IEEE Trans. Power Deliv.* **2020**, *35*, 2700–2710. [CrossRef]
110. Buzau, M.; Tejedor-Aguilera, J.; Cruz-Romero, P.; Gómez-Expósito, A. Hybrid Deep Neural Networks for Detection of No Technical Losses in Electricity Smart Meters. *IEEE Trans. Power Syst.* **2020**, *35*, 1254–1263. [CrossRef]
111. International Renewable Energy Agency. *Demand-Side Flexibility for Power Sector Transformation—Analytical Brief*; Technic Report; IRENA: Abu Dhabi, UAE, 2019; ISBN 978-92-9260-159-1.
112. Shoreh, M.H.; Siano, P.; Shafie-khah, M.; Loia, V.; Catalão, J.P.S. A survey of industrial applications of Demand Response. *Ele Power Syst. Res.* **2016**, *141*, 31–49. [CrossRef]
113. Bilton, M.; Aunedi, M.; Woolf, M.; Strbac, G. *Smart Appliances for Residential Demand Response*; Technical Report A10 for the "Lo Carbon London" LCNF Project; Imperial College London: London, UK, 2014.

4. Neupane, B.; Pedersen, T.B.; Thiesson, B. Towards Flexibility Detection in Device-Level Energy Consumption. In *Data Analytics for Renewable Energy Integration*; Number 8817 in Lecture Notes in Computer Science; Springer: Berlin/Heidelberg, Germany, 2014; pp. 1–16.
5. D'hulst, R.; Labeeuw, W.; Beusen, B.; Claessens, S.; Deconinck, G.; Vanthournout, K. Demand response flexibility and flexibility potential of residential smart appliances: Experiences from large pilot test in Belgium. *Appl. Energy* **2015**, *155*, 79–90. [CrossRef]
6. Afzalan, M.; Jazizadeh, F. Residential loads flexibility potential for demand response using energy consumption patterns and user segments. *Appl. Energy* **2019**, *254*, 113693. [CrossRef]
7. Pecan Street Inc. Dataport—Researcher Access to Pecan Street's Groundbreaking Energy and Water Data. 2020. Available online: https://www.pecanstreet.org/dataport/ (accessed on 21 January 2021).
8. Mesarić, P.; Krajcar, S. Home demand side management integrated with electric vehicles and renewable energy sources. *Energy Build.* **2015**, *108*, 1–9. [CrossRef]
9. Li, P.H.; Pye, S. Assessing the benefits of demand-side flexibility in residential and transport sectors from an integrated energy systems perspective. *Appl. Energy* **2018**, *228*, 965–979. [CrossRef]
10. Fanitabasi, F.; Pournaras, E. Appliance-Level Flexible Scheduling for Socio-Technical Smart Grid Optimization. *IEEE Access* **2020**, *8*, 119880–119898. [CrossRef]
11. Ciabattoni, L.; Comodi, G.; Ferracuti, F.; Foresi, G. AI-Powered Home Electrical Appliances as Enabler of Demand-Side Flexibility. *IEEE Consum. Electron. Mag.* **2020**, *9*, 72–78. [CrossRef]
12. Lucas, A.; Jansen, L.; Andreadou, N.; Kotsakis, E.; Masera, M. Load Flexibility Forecast for DR Using Non-Intrusive Load Monitoring in the Residential Sector. *Energies* **2019**, *12*, 2725. [CrossRef]
13. Zhai, S.; Zhou, H.; Wang, Z.; He, G. Analysis of dynamic appliance flexibility considering user behavior via non-intrusive load monitoring and deep user modeling. *CSEE J. Power Energy Syst.* **2020**, *6*, 41–51. [CrossRef]
14. Yue, H.; Yan, K.; Zhao, J.; Ren, Y.; Yan, X.; Zhao, H. Estimating Demand Response Flexibility of Smart Home Appliances via NILM Algorithm. In Proceedings of the 2020 IEEE 4th Information Technology, Networking, Electronic and Automation Control Conference (ITNEC), Chongqing, China, 12–14 June 2020; Volume 1, pp. 394–398. [CrossRef]
15. Gajowniczek, K.; Ząbkowski, T. Electricity forecasting on the individual household level enhanced based on activity patterns. *PLoS ONE* **2017**, *12*, e0174098. [CrossRef]
16. Pirbazari, A.M.; Farmanbar, M.; Chakravorty, A.; Rong, C. Short-Term Load Forecasting Using Smart Meter Data: A Generalization Analysis. *Processes* **2020**, *8*, 484. [CrossRef]
17. Amin, P.; Cherkasova, L.; Aitken, R.; Kache, V. Automating Energy Demand Modeling and Forecasting Using Smart Meter Data. In Proceedings of the 2019 IEEE International Congress on Internet of Things (ICIOT), Milan, Italy, 8–13 July 2019; pp. 133–137. [CrossRef]
18. Dinesh, C.; Makonin, S.; Bajic, I.V. Residential Power Forecasting Based on Affinity Aggregation Spectral Clustering. *IEEE Access* **2020**, *8*, 99431–99444. [CrossRef]
19. Hashmi, M.U.; Mukhopadhyay, A.; Bušić, A.; Elias, J. Optimal control of storage under time varying electricity prices. In Proceedings of the 2017 IEEE International Conference on Smart Grid Communications (SmartGridComm), Dresden, Germany, 23–26 October 2017; pp. 134–140. [CrossRef]
20. Hashmi, M.U.; Pereira, L.; Bušić, A. Energy storage in Madeira, Portugal: co-optimizing for arbitrage, self-sufficiency, peak shaving and energy backup. In *IEEE Milan PowerTech*; IEEE: Milan, Italy, 2019; pp. 1–6.
21. Kiedanski, D.; Hashmi, M.U.; Bušić, A.; Kofman, D. Sensitivity to Forecast Errors in Energy Storage Arbitrage for Residential Consumers. In Proceedings of the 2019 IEEE International Conference on Communications, Control, and Computing Technologies for Smart Grids (SmartGridComm), Beijing, China, 21–23 October 2019; pp. 1–7. [CrossRef]
22. Xydas, E.S.; Marmaras, C.E.; Cipcigan, L.M.; Hassan, A.S.; Jenkins, N. Forecasting Electric Vehicle charging demand using Support Vector Machines. In Proceedings of the 2013 48th International Universities' Power Engineering Conference (UPEC), Dublin, Ireland, 2–5 September 2013; pp. 1–6. [CrossRef]
23. Ai, S.; Chakravorty, A.; Rong, C. Household EV Charging Demand Prediction Using Machine and Ensemble Learning. In Proceedings of the 2018 IEEE International Conference on Energy Internet (ICEI), Beijing, China, 21–25 May 2018; pp. 163–168. [CrossRef]
24. Fachrizal, R.; Munkhammar, J. Improved Photovoltaic Self-Consumption in Residential Buildings with Distributed and Centralized Smart Charging of Electric Vehicles. *Energies* **2020**, *13*, 1153. [CrossRef]
25. Vassileva, I.; Campillo, J. Adoption barriers for electric vehicles: Experiences from early adopters in Sweden. *Energy* **2017**, *120*, 632–641. [CrossRef]
26. Svangren, M.K.; Jensen, R.H.; Skov, M.B.; Kjeldskov, J. Driving on sunshine: aligning electric vehicle charging and household electricity production. In Proceedings of the 10th Nordic Conference on Human-Computer Interaction, Oslo, Norway, 1–3 October 2018; Association for Computing Machinery: New York, NY, USA, 2018; pp. 439–451. [CrossRef]
27. Barros, L.; Barreto, M.; Pereira, L. Understanding the challenges behind Electric Vehicle usage by drivers—A case study in the Madeira Autonomous Region. In Proceedings of the 7th International Conference on ICT for Sustainability, Bristol, UK, 21–27 June 2020; Association for Computing Machinery: New York, NY, USA, 2020; pp. 88–97. [CrossRef]

138. Vasudevarao, B.V.M.; Stifter, M.; Zehetbauer, P. Methodology for creating composite standard load profiles based on real load profile analysis. In Proceedings of the IEEE PES Innovative Smart Grid Technologies Conference Europe (ISGT-Europe), Ljubljana, Slovenia, 9–12 October 2016; pp. 1–6.
139. Marlen, A.; Maxim, A.; Ukaegbu, I.A.; Kumar Nunna, H.S.V.S. Application of Big Data in Smart Grids: Energy Analytics. In Proceedings of the 2019 21st International Conference on Advanced Communication Technology (ICACT), PyeongChang, Korea, 17–20 February 2019; pp. 402–407. [CrossRef]
140. Røsok, J.M. Combining smart energy meters with social media: Increasing energy awareness using data visualization and persuasive technologies. In Proceedings of the 2014 International Conference on Collaboration Technologies and Systems (CTS), Minneapolis, MN, USA, 19–23 May 2014; pp. 27–32. [CrossRef]
141. Toma, A.R.; Gheorghe, C.M.; Neacșu, F.L.; Dumitrescu, A. Conversion of smart meter data in user-intuitive carbon footprint information. In Proceedings of the 2017 5th International Symposium on Electrical and Electronics Engineering (ISEEE), Galați, Romania, 20–22 October 2017; pp. 1–6. [CrossRef]
142. Park, K.; Son, S. A Novel Load Image Profile-Based Electricity Load Clustering Methodology. *IEEE Access* **2019**, *7*, 59048–59058. [CrossRef]
143. Lee, E.; Kim, J.; Jang, D. Load profile segmentation for effective residential demand response program: Method and evidence from Korean pilot study. *Energies* **2020**, *16*. [CrossRef]
144. Zhan, S.; Liu, Z.; Chong, A.; Yan, D. Building categorization revisited: A clustering-based approach to using smart meter data for building energy benchmarking. *Appl. Energy* **2020**, *269*, 114920. [CrossRef]
145. Choksi, K.A.; Jain, S.; Pindoriya, N.M. Feature based clustering technique for investigation of domestic load profiles and probabilistic variation assessment: Smart meter dataset. *Sustain. Energy Grids Netw.* **2020**, *22*, 100346. [CrossRef]
146. Alonso, A.M.; Nogales, F.J.; Ruiz, C. Hierarchical Clustering for Smart Meter Electricity Loads Based on Quantile Autocovariances. *IEEE Trans. Smart Grid* **2020**, *11*, 4522–4530. [CrossRef]
147. Varga, E.D.; Beretka, S.F.; Noce, C.; Sapienza, G. Robust Real-Time Load Profile Encoding and Classification Framework for Efficient Power Systems Operation. *IEEE Trans. Power Syst.* **2015**, *30*, 1897–1904. [CrossRef]
148. Ullah, A.; Haydarov, K.; Haq, I.U.; Muhammad, K.; Rho, S.; Lee, M.; Baik, S.W. Deep Learning Assisted Buildings Energy Consumption Profiling Using Smart Meter Data. *Sensors* **2020**, *20*, 873. [CrossRef]
149. Sun, M.; Wang, Y.; Teng, F.; Ye, Y.; Strbac, G.; Kang, C. Clustering-Based Residential Baseline Estimation: A Probabilistic Perspective. *IEEE Trans. Smart Grid* **2019**, *10*, 6014–6028. [CrossRef]
150. Haben, S.; Singleton, C.; Grindrod, P. Analysis and Clustering of Residential Customers Energy Behavioral Demand Using Smart Meter Data. *IEEE Trans. Smart Grid* **2016**, *7*, 136–144. [CrossRef]
151. Sun, M.; Konstantelos, I.; Strbac, G. C-Vine Copula Mixture Model for Clustering of Residential Electrical Load Pattern Data. *IEEE Trans. Power Syst.* **2017**, *32*, 2382–2393. [CrossRef]
152. Hurst, W.; Montañez, C.A.C.; Shone, N. Time-Pattern Profiling from Smart Meter Data to Detect Outliers in Energy Consumption. *IoT* **2020**, *1*, 92–108. [CrossRef]
153. Donaldson, D.L.; Jayaweera, D. Effective solar prosumer identification using net smart meter data. *Int. J. Electr. Power Energy Syst.* **2020**, *118*, 105823. [CrossRef]
154. Berges, M.E.; Goldman, E.; Matthews, H.S.; Soibelman, L. Enhancing Electricity Audits in Residential Buildings with Nonintrusive Load Monitoring. *J. Ind. Ecol.* **2010**, *14*, 844–858. [CrossRef]
155. Çimen, H.; Çetinkaya, N.; Vasquez, J.C.; Guerrero, J.M. A Microgrid Energy Management System based on Non-Intrusive Load Monitoring via Multitask Learning. *IEEE Trans. Smart Grid* **2020**. [CrossRef]
156. Uihlein, A.; Caramizaru, A. *Energy Communities: An Overview of Energy and Social Innovation*; Technical Report EUR 30083 EN; Publications Office of the European Union: Luxembourg, 2020.
157. IEC. *Sensing Devices for Non-Intrusive Load Monitoring (NILM) Systems*; Norm DIN IEC/TS 63297; International Electrotechnical Commission: Geneva, Switzerland, 2020.
158. Hart, G.W. *Prototype Nonintrusive Appliance Load Monitor*; Technical Report; MIT Energy Laboratory and Electric Power Research Institute: Cambridge, MA, USA, 1985.
159. Reinhardt, A.; Klemenjak, C. How does load disaggregation performance depend on data characteristics? Insights from a benchmarking study. In Proceedings of the Eleventh ACM International Conference on Future Energy Systems, Melbourne, Australia, 22–26 June 2020; pp. 167–177.
160. European Union. Regulation (EU) 2016/679 of the European Parliament and of the Council of 27 April 2016 on the protection of natural persons with regard to the processing of personal data and on the free movement of such data, and repealing Directive 95/46/EC (General Data Protection Regulation). *Off. J. L110* **2016**, *59*, 1–88.
161. Egarter, D.; Prokop, C.; Elmenreich, W. Load hiding of household's power demand. In Proceedings of the IEEE International Conference on Smart Grid Communications (SmartGridComm), Venice, Italy, 3–6 November 2014; pp. 854–859.
162. Reinhardt, A.; Konstantinou, G.; Egarter, D.; Christin, D. Worried About Privacy? Let Your PV Converter Cover Your Electricity Consumption Fingerprints. In Proceedings of the IEEE International Conference on Smart Grid Communications (SmartGridComm) Symposium on Cyber Security and Privacy, Miami, FL, USA, 2–5 November 2015; pp. 25–30.

Okta. The Cost of Privacy: Reporting on the State of Digital Identity in 2020. Available online: https://www.okta.com/cost-of-privacy-report/2020/ (accessed on 28 December 2020).

Jacobsen, R.; Torring, N.; Danielsen, B.; Hansen, M.; Pedersen, E. Towards an app platform for data concentrators. In Proceedings of the IEEE PES Innovative Smart Grid Technologies Conference, ISGT 2014, Washington, DC, USA, 19–22 February 2014; pp. 1–5.